Mucopolysaccharides

Rajendra Varma
Ranbir S. Varma

Mucopolysaccharides – Glycosaminoglycans – of Body Fluids in Health and Disease

Walter de Gruyter
Berlin · New York 1983

About the Authors

Rajendra Varma received undergraduate education at the University of Delhi, India and graduate education at the University of New South Wales, Sydney, Australia with a Ph.D. in 1966. She did postdoctoral work at the Iowa State (1967), Purdue (1968) Universities in U.S.A. and thereafter worked as an assistant professor (Alliance College, Pa.) and as an associate professor (Edinboro State College, Pa.) and from 1971 to the present at Warren State Hospital, Warren, Pennsylvania 16365, U.S.A. She is the director of the biochemistry department at this hospital, a fellow of the American Institute of Chemists, and a member of the New York Academy of Sciences, Sigma Xi, and the American Chemical Society.

Ranbir S. Varma earned two graduate degrees at Glancy Medical College, Amritsar, and Panjab University, India, and received postgraduate degrees from the Institute of Basic Medical Sciences, Panjab University, Chandigarh, India, and also the School of Medicine of the State University of New York at Buffalo, N. Y. He worked as an assistant research officer at the All-India Institute of Medical Sciences, New Delhi, India, and as a clinical biochemist/medical scientist in different hospitals from 1961 to 1970 and at Warren State Hospital, Warren, Pennsylvania 16365, U.S.A. since 1971. He is a bioanalyst certified by the American Board of Bioanalysis, a fellow of the American Institute of Chemists, and a member of the Biochemical Society (U.K.), and Sigma Xi.

CIP-Kurztitelaufnahme der Deutschen Bibliothek

Varma, Rajendra: Mucopolysaccharides – Glycosaminoglycans – of body fluids in health and disease / Rajendra Varma ; Ranbir S. Varma. – Berlin ; New York : de Gruyter, 1983.
ISBN 3-11-008471-6

NE: Varma, Ranbir S.:

Library of Congress Cataloging in Publication Data

Varma, R. (Rajendra), 1942 – Mucopolysaccharides (Glycosaminoglycans) of body fluids in health and disease.

Bibliography: p.
Includes index.
1. Mucopolysaccharides. 2. Glycosaminoglycans.
3. Mucopolysaccharides -- Metabolism -- Disorders. 4. Body fluids.
I. Varma, R. S. (Ranbir S.), 1938 – . II. Title [DNLM: 1. Body fluids -- Metabolism. 2. Mucopolysaccharides -- Metabolism. 3. Glycosaminoglycans -- Metabolism. 4. Body fluids -- Metabolism. QU 83 V316m]
QP 702.M8V37 1983 612'.015782 83-2021
ISBN 3-11008471-6

 Printing: Druckerei Gerike GmbH, Berlin. – Binding: Dieter Mikolai, Berlin. – Printed in Germany

Dedicated with love to our sons
Rajeev and *Sunil*

Preface

In recent years there has been a rapid advancement in the knowledge of the structure and function of connective tissues and the importance of the acid mucopolysaccharides (Glycosaminoglycans) in physiological and pathological states. This has been reflected in an increased success in the search for means of amelioration of distressing conditions like coronary arteriosclerosis, rheumatoid arthritis, mucopolysaccharidoses, etc. Also, the measurement of the changes of glycosaminoglycan metabolism in connective tissue reflected in the biological fluids have provided useful parameters for diagnosis and prognosis of the patients. These studies have been approached on a multidisciplinary basis and the contribution to the literature has been enormous and scattered in original articles and reviews in different specialty journals. Since no recent single volume that covers the various aspects of all the biological fluids was available, this monograph was prompted to meet this pressing need. The comprehensive, up-to-date review of the literature presented in this monograph will not only provide a starting point for a new investigator entering the field but will also serve the expert in broadening his/her interest and acquiring a greater depth of knowledge.

Because of the broad scope of the field, the subject matter has been organized systematically in a logical sequence of chapters to be of use to the multidisciplinary readers. The first two chapters cover the basic background without involvement of too much of chemistry. The chapter 3 reviews the clinical therapeutic uses and biological roles of glycosaminoglycans. The succeeding chapters relate to the information of glycosaminoglycan metabolism in connective tissues and how these changes in tissue glycosaminoglycans reflected in biological fluids are monitored and interpreted for understanding the mechanisms underlying the physiology and pathology of the tissues and for diagnosis and prognosis of diseases. Each biological fluid has been treated separately and virtually all the disease entities reported in literature have been dealt to create a reference work in depth.

The authors have tried to be comprehensive and systematic but at some places, where changes in glycosaminoglycans in a disease occur simultaneously in two or more biological fluids, overlap became inevitable and we tried to be repetitious rather than risking omission of a related information. The authors hope that this will give the readers a more complete picture of metabolic changes in glycosaminoglycans occurring in a disease. To ensure an up-to-date bibliography a reference list with title and the first and last pages of the articles has been presented at the end of each chapter.

In spite of the tremendous amount of work carried out in the field of connective tissue metabolism, information on glycosaminoglycans in some of the biological fluids is incomplete. Although strictly speaking glycosaminoglycans have not been identified in some minor biological fluids, the authors have digressed to present the information available on other glycoconjugates, like glycoproteins, reported in those biological fluids to serve as a starting point for further exploration. It is important to clarify that the glycosaminoglycans present in the biological fluids are actually the degradation products of the proteoglycans and their aggregates (Chapter 1) in tissues, although in a few instances proteoglycans per se may appear in a biological fluid. Thus a proteoglycan of hyaluronic acid was detected in cyst fluid of cystic mucoid degeneration (Chapter 13) and chondroitin sulfate proteoglycans were found in plasma (Chapter 16) and in synovial fluid from rheumatoid arthritis patients (Chapter 12). In addition to the proteoglycans, aggregates between glycosaminoglycans and glycoproteins were observed in urine (Chapter 17).

The connective tissue and glycosaminoglycan metabolism problems are the concerns of people in many scientific specialties. The authors sincerely believe that the integrated, comprehensive survey of the information presented in this monograph will be very useful to diversified groups of individuals, including chemists, biochemists, physiologists, pharmacologists, pathologists, physicians and individuals in the speciality of neurological sciences, orthopedics and laboratory medicine.

R. Varma

R. S. Varma

Contents

Chapter 1

Chemistry and Metabolism of Glycosaminoglycans and Proteoglycans

1.1 Nature and classification

The acid mucopolysaccharides (or glycosaminoglycans) are high molecular weight linear heteropolysaccharides formed by polymerization of repeating disaccharide units made up of a hexosamine (glucosamine or galactosamine) and a hexuronic acid (D-glucuronic or/and L-iduronic acid) linked by glycosidic bonds. Acetyl and sulfate substituent groups may be present on either or both residues of the polymer. The glycosaminoglycans (GAG) are present as a component of the extracellular ground substance of the connective tissue. These are synthesized by the connective tissue cells. Besides the cells and the ground substance, the connective tissue also contains collagen and elastin fibers. The connective tissue originates from the differentiation of the embryonic mesenchymal cells. The relative proportions of the cells, fibers and the extracellular ground substance vary in different connective tissues. These variations produce differences in the texture of the connective tissues, as an adaptation to the function performed by different connective tissues. Various GAG show heterogeneity with respect to replacement of a glycosyl residue by another, differences in linkage regions and presence of hybrid polymers. Also, these may show variability in the number and sequence of groups (like sulfate and acetyl) present on the outer parts of the polymer chain.

The glycosaminoglycans (GAG) may be classified into groups in different ways:
On the basis of the presence or absence of sulfate gorups, these can be divided into:
a) *Non-sulfated GAG* include hyaluronic acid which contains equimolar amounts of N-acetylglucosamine and D-glucuronic acid and chondroitin, which is the galactosamine-containing counterpart of hyaluronic acid.

b) *Sulfated GAG* contain sulfate groups. Examples are the chondroitin sulfates, heparin, heparan sulfate, dermatan sulfate and keratan sulfate.

Another classification based on the nature of the hexosamine divides GAG into:

a) *Glucosaminoglycans* are those containing glucosamine e. g., heparin, heparan sulfate and hyaluronic acid and keratan sulfate.

b) *Galactosaminoglycans* are those containing galactosamine. Examples are chondroitin 4- and 6-sulfates and dermatan sulfate.

On the basis of the polyanionic character and charge densities, the GAG are classified into:

a) *Polycarboxylates* containing uronic acid but no sulfate groups, e. g., hyaluronic acid and chondroitin

b) *Polysulfates* contain sulfate groups but no hexuronic acid, e. g., keratan sulfate in which D-galactose replaces hexuronic acid.

c) *Polycarboxylsulfates* have both carboxyl groups of hexuronic acid and sulfate groups. The chondroitin sulfates, heparan sulfate, heparin and dermatan sulfate belong to this group.

1.2 Modern nomenclature

Throughout the different sections of this monograph certain old and new terms related to mucopolysaccharides (glycosaminoglycans) used by different investigators will be mentioned. A survey of both the modern, systematic and old, trivial terminology will be presented first.

Meyer (1938) coined the word mucopolysaccharides to describe hexosamine-containing polysaccharides of animal origin present in free state or as protein salts and which gave viscous (= muco) solutions and like the epithelial secretions had the ability to form "mucin clots" with proteins at a low pH. Since then several investigators have used the prefix "muco" for several types of carbohydrate-containing macromolecules in several different ways as: mucoproteins, mucolipids, mucoids, mucoitin, etc., which created a great confusion. To ease up the confusing state of nomenclature, Jeanloz (1960) proposed the name glycosaminoglycuronans for mucopolysaccharides containing a hexosamine and hexuronic acid as repeating disaccharide units and other systematic names for members of the glycosaminoglycan family. The old names and systematic names along with their abbreviations are given in Table I. A more detailed systematic nomenclature can be found elsewhere (Jeanloz, 1960; Balazs and Jeanloz, 1965; Balazs, 1970). Although the proposed systematic names are being widely used, the term glycosaminoglycuronans is cumbersome and excludes keratan sulfate which lacks uronic acid. For this reason the term has been replaced by "glycosaminoglycans".

In the medical literature the term "mucopolysaccharides" is still used extensively and the genetic defects of degradation of mucopolysaccharides are called "mucopolysaccharidoses" and not "glycosaminoglycanoses" or "glycosaminoglycuronoses". For this reason we have included the term mucopolysaccharides in the title of this monograph along with the term glycosaminoglycans. The former term is related to the physical properties of the solutions and the latter to the chemical structure and composition of these macromolecules.

Although the terms chondroitin sulfates A and C have been replaced by chondroitin 4- and 6-sulfates, the bacterial enzymes for their degradation are still called chondroitinases AC and ABC. Also, there is no systematic name for the oversulfated chondroitin sulfate types D and E. The trend is changing; it appears that the terms like "hypermucopolysacchariduria" will eventually be replaced by "hyperglycosaminoglycanuria".

1.3 Development of terminology

Before the systematic modern nomenclature was suggested (Jeanloz, 1960), various other terms were used for the acidic mucopolysaccharides (acidic glycosaminoglycans). Many of these terms have become obsolete and were actually misleading. The nomenclature went through several changes. A survey of these is presented below not with a view to confuse the reader but to help in interpretation of the older literature.

The amino sugar-containing heteropolysaccharides have been known for a long time. Levene (1925) reviewed the earlier work in a monograph. He considered mucoproteins as conjugated proteins with chondroitin sulfate or mucoitin sulfate as prosthetic groups. Karl Meyer of the Columbia University coined the name "mucopolysaccharides" in 1920. These complexes were later classified and reviewed (Meyer, 1938; Stacey, 1946). Originally the term mucopolysaccharide was chosen to indicate the presence of these macromolecules in viscous solutions. Later it was realized that these macromolecules were not only present in biological fluids or excretions but were also widely distributed in mammalian tissues. Meyer (1938) then defined mucopolysaccharides as "hexosamine-sugar-containing polysaccharides of animal origin occurring in a pure state or as their salts with proteins". Blix (1940) used the term "acidoglucoproteins" for the heteropolysaccharides which contained acid carbohydrate radicals: uronic acid, sialic acid and ester sulfate, in comparison to "neutroglucoproteins" which lacked acidic groups and contained hexosamines and neutral sugars. Meyer (1953) in discussing the amino sugar-containing compounds defined mucopolysaccharides as those containing hexosamine. The mucopolysaccharides were present in mucoproteins and glycoproteins in firm chemical union with protein and contained over 4 % hexosamine in mucoproteins and less than 4 % hexosamine in glycoproteins. Stacey (1946) divided the protein-carbohydrate compounds into mucopolysaccharides and mucoproteins. The term mucopolysaccharide was used for heteropolysaccharide containing a low protein content and with or without hexosamine which gave viscous solutions. The name "mucoproteins" was then used for the carbohydrate-protein compounds in which the protein content predominated. Meyer (1953) gave the following classification on the basis of relative percentage composition of carbohydrates and proteins:

Glycoproteins	trace-4 % carbohydrate
Mucoids	4-15 % carbohydrate
Mucoproteins	10-85 % carbohydrate
Mucopolysaccharides	100 % carbohydrate.

Meyer (1953) considered mucoproteins as ionic or easily split complexes between acid aminopolysaccharides and protein. The impact of Meyer's definition of mucopolysaccharides as 100 % carbohydrates was so strong that the earlier workers used and devised drastic procedures for isolation of "pure" mucopolysaccharides from tissues and that any presence of protein was considered an impurity.

Kent and Whitehouse (1955) used "aminopolysaccharide" for nitrogenous polysaccharides which contained no protein and in which aminosugars (hexosamines) accounted for all the nitrogen. The mucopolysaccharides, though solely polysaccharides in character, conjugate with proteins and lipids to form complexes which may be called mucoproteins and mucolipids, respectively. They called mucoproteins and mucopolysaccharides collectively as mucocomplexes or mucosubstances and these were complexes of protein with carbohydrates containing hexosamine but no glucuronic acid or sulfate. They distinguished mucopolysaccharides from mucoproteins in having aminosugar as the major component in the former and with protein as the major component in mucoproteins.

Masamune (1956) used the name glucidamins for proteins having 10-50 % neutral sugars and hexosamines and glycoproteins for the complexes of proteins with acidic polysaccharides.

Bettelheim-Jevons (1958) described the complexes of acidic aminopolysaccharides with protein which he believed were through ionic linkages and could be easily split at high pH or ionic strength.

From the aforementioned survey of nomenclature of protein-carbohydrate complexes, it is clear that several systems have been proposed and different names have been used for the same compounds. This added a great deal of confusion. To overcome this problem Jeanloz (1960) suggested semi-systematic names for the family of mucopolysaccharides as "glycosaminoglycuronans" and the individual names for other members of the family except hyaluronic acid, chondroitin and heparin. This systematized nomenclature has been further elaborated (Balazs and Jeanloz, 1965; Balazs, 1970). In general this nomenclature (Table I) has been widely accepted. However, for the sake of simplicity and to include keratan sulfate, which contains no glucuronic acid "glycosaminoglycans" is the preferred term instead of "glycosaminoglycuronans".

Table I. Nomenclature and Chemical Composition of Acid Mucopolysaccharides (Glycosaminoglycans)

Common Old Name and Abbreviation	Modern Systematic Name and Abbreviation	Monosaccharide Composition of repeating disaccharide	N-Acetyl Group	0-sulfate Group	N-sulfate sulfamate Group	Carbohydrate-protein Linkage Region
Acid Mucopolysaccharides (AMPS)	Glycosaminoglycuronoglycan (GAG)	D-GlcUA and/L-IdUA D-GlNH$_2$/D-GalNH$_2$	+	±	±	
Chondromucoprotein (CMP) or Protein-polysaccharide complexes (PPC)	Proteoglycan (PG)					
Hyaluronic Acid (HA)	Hyaluronic Acid (HA)	D-GlNH$_2$ and D-GlcUA	+	–	–	?
Chondroitin (Ch)	Chondroitin (Ch)	D-GalNH$_2$ and D-GlcUA	+	–	–	GlcUA-Gal-Gal-Xyl-Ser
Chondroitin sulfate A (CSA)	Chondroitin 4-sulfate (C-4-S)	D-GalNH$_2$ and D-GlcUA	+	+	–	GlcUA-Gal-Gal-Xyl-Ser
Chondroitin sulfate B (CSB), β-Heparin	Dermatan sulfate (DS)	D-GalNH$_2$ and L-IdUA/D-GlcUA	+	+	–	GlcUA-Gal-Gal-Xyl-Ser
Chondroitin sulfate C (CSC)	Chondroitin 6-sulfate (C-6-S)	D-GalNH$_2$ and D-GlcUA	+	+	–	GlcUA-Gal-Gal-Xyl-Ser
Chondroitin sulfate D (CSD)	Oversulfated Chondroitin 6-sulfate (C-6-S)	D-GalNH$_2$ and D-GlcUA	+	++	–	GlcUA-Gal-Gal-Xyl-Ser
Heparitin sulfate (HS) or Heparin monosulfate (HMS)	Heparan sulfate (HS)	D-GlNH$_2$ and D-GlcUA/L-IdUA	+	+	+[1]	GlcUA-Gal-Gal-Xyl-Ser
Heparin (Hep)	Heparin (Hep)	D-GlNH$_2$ and D-GlcUA/L-IdUA	+	+	++[1]	GlcUA-Gal-Gal-Xyl-Ser
Corneal keratosulfate (KS)	Keratan sulfate I (KS-I)	D-GlNH$_2$ and D-Gal	+	+	–	GlcNAC-Asp
Skeletal keratosulfate (KS)	Keratan sulfate II (KS-II)	D-GlNH$_2$ and D-Gal	+	+	–	GalNAc-Thr | Gal-NANA

1. The repeating units of HS and heparin contain an average of 1.5 and 2.5 sulfate, respectively.

Legends for Table I

+ = Present, – = Absent, ± = Present or absent, ++ = Present more than once in a repeating disaccharide unit, D-GlNH$_2$ = D-Glucosamine, D-GalNH$_2$ = D-Galactosamine, D-GlcUA = D-Glucuronic acid, L-IdUA = L-Iduronic acid, GlcNAc = N-Acetyl-D-glucosamine, GalNAc = N-Acetyl-D-galactosamine, Xyl = D-Xylose, Gal = D-Galactose, NANA = N-Acetylneuraminic acid, Ser = Serine, Thr = Threonine, Asp = Aspartic acid

With the concept of "pure" mucopolysaccharides as those containing 100 % carbohydrates, earlier workers used drastic conditions like lengthy proteolytic digestion, treatment with alkali, extraction with strong salt solutions to obtain protein-free preparations of polysaccharides. The material so isolated was considered to be "pure" glycosaminoglycan and the presence of any protein an "impurity". However, it has been clearly established that the glycosaminoglycans isolated by the drastic procedures are actually artifacts and represent products formed by degradation of the native macromolecules called proteoglycans (PG). In the native state the glycosaminoglycans are not present as such but as protein-carbohydrate complexes formed by stable covalent linkages. For such complexes Balazs in 1967 suggested the name "proteoglycans". Proteoglycans containing a single GAG are named by adding the prefix "proteo" before the name of glycosaminoglycan like proteoheparan sulfate, proteokeratan sulfate, etc. When two different types of polysaccharides are attached to protein, e. g., keratan sulfate and chondroitin 4-sulfate, the protein-polysaccharide complex is named as proteokeratan chondroitin 4-sulfate. A detailed structure of the proteoglycans will be presented later in section 1.7. The mucopolysaccharides (glycosaminoglycans) present in biological fluids are actually the degradation products formed by the catabolism of proteoglycans in the amorphous ground substance of the connective tissue.

Another new term "glycoconjugates" has been introduced by the European workers. This term covers three types of macromolecules: glycoproteins, glycolipids and proteoglycans (Horowitz and Pigman, 1977; Gregory and Jeanloz, 1979).

1.4 Other related terms in literature

Besides the terms described in section 1.3, some other terms have been used in literature and may be used at some places in this monograph during review of the older literature. A survey of these terms follows. *Mucopolysaccharide-protein complexes and polysaccharide-protein* were used in older literature to describe association of mucopolysaccharides with protein.

Proteoglycosaminoglycan is synonymous with proteoglycan and refers to mucopolysaccharide (glycosaminoglycan) linked covalently to protein as a complex.

Chondromucoprotein refers to proteoglycan isolated from cartilage.

Glycoproteins are macromolecules which have the characteristics of a protein and carry covalently bound oligosaccharide. It is a protein molecule to which an oligosaccharide chain with less than 15-20 monosaccharide residues is attached. Originally, the protein nomenclature committee of the American Society of Biological Chemists and the American Physiological Society introduced this term in 1908 for "compounds of the molecule with a substance containing a carbohydrate group other than nucleic acid".

Mucin is a term used in a physiological sense to denote a slimy viscous secretion. An example is gastric mucin.

Mucous glycoprotein is used to describe mucus of epithelial secretions.

Glycopeptide is the term used for a peptide chain consisting of a few amino acids having a covalently-linked polysaccharide chain. Glycopeptides are formed by the action of proteolytic enzymes on glycoproteins.

Peptidoglycan refers to macromolecules which have one or more peptides attached covalently to one or more polysaccharide chains, form a bridge between two points of the same chain or be attached to the end of a polysaccharide chain or at branch points.

Proteoglycan complex is used for macromolecular complexes having one or more proteoglycans alone or with one or more protein, lipid or polysaccharide molecules linked together by covalent bonds.

Protein-glycosaminoglycan complex is the same as protein-polysaccharide and stands for a complex of protein with glycosaminoglycans linked by noncovalent bonds. An example of this is a collagen-chondroitin sulfate complex formed by an ionic interaction.

Prosthetic group is the oligosaccharide group attached to a protein in a glycoprotein or to a lipid in a glycolipid.

Acid mucopolysaccharide is synonymous with acidic mucopolysaccharides and acidic glycosaminoglycans. It is used for glycosaminoglycans isolated from tissue proteoglycans or biological fluids which may or may not have a small peptide attached to it covalently, but always have an acidic character due to presence of sulfate and/or hexuronic acid.

Acidic polysaccharides represent any polysaccharide containing acidic groups

Oligosaccharide groups refer to carbohydrate chains (usually of less than 15-20 monosaccharide residues) present covalently-linked in glycoproteins or glycolipids.
Aminopolysaccharides was the term used in old literature to include the hexosamine-containing compounds: acidic mucopolysaccharides, blood group substances, chitin, etc.

The majority of the names used for protein-polysaccharide compounds were based on whether the compound had the properties of a protein or a polysaccharide, which depended on the relative percentage contents of the two components. Meyer (1970) has pointed out if a protein-polysaccharide contains over 70 % carbohydrate it has the properties of a polysaccharide.

1.5 Occurrence

The glycosaminoglycans occur in the extracellular amorphous ground substance of all the connective tissues and biological fluids. With the exception of heparin, these macromolecules are present in the tissues as covalently-linked protein-complexes called proteoglycans and as proteoglycan aggregates of very high molecular weights. These are responsible for the physical characteristics of the tissues like shape and compactness. In the biological fluids these macromolecules are present as glycosaminoglycans. However, proteoglycans having a single type of GAG have been isolated from some biological fluids. Two such examples are proteohyaluronic acid from ovarian cyst fluid (section 13.4) and proteochondroitin sulfate A from the euglobulin fraction of human plasma (section 16.5).

The connective tissue consists of 3 components: the extracellular fibers (collagen and elastin), the cells (fibroblasts, chondroblasts, neuroblasts) and the extracellular ground substance which has no shape or form (= amorphous) and contains the glycosaminoglycans. Heparin is an exception and does not occur in the extracellular fluid but occurs intracellularly as granules in mast cells and is released from these intracellular granules under the influence of certain stimuli. The amorphous matrix is present as a viscous gel and fills the space between the cells and the fibers

Table II. Occurrence of Various Glycosaminoglycans

Glycosaminoglycans	Sources	Molecular Weight Range X 10^3	Sulfate Disaccharide		Unit anionic charge Dissacharide
			Average	Range	
Keratan sulfate I (corneal)	cornea	4–19	(1.0)	0.9–1.3	2
Keratan sulfate II (cartilage or skeletal)	human cartilage nucleus pulposus, annulus shark cartilage ox cartilage cartilage from fish, squid nasal septum bovine intervertebral disc morquio syndrome urine	8–12	(1.0)	0.9–1.8	2
Hyaluronic acid	vitreous humor aqueous humor subretinal fluid umbilical cord aorta synovial fluid rooster comb streptococci urine blood in cancer skin	50–8,000		0.0	1
Heparin	lung, liver, skin	7–16	(2.5)	2–3	3–4
Heparan sulfate	ox lung, spleen, thymus, granulation tissue bovine and human aorta, cervix uteri human amyloid liver, hen oviduct, urine in mucopolysaccharidoses	15	(0.5)	0.1–2.8	1–3.5
Dermatan sulfate	skin, umbilical cord, urine in mucopolysaccharidoses, beef lung, tendon, blood vessels, aorta	15–40	(1.0)	1–2	2
Chondroitin 4-sulfalte	cartilage, bone, cornea, skin, notocord, blood plasma	5–50	(1.0)	0.05–1.0	2
Chondroitin 6-sulfate	umbilical cord, aorta, skin, cartilage	5–50	(1.0)	0.2–1.3	2

and acts as a cementing substance. Because of differences in the relative proportions of the ground substance, cells and fibers, the connective tissues have different physical appearances and shapes and varied physicochemical properties which make them suitably adapted for special physiological functions. The composition of individual GAG components also varies in different connective tissues in relation to the function performed by that tissue. The occurrence of individual GAG components in different tissues has been presented in Table II.

1.6 Chemical structures

All the GAG components except keratan sulfate are glycosidically linked polymers of repeating disaccharide units containing glucosamine or galactosamine as the basic sugar residue and a sugar acid, D-glucuronic acid or L-iduronic acid or a combination of the two acids. Hybrid structures containing both glucosamine and galactosamine are not known. However, keratan sulfate from cartilage (keratan sulfate II), which has physical and structural characteristics of glycosaminoglycans but lacks hexuronic acid, does contain a small amount of D-galactosamine.

A brief account of the structural features of the individual GAG along with heterogeneity and variability follows. For details of how these structures were arrived at, the reader is advised to consult several reviews and monographs available (Clark and Grant, 1961; Wolstenholme and O'Connor, 1958; Brimacombe and Weber, 1964; Muir, 1964; Quintarelli, 1968; Jeanloz, 1963, 1970).

1.6.1 Chondroitin sulfate isomers

Chondroitin 4-sulfate, dermatan sulfate and chondroitin 6-sulfate are galactosamine containing glycosaminoglycans (galactosaminoglycuronans). These are linear unbranched polymers of the repeating disaccharide units of an N-acetyl-O-sulfated-D-galactosamine and a hexuronic acid. The disaccharide units are formed by β-(1→4)-hexosaminidic bonds and the disaccharide units are linked by β-(1→3)-uronidic bonds. Therefore, the β-(1→3)-uronidic and β-(1→4)-hexosaminidic bonds alternate in these polymers. Dermatan sulfate and chondroitin 4-sulfate carry an ester sulfate group at C-4 in the axial position, whereas a sulfate group is present on C-6 in the equatorial position in case of chondroitin 6-sulfate. The three chondroitin sulfate isomers exist as proteoglycans in tissues covalently linked to protein through the sequence (-GlcUA-Gal-Gal-Xyl-Ser). The hexuronic acid in chondroitin 4- and 6-sulfates is D-glucuronic acid, whereas dermatan sulfate contains mostly L-iduronic acid in addition to a small amount of D-glucuronic acid. Dermatan sulfates from different sources vary in relative proportions of the two hexuronic acids and in degrees of sulfation.

The repeating disaccharide units of the chondroitin 4- and 6-sulfates and dermatan sulfates are 10 Å (1 nm) long and have molecular weights of 450-500 daltons. The chondroitin sulfate chains are 500-600 Å long and are 20,000-30,000 in molecular weight. In the chondroitin 6-sulfate from cartilage, the chains consist of 40-60 repeating disaccharide units.

The three chondroitin sulfate isomers can be distinguished by their optical rotations, solubilities in alcohol, infra-red spectra and by the action of bacterial chondroitinases ABC and AC. Also, chondroitin 4- and 6-sulfates are susceptible to testicular hyaluronidase (EC 3.2.1.35) while dermatan sulfate is resistant to this enzyme. However, evidence is now available that dermatan sulfate does actually undergo hydrolysis by the enzyme to a limited extent due to the presence of a small amount of D-glucuronic acid.

A chondroitin 6-sulfate isolated from shark cartilage is oversulfated and has sulfate/hexosamine ratios of 1.3 or higher. It has been named as chondroitin sulfate D. It has been found to be disulfated and carries the two sulfated groups at C-4 and C-6 positions of galactosamine residue. The name chondroitin sulfate E has been assigned to a chondroitin 4-sulfate or chondroitin 6-sulfate with an additional sulfate group at C-2 position of glucuronic acid residues. A chondroitin 4-sulfate with sulfate/hexosamine ratio less than unity has been isolated from cornea. Similarly, chondroitin 4- and 6-sulfates with low degrees of sulfation have been isolated from tadpole carti-

Fig. 1-1. Glycosaminoglycans and Their Linkages

GlcNAc(β1→4)GlcUA(β1→3)GlcNAc(β1→4)GlcUA (Linkage to protein?)

Hyaluronate

GalNAc(β1→4)GlcUA(β1→3)GalNAc(β1→4)GlcUA(β1→3)Gal(β1→3)Gal(β1→4)Xyl→Ser

Chondroitin

OSO_3^- ↓ 4 OSO_3^- ↓ 4

GalNAc(β1→4)GlcUA(β1→3)GalNAc(β1→4)GlcUA(β1→3)Gal(β1→3)Gal(β1→4)Xyl→Ser

Chondroitin 4-sulfate

OSO_3^- ↓ 6 OSO_3^- ↓ 6

GalNAc(β1→4)GlcUA(β1→3)GalNAc(β1→4)GlcUA(β1→3)Gal(β1→3)Gal(β1→4)Xyl→Ser

Chondroitin 6-sulfate

OSO_3^- ↓ 4 OSO_3^- ↓ 4

GalNAc(β1→4)IdUA(α1→3)GalNAc(β1→4)GlcUA(β1→3)Gal(β1→3)Gal(β1→4)Xyl→Ser

Dermatan sulfate

OSO_3^- ↓ 6 OSO_3^- ↓ 6 OSO_3^- ↓ 6 OSO^- ↓ 6

Gal(β1→4)GlcNAc(β1→3)Gal(β1→4)GlcNAc(β1→6)GlcNAc→Asp

3 β↓ 1

Gal(4←1α)NeuAc

Keratan sulfate I

OSO_3^- ↓ 6 OSO_3^- ↓ 6 OSO_3^- ↓ 6 OSO^- ↓ 6

Gal(β1→4)GlcNAc(β1→3)Gal(β1→4)GlcNAc(β1→6)GalNAc→Thr

3 β↑ 1

Gal(4←1α)NeuAc

Keratan sulfate II

OSO_3^- → 6 (first residue); OSO_3^- → 6 (third residue)

GlcNAc(α1→4)IdUA(α1→4)GlcNAc(α1→4)GlcUA(β1→3)Gal(β1→3)Gal(β1→4)Xyl→Ser

Heparan sulfate

OSO_3^- → 6 (first residue); OSO_3^- → 2 (second residue); OSO_3^- → 6 (third residue)

GlcNSO$_3^-$(α1→4)IdUA(α1→4)GlcNAc(α1→4)GlcUA(β1→3)Gal(β1→3)Gal(β1→4)Xyl→Ser

Heparin

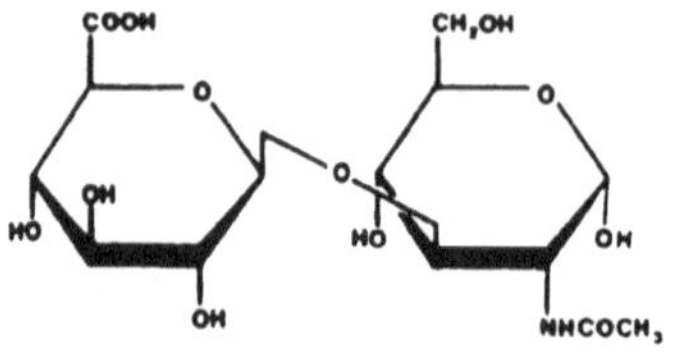

HYALURONIC ACID

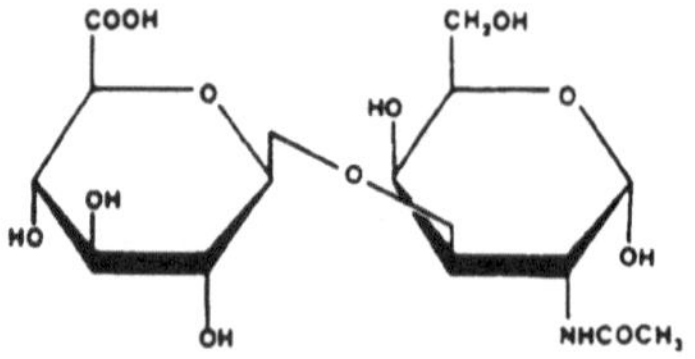

CHONDROITIN

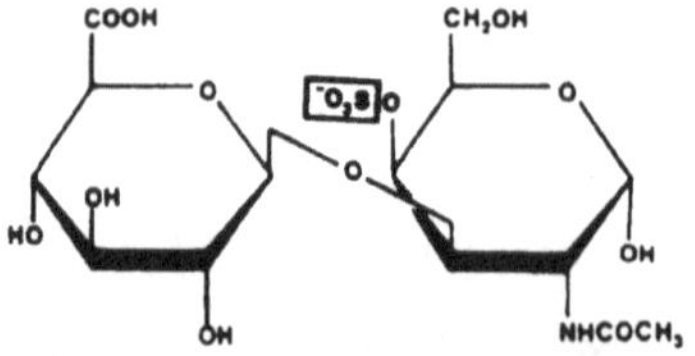

CHONDROITIN 4-SULFATE

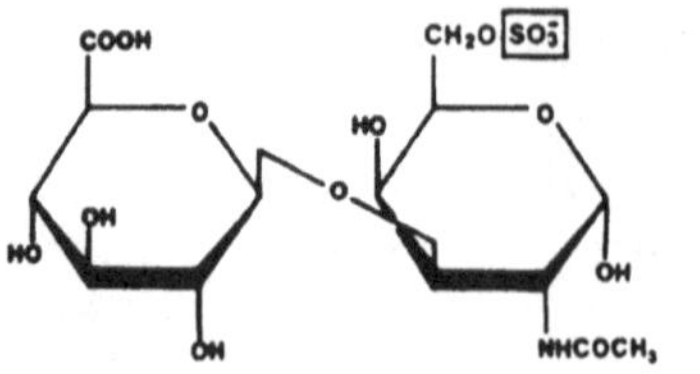

CHONDROITIN 6-SULFATE

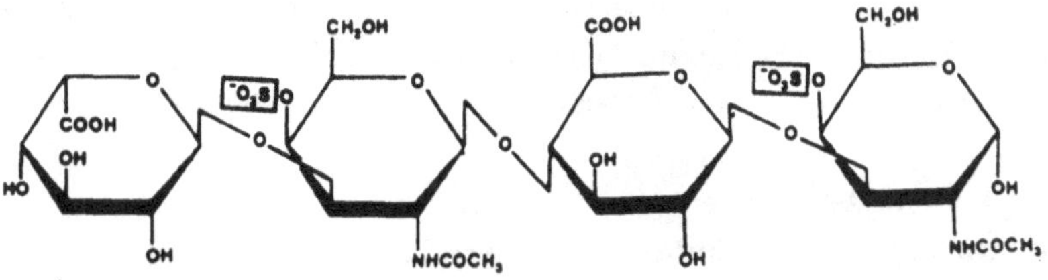

DERMATAN SULFATE TETRASACCHARIDE

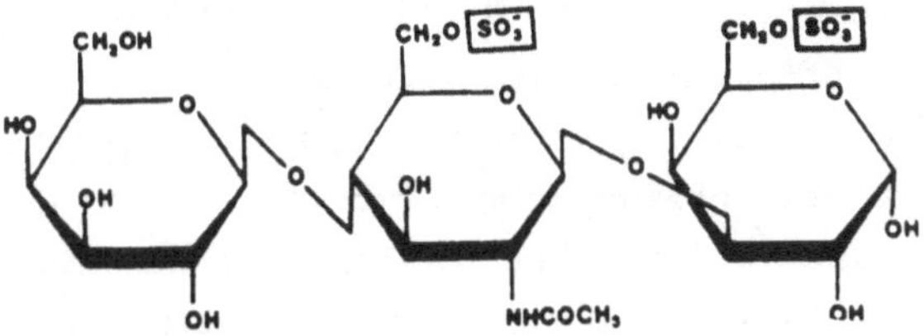

KERATAN SULFATE (TRISACCHARIDE)

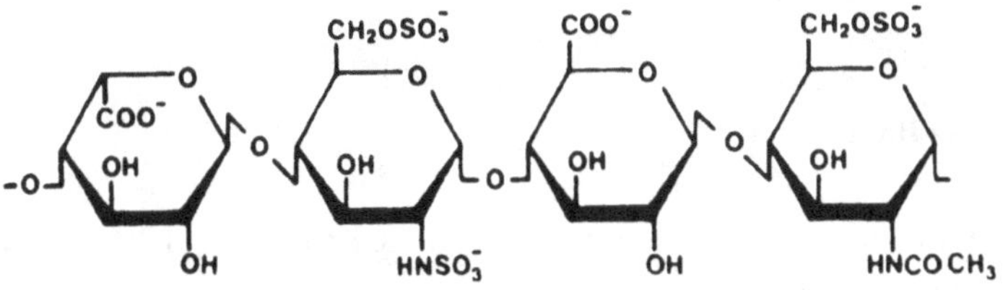

HEPARAN SULFATE (TETRASACCHARIDE)

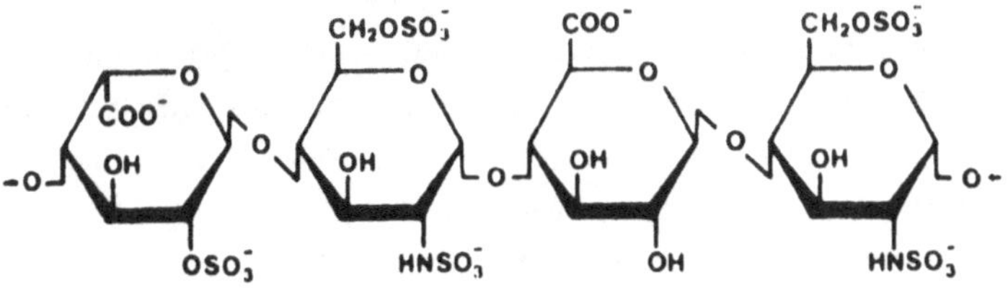

HEPARIN (TETRASACCHARIDE)

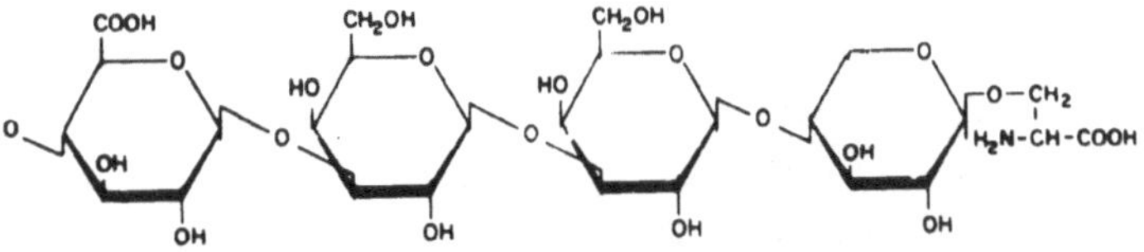

CARBOHYDRATE-PROTEIN LINKAGE REGION

lage. Plasma contains an undersulfated chondroitin 4-sulfate as the major GAG along with fully sulfated and oversulfated forms as minor components (see section 16.5).

Dermatan sulfates isolated from some tissues including arterial tissue, skin and umbilical cord are hybrid structures. In the proteoglycan from bovine tissue about 80 polysaccharide chains (each of M_r 15,000-20,000) are present attached to protein by O-seryl (threonyl) bonds. Each polysaccharide chain is a hybrid chain and consists of 75 % chondroitin 4-sulfate and 25 % dermatan sulfate (Buddecke et al., 1971). Buddecke and von Figura (1975) have reviewed the different types of chondroitin sulfate-dermatan sulfate hybrids in ox aorta. Dermatan sulfate from skin is a mixed copolymer of chondroitin 4-sulfate disaccharides and iduronic acid containing 4-sulfated disaccharides. The umbilical cord dermatan sulfate is a copolymer of iduronic acid containing 4- and 6-sulfated disaccharides and chondroitin 6-sulfate disaccharides.

1.6.2 Keratan sulfate

Although keratan sulfate does not contain any hexuronic acid it is classified among the glycosaminoglycans because of the repeating sequence of the disaccharide units and of the sulfation at C-6 position of both the monosaccharide residues. Keratan sulfate is a linear polymer of repeating disaccharide structure composed of alternating galactose and N-acetylglucosamine residues linked β-1,4 and β-1,3, respectively. In keratan sulfates from most of the sources there is a molar excess of galactose over N-acetylglucosamine and these extra galactose residues are believed to be present as side branches from the keratan sulfate chains (Bhavanandan and Meyer, 1968). Keratan sulfates from other sources have also been reported to contain fucose, sialic acid, mannose and N-acetylgalactosamine as components of their structures.

Keratan sulfate is a small polymer containing only 10-20 repeating disaccharide units, 100-200 Å long and 5,000-10,000 daltons in molecular weight. In the keratan sulfate from cartilage the protein-carbohydrate linkage is an

O-glycosidic linkage between N-acetylgalactosamine on the keratan sulfate chain and the serine or threonine residues on the protein core (Kieras, 1974). Sialic acid and neutral sugar residues are attached to this N-acetylgalactosamine involved in the carbohydrate-protein linkage. Depending on the source of keratan sulfate the neutral sugar residues vary. In keratan sulfate from cartilage, galactose is present as the neutral sugar (Hopwood and Robinson 1974, 1974 a). Thus keratan sulfate resembles glycoproteins in having the sugar residues; fucose, mannose, sialic acid, branched nature and O-glycosidic linkage. On the other hand, this macromolecule as mentioned above possesses structural characteristics and physical properties of glycosaminoglycans. For these reasons keratan sulfate is regarded as a macromolecule intermediate between glycosaminoglycans and glycoproteins.

From a comparative analysis of a wide variety of vertebrate species, two types of keratan sulfates were recognized (Mathews and Cifonelli, 1965). The corneal keratan sulfate is basically similar to the skeletal analogue, cartilage keratan sulfate but is simpler in structure and differs in the protein-carbohydrate linkage from the latter. The major difference between the two types is the presence of a substantial amount of galactosamine, in the absence of chondroitin sulfate, in case of cartilage keratan sulfate and the higher degree of sulfation in cartilage keratan sulfate. The carbohydrate-protein linkage in cartilage keratan sulfate is an O-glycosidic linkage between N-acetylgalactosamine and threonine which is cleaved by treatment with alkali (Hopwood and Robinson, 1974) in case of keratan sulfate from adult bovine intervertebral disc. Two different types of linkages exist between skeletal keratan sulfate from bovine intervertebral disc and bovine nasal septum (Hopwood and Robinson, 1974 a).

The corneal keratan sulfate from mammals, fish and birds contains an equimolar amount of galactose and glucosamine, sulfate/glucosamine molar ratio of 1.0-1.3, and average molecular weight of 10,000. Aspartic acid is the principal amino acid residue and is involved in the carbohydrate-protein linkage.

Keratan sulfate is the most heterogeneous of the connective tissue glycosaminoglycans. Depending upon the tissue of origin, it shows heterogeneity with respect to protein-carbohydrate linkage, degree of sulfation on the

D-galactose residues and in nature and number of the additional neutral sugars and sialic acid. Even with keratan sulfate preparations from a single tissue there is a wide variation in the contents of fucose and sialic acid. Tissues at different ages also show similar variations (Seno et al., 1965; Mathews and Cifonelli, 1965; Toda and Seno, 1970). The C-6 of N-acetylglucosamine carries a sulfate ester group in all keratan sulfates; the variability in the degree of sulfation is restricted to O-galactose residues.

1.6.3 Heparan sulfate and heparin

Heparin is not found in the extracellular matrix of connective tissue but resembles heparan sulfate structurally. Both heparin and heparan sulfate contain equimolar proportions of glucosamine and hexuronic acid linked by 1,4 linkages. Heparan sulfate contains a mixture of N-acetylglucosamine and N-sulfated glucosamine, while heparin contains mainly N-sulfated glucosamine. The hexuronic acid in heparan sulfate is mainly D-glucuronic acid with some L-iduronic acid. Structurally heparin differs from heparan sulfate in higher contents of L-iduronic acid (70-90 %; 10-30 % being glucuronic acid) and sulfate groups. The glucosamine and glucuronic acid residues are alpha-linked, while the glucosamine and L-iduronic acid residues are beta-linked. The O-sulfate groups are present on C-6 of N-acetylglucosamine residues and on C-2 of most of the iduronic acid residues, but glucuronic acid residues carry no sulfate groups. The sulfate/hexosamine ratios for heparan sulfate and heparin are 0.5 and 2.5, respectively. Different sulfate preparations vary in relative contents of N-acetylglucosamine and N-sulfated glucosamine. These two moieties are arranged as clusters or bunches with N-sulfated being present in the inner core and N-acetylatedglucosamine on the outer branches (Meyer, 1969). Both heparin and heparan sulfate are linked O-glycosidically to the serine residues on protein through the sequence: -glucuronic acid-galactose-galactose-xylose-serine-protein, which suggests these are synthesized as proteoglycans. Heparan sulfate (M_r 40,000-45,000) is a much larger macromolecule as compared to heparin (M_r 5,000-15,000).

Heparin and heparan sulfate differ from all other glycosaminoglycans in having sulfoamino groups (N-sulfoglucosamine), which makes them amenable to deamination with nitrous acid. The determination of the 2,5-anhydromannosyl residues so liberated provides a method for the analysis of heparan sulfate/heparin in mixtures with other glycosaminoglycans (Lagunoff and Warren, 1962).

Heparan sulfate and heparin preparations show heterogeneity with respect to the molecular size, variations in the component residues of D-glucosamine, L-iduronic acid, D-glucuronic acid, N-sulfate and O-sulfate. Although glucosamine is always the only hexosamine present, yet the relative proportions of L-iduronic acid and D-glucuronic acid vary. The relative proportions of N-acetyl and N-sulfated glucosamine and also O-sulfate/N-sulfate groups are variable. These variations affect the physicochemical properties and create problems in the interpretation of anion-exchange chromatographic profiles and electrophoretic patterns.

The polysaccharide portions of heparin and heparan sulfate are synthesized in a similar manner. Some investigators have suggested that heparan sulfate may be a precursor of heparin, being converted to the latter by sulfation and C-5 epimerization of the glucuronic acid. However, others (Silbert et al., 1975) consider them as separate and different macromolecules, distinguished by source and location of synthesis and function. Heparin is localized and synthesized intracellularly in mast cells but functions extracellularly when being released into the extracellular matrix and blood in response to a specific stimulus. Heparin has an anticoagulant action and is an activator of the plasma lipoprotein lipase, which brings about clearing of plasma. On the other hand, heparan sulfate is synthesized by connective tissue cells other than mast cells (e. g., fibroblasts) and is a component of external surface of all normal cells and may act as an inhibitor of fibrinolysis as a check in the spread of normal cells as opposed to fibrinolysis effect of the malignant cells (Silbert et al., 1975).

1.6.4 Hyaluronic acid

Hyaluronic acid is the nonsulfated glycosaminoglycan and is a very high molecular weight polymer of a repeating disaccharide unit called hyalbiuronic acid. The hyalbiuronic acid consists of equimolar amounts of D-glucuronic acid residues in β-(1→3) linkage to the N-acetylglucosamine residues. The hyalbiuronic acid repeating units carry β-(1→4) linkages in hyaluronic acid. This GAG was first isolated from vitreous humor (Meyer and Palmer, 1934) followed by its isolation from umbilical cord (Meyer and Palmer, 1936). Hyaluronic acid is known to be present in the mesenchymal fluid-filled cavities or the interfibrillar sacs like synovial fluid, aqueous and vitreous humors of the eye and the labyrinthine fluids (perilymph and endolymph) in the internal ear. Only a small amount of hyaluronic acid is present in cartilage where it acts as a filamentous backbone for making non-covalent associations with several proteoglycan subunits (consisting of chondroitin sulfate and keratan sulfate chains attached to a protein core) to form giant proteoglycan aggregates.

Hyaluronic acids isolated from different sources vary in size (molecular weight) and with respect to its association with protein. Since this macromolecule can be isolated from some tissues and biological fluids without proteolysis, the presence of a stable, covalent linkage to protein appears improbable. However, in some biological fluids it may be associated with protein. For an example, a proteohyaluronic acid has been isolated from ovarian cyst fluid (see section 13.4).

Despite any evidence for the presence of a carbohydrate-protein linkage it is impossible to obtain a hyaluronic acid preparation free of amino acid residues. Hyaluronic acid preparations from different sources are always associated with a small amount of amino acids and neutral sugars. The significance and role of these amino acids and neutral sugars is not clear. One neutral sugar, arabinose, has been consistently found as a component of hyaluronate from bovine and human brain (Stary et al., 1965; Wardi et al., 1966; Margolis, 1967; Varma et al., 1977), in hyaluronate from urine along with glucose (Chakrapani and Bacchawat, 1968); along with fucose and glucose in hyaluronates from vitreous humor (Varma et al., 1974, 1974 a), from aqueous humor (Varma et al., 1975), from human umbilical cord (Varma et al., 1975 a). The hyaluronate from synovial fluid has been reported to contain glucose and chemical and immunological evidence was provided for binding of hyaluronate to protein (Hamerman et al., 1966). While some workers have also indicated a carbohydrate-protein linkage, others have presented evidence against such a chemical combination (for review see Varma et al., 1975 a). Some investigators have suggested the presence of protein in hyaluronate samples to be due to mutual interaction of these polymers. Varma et al. (1974, 1975, 1975 a) based on physicochemical studies and chemical analysis of hyaluronates before and after reduction with sodium-palladium chloride, suggested silanolate cross-links between the hyaluronate chains through minority neutral sugars and/or suitable hydroxyl groups of glucuronyl residues.

1.7 Organization as proteoglycans and aggregates

Although the functional properties of the connective tissues are due to the glycosaminoglycans species, these are by themselves unable to function unless these are arranged into ordered aggregates of very large molecular weights. The macromolecular glycosaminoglycans serve only as the basic structural units, which are assembled into functional aggregates, which can be described at different levels of organization as the primary, secondary and tertiary structures. These levels of organization in cartilage would be a typical example and are described below:

1.7.1 Primary structure

The arrangement of the alternating, repeating units of hexosamine and hexuronic acid (40-60 for chondroitin 6-sulfate and 10-20 repeating units for keratan sulfate) along with the neutral sugars of the carbohydrate-protein linkage region (GlcU-Gal-Gal-Xyl) constitutes as primary structure of the glycosaminoglycans. The chondroitin sulfate chains are 500-600 units long and have molecular weights of 20,000-30,000 while keratan sulfates are short 100-200 Å long and 5,000-10,000 in molecular weights. In the tissues GAG are present as protein-polysaccharide complexes called proteoglycans. The linkage of the neutral sugar (xylose) end of chondroitin 6-sulfate and of N-acetylglucosamine of keratan sulfate to the threonine or serine residues on the protein chain is described as the primary structure of the proteoglycan.

1.7.2 Secondary structure

Like proteins and nucleic acids, the glycosaminoglycans (GAGs) exist as right-handed (e. g. chondroitin sulfates) or left-handed (e. g. heparan sulfate, heparin) helical structures both in crystalline and solution states. These have been studied by chiroptical properties (ORD, CD) in the far UV region and by X-ray analysis of their axial and lateral periodicity in their stretched crystalline films. The dimensions of the elementary cell in X-ray diffraction give information on the number of chains in the helix and the relative arrangement of saccharides forming the helix. Differences in the substituent groups on the chains produce varying H-bonds and packing and conformation changes in solutions containing divalent cations. Thus hyaluronic acid changes from a 4-fold to a 3-fold helix, while chondroitin 4-sulfate goes from 3- to a 2-fold helix in solution.

1.7.3 Tertiary structure

Differences in helical and stereochemical molecular forms affect the packing, aggregations and interactions of GAGs with other molecules. The transitions in packing, conformations have been studied by equilibrium dialysis binding studies, ORD, CD, and X-ray diffraction studies. The differences in helical forms (secondary structures) and packing and aggregations among the various GAGs are responsible for the differences in their biological interactions with collagen, lipoproteins and other properties.

The proteoglycans at the tertiary structure level vary in size and composition depending on the tissue source. This variation in size and GAG components of the aggregate accounts for the differences in the mechanical properties of the different tissues and actually represents an adaptation of the tissue to a specific function.

1.7.4 Chemical and molecular heterogeneity

Cartilage proteoglycans, as an example, have differences not only in the core proteins but also in the relative proportions of chondroitin sulfate and keratan sulfate chains attached to the protein core. These molecular species may be dissociated from the cartilage "proteoglycan aggregates" with 4 M guanidinium chloride and could be fractionated by gel-permeation chromatography, and density gradient centrifugation. The proteoglycan fractions so isolated exhibit variations in length of the protein core, type, distribution and number of carbohydrate chains attached to it, thus giving rise to four types of variations (Tsiganos and Muir, 1970; Sajdera and Hascall, 1969).

The proteoglycans extracted from pig cartilage or bovine nasal septum with sodium acetate or 2 M $CaCl_2$ show changes in size, chemical composition and extractability. The ease of extraction of proteoglycans declines with the age of the tissue. Proteoglycans contain increasing proportions of keratan sulfate with age (Simunek et al., 1971).

1.8 Role of silicon in molecular organization of glycosaminoglycans

Silicon has been reported as a component of glycosaminoglycans: chondroitin 4-sulfate, dermatan sulfate and heparan sulfate, which contain about one atom of silicon per 130-180 repeating disaccharide units or approximately 0.04 % of their weights (Schwarz, 1973). Other studies have indicated that this element is an essential micronutrient for growth and development (Hoekstra et al., 1974) and is present in tissues, being more (about 0.01 %/weight) in cartilage, calcification sites, ligaments and skin than in the internal organs (Carlisle, 1972). Suggestions have been made that silicon may function as a biological cross-linking agent in connective tissue, where it may form a bridge between glycosaminoglycan chains through ether linkages formed by reaction of silicic acid, $Si(OH)_4$ with hydroxyl groups on the GAG. Further cross-linkages to additional GAG chains may occur through cross-linkages formed between the hydroxyl groups present on these ether-linked chains (Schwarz, 1973). The presence of Si linkages even in the fragments obtained by hyaluronidase digestion of GAG preparations further suggests this possible role of silicon.

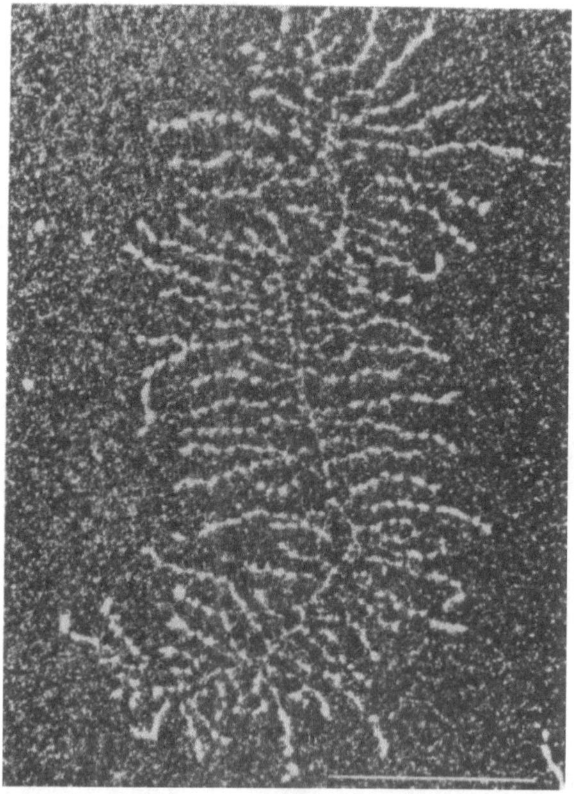

Fig. 1-2. Dark field electron micrograph of a proteoglycan aggregate (x 120,000). The bar at the lower right equals 0.5 μm. Reproduced with permission from Rosenberg, L. et al. Electron microscopic studies of proteoglycan aggregates from bovine articular cartilage. J. Biol. Chem. 250, (1975) 1877-1883

PROTEOGLYCAN AGGREGATE

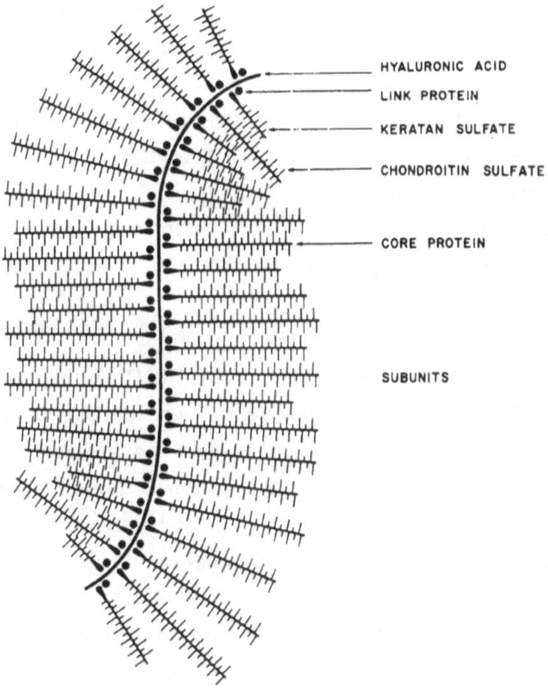

Fig. 1-3. Diagrammatic representation of proteoglycans aggregate. Reproduced with permission from Rosenberg, L. et al.: Organization of extracellular matrix in bovine articular cartilage. In: Slavkin, H. C. and Greulich, R. C. (Eds.) Extracellular Matrix Influences on Gene Expression. Academic Press, New York, pp. 415-421 (1975)

Varma et al. (1974, 1975, 1975 a) have reported the presence of bound silicon in hyaluronic acid preparations from bovine aqueous and vitreous humors and umbilical cord. Based on the physicochemical data, analyses of carbohydrates, amino acids and silicon contents of these hyaluronates before and after reduction with $NaBH_4/PdCl_2$, it was suggested that the silicon may bind two chains of hyaluronate through hydroxyl groups on the hexuronic acid and/or the neutral sugars.

1.9 Interactions with basic compounds and cations

Because of the polyanionic nature, the glycosaminoglycans interact with a number of basic dyes, like toluidine blue, Alcian blue, methylene blue, acridine orange, ruthenium red, etc., to form insoluble precipitates. This forms the basis for the use of these dyes for detection and localization of the glycosaminoglycans in histochemical, cytochemical and electrophoretic studies. Formation of metachromasia with some basic dyes like toluidine blue and methylene blue and fluorescent complexes with acridine orange have been used for detection and determination of GAG by spectrometry and fluorescent microprobe analysis.

The negative electrical field of a polyanion (GAG) attracts the positively charged basic dyes and cations forming a colored complex with basic dyes (Hopfinger, 1977). When both the basic dye and an uncolored cation are present together or when an uncolored cation is added to the GAG-basic dye colored precipitate, a competition occurs between the dye and the cation for anionic sites on the glycosaminoglycans until at a sufficiently high cation concentration called the "critical electrolyte concentration" all the bound dye is released and can be measured (Scott, 1973). This principle has been used as the basis for analysis of GAG in biological fluids.

The glycosaminoglycans form a precipitate with cetylpyridinium salts like cetylpyridinium chloride (CPC) and cetyltrimethylammonium bromide (CTAB). This property has been exploited in several methods for isolation and purification and turbidimetric measurements of GAG from tissue and biological fluids (Scott, 1960).

Glycosaminoglycans and proteoglycans, both in the aggregated and disaggegated forms, bind cations. The role of these macromolecules as ion-exchange resins has been described elsewhere (see section 3.12). These macromolecules interact with calcium and phosphate ions (MacGregor and Bowness, 1971), with similar binding properties as glycosaminoglycans and disaggregated proteoglycans but with greater binding in case of the aggregated proteoglycan. The implication of calcium binding by these marcromolecules in the biological process of calcification has been discussed in section 3.13. In the presence of cations (counterions), the macromolecular properties of these macromolecules are changed. Also, these changes vary with different counterions. Heparan sulfate contracts in the presence of calcium ions due to greater binding of counterions than with sodium ions. This contraction of the macromolecule produces differences in charge densities and differences in the sedimentation coefficients and diffusion coefficients in the presence of sodium ions ($S^{\circ}_{20,w} = 2.28 \cdot 10^{-13}$ s and $D^{\circ}_{20,w} = 2.7 \cdot 10^{-7}$ cm^2/s) and in the presence of calcium ions ($2.70 \cdot 10^{-13}$ s and $3.1 \cdot 10^{-7}$ cm^2/s respectively). Also, the concentration dependence is much less in the presence of calcium ions than in sodium chloride (Öbrink et al., 1975 a). Therefore, there are advantages in determining molecular weights of GAG in the presence of calcium ions. Changes in optical rotatory dispersion and circular dichroism data in the presence of counterions indicated conformational changes in hyaluronic acid in the presence of different counterions (Chakrabarti, 1977).

1.10 Interaction with other macromolecules

The glycosaminoglycans and proteoglycans interact with macromolecules including collagen, lipoproteins, lectins, basic peptides, basic proteins and platelet factor IV. These interactions have been reviewed (Laurent, 1977; Lindahl and Höök, 1978; Rosenberg et al., 1979).

Both the protein-free sulfated GAG (Wood, 1960) and intact proteoglycans (Mathews, 1965, Mathews and Decker, 1968) interact ionically with tropocollagen. Such interactions with collagen depend mainly on electrostatic forces and are of a great significance in the molecular organization of collagen in connective tissue (Öbrink et al., 1975). These involve ionic interactions between $-SO_3$ on proteoglycan or GAG and $-NH_3^+$ groups on collagen (Steven, 1971). Due to this, the ability of GAG to interact with collagen is directly related to their chain lengths and charge densities. However, the presence of iduronic acid in dermatan sulfate and heparan sulfate is probably responsible for their stronger interaction with collagen than chondroitin 4- or 6-sulfates of comparable charge densities (Öbrink and Wasteson, 1971). Meyer (1970) has suggested a special role for dermatan sulfate in vivo for formation of coarse and thick collagen fibers, since dermatan sulfate is absent from tissues like cornea or cartilage where the collagen fibers need to be very fine. In fact, the differences in distribution of the relative amounts of various GAG in various tissues may be related to the thickness, distribution and orientation of the collagen fibers needed as an adaptation to functions in those tissues. Also see section 3.2.

The interaction of a GAG with collagen depends upon its chemical structure and also probably on its chain conformation. Hyaluronic acid, keratan sulfate, dermatan sulfate and chondroitin sulfate changed the denaturation behavior of the acid-soluble collagen in dilute aqueous solution at pH 4. However, a different proportion of each GAG was necessary to increase the melting temperature from 38 to 46 °C (Gelman and Blackwell, 1974).

The free glycosaminoglycan fraction of plasma which is highly sulfated interacts with plasma lipoproteins and this interaction is significant in the pathogenesis of atherosclerosis (Iverius, 1972; Nakashima et al., 1975; Robinson et al., 1975). The protein-bound GAG fraction of plasma, which consists mostly of undersulfated chondroitin 4-sulfate prevents this interaction (Singh et al., 1977). Like the interactions with collagen, the interactions of GAG with lipoproteins involve electrostatic forces, are directly related to charge densities on the GAG macromolecules and are stronger with iduronic acid-containing dermatan sulfate and heparan sulfate than with chondroitin sulfates of comparable charged densities (Iverius, 1972). The arrangement of iduronic acid and total iduronic acid contents, along with the proportion and distribution of sulfate groups are related to the stronger interactions of heparan sulfate and dermatan sulfate with low density lipoproteins. An increase in the amounts of these two macromolecules in aging and atherosclerosis may possibly be responsible for deposition of lipoproteins in fatty streaks. In fact, the dermatan sulfate contents of atherosclerotic human aortas were higher than non-atherosclerotic controls (Robinson et al., 1975; Engelberg, 1978).

Platelet factor IV, is the heparin neutralizing factor released from the blood platelets during the early phase of blood coagulation. This factor interacts and strongly binds isolated chondroitin sulfate chains in a molar ratio of 1:1. Chondroitin 4-, 6-sulfates and dermatan sulfate interact equally, while heparan sulfate and heparin interact more strongly and can displace the three chondroitin sulfate isomers from their complexes with platelet factor IV (Barber et al., 1972). The platelet factor IV is present in the cytoplasmic granules of blood platelets complexed to the chondroitin 4-sulfate proteoglycan (M_r 350,000). Each proteoglycan consists of four chondroitin sulfate chains of molecular weight 12,000 each bound to the protein core and binds four moles of platelet factor IV. The whole complex is present as a dimer in the platelet granules (Barber et al., 1972). During the early phase of blood coagulation, the chondroitin 4-sulfate proteoglycan and platelet factor IV are released from the broken platelets (see section 16.6, chapter 16).

The proteoglycans interact with some lectins. Also, the proteoglycans are antigenic because of the presence of the protein and give precipitation reaction with antisera in double diffusion. By this procedure the identity or nonidentity of the protein cores of two proteoglycans can be checked (Muir et al., 1973). The proteoglycans and glycosaminoglycans also interact with basic proteins like protamine and myelin basic protein and basic peptides like poly-L-lysine and poly-L-arginine. Such interactions change the extended charged coil conformation of glycosaminoglycans into an α-helical conformation and this conformation-directing effect of a cationic polypeptide can be measured by the extent of ellipticity found on circular dichroism spectroscopy in dilute aqueous solutions at a neutral pH. The conformation-directing effects of poly-L-lysine or poly-L-arginine are weakened and lost with the rise of temperature (Rosenberg et al., 1979). Also, the conformation-directing effects exerted by a cationic polypeptide vary with different types of glycosaminoglycans (Hopfinger, 1977). Both the stoichiometry and the strength of the interactions depend upon the position, number and type of anionic groups on the polysaccharide backbone. The strength of interactions vary as: hyaluronic acid < chondroitin 4-sulfate < heparan sulfate < chondroitin 6-sulfate < dermatan sulfate (Gelman and Blackwell, 1974 a). Similarly, differences in the interactions of four proteoglycans with poly-L-arginine have been reported (Schodt and Blackwell, 1976). The position of the sulfate group is an important factor in such interactions. Studies using CD spectroscopy showed that at neutral pH poly-L-lysine instead of charged coil form was 80 % and about 20 % alpha helical in the presence of chondroitin 6-sulfate and chondroitin 4-sulfate, respectively (Gelman and Blackwell, 1973).

1.11 Biosynthesis of glycosaminoglycans and proteoglycans

The biosynthesis of glycosaminoglycans occurs as protein-linked glycosaminoglycans in different types of connective tissue cells, mesenchymal cells, mast cells, fibroblasts, chondrocytes. Cells other than those present in connective tissue, e. g., lymphocytes and leucocytes, also synthesize these proteoglycans. The mechanism and steps involved in the biosynthesis of different glycosaminoglycans and proteoglycans have been the subject of several excellent reviews (Stoolmiller and Dorfman, 1970; Silbert, 1970; Dorfman, 1970; Roden et al., 1972; Roden, 1973; Silbert, 1973; Phelps, 1973; Roden and Schwartz, 1974; Roden and Horowitz, 1978; Jourdian, 1979). A brief introduction to the biosynthesis steps is given below. The reader is referred to the reviews for other details or for references to the original articles.

Information on biosynthesis is still incomplete. However, the available facts can be summarized. The glycosaminoglycans are synthesized as proteoglycans for which the core protein is synthesized by established routes for protein synthesis. The carbohydrate chains are formed on this core protein by stepwise transfer of monosaccharides from the corresponding nucleotide sugars. The transfer of each monosaccharide is catalyzed by a transferase which is specific for the donor nucleotide sugar and the acceptor. The epimerization of D-glucuronic acid at C-5 to form L-iduronic acid occurs in the intact macromolecule. Further modification of the polysaccharide chain occurs by sulfation in which the 3-phosphoadenosine 5'-phosphosulfate (PAPS) acts as the donor for the active sulfate and the transfer is catalyzed by a sulfotransferase.

The GAG present extracellularly in connective tissue are first synthesized intracellularly and then are exported outside the cells by the Golgi apparatus, when they appear close to the surface and then spread into the extracellular space. The first step is the synthesis of core protein on the ribosomes, which then serve as a chain initiation point for addition of the carbohydrate-protein region neutral sugars and the glucuronyl residue. The residues for transfer to the protein core are used as their uridine dinucleotides. These transfers are catalyzed by specific transfer enzymes (transferases), which transfer the sugar residue to the protein core or the growing polysaccharide chain from the donor UDP-nucleotide sugars. All the monosaccharide precursors are formed from D-glucose. The formation and utilization of UDP-sugar derivatives as precursors for GAG synthesis involve at least 7 transferases and a number of oxidases and deaminases are used to synthesize sugar residues from glucose (Dorfman, 1970). These enzyme activities for formation of sugar nucleotides and the active sulfate donor "PAPS" are present in the cytoplasm (cytosol) fraction obtained as the high speed nonparticulate supernatent fraction.

Before the polysaccharide chain synthesis can begin, the protein core needs to be formed, since puromycin or cycloheximide prevents GAG synthesis. Hyaluronic acid which appears to exist free of protein may be an exception to this rule. The first step in the formation of the proteoglycans is the synthesis of the core protein on the ribosome, which then serves as a chain initiation point for the polysaccharide assembly. The enzymes involved in the formation of the protein core and for the polysaccharide assembly are found in the crude microsomal fraction attached to the membranous endoplasmic reticulum in the cell. The formation of the protein-carbohydrate linkage region and the formation of the repeating disaccharide sequence of the polysaccharide chain takes place by the specific glycosyl transferases catalyzed transfers of monosaccharide residues from their UDP-derivatives after the protein core has left the ribosome and has become localized in the cisternal space of the endoplasmic reticulum. The first step in the chain initiation is the xylosyl transferase catalyzed transfer of xylosyl residues from UDP-D-xylose to the protein core to form the xylose-serine linkage. Following this the linkage region (-GlcUA-Gal-Gal-Xyl-Ser-) is completed by the sequential transfers of two galactosyl residues by two specific galactosyl-transferases and a glucuronosyl residue transfer catalyzed by the UDP-D-glucuronic acid: 3-O-β-D-galactosyl-D-galactose glucuronosyltransferase. These enzymes are concentrated in the rough endoplasmic reticulum.

The formation of the repeating disaccharide sequence of the main polysaccharide chain takes place on the smooth membranes as the growing molecule is moving through the smooth membranes to the Golgi complex. This occurs by the alternating actions of an N-acetylglucosaminyltransferase in case of biosynthesis of hyaluronic acid, heparan sulfate and heparin (or N-acetylgalactosaminyltransferase for biosynthesis of galactosaminoglycans) and a glucuronosyltransferase different from that involved in the transfer of glucuronosyl residue next to galactose in the linkage region. These glycosyltransferases are probably distributed equally in the rough and smooth membrane fractions for chondroitin sulfate synthesis.

After the completion of the main polysaccharide chain, modifications in this completed polymer are catalyzed by specific enzymes. These modifications vary with the nature of the final glycosaminoglycan molecule. These include N-acetylation, N-sulfation of the amino groups of hexosamines in the polymer,C-5 epimerization of D-glucuronic acid residues to form L-iduronic acid residues and O-sulfation at C-4, C-6 position of the hexosamine residues and at the C-2 position of L-iduronic acid. The epimerization of D-glucuronic acid to L-iduronic acid in the intact polymer is catalyzed by a specific epimerase. The sulfate transfers from the sulfate donor "PAPS" (3-phosphoadenosine 5'-phosphosulfate) are catalyzed by the sulfotransferases, which are concentrated in the smooth membrane fraction. The N-sulfation of the amino groups of hexosamines involves their deacetylation followed by transfer of the sulfate groups to the freed amino groups. The completed polymer is released by the proteolytic degradation of the core protein. In case of heparin, the product so released is a macromolecular product which is biologically inactive. Cleavage of the inactive macromolecular heparin by a specific endoglucuronidase produces biologically active heparin chains.

The biosynthesis of GAG is regulated at the level of precursor synthesis and by several other factors like vitamins, hormones, etc. (section 1.13). The feedback inhibition exerts control of biosynthesis at two levels. The excessive accumulation of UDP-N-acetylglucosamine inhibits the enzyme L-glutamine-D-fructose-6-phosphate aminotransferase, which turns off the conversion of glutamine and fructose-1-phosphate into glucosamine-1-phosphate. Similarly, UDP-xylose inhibits the enzyme UDP-glucose dehydrogenase and this turns off the formation of UDP-glucuronic acid from UDP-glucose. These two feedback reactions, can control the availability of hexosamine and hexuronic acid for formation of the polysaccharide chain.

1.12 Degradation of proteoglycans and glycosaminoglycans

Since the proteoglycans consist of a protein core with several glycosaminoglycans attached to this protein, the complete breakdown of the proteoglycans requires a combination of the action of proteases and different types of glycosidases. These proteases and glycosidases have different specificities, attack at different points on the protein-GAG complexes and produce various types of end products. Some of the enzymes involved in the degradation of the complex proteoglycans of the connective tissues have been found in mammalian tissues and some of these have been purified. However, the information on the degradation process in the body is incomplete. The delineation of the defective degradation of the GAG in the patients with various mucopolysaccharidoses has helped to understand partly the importance of these degradative enzymes in the connective tissue. All the proteases and glycosidases involved in the degradation of proteoglycans and GAG are localized in the lysosomes.

Proteolytic enzymes degrade the proteoglycans producing polysaccharide chains attached to protein or peptides. Cathepsins A, B_1, C and D, the proteases of lysosomal origin, are important in the degradation of tissue proteoglycans in vivo. Cathepsins B_1 and D are endopeptidases and are more important in the initial breakdown of cartilage matrix and the release of GAG. The acid protease, cathepsin D, has been identified by the use of specific antiserum to be principal enzyme involved in the breakdown of cartilage proteoglycans (Morrison, 1970; Woessner, 1973). The cathepsins A and C are exopeptidases and act on the degradation products formed by cathepsins B_1 and D. Degradation by trypsin produces "doublets" consisting of two polysaccharide chains linked together by a trypsin-resistant polypeptide while pepsin or cathepsin D gives a triplet and papain and cathepsin B_1 produce single polysaccharide chains (Morrison, 1974).

Injections of two exogeneous proteases, papain (EC 3.4.4.10) and plasmin (EC 3.4.4.14) produce degradation of the cartilage proteoglycans with liberation of GAG into blood circulation, which results in increased uronic acid in blood and urine. Hypovitaminosis A increases the synthesis of cathepsin D in the lysosomes of cartilage and liver bringing about breakdown of the intercellular matrix of the cartilage. Injection of papain into the rabbit ear or therapeutic injection of this protease into the area of prolapsed intervertebral disc (herniation) results in loss of cartilage rigidity and dissolution of the prolapsed intervertebral disc, respectively, with an increase in urinary GAG excretion (Stern et al., 1968).

Lysozyme found in the cartilage may bring about disaggregation of the cartilage proteoglycans by degradation of the hyaluronic acid core (Kuettner et al., 1974). Collagenolytic enzymes are found in mammalian systems and are of a significance in the connective tissue breakdown in human rheumatoid arthritis. Leukocytes also contain collagenase that degrades insoluble collagen in cartilage (Anderson et al., 1977). The therapeutic effectiveness of phenylbutazone is due to its effect on partial inhibition of collagenase, lysosomal hyaluronidase and β-glucuronidase.

β-Xylosidase (EC 3.2.1.37) is present in mammalian tissues but being an exoglycosidase is unable to act on xylosyl-O-serine linkages in the intact chondroitin sulfate proteoglycans, which accounts for excretion of a significant quantity of xylosyl-serine in human urine. However, it can act in later stages of degradation. The mammalian tissue also contains β-galactosidases (EC 3.2.1.23) which may also be effective in later stages of degradation.

Lysosomes in several mammalian tissues contain hydrolytic enzymes for degradation of glycosaminoglycans. Two types of processes are involved in the release of hydrolytic enzymes from lysosomes. In one type the rupture of the lysosomes may destroy the entire cells and adjacent intercellular substance. Such is the case in erosion of bone by osteoclasts. Alternatively, the enzymes may be secreted into the intercellular substance or the substance to be hydrolyzed may be taken into the cell or within the lysosome. The latter may be the case in the case of mucopolysaccharidoses where the GAG enter the cells but due to deficiencies of the lysosomal enzymes these GAG are not degraded and therefore accumulate in the tissues.

The average half life of the glycosaminoglycan chain is believed to be the same as in experimental animals and is 2 $^{1}/_{2}$ days. On the basis of this half life and the total GAG of the body it has been estimated that in an adult approximately 250 mg of GAG are disposed off daily (Leaback, 1970). Out of this only 3-6 mg are excreted in the urine per day while the remaining are degraded in the body. In rats (Wood et al., 1973) the majority of the injected chondroitin 4-sulfate is broken down by the liver. Also since chondroitin 4-sulfate is the major GAG of plasma (see section 16.5) it appears that in the normal state, cleavage of chondroitin 4-sulfate from the protein moiety in proteoglycans in tissues like cartilage and aorta may be one major catabolic activity of tissues. Also, following injection of ^{35}S-labeled chondroitin sulfate, the radioactivity in urine is first in the macromolecular fraction and then increases with time in inorganic sulfate (Kaplan and Meyer, 1962). This suggests the presence of sulfatases, the enzymes that cleave ester sulfate groups from GAG. The degradation capacities of a tissue for different, individual GAG have been noted. When acetates, sulfate and glucose were employed in a type of triple labeling, hyaluronic acid and dermatan sulfate had different half lives (Schiller et al., 1956).

Enzymes having hyaluronidase (EC 3.2.1.30) activity have been found, besides the testis, in several tissues including rat bones, kidney, spleen, liver, lung, submandibular gland, synovial fluid, artery, urine and serum. The enzyme is an endo-poly-β-hexosaminidase and cleaves the β-1→4 hexosaminide bonds between N-acetyl-D-glucosamine and D-glucuronic acid in hyaluronic acid. This enzyme also acts on chondroitin 4- and 6-sulfates, bringing about a random splitting of the polysaccharide chain to mainly the tetrasaccharides. Due to the transglycosylase activity present in hyaluronidase, higher oligosaccharides are also formed to some extent.

Mammalian tissues contain β-glucuronidase (EC 3.2.1.31) which can cleave the β-1→3 hexuronide linkages at the terminal non-reducing ends of even-numbered oligosaccharides produced by hyaluronidase action. The result is removal of a D-glucuronic acid residue producing odd-numbered oligosaccharides having N-acetylhexosamine moieties at both reducing and non-reducing terminal ends. The β-N-acetylhexosaminidase (EC 3.2.1.30) present in liver can cleave these oligosaccharides with non-reducing terminal N-acetyl-D-glucosamine residues liberating free N-acetyl-D-glucosamine as one of the products.

Enzymes for in vivo depolymerization of dermatan sulfate and heparin are known in mammalian tissues. The deficiency of α-L-iduronidase acitivity is the cause of storage of heparan sulfate and its increased urinary excretion in patients with Sanfilippo syndrome. A limited depolymerization of GAG may also occur by the action of exoglycosidases like α-fucosidase, β-galactosidase, etc. Only 11 % of the chondroitin 6-sulfate injected into a dog (at 5 mg/kg) appeared in the urine either as intact GAG or as degradation products, whereas 39 % of heparan sulfate and 30 % of dermatan sulfate were excreted after separate prolonged injections (Kaplan and Meyer, 1962). Human tissues and serum contain enzymes for degradation of GAG (Öckerman et al., 1969).

Sulfatases for desulfation of GAG are present in mammalian tissues, since the labeled sulfated glycosaminoglycans introduced subcutaneously or intraperitoneally in dog and man are removed from blood circulation with a concomitant increase of radioactivity in the urinary inorganic sulfate (Kaplan and Meyer, 1962). The enzyme 4-L-aspartylglycosylamine amidohydrolase (EC 3.5.1.37) cleaves the linkage regions that occur in some keratan sulfates. This enzyme is present in human tissues and the deficiency of this enzyme is seen in patients with aspartylglycosylaminuria (also see section 16.7, 13.4).

1.13 Effects of vitamins, hormones and minerals, etc.

The genetic information can control biosynthesis by the effects on formation of the core proteins or the enzymes for the biosynthetic steps. The metabolism is modified by several other factors including vitamins, hormones, drugs and minerals.

Vitamin A deficiency or food deprivation decrease the ATP-sulfurylase thereby producing less "PAPS" to act as a sulfate donor. The vitamin A deficient or food deprived animal therefore contained a higher proportion of nonsulfated and less of sulfated GAG in the tissues. Hypervitaminosis had an opposite effect. Also, excess vitamin A appeared to accelerate catabolism of GAG by activation of a lysosomal protease. Addition of vitamin A to rat liver lysosomes releases cathepsins probably by damaging the lysosomal membrane. Vitamin A injection into rabbits produces the same effect as that produced by papain injection in the ear cartilage.

Vitamin C deficiency has been found to affect changes in the connective tissue either secondary to collagen formation or in decreased healing of granulation tissue due to a decreased incorporation of galactosamine and sulfate into GAG. However, the mechanism involved in these may be complicated. In vitro, vitamin C causes partial depolymerization of hyaluronic acid by an oxidation-reduction reaction. This effect may operate in vivo, since intradermal injection of the vitamin C increases permeability.

Administration of vitamin D to rachitic animals increases the galactosamine content of cartilage and the GAG content of the cartilage in normal subjects was more. These observations suggest that vitamin D affects biosynthesis of GAG by some yet unknown mechanism.

Insulin affects biosynthesis. In skin from diabetic rats, the incorporation of labeled acetate and sulfate was diminished and this was reversed by insulin administration. This was probably related to increased degradation of GAG and of UDP-glucuronic acid in skin from diabetic animals.

Thyrotropin released from the anterior pituitary gland (adenohypophysis) increases the hyaluronic acid in the retrobulbar tissue causing proptosis (protrusion of the eyeball). Experimental exophthalmos can be produced in guinea pigs by administration of thyrotropin. Patients with exophthalmos excrete increased amounts of GAG in urine. Administration of growth hormone or thyroxine increase GAG synthesis. A shift in the synthesis from dermatan sulfate to hyaluronic acid occurs in the localized myxedema in Grave's disease.

Androgens produce increased synthesis and accumulation of hyaluronic acid in the comb of capons. Estrogens also cause an increase in biosynthesis and are responsible for tumescence of perineal sex skin and of increased cervical mucus during the estrous cycle.

Administration of methylprednisolone in experimental animals produces an increase of hyaluronic acid in the vitreous humor. In other tissues, this decreases the GAG contents. Adrenalectomy and stress decrease GAG biosynthesis due to decreased formation of UDP-glucuronic acid.

Anti-inflammatory drugs like salicylates, chloroquine and phenylbutazone reduce incorporation of labeled sulfate or glucosamine into GAG due to decreased formation of PAPS or uncoupling of oxidative phosphorylation. Gold salts used in treatment of rheumatoid arthritis decrease degradation of GAG by inhibiting cathepsins and β-N-acetylglucosaminidase of human synovial fluid.

Manganese appears to be involved as a cofactor for protein-carbohydrate linkage formation and for polymerization of chondroitin sulfate. In guinea pigs reduced amounts of GAG were present in cartilage in manganese deficiency. Deficiency of copper ions had the opposite effect in aortas and large blood vessels in swine. In this case, whereas heparan sulfate remained unchanged, dermatan sulfate and chondroitin sulfate increased.

Other factors such as hypoxia, food deprivation and starvation, cancer, chemical agents, etc., also modify the metabolism of GAG and so do pathological states, particularly the connective tissue disorders.

1.14 Bibliography

1. Anderson, J. C., Labedz, R. I. and Kewley, M. A. (1977): The effect of bovine tendon glycoprotein on the formation of fibrils from collagen solutions. Biochem. J. 167, 345-351

2. Balazs, E. A., Ed. (1970): Guide to nomenclature. In: E. A. Balazs (Ed.) Chemistry and Molecular Biology of the Intercellular Matrix. Academic Press, New York, pp. XXXIX-XXXI

3. Balazs, E. A. and Jeanloz, R. W. Eds. (1965): A guide to the nomenclature. The Amino Sugars, Vol. II A. Academic Press, New York, pp. XIII-XIV

4. Barber, A. J., Kaser-Glanzmann, R., Jakabova, M. and Luscher, E. F. (1972): Characterization of a chondroitin 4-sulfate proteoglycan carrier for heparin neutralizing activity (Platelet factor 4) released from human blood platelets. Biochim. Biophys. Acta 286, 312-329

5. Bettleheim-Jevons, F. R. (1958): Protein-carbohydrate complexes. Adv. Protein Chem. 13, 35-105

6. Bhavanandan, V. P. and Meyer, K. (1968): Studies on keratosulfates. Methylation, desulfation and acid hydrolysis studies on old human rib cartilage keratosulfate. J. Biol. Chem. 243, 1052-1059

7. Blix, G. (1940): Carl Thore Morner. Svenska lak-tidning 37, 1569-1572

8. Brimacombe, J. S. and Weber, J. M., Eds. (1964): Mucopolysaccharides. Elsevier Publishing Co., Amsterdam

9. Buddecke, E., Kresse, H. and Segeth, G. (1971): Chemical and metabolic heterogeneity of ^{14}C and ^{35}S sulfate labelled glycosaminoglycans of bovine arterial tissue. Z. Klin. Chem. Klin. Biochem. 9, 74

10. Buddecke, E. and Von Figura, K. (1975): Different types of chondroitin sulfate-dermatan sulfate hybrids in arterial tissue. Protides Biol. Fluids, Proc. Colloq. 22, 87-90

11. Carlisle, E. M. (1972): Silicon: an essential element for the chick. Science 178, 619-621

12. Chakrabarti, B. (1977): Effect of counterions on the conformation of hyaluronic acid. Arch. Biochem. Biophys. 180, 146-150

13. Chakrapani, B. and Bacchawat, B. K. (1968): Glycosaminoglycans of human urine. Part I. Protein-polysaccharide linkage region in the nonsulfated glycosaminoglycans of normal urine. Indian J. Biochem. 5, 9, 12

14. Clark, F. and Grant, J. K. (1961): The Biochemistry of Mucopolysaccharides of Connective Tissue, Biochem. Soc. Sympos. No. 20, Cambridge University Press, Cambridge

15. Dorfman, A. (1970): The biosynthesis of acid mucopolysaccharides of connective tissues. In: W. Montagna, J. P. Benteley and R. L. Dobson (Eds.) Advances in Biology of Skin, Vol. X: The Dermis. Appleton-Century-Crofts, New York, Chapter 8

16. Engelberg, H. (1978): Heparin. Monographs on Atherosclerosis. Vol. 8, S. Karger, Basel, pp. 1-72

17. Gelman, R. A. and Blackwell, J. (1973): Mucopolysaccharide-polypeptide interactions. Effect of the position of the sulfate group. Biochim. Biophys. Acta 297, 425-455

18. Gelman, R. A. and Blackwell, J. (1974): Collagen mucopolysaccharide interactions at acid pH. Biochim. Biophys. Acta 342, 254-261

19. Gelman, R. A. and Blackwell, J. (1974 a): Interactions between mucopolysaccharides and cationic polypeptides in aqueous solution: hyaluronic acid, heparitin sulfate, and keratan sulfate. Biopolymers 13, 139-145

20. Gregory, J. D. and Jeanloz, R. W. Eds. (1979): Glycoconjugate Research, Vol. I and II. Academic Press, New York

21. Hamerman, D., Rojkind, H. and Sandson, J. (1966): Protein bound to hyaluronate. Chemical and immunological studies. Fed. Proc. 25, 1040-1045

22. Hoekstra, W. H., Suttie, J. W., Ganther, H. E. and Mertz, W. Eds. (1974): Trace Element Metabolism in Animals, Vol. 2. University Park Press, Baltimore

23. Hopfinger, A. J. (1977): Intermolecular Interactions and Bimolecular Organization, Wiley Interscience, New York, 1977

24. Hopwood, J. J. and Robinson, H. C. (1974): The alkali-labile linkage between keratan sulfate and protein. Biochem. J. 141, 57-69

25. Hopwood, J. J. and Robinson, H. C. (1974 a): The structure and composition of cartilage keratan sulfate. Biochem. J. 141, 517-526

26. Horowitz, M. and Pigman, W., Eds. (1977): The Glycoconjugates: Mammalian Glycoproteins and Glycolipids. Vol. I. Academic Press, New York

27. Iverius, P. H. (1972): The interaction between human plasma lipoproteins and connective tissue glycosaminoglycans. J. Biol. Chem. 247, 2607-2613

28. Jeanloz, R. W. (1960): The nomenclature of mucopolysaccharides. Arthritis Rheumat. 3, 233-237

29. Jeanloz, R. W. (1963): Mucopolysaccharides (acidic glycosaminoglycans). In: M. Florkin and E. H. Stotz (Eds.) Comprehensive Biochemistry Vol. V. Academic Press, New York, pp. 262-296

30. Jeanloz, R. W. (1970): Mucopolysaccharides of higher animals. In: W. Pigman and D. Horton (Eds.): The Carbohydrates. Academic Press, New York, 1970, Vol. 2B, pp 589-625

31. Jourdian, G. W. (1979): Biosynthesis of glycosaminoglycans. In: R. U. Margolis and R. K. Margolis (Eds.) Complex Carbohydrates of Nervous Tissue. Plenum Publ. Corp., New York, Chapter 5, pp. 103-125

32. Kaplan, D. and Meyer, K. (1962): The fate of injected mucopolysaccharides. J. Clin. Invest. 41, 743-749

33. Kent, P. W. and Whitehouse, M. W. (1955): Biochemistry of the Amino Sugars. Butterworths, London, Chapter I, pp. 1-9

34. Kieras, F. (1974): Linkage region of cartilage keratan sulfate to protein. J. Biol. Chem. 249, 7506-7513

35. Kuettner, K. E., Croxen, R. L., Eisenstein, R. (1974): Proteinase inhibitor activity in connective tissues. Experientia 30, 595-597

36. Lagunoff, D. and Warren, G. (1962): Determination of 2-deoxy-2-sulfaminohexose content of mucopolysaccharides. Arch. Biochem. Biophys. 99, 396-400

37. Laurent, T. C. (1977): Interaction between proteins and glycosaminoglycans. Fed. Proc. 36, 24-27

38. Leaback, D. H. (1970): The metabolic hydrolysis of hexosaminide linkages. In: W. H. Fishman (Ed.) Metabolic Conjugation and Metabolic Hydrolysis, Vol. 2, Academic Press, New York, pp 443-517

39. Levene, P. A. (1925): Hexosamine and Mucoproteins. Longmans, Green and Co., New York

40. Lindahl, U. and Hook, M. (1978): Glycosaminoglycans and their binding to biological macromolecules. Annu. Rev. Biochem. 47, 385-417

41. MacGregor, E. A. and Bowness, J. M. (1971): Interaction or proteoglycans and chondroitin sulfates with calcium or phosphate ions. Can. J. Biochem. 49, 417-425

42. Margolis, R. U. (1967): Acid mucopolysaccharides and proteins of bovine whole brain, white matter and myelin. Biochem. Biophys. Acta 141, 91-102

43. Masamune, H. (1956): On the mode of combination of carbohydrate with amino acid grouping in mucopolysaccharides and glycoproteins. Proc. 3rd. Intern. Congr. Biochem., Brussels, 1955, p. 72

44. Mathews, M. B. (1965): The interaction of collagen and acid mucopolysaccharides. A model for connective tissue. Biochem. J. 96, 710-716

45. Mathews, M. B. and Cifonelli, J. A. (1965): Comparative biochemistry of keratosulfates. J. Biol. Chem. 240, 4140-4145

46. Mathews, M. B. and Decker, L. (1968): The effect of acid mucopolysaccharides and acid mucopolysaccharide-proteins on fibril formation from collagen solutions. Biochem. J. 109, 517-526

47. Meyer, K. (1938): The chemistry and biology of mucopolysaccharides and glycoproteins. Cold Spring Harbor Symp. Quant. Biol. 6, 91-102

48. Meyer, K. (1953): Mucoproteins and mucoids. In: H.W.Cole, (Ed.) Symposium on Some Conjugated Proteins. Ann. Conf. Protein Metabolism, Bur. Biol. Research, Rutgers University, 9, 64-73

49. Meyer, K. (1969): Biochemistry and biology of mucopolysaccharides. Am. J. Med. 47, 664-672

50. Meyer, K. (1970): Reflections on "Mucopolysaccharides" and their protein polysaccharide complexes. In: E. A. Balazs, Ed. The Chemistry and Molecular Biology of the Intercellular Matrix. Academic Press, London, Vol. 1, pp. 5-24

51. Meyer, K. and Palmer, J. W. (1934): The polysaccharide of the vitreous humor. J. Biol. Chem. 107, 629-634

52. Meyer, K. and Palmer, J. W. (1936): On glycoproteins II. The polysaccharides of vitreous humor and of umbilical cord. J. Biol. Chem. 114, 689-703

53. Morrison, R. I. G. (1970): The breakdown of proteoglycans by lysosomal enzymes and its specific inhibition by an antiserum to cathepsin D. In: E. A. Balazs (Ed.) Chemistry and Molecular Biology of Intercellular Matrix. Academic Press, New York, Vol. 3, pp. 1683-1706

54. Morrison, R. I. G. (1974): The enzymatic degradation of proteoglycans. In: R. Fricke and F. Hartman (Eds.) Connective Tissues: Biochemistry and Pathophysiology, Springer Verlag, New York, pp. 150-157

55.Muir., H. (1964): Chemistry and metabolism of connective tissue glycosaminoglycans (mucopolysaccharides). Int. Rev. Connec. Tissue Res. 2, 101-154

56. Muir, H., Baxter, E. and Brandt, K. D. (1973): The antigenicity of extracellular proteoglycans: implications for structure and biosynthesis. Biochem. Soc. Trans. 1, 223-226

57. Nakashima, Y., Di Ferrante, N., Jackson, R. L. (1975): The interaction of human plasma glycosaminoglycans with plasma lipoproteins. J. Biol. Chem. 250, 5386-5392

58. Öbrink, B., Laurent, T. C. and Carlsson, B. (1975): The binding of chondroitin sulfate to collagen. FEBS Lett. 56, 166-169

59. Öbrink, B., Pertoft, H. and Iverius, P. H. (1975 a): The effect of calcium on the macromolecular properties of heparan sulfate. Connective Tissue Res. 3, 187-193

60. Öbrink, B. and Wateson, A. (1971): Nature of the interaction of chondroitin 4-sulphate and chondroitin sulphate-proteoglycan with collagen. Biochem. J. 121, 227-233

61. Öckermann,P. A., Hultberg, B. and Erickson, O. (1969): Enzyme patterns in tissue and body fluids in the mucopolysaccharidoses. Clin. Chim. Acta 25, 97-102

62. Phelps, C. F. (1973): The biosynthesis of glycosaminoglycans. Biochem. Soc. Trans. 1, 814-819

63. Quintarelli, G. (1968): The Chemical Physiology of Mucopolysaccharides. Little Brown and Company, Boston

64. Robinson, R. W., Likar, I. N. and Likar, L. J. (1975): Glycosaminoglycans and arterial disease. Monographs on Atherosclerosis (S. Karger, Basel) 5, 1-136

65. Roden, L. (1973): Biosynthesis of connective tissue polysaccharides. Pure and Applied Chem. 35, 181-193

66. Roden, L., Baker, J. R., Schwartz, N. B., Stoolmiller, A. C., Yamagata, S. and Yamagata, T. (1972): Some aspects of the structure and biosynthesis of connective tissue proteoglycans. In: R. Piras and H. G. Pontis (Eds.). Biochemistry of the Glycosidic Linkage: An Integrated View, PAABS. Symposium, Vol. 2, Academic Press, New York, pp. 345-386

67. Roden, L. and Horowitz, M. I. (1978): Structure and Biosynthesis of Connective Tissue Proteoglycans. In: M. I. Horowitz and W. Pigmans (Eds.) The Glycojugates, Vol. II, Academic Press, New York, pp. 3-71

68. Roden, L. and Schwartz, N. B. (1974): Biosynthesis of connective tissue proteoglycans. In: R. Fricke and Hartman, F. (Eds.). Connective Tissues and Pathophysiology. Springer Verlag, New York, pp. 73-84

69. Rosenberg, L. (1979): Structure and function of proteoglycans. In: D. J. McCarty (Ed.). Arthritis and Allied Conditions. A Text Book of Rheumatology. 9th Ed., Lea and Febiger, Philadelphia, pp. 189-200

70. Rosenberg, L., Choi, H., Pal, S. and Tang, L. (1979): Carbohydrate-protein interactions in proteoglycans. In: I. J. Goldstein (Ed.). Carbohydrate-Protein Interactions. ACS Sympos. Ser. 88, ACS, Washington, D. C., pp. 186-216

71. Sajdera, S. W. and Hascall, V. C. (1969): Proteinpolysaccharide complex from bovine nasal cartilage. J. Biol. Chem. 244, 77-87

72. Schiller, S., Mathews, M. B., Cifonelli, J. A. and Dorfman, A. (1956): Metabolism of mucopolysaccharides in animals; further studies on skin utilizing C^{14}-glucose, C^{14}-acetate and S^{35}-sodium sulfate. J. Biol. Chem. 218, 139-145

73. Schodt, K. P. and Blackwell, J. (1976): Comparison of four proteoglycans in terms of their interactions with poly (L-arginine). Biopolymers 15, 469-482

74. Schwarz, K. (1973): A bound form of silicon in glycosaminoglycans and polyuronides. Proc. Natl. Acad. Sci. USA 70, 1608-1612

75. Scott, J. E. (1960): Aliphatic ammonium salts in the assay of acidic polysaccharides from tissues. Methods Biochem. Anal. 8, 145-197

76. Scott, J. E. (1973): Affinity, competition and specific interactions in the biochemistry and histochemistry of polyelectrolytes. Biochem. Soc. Trans. 1, 787-806

77. Seno, N., Meyer, K. and Anderson, B. (1965): Variations in keratosulfates. J. Biol. Chem. 240, 1005-1010

78. Silbert, J. E. (1970): Biosynthesis of heparin. In: E. A. Balazs, Ed. Chemistry and Molecular Biology of the Intercellular Matrix, Academic Press, New York, 1970, Vol. 2, pp. 991-1004

79. Silbert, J. E. (1973): Biosynthesis of mucopolysaccharides and protein-polysaccharides. In: R. Perez-Tamayo and M. Rojkind (Eds.). The Biochemistry of Disease. Vol. 3: Molecular Pathology of Connective Tissue. Marcel Dekker, Inc., New York, Chap. 6, pp. 323-353

80. Silbert, J. E., Kleinman, H. K. and Silbert, C. K. (1975): Heparin and heparin-like substances of cells. Adv. Exptl. Med. Biol. 52, 51-60

81. Simunek, Z., Tsiganos, C. P. and Muir, H. (1971): Chemical and Molecular heterogeneity of proteoglycans. Z. Klin. Chem. Klin. Biochem. 9, 72-73

82. Singh, J., Di Ferrante, N., Gyorkey, F. and Wilson, N. (1977): Plasma glycosaminoglycans in normal individuals of various age. Atherosclerosis. 28, 319-324

83. Stacey, M. (1946): The chemistry of mucopolysaccharides and mucoproteins. Adv. Carbohydrate Chem. 2, 161-201

84. Stary, Z., Wardi, A. H., Turner, D. L. and Allen, W. S. (1965): Arabinose as a mucopolysaccharide component in human and animal brain tissue. Arch. Biochem. Biophys. 110, 388-394

85. Stern, I. J., Cosmas, F. and Smith, L. (1968): Urinary polyuronide excretion in men after enzymic dissolution of the chondromucoprotein of the intervertebral disc or surgical stress. Clin. Chim. Acta 21, 181-190

86. Steven, F. S. (1971): The influence of proteoglycans on the molecular organization of collagen fibrils. Z. Klin. Chem. Klin. Biochem. 9, 75-76

87. Stoolmiller, A. C. and Dorfman, A. (1970): The biosynthesis of hyaluronic acid in Group A streptococcus. In: E. A. Balazs (Ed.). Chemistry and Biology of Intercellular Matrix. Academic Press, New York, pp. 783-794

88. Toda, N. and Seno, N. (1970): Sialic acid in the keratan sulfate fraction from whale cartilage. Biochim. Biophys. Acta 208, 227-235

89. Tsiganos, C. P. and Muir, H. (1970): The natural heterogeneity of proteoglycans of porcine and human cartilage. In: E. A. Balazs (Ed.). Chemistry and Molecular Biology of the Intercellular Matrix, Vol. 2, Academic Press, 859-866

90. Varma, R., Varma, R. S., Allen, W. S. and Wardi, A. H. (1974): On the carbohydrate-protein linkage in vitreous humor hyaluronate. Biochim. Biophys. Acta 362, 584-588

91. Varma, R., Varma, R. S., Allen, W. S. and Wardi, A. H. (1974 a): Neutral sugars from the hyaluronate-peptide of vitreous humor. Carbohyd. Res. 32, 386-395

92. Varma, R., Varma, R. S., Allen, W. S. and Wardi, A. H. (1975): Human umbilical cord hyaluronate: Neutral sugar content and carbohydrate-protein linkage studies. Biochim. Biophys. Acta 399, 139-144

93. Varma, R., Varma, R. S., Allen, W. S. and Wardi, A. H. (1975 a): Hyaluronate from aqueous humor: Some physicochemical studies. Biochem. Exp. Biol. 11, 407-417

94. Varma, R., Vercellotti, J. R. and Varma, R. S. (1977): On arabinose as a component of brain hyaluronate. Confirmation by chromatographic, enzymatic and chemical ionization-mass spectrometric analyses. Biochim. Biophys. Acta 497, 608-614

95. Wardi, A. H., Allen, W. S., Turner, D. L. and Stary, Z. (1966): Isolation of arabinose-containing hyaluronate peptide and xylose-containing chondroitin sulfate-peptides from protease digested brain tissue. Arch. Biochem. Biophys. 117, 44-53

96. Woessner, J. F., Jr. (1973): Cartilage cathepsin D and its action on matrix components. Fed. Proc. 32, 1485-1488

97. Wolstenholme, G. E. W. and O'Connor, M. (1958): Ciba Symposium on Chemistry and Biology of Mucopolysaccharides. Little Brown and Company, Boston

98. Wood, G. C. (1960): The formation of fibrils from collagen solutions. 3. effect on chondroitin sulfate and some other naturally occurring polyanions on the rate of formation. Biochem. J. 75, 605-612

99. Wood, W. M., Wusteman, F. S. and Curtis, C. G. (1973): The degradation of intravenously injected chondroitin 4-sulfate in the rat. Biochem. J. 134, 1009-1013

Chapter 2

General Methods for Isolation, Fractionation and Analysis

2.1 Introduction

For understanding the physicochemical behavior of a macromolecule, it is essential to find out what kinds of molecules it is composed of in the native state in a tissue or in a biological fluid. The first step toward this goal would require that the mixture of various macromolecules present must be isolated from the biological material and then separated into their individual molecular species for study of their individual properties. This is achieved by use of a number of fractionation procedures. The next step would be the analysis of the components of individual macromolecules separated by the fractionation procedures. This would involve determination of the contents of the components: hexosamines, hexuronic acids, sulfate and acetyl groups and neutral sugars, and the manner in which these components are linked or positioned to make the macromolecule. The present discussion would be restricted to the isolation of the glycosaminoglycans from biological fluids, their fractionation into the component members, physical characterization for parameters such as molecular weight determination and analysis of the constituents of these macromolecules.

Sometimes it is possible to obtain information about the nature of GAG components in the mixture without fractionation from the chemical analysis. For example, orcinol/carbazole ratios provide information about the presence of iduronic acid-containing GAG components. Similarly, N-sulfohexosamine and N-sulfohexosamine/total sulfate ratios give information about the amounts of heparan sulfate and heparin in the mixture.

All the isolation and fractionation methods for glycosaminoglycans are based upon polyanionic character of these macromolecules. The differences in the distribution of anionic charges on the various glycosaminoglycans permit their separation and characterization by differential staining and precipitation, anion exchange chromatography and electrophoretic methods (Taniguchi, 1976).

2.2 Isolation procedures

The methods for fractionation and analysis of glycosaminoglycans (GAG) isolated from biological fluids are similar to those for the macromolecules isolated from the tissues. However, the isolation procedures from the two sources vary, since in tissues the GAG are present as protein-polysaccharide complexes called proteoglycans, whereas in biological fluids these exist generally as glycosaminoglycans, which are actually the degradation products of tissue proteoglycans. In the biological fluids the GAG are present in a solubilized form and the steps involving homogenization with acetone for defatting and for making acetone powders for isolating GAG from the tissues are unnecessary. In some of the biological fluids such as blood plasma, a part of the GAG may be present in the free fraction unbound to proteins and this fraction can be selectively removed and separated from the protein-bound fraction by chromatography on a weak anionic resin, whereas the protein-bound fraction can be isolated from the eluate from the weak anion-exchange column either directly or after digestion with a proteolytic enzyme such as pronase, papain, trypsin, etc. (chapter 16, section 16.5).

The most commonly used procedure for isolating GAG from the biological fluids involve precipitation of all the GAG fractions together using quaternary ammonium salts such as cetylpyridinium chloride (CPC) or cetyltrimethylammonium bromide (CTAB). Various aspects of the use of quaternary ammonium salts have been reviewed (Scott, 1960, 1965, 1970). Anion-exchange chromatography can also be used for isolation of the GAG from diluted fluids. Following isolation the analysis of the total GAG can be carried out on the basis of analysis of the intact macromolecule by interaction with a cationic agent or on the basis of analysis of its components.

For isolation of GAG from biological fluids such as blood plasma or serum where some GAG are present protein-bound, proteolytic digestion is necessary before precipitation of the total GAG. For this purpose digestion with papain at 65-70 °C or with pronase at pH 7.8-8.0 and 37-40 °C are commonly used. Pronase is more useful and efficient for cleaving the peptide bonds near the carbohydrate-peptide linkage region. Also, papain glycopeptides appear as contaminants in the hyaluronic acid fraction of the proteolytic digest. Pepsin and trypsin are usually not used; the low pH of 1.5 needed for pepsin may produce N-desulfation in heparan sulfate and heparin.

Following proteolytic digestion the residual protein and the low molecular weight fragments released upon proteolysis and those present prior to digestion are removed by shaking with chloroform-butanol (1:1v/v, Staub, 1965), addition of trichloroacetic acid to a final concentration of 10% or acidification and immersing in boiling water for 5-10 minutes, followed by centrifugation or filtration. The supernatent or filtrate is neutralized and dialyzed. Alternatively, the GAG components can be precipitated with 3-5 volumes of ethanol. In both dialysis and ethanol precipitation the low molecular weight fractions are lost and to overcome this ultrafiltration through Amicon Diaflo membranes of different pore sizes or gel filtration through Sephadex of various mesh sizes may be used as an alternative. Dialysis in cellophane casings preheated to reduce the pore size have also been used in some studies. Precipitation with quaternary ammonium salts (CPC 3 mg/mg GAG at neutral or slightly acidic pH or CTAB 0.05-0.1%) may be used as an alternative to ethanol precipitation. The CPC-GAG complexes obtained by centrifugation may be freed from the bound CPC by washing the precipitate with 0.1% CPC and dissolving it first in 0.6 M NaCl for bringing GAG into solution leaving glycoproteins in the precipitate. Complete dissociation of the complexes can also be carried out with 2 M NaCl and the CPC can be precipitated from the supernatent with a saturated solution of potassium thiocyanate (Korn, 1959) for several hours or overnight at 4 °C, followed by centrifugation at 4 °C and dialysis. An immiscible organic solvent like chloroform or amyl alcohol may also be used for removal of excess CPC (Scott, 1960) or it may be removed by passing the solution through a column of cationic resin such as AG 50W-X8 (H^+). The GAG can be precipitated from the CPC-free solution by addition of 4 volumes of ethanol.

2.3 Fractionation of GAG mixtures

The GAG mixture in the deproteinized pronase/papain digest may be fractionated directly by anion-exchange chromatography using AG 1-X2 (Dowex 1-X2) or another anionic resin. Alternatively, the GAG obtained in the ethanol precipitate may be separated into various fractions by using one of the following procedures.

2.3.1 Fractionation with alcohol

Fractionation of the GAG mixture can usually be carried out by adding varying concentrations of ethanol to a 1-2 % solution of the mixture in an acidic medium containing divalent metal ions (5 % $CaCl_2$ solution in 0.5 M acetic acid). By this procedure dermatan sulfate precipitates with 30-40 % ethanol, chondroitin 6-sulfate with 40-50 % ethanol and keratan sulfate with 45-65 % ethanol. Heparan sulfate and heparin require higher concentrations of alcohol.

Gardell (1957) presented an improved procedure for fractional elution of GAG precipitated with ethanol on a column packed with Whatman cellulose powder. A 0.5 ml of a 10 % GAG solution in 0.3 % barium acetate is applied to the cellulose column pre-equilibrated with 80 % ethanol containing 0.3 % barium acetate and washed with a small amount of a 0.3 % barium acetate solution. The column is then eluted stepwise with decreasing 80 → 20 % alcohol containing 0.3 % barium acetate. This column procedure has been found very useful for isolation of keratan sulfate and dermatan sulfate from various sources.

2.3.2 Precipitation with inorganic ions

The GAG form precipitates with inorganic ions like copper in alkaline medium (Benedict solution), scandium ions (Wardi et al., 1974), zirconium ions (Wardi and Michos, 1973), zinc ions (Sato and Gyorkey, 1976) in an acidic medium. Precipitation with Benedict's reagent separates dermatan sulfate from other GAG components, zirconium and scandium ions separate the total GAG from proteins or glycoproteins and complexes of various GAG components with zinc ions can be separated by countercurrent distribution (Sato and Gyorkey, 1977).

2.3.3 Fractionation from quaternary ammonium precipitates

Fractional precipitation methods using quaternary ammonium salts have been reviewed (Scott, 1960, 1965). The complex formation between the GAG and quaternary ammonium salt (CPC or CTAB) depends on the charge densities of the GAG components. The GAG components are released from the complex when the salt concentration of the solution reaches the critical electrolyte concentration for a particular charge density on the GAG molecules, thus by using varying salt concentrations the GAG components can be separated into three groups: hyaluronic acid and chondroitin; chondroitin 4- and 6-sulfate, dermatan sulfate and heparan sulfate; and heparin (Scott, 1960, 1965). Several modifications of this principle have been used by different investigators. One procedure involves gradient elution of the GAG-CPC complex from the inert support Whatman cellulose powder packed in a column (Antonopoulos et al., 1961, 1964). Applications of this procedure on a microgram scale have been reported (Szirmai et al., 1967; Svejcar and Robertson, 1967; Dorner et al., 1968). A 1-2 ml of GAG solution in 0.05 M sodium sulfate per cm square cross section area of the column is drained into the column and washed sequentially with water and 2 bed volumes of 1 % CPC. The column is then eluted with increasing concentrations of salt solutions containing 0.05 % CPC. The fractions are monitored by carbazole reaction in the effluent from which GAG components can be recovered by concentration to a small volume and precipitation with ethanol (Thunell et al., 1967). Gardell (1965) has reported separation of a mixture of hyaluronic acid, chondroitin sulfate and heparin on a column packed with cellulose powder slurry by first eluting with 1 % CPC and then sequentially with 0.15, 0.5 and 1.0 M $MgCl_2$ containing 0.05 % CPC.

In another modification the GAG-CPC complex is dissolved in 2 M NaCl and the GAG present in supernatents are obtained by stepwise dilution with varying concentrations (1.4, 1.2, 0.5 and 0.1 M) of NaCl containing 0.05 % CPC. The GAG fractions at various dilutions are recovered from the supernatent by precipitation with ethanol. The various fractions obtained at various salt dilutions depend upon the sulfate content of the GAG components. Thus hyaluronic acid is recovered at 0.1 M NaCl, the mixture of the three chondroitin sulfate isomers and heparan sulfate at 0.5 M, heparan sulfate and heparin of a lesser sulfate content at 1.2 M and heparin at 1.4 M NaCl.

Some investigators have used the reverse approach of bringing the various GAG fractions into solution by use of increasing NaCl concentration.

2.3.4 Paper and thin-layer chromatography

The mixture of GAG can be separated into the individual members by paper chromatography or thin-layer chromatography using ammonium formate buffer-isopropanol (65:35, v/v), heparin could be separated from chondroitin sulfate and commercial heparin could be separated into two principal components on Whatman no. 1 filter paper (Spolter and Marx, 1959). Ascending paper chromatography using acetate buffer-ethanol was used for separation of sulfated GAG (Good, 1967). Similarly, hyaluronic acid could be separated from heparin without protein interference using 0.04 M ammonium carbonate and methanol (45:55, v/v) (Castor and Dorstewitz, 1964). Displacement chromatography on silica gel-treated glass papers with different proportions of ethanol-$CaCl_2$ separated 7 GAG components upon detection by charring with sulfuric acid (Berenson and Dalferes, 1962).

Chondroitin 4-sulfate could be separated from chondroitin 6-sulfate by thin-layer chromatography on plates precoated with MN-Polygram cellulose 400 using 0.04 M $HCOONH_4$-isopropanol (50:50, v/v). On these plates the solvent system 0.04 M $HCOONH_4$-methanol (45:55, v/v) containing 1.1 mM complexone-III gave R_F of 0.31 for chondroitin 4-sulfate and chondroitin 6-sulfate and R_F of 0.14 to 0.22 for heparan sulfate, dermatan sulfate and hyaluronic acid, when the spots were detected with Azure A (Havass and Szabo, 1972). Marzullo and Lash (1967) have reported separation of the GAG components on plain silica gel in 1 μg amounts using the solvent system n-propanol-NH_4OH-water (40:60:5, v/v) for 2 hours followed by staining of the sulfated GAG with 1% toluidine blue and H_2SO_4 charring for hyaluronic acid. Thin-layer chromatography on cellulose powder or Sephadex has also been used for separation of < 1 μg of GAG following metachromatic detection with toluidine blue (Bischel et al., 1966). Reversed-phase chromatography on kieselguhr, using fatty amines as carriers to increase the solubility of GAG, separated hyaluronic acid and chondroitin sulfate (Berenson et al., 1955). Microquantities of the chondroitin 4-sulfate, chondroitin 6-sulfate, dermatan sulfate and keratan sulfate in mixtures may be separated either on acidic or alkaline silicated glass fiber sheets by sequential development with a series of acidic ethanolic calcium-containing solvents (Lippiello and Mankin, 1971).

2.3.5 Gel filtration

Sephadex of various pore sizes packed in columns or applied on a plate support as thin layer acts as a molecular sieve and separates the various GAG components on the basis of the molecular sizes of the diffusing species. These technics (Lee and Montgomery, 1965) have been used both for fractionation and molecular weight determination. Methods for determination of molecular weights of GAG components isolated from biological fluids on Sephadex columns have been described (Constantopoulos et al., 1969).

Gel filtration on thin layers of superfine Sephadex G-200 have been used for fractionation and molecular weight determination of GAG from urine (Taniguchi, 1970). Thin-layer gel filtration methods for molecular weight determination require microamounts of GAG. The sulfated GAG components were separated on thin layers of a 4:1 mixture of Sephadex G-100 and cellulose and molecular weights were determined from the plot of migration distance and molecular weights for standards (Tortolani and Romagnoli, 1975). Thin layer plates of Sephadex G-25 fine, G-50 fine, G-100 and G-200 developed with 0.05 M KCl have been used for gel filtration of mucopolysaccharides (Roberts, 1966).

If gel filtration is carried out in columns, the eluates from the chromatographic columns can be continously monitored by nephelometry of the precipitate formed with CPC (Blau and Dodge, 1974) or by use of a colorimetric reaction such as carbazole reaction.

2.3.6 Ion-exchange chromatography

Because of the polyanionic character and differences in charge densities on the molecules, the various GAG can be fractionated by using different types of anionic resins (Pearce et al., 1968; Kennedy, 1974). Methods for fractionation and analysis on columns packed with DEAE-cellulose (Lloyd et al., 1960; Neukom and Kuendig, 1965), or DEAE-Sephadex (Schmidt, 1962) have been described. Ion-exchange chromatography on ECTEOLA-cellulose columns (Ringertz and Reichard, 1960; Antonopoulos et al., 1967) gives a good separation of chondroitin plus hyaluronic acid, chondroitin sulfate plus heparin using a gradient elution with NaCl solution. Anion-exchange resin, Dowex 1 was used for the first time for fractionation of chondroitin and chondroitin sulfate from cornea (Davidson and Meyer, 1954). Anion-exchange chromatography on AG 1-X2 or Dowex 1 (Cl^-) Schiller et al., 1961; Pearce et al., 1968), using stepwise elution with 0.5-2.0 M NaCl, has been widely used for fractionation of the GAG mixtures.

Anion-exchange chromatography on ECTEOLA-cellulose or Dowex 1-X2 (AG 1-X2) fractionates hyaluronic acid with 0.5 M NaCl, heparan sulfate with 1.25 M NaCl. However, neither of the two resins gives a good separations of chondroitin 4-sulfate, chondroitin 6-sulfate and dermatan sulfate. Also, with both resins the elution of chondroitin sulfate overlaps keratan sulfate.

Micromethods using Deacidite FF resin have been used in conjunction with radioactive sulfate labeling for characterization of individual GAG components (Barker et al., 1965). DEAE-Sephadex A-25 columns could be used for fractionation of GAG by stepwise elution with different NaCl concentrations (Makharadze et al., 1977).

2.3.7 Adsorption and ligand chromatography

Adsorption chromatography on a column packed with a mixture of calcium phosphate and celite using pH 6.5 phosphate buffers with increasing ionic strengths up to 0.2 M gives a good separation of chondroitin sulfate from hyaluronate and separates a hyaluronate preparation into fractions of differing molecular weights (Bowness, 1965). The GAG liberated from tissues by proteolysis or those present in the biological fluids have a peptide or serine residue at the reducing end of the polysaccharide chain. The free amino group in these preparations can be attached covalently to agarose beads (Sepharose 4B) activated by cyanogen bromide. Such matrix-bound glycosaminoglycans have been used for affinity chromatography for determination of enzyme activities in biological fluids (Iverius, 1971). The GAG-bound AH-Sepharose 4B (previously coated non-covalently with GAG) was used as the matrix for isolation of chondroitinases C, AC, heparitinase and heparinase from Flavobacterium herparinum by affinity chromatography (Ototani and Yosizawa, 1978). Affinity chromatography on heparin-Sepharose 4B was used for purification of a collagenase (Sakamoto et al., 1978).

2.4 Molecular weight determination

The average molecular weight of the glycosaminoglycans can be determined by using one of the several physical methods available. Density gradient centrifugation (Franek and Dunstone, 1966; Silpanata et al., 1968; Scher and Hamerman, 1971); ultracentrifugal sedimentation (Preston et al., 1965); equilibrium gradient centrifugation, which involves two interrelated equilibria, the equilibrium of the neutral salt (density gradient) and the banding of the macromolecule at its buoyant density in the gradient (Mashburn et al., 1974); polyacrylamide gel electrophoresis (Hillborn and Anastassiadis, 1969, 1971; Mathews and Decker, 1971; Hsu et al., 1973; Mathews, 1976); Sephadex gel filtration (Constantopoulos et al., 1969); thin layer Sephadex cellulose chromatography (Tortolani and Ramagnoli, 1975) using GAG of known molecular weights for calibration give good results. Intrinsic viscosity measurements can be used for calculation of number average molecular weights

(Mathews, 1956; Laurent et al., 1960; Preston et al., 1965). These physical methods and others including diffusion method, light scattering and osmotic pressure measurements for molecular weight determination are given (Whistler, 1964). Bettelheim (1977) has reviewed these technics.

Chemical end-group analysis after cleavage of the carbohydrate-protein linkage can also be used. The free end-group can be labeled with ^{14}C by reaction with $Na^{14}CN$, tritiated by reduction with NaB^3H_4 or the free reducing activity could be measured. For a review of these methods see Varma and Varma (1977). The terminal reducing sugar, xylose exposed by the beta-elimination reaction in the alkaline medium can be determined by gas-liquid chromatography of the hydrolyzate as its aldononitrile acetate derivatives (Varma and Varma, 1977) or the terminal xylose may be reduced with borohydride and determined as its alditol acetate (Wasteson and Wessler, 1971; Yamaguchi and Makino, 1977). The degree of polymerization and chain length can be calculated from the xylose content on the basis of the weight of the polymer that would contain one mole of xylose.

2.5 Fractionation by preparative electrophoresis

The glycosaminoglycans are polyanions and migrate to the anode in an electric field. These polyelectrolytes have been separated by free electrophoresis in Tiselius apparatus and by zone electrophoresis on different supports. The various electrophoretic technics for glycosaminoglycans have been recently reviewed in a monograph (Gaal et al., 1980). Preparative fractionation of glycosaminoglycans has been carried out on several different support media such as Pevikon blocks (Wessler, 1968), Agarose gel (Horner, 1967), Hyflo Supercel or Celite (Gardell et al., 1950), on starch gel (Brookhart, 1965). Kieselguhr slab has also been used for GAG quantities of 20 mg or more (Gardell et al, 1950). Difficulties sometimes arise in the identification of the separated components on the basis of mobilities with respect to those of the reference compounds, since the migration distances, among other factors, also depend upon the degree of sulfation and molecular size.

Methods for preparative separation of GAG mixtures into individual glycosaminoglycans by isoelectric focusing have been reported (McDuffie et al., 1975; Nadar et al., 1978; Coppa et al., 1979).

2.6 Characterization by microscopic, optical and spectrometric methods

Electron microscopy has been used for determination of shape and dimensions of various GAG (Fessler and Fessler, 1966). For example, the electron microscopic study indicated that heparin had an unbranched linear structure with a length of 160 ± 40 Å (Hirano, 1972). An estimate of molecular weight can be made from the molecular dimensions, since one repeating disaccharide unit (M_r 450-500) is 10 Å long.
Optical rotation measurements have been used for characterization of the pure individual GAG. The three chondroitin sulfate isomers (Mathews, 1958) and other GAG components have characteristic specific rotations (Brimacombe and Webber, 1964). Optical rotations and circular dichroism spectra of GAG and their complexes with dyes provide information about the helical content and molecular conformation from the Cotton effect induced upon binding with a basic dye (Stone, 1965; Hirano, 1972; Chakrabarti and Balazs, 1973; Buffington et al., 1977; Mathews and Decker, 1977). Metachromatic binding of methylene blue to neighboring anionic sites on the polysaccharide molecule induced Cotton effects which could be varied by changes in salt or dye concentrations producing different effects for various dye-polymer systems. Thus, the ultraviolet ORD and CD spectra for keratan sulfate and heparan sulfate and for heparan sulfate and dermatan sulfate differed (Stone, 1976). This can help in their identification.

Nuclear magnetic resonance (N.M.R.) spectrometry has been found useful in characterization of GAG, their derivatives and their degradation products produced by enzymatic and chemical reactions. It has also been used for studies on interactions of GAG with cations or proteins. Variations in the nature of monosaccharide units and the nature, number and position of the exo-groups (sulfate, carbonyl, acetyl) and the position and configuration of the glycosidic bonds give rise to distinctive chemical shifts and spin-spin coupling patterns. Because of the variability in these factors contributing to the N.M.R. spectra, distinction between various polymers is possible. For an example, heparin gives ^{1}H and ^{13}C N.M.R. spectra distinguishable from other GAG (Perlin et al., 1970; Perlin 1975, 1976, 1978). Similarly, the chemical shift of acetate-methyl signals in the N.M.R. spectra of peracetylated heparin indicated 1 C conformation of the hexuronide moieties (Hirano, 1972). N.M.R. characterization of heparan sulfate and dermatan sulfate isolated from the urines of patients with mucopolysaccharidoses type II, III and VI have been reported (Tira et al., 1979). N.M.R. spectra also provide a satisfactory quantitation and serve as a non-destructive and convenient procedure for this purpose (Perlin, 1978).

Raman spectroscopy has been used as a probe for structural studies of glycosaminoglycans (Bansil et al., 1978).

Infra-red spectroscopy has been very useful for detection of functional groups α or β anomeric conformation and even sometimes for information about the point of attachment of glycosidic linkages. The assignment of bands to different groups in polysaccharides (Orr et al., 1952; Orr, 1954) provided a basis for identification of glycosaminoglycans isolated from biological fluids and tissues. The differentiation and identification of the three isomeric chondroitin sulfates has been reported (Mathews, 1958). All the three isomers show a strong absorption at 1240 cm^{-1} due to S-O stretching vibrations. However, these give different absorbances in the fingerprint region (700-1000 cm^{-1}). The dermatan sulfate and chondroitin 4-sulfate contain the sulfate groups in the axial position and give typical absorbance bands at 725 cm^{-1}, 852 cm^{-1} and 928 cm^{-1}. On the other hand, in chondroitin 6-sulfate the sulfate group is in the equitorial position and it has characteristic bands at 775 cm^{-1}, 820 cm^{-1} and 1000 cm^{-1}. The free acids but not the salts of the GAG have an additional band at 1736 cm^{-1} attributable to the C=O stretching of the -COOH group. This is in addition to the C=O stretching and N-acetyl deformation of the N-acetyl groups at 1648 cm^{-1} and 1560 cm^{-1} respectively. Heparin can be differentiated from heparan sulfate by the absence in heparin of infrared absorption at 1560-1508 cm^{-1} attributable to the N-H deformation mode of N-acetyl group.

Thermal analysis has been used for qualitative and quantitative analyses of the macromolecules. The characterization and quantitation of glycosaminoglycans and proteoglycans for elucidating the age-related and pathological changes have been carried out by thermal decomposition of these macromolecules using a Paulik-Paulik-Erdey MOM derivatograph programmed at 2 °C/min up to 900 °C. The thermal decomposition curves for the macromolecules vary with the degree of polymerization, structural differences and the substituent groups which help their characterization and quantitation (Bihari-Varga, 1971; Bihari-Varga et al., 1975).

2.7 Analysis without hydrolysis

The methods based on analysis of the intact, unhydrolyzed glycosaminoglycans include turbidity measurement upon adding quaternary ammonium salts (Di Ferrante, 1954) or albumin at an acidic pH (Dorfman and Matalon, 1972), spectrometric titration with acridine orange (Stone and Bradley, 1967), precipitation with protamine or quinacrine (Mitra and Blau, 1978), zirconium chloride (Wardi and Michos, 1973), scandium chloride (Wardi et al., 1974), N,N-diethylpseudoisocyanin halide (Vilter et al., 1974; Suschke, 1973), or Alcian blue (Whiteman, 1973, 1973 a). In these procedures, the precipitating agent may be measured spectrophotometrically in the supernatent or in the extract of the complex (Rosenquist and Rosenquist, 1974) or alternatively if the complex is uncolored, hexuronic acid analysis can be carried out directly on the precipitate. A method based on a characteristic absorption of the Alcian blue-GAG complexes at 480 nm has been reported (Gold, 1979). Methods based on measurement of the precipitating agent are less accurate, since the different individual glycosaminoglycans have different charges and therefore bind different amounts of the precipitating agent. Also, the value needs to be read from a standard curve made with a single GAG component usually chondroitin sulfate, although a mixture of several different macromolecules may be present in a biological fluid. The methods based on analysis of hexuronic acid content of the precipitate such as those formed with quaternary detergents, protamine sulfate or zirconyl chloride are quite reliable.

2.8 Analysis of components in hydrolysates

2.8.1 Hexuronic acid

Since hexuronic acid is characteristically present in all GAG (except keratan sulfate) its measurement provides the most rapid and sensitive analysis of the polyanionic GAG. However, it is difficult to determine very low concentrations of hexuronic acid in the presence of a substantial amount of protein or glucose, which interfere with the colorimetric methods. Only two hexuronic acids, glucuronic acid and iduronic acid, have been identified as components of the GAG from connective tissues. Galacturonic acid was reported as a component of GAG from rabbit and human brains (Stary et al., 1964).

The isolation and identification of the hexuronic acids from GAG involve several problems. These are very labile and are usually destroyed during acid or alkaline hydrolysis (Marsh, 1966). The glucuronic acid and galacturonic acid form lactones during hydrolysis and as a result each of these hexuronic acids gives two spots on chromatography. Also, epimerization of D-glucuronic acid to L-iduronic acid occurs during deaminative cleavage of the (1 → 4) glycosidic bonds between D-glucosamine and D-glucuronic acid. Some of these problems can be avoided by hydrolysis with a resin such as Dowex 50 (H^+) at 100 °C instead of acid hydrolysis (Hoffman et al., 1956) and by saponification of any lactones formed during this hydrolysis prior to analysis by chromatography. The hexuronic acids so isolated may be separated by paper chromatography (Hoffman et al., 1956; Masamune and Satake, 1959; Yosizawa, 1960) or by electrophoresis (St. Cyr, 1970) or by ion-exchange chromatography (Dziewiatkowski, 1962) on Dowex 1 (AG 1-X2) acetate form and elution with 0.15 M acetic acid (Khym and Doherty, 1952). An automated method for ion-exchange chromatographic determination of glucuronic acid and iduronic acid (Radhakrishnamurthy et al., 1976) or paper chromatography (Radhakrishnamurthy and Berenson, 1963) after hydrolysis of GAG with formic acid are available. A radiochemical method for uronic acid analysis has been described (Shively et al., 1976). Identification and quantification of relative amounts of glucuronic acid and iduronic acid in sulfated GAG may be carried out by gas-chromatographic procedures (Radhakrishnamurthy et al., 1968; Inou and Miyaki, 1975) following methanolysis, deamination-methanolysis and other hydrolytic methods.

High voltage paper electrophoresis using pyridine-acetic acid-water at pH 2.7, 100 V/cm for 180 minutes gives a good separation of hexuronic acid mixture (Kosaki and Yosizawa, 1975).

The total hexuronic acid contents of the GAG are ususally determined by the carbazole reaction (Dische, 1947) or modification of this reaction in the presence of borate (Bitter and Muir, 1962) or borate and sulfamate (Galambos, 1967, 1967 a). Other methods used for this purpose include the orcinol reaction (Brown, 1946), the naphthoresorcinol reaction (Nir, 1964). Other methods measure CO_2 liberated by decarboxylation of hexuronic acid with 12 % HCl by manometry (Wollin, 1978) or with 57 % HI followed by gas chromatography (Kosheleva et al., 1976), reaction with harmine (Wardi et al., 1974 a) and reaction with m-hydroxydiphenyl using the manual (Blumenkrantz and Asboe-Hansen, 1973) or the automated (Blumenkrantz and Asboe-Hansen, 1974) assay procedure.

A method for differential analysis of a mixture of glucuronic acid in the free, conjugated and GAG-bound forms is based on hypoiodite oxidation and Dische carbazole reaction (Yuki and Fishman, 1963). Periodate oxidation and pK_a analysis give information on the nature and conformation of the hexuronic acids in the glycosaminoglycans (Scott, 1968).

All the aforementioned colorimetric reactions can be carried out directly on the polymeric GAG. Heating with acid hydrolyzes the GAG and converts the free hexuronic acid so formed into a furfural type compound, which gives the color with the reagent. However, the various hexuronic acids give differing color yields in the carbazole reaction of Dische. Whereas the color yield for L-iduronic acid is less than that of D-glucuronic acid in the carbazole reaction; in the orcinol reaction (Brown, 1946) the two hexuronic acids yield the same amount of color. When the two reactions are simultaneously carried out the carbazole/orcinol ratios obtained are: L-iduronic acid, 0.22; D-glucuronic acid, 1.0; D-galacturonic acid, 0.95. The determination of carbazole/ orcinol ratio on a mixture of GAG gives semiquantitative measure of iduronic acid-containing compounds (dermatan sulfate, heparan sulfate and heparin) without fractionation of the mixture into these individual glycosaminoglycans.

2.8.2 Hexosamine

The glucosamine and galactosamine present in the acid hydrolyzates of GAG may be identified by paper chromatography either directly (Mukerjee and SriRam, 1964), after oxidation with ninhydrin to arabinose and xylose, respectively, after derivatization with fluorodinitrobenzene or this derivatization following borohydoride reduction to hexosaminitols. Gardell (1961) has reviewed these methods.

Ion-exchange chromatographic methods for determination of the relative amounts of glucosamine and galactosamine include the differential elution of these amino sugars from a Dowex-50 or Zeokarb 225 column with 0.3 M hydrochloric acid (Gardell, 1961). Hexosamines may be separated from a Dowex-50 column with 0.1 M citrate buffer pH 5.0 or from an Amberlite CG-120 column and monitoring the elution by ninhydrin or Elson-Morgan reaction (Pearson, 1963). The differential determination of hexosamine together with hexosaminitols may be easily achieved by automated ion-exchange chromatography (Walborg et al., 1963; Downs and Pigman, 1976; Valle and Shively, 1979).

Spectrometric determination of the total hexosamine is easily carried out by one of the several manual or automated modifications of the Elson-Morgan reaction (Boas, 1953; Swann and Balazs, 1966). The original Elson-Morgan reaction, involving reaction with an alkaline solution of acetylacetone followed by reaction with an acidic solution of p-dimethylaminobenzaldehyde, gives the same amount of color with glucosamine and galactosamine. By carrying out the acetylation at 0-1 ^{o}C instead of 25 ^{o}C (Wagner, 1979) and by carrying out the reaction in the presence of tetraborate (Ludowieg and Benmaman, 1967) and by further reducing the tetraborate concentration to 0.175 *M* (Oguchi and Oguchi, 1979), the color yield from glucosamine is considerably reduced. These modifications help differential determination of the two hexosamines spectrophotometrically (Blumenkrantz and Asboe-Hansen, 1976).

Methods for fluorometric determination of the amino sugars in very small samples by reaction with pyridoxal and zinc ions (Maeda et al., 1970) or with O-phthaldehyde (Carroll and Nelson, 1979) are available.

The hexosamines have also been measured by alkaline decomposition and determination of the liberated ammonia (Tracey, 1952) or estimation of the decomposition product at 302 nm (Lewandowski, 1973). Other methods are based on deamination with nitrous acid and determination of the liberated anhydrosugars by reaction with indole-sulfuric acid (Dische and Borenfreund, 1950), with Pyrrole-HCl (Exley, 1957), with 3-methyl-2-benzothiazolone hydrazone hydrochloride (Smith and Gilkerson, 1979), by gas chromatography following derivatization to aldononitrile acetates (Varma et al., 1974; Varma and Varma, 1976) or upon conversion to alditol acetates (Niedermeier, 1971). Gas chromatographic methods as trimethylsilyl derivatives (Stimson, 1971) or as other derivatives (Dutton, 1973, 1974) have also been used.

2.8.3 Neutral sugars

Xylose and galactose present in the carbohydrate-protein linkage region and galactose, fucose, mannose, etc. in keratan sulfate can be measured by spectrophotometric methods, but interferences are observed from other components. Tsignos and Muir (1966) and Stevenson et al. (1967) have reported methods for determination of xylose in GAG with less interference. Pentoses and hexuronic acid may be determined simultaneously by reaction with harmine (Wardi et al., 1974 a). The neutral sugars and hexosamines may be determined simultaneously as trimethylsilyl derivatives (Oates and Schrager, 1967), and also as several other volatile derivatives (Dutton, 1973, 1974) or as aldononitrile acetates (Varma and Varma, 1976). Mass spectrometry coupled to gas-liquid chromatography using aldononitrile acetates (Varma et al., 1977) or other derivatives (Kochetov and Chizov, 1972; Lönngren and Svensson, 1974) helps in positive identification of the neutral sugars from the fragmentation pattern corresponding to a peak for a neutral sugar. Paper or thin-layer chromatography may be used for isolation of neutral sugars on a preparative scale or for their identification (Smith, 1976).

High voltage paper electrophoresis in 0.05 M borate buffer of pH 9.2 or 0.05 M acetic acid-pyridine buffer of pH 6.0 separates the neutral sugars (Smith, 1976).

2.8.4 N-Acetylhexosamine

The N-acetylhexosamine reducing groups released during hyaluronidase digestion or those present in GAG are determined by the Morgan-Elson reaction as modified by Reissig et al., (1955). This reaction consists of a brief boiling with potassium tetraborate followed by treatment with p-dimethylaminobenzaldehyde and spectrophotometric determination of the red-violet color obtained.

2.8.5 N-sulfated hexosamine

The analysis of N-sulfated hexosamine is an indirect measure of the N-sulfate content of heparan sulfate and heparin and a measure of these macromolecules in the GAG mixtures. Treatment of heparin or heparan sulfate with nitrous acid causes deamination of the N-sulfated D-glucosamine residues producing degradation products with reducing terminal 2,5-anhydromannose groups which are measured by reaction with indole (Lagunoff and Warren, 1962; Cifonelli, 1976). The molar ratio of N-sulfate/N-acetyl groups in an isolated GAG helps its identification as heparin or heparan sulfate, the ratio being much higher in the former.

As compared to 0-sulfate groups the N-sulfate groups are very labile and are very easily released by heating with 0.04 M HCl at 100 °C for 90 minutes and their determination (Margolis and Margolis, 1972) along with total sulfate can be used for calculation of 0-sulfate by difference.

2.8.6 N-acetyl groups

The N-acetyl groups may be measured indirectly from N-acetylhexosamine determination (section 2.8.4) by the method of Reissig et al., (1955). A gas-chromatographic method for N-acetyl groups has also been reported (Radhakrishnamurthy et al., 1968 a).

2.8.7 Sulfate content

The methods for sulfate determination can be divided into two groups:
a) Those involving determination on the intact GAG macromolecules include the gas chromatographic procedure of Srinivasan et al., (1970) in which n-butylamine formed a butylammonium sulfate upon reaction with the free sulfuric ester (-OSO_3H and -$NHSO_3H$) groups on the GAG and butylamine was then liberated from this complex with an alkali and determined by gas chromatography. Similarly, the sulfate content can be estimated by electrophoresis in 0.1 M HCl. Under these conditions, pH 1.0, the carboxyl groups remain unionized and the electrophoretic mobility varies directly with the sulfate content (Wessler, 1971).
b) Methods analyzing inorganic sulfate liberated by hydrolysis in 1 M HCl for 4 hr. at 100 °C are commonly used. The liberated inorganic sulfate may be determined by precipitation with barium chloride, benzidine or barium chloranilate (Nalefski and Takano, 1950; Spencer, 1960; Dodgson and Price, 1962; Ginsberg and Di Ferrante, 1977). Other modified micromethods are based on radioactivity (Braselman, 1965), flame photometry (Barker et al., 1968), chromatographic separation of inorganic sulfate and complexation with toluidine blue (Nader and Dietrich, 1978), measurement of barium in EDTA solution of barium sulfate by atomic absorption spectrophotometry (Wollin, 1978 a) and other modifications (Khan et al., 1979).

2.9 Microfractionation and microanalysis of glycosaminoglycans in mixtures

For working with small volumes of biological fluids or with limited amounts of GAG isolated from the biological fluids, several methods on microscale have been devised. The various methods for microscale identification, fractionation include those using microcolumn fractionation on cellulose (Svejcar and Robertson, 1967), ion-exchange resins (Lakatos and Di Ferrante, 1974; Applegarth et al., 1975), electrophoresis on agarose gel (Dietrich and Dietrich, 1972; Dietrich et al., 1977) or polyacrylamide (Rivkina et al., 1974) and other microprocedures (Brown et al., 1965; Svejcar and Robertson, 1967; Barker et al., 1968; Nikata et al., 1971; Keleti and Lederer, 1974; Blumenkrantz and Asboe-Hansen, 1976).

2.10 Keratan sulfate determination

Keratan sulfate analysis presents several problems. Firstly, it cannot be determined in mixture with other GAG due to the absence of glucuronic acid in this macromolecule. Secondly, it behaves differently than other GAG during precipitation with quaternary ammonium salts, since keratan sulfate fractions redissolve in the presence of an excess of these precipitating agents. Also, the testicular hyaluronidase and bacterial chondroitinases have no hydrolytic effect on keratan sulfate.

Since the repeating disaccharide unit of keratan sulfate consists of D-glucosamine and D-galactose, the measurement of galactose is used for quantitation of keratan sulfate in a GAG using the cysteine-sulfuric acid reaction (Dische, 1955). Keratan sulfate is measured as galactose in the fraction isolated by anion-exchange chromatography on ECTEOLA-cellulose (Wortman, 1962). A modification of this procedure using GAG isolated from 50 ml of urine by adsorption and elution from ECTEOLA-cellulose followed by galactose determination on the ethanolic precipitate has been described (Robertson and Harvey, 1972). A microprocedure for determination of keratan sulfate as galactose in the 3 M NaCl fraction isolated from Dowex 1-X2 column using only 10 ml of urine has been reported (Humbel, 1974). Thin-layer chromatography on silica gel with n-propanol-25 % ammonia-water (40:60:5, v/v) or on cellulose plates can be used for detection and quantitation of keratan sulfate in pathological urine. No keratan sulfate is detectable in urine from normal adults and children (Humbel, 1974).

2.11 Hyaluronic acid determination

Several micromethods reported for analysis of this macromolecule include a capillary turbidity test (Ibrahim and Streitfeld, 1973), selective elution from microcolumns of anion-exchange resin (Schiller et al., 1961) or CPC-cellulose (Antonopoulos et al., 1964; Svejcar and Robertson, 1967), cellulose acetate electrophoresis (Breen et al., 1976). Specific enzymatic hydrolysis using a novel hyaluronate lyase from Streptomyces hyalurolyticus forms oligosaccharides terminating in $\Delta^{4,5}$-unsaturated glucuronic acid units which can be measured by its characteristic absorption maxima at 232 nm (Ohya and Kaneko, 1970; Greiling, 1974) or by colorimetric measurement of the N-acetylglucosamine present at the reducing termini by reaction with p-dimethylaminobenzaldehyde and measurement at 585 nm (Greiling, 1974; Hatae and Makita, 1975). The unsaturated residues at the non-reducing termini of the oligosaccharides produced by bacterial hyaluronate lyase on periodate oxidation form formyl pyruvic acid which gives a chromogen with thiobarbituric acid with absorption maxima at 549 nm (Jourdian et al., 1979).

2.12 Determination of dermatan and chondroitin sulfates

The fractionation of the three isomeric chondroitin sulfates is not possible by chromatographic, electrophoretic or ion-exchange chromatography. However, microanalytical technics for colorimetric determination of dermatan sulfate in the presence of chondroitin 4- and 6-sulfates have been developed (Di Ferrante et al., 1971; Humbel, R., 1972; Kinoshita et al., 1975). Paper electrophoresis in zinc salts separates dermatan sulfate from the other two isomers (Haruki and Kirk, 1967). Bidimensional electrophoresis on cellulose acetate is helpful in this respect (Hata and Nagai, 1973; Sato and Gyorkey, 1977). In veronal buffer pH 8.6, μ 0.05, dermatan sulfate and heparan sulfate were resolved but chondroitin 4- and 6-sulfates and heparin had the same electrophoretic mobility (Kimura and Tsurumi, 1969). A better electrophoretic procedure (Breen et al., 1976) has been reported but cannot resolve dermatan sulfate and chondroitin 4- and 6-sulfates. Recently a single dimensional electrophoretic method which resolves all the known GAG has been reported (Cappelletti et al., 1979). In these electrophoretic procedures the components separated can be quantified by elution of the dye from the strips (Hsu et al., 1972; Newton et al., 1974; Endo and Yosizawa, 1975).

Enzymatic micromethods for determination of small quantities of the three isomeric chondroitin sulfates have been developed (Saito et al., 1968). In these procedures the unsaturated disaccharides produced by digestion of the various glycosaminoglycans with chondroitinases ABC and AC are analyzed by paper chromatography (Saito et al., 1968), paper electrophoresis (Elliott and Gardner, 1978), thin layer chromatography (Wasserman et al., 1977), anion exchange chromatography on Dowex 1 using a lithium gradient (Greiling and Eberhard, 1974), by high performance liquid chromatography (Lee et al., 1978, 1978 a; ref. 212) or by use of multienzyme systems (Greiling, 1974 a).

Murata (ref. 213) has recently reviewed the glycosaminoglycan analyses using electrophoresis; enzymic digestion with Streptomyces hyaluronidase; chondroitinase ABC and AC, and chondrosulfatases followed by analyses of the resulting unsaturated disaccharides by paper chromatography, high voltage electrophoresis and also by TLC of hexosamines. Dietrich et al. (ref. 214) have reviewed the use of mucopolysaccharidases from Flavobacterium heparinum for structural analyses of chondroitin sulfates, heparan sulfate and heparin.

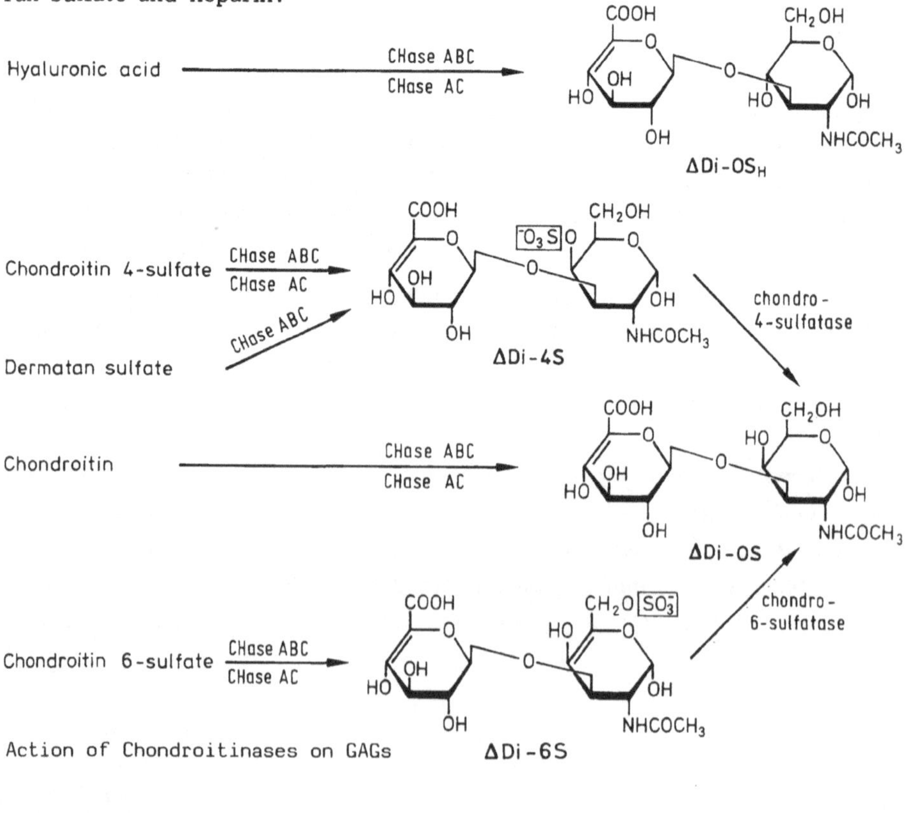

Action of Chondroitinases on GAGs ΔDi-6S

2.13 Bibliography

1. Antonopoulos, C. A., Borlius, E., Gardell, S., Hamnstrom, B. and Scott, J. E. (1961): The precipitation of polyanions by long-chain aliphatic ammonium compounds. IV. Elution in salt solutions of mucopolysaccharide-quaternary ammonium complexes adsorbed on a support. Biochim. Biophys. Acta 54, 213-226

2. Antonopoulos, C. A., Fransson, L. A., Heinegard, D. and Gardell, S. (1967): Chromatography of glycosaminoglycans on ECTEOLA-cellulose columns. Biochim. Biophys. Acta 148, 158-163

3. Antonopoulos, C. A., Gardell, S., Szirmai, J. A. and Tyssonsk, E. R. (1964): Determination of glycosaminoglycans (mucopolysaccharides) from tissues on the microgram scale. Biochim. Biophys. Acta 83, 1-19

4. Applegarth, D. A., Bozoian, G., and Lowry, R. B. (1975): Analysis of urinary mucopolysaccharides using small ion exchange columns. Clin. Biochem. 8, 151-160

5. Bansil, R., Yannas, I. V. and Stanley, H. E. (1978): Raman Spectroscopy: A structural probe of glycosaminoglycans. Biochim. Biophys. Acta 541, 535-542

6. Barker, S. A., Cruickshank, C. N. D. and Webb, T. (1965): Mucopolysaccharides in rat skin. Carbohydr. Res. 1, 52-61

7. Barker, S. A., Kennedy, J. F., Somers, P. J. and Stacey, M. (1968): A new method for the quantitative, microscale determination of the sulfate content of the carbohydrates. Carbohydr. Res. 7, 361-368

8. Berenson, G. S. and Dalferes, E. R. (1962): Identification of acid mucopolysaccharides by glass paper chromatography. Biochim. Biophys. Acta 58, 34-40

9. Berenson, G. S., Roseman, S. and Dorfman, A. (1955): A chromatographic method for the separation of acid mucopolysaccharides. Biochim. Biophys. Acta 17, 75-80

10. Bettelheim, F. A. (1977): Molecular weight determination and physical properties. In: M. I. Horowitz and W. Pigman (Eds.). The Glycoconjugates. Vol. I. Academic Press, New York, 111-128

11. Bihari, Varga, M. (1971): Thermoanalytical assay of glycosaminoglycans. Acta Biochim. Biophys. Acad. Sci. Hung. 6, 271-275

12. Bihari-Varga, M., Sepulchre, C. and Moczar, E. (1975): Thermoanalytical studies on protein-polysaccharide complexes of connective tissues. J. Thermal. Anal. 7, 675-683

13. Bischel, M. D., Austin, J. H., Kemeny, M. D., Hubble, C. M. and Lear, R. K. (1966): Separation and identification of acid polysaccharides by thin-layer chromatography. J. Chromatogr. 21, 40-45

14. Bitter, T. and Muir, H. M. (1962): A modified uronic acid carbazole reaction. Anal. Biochem. 4, 330-334

15. Blau, K. and Dodge, C. S. (1974): Continuous nephelometric monitoring of glycosaminoglycans eluted from chromatographic columns. Anal. Biochem. 58, 650-654

16. Blumenkrantz, N. and Asboe-Hansen, G. (1973): New method for quantitative determination of uronic acids. Anal. Biochem. 54, 484-489

17. Blumenkrantz, N. and Asboe-Hansen, G. (1974): An automated quantitative assay for uronic acids. Biochem. Med. 11, 60-66

18. Blumenkrantz, N. and Asboe-Hansen, G. (1976): An assay for total hexosamine and a differential assay for glucosamine and galactosamine. Clin. Biochem. 9, 269-274

19. Blumenkrantz, N. and Aboe-Hansen, G. (1976 a): Micromethod for fractionation of acid mucpolysaccharides. Clin. Biochem. 9, 9-15

20. Boas, N. (1953): Method for the determination of hexosamines in tissues. J. Biol. Chem. 204, 553-563

21. Bowness, J. M. (1965): Fractionation of acidic mucopolysaccharides on columns containing celite and calcium phosphate. Methods Carbohydr. Chem. 5, 18-20

22. Braselmann, H. (1965): Microdetermination of the sulfate content and ^{35}S-sulfate activity of mucopolysaccharides. Acta Biol. Med. German. 15, 173-179 (Ger)

23. Breen, M., Weinstein, H. G., Blacik, L. J., Borcherdang, M. S. and Sittig, R. A. (1976): Microanalysis and characterization of glycosaminoglycans from human tissue via zone elctrophoresis. Methods Carbohydr. Chem. 7, 101-115

24. Brimacombe, J. S. and Webber, J. M. (1964): Mucopolysaccharides. Elsevier, Amsterdam

25. Brookhart, J. H. (1965): Separation of acid polysaccharides by starch gel electrophoresis. J. Chromatogr. 20, 191-193

26. Brown, A. H. (1946): Determination of pentose in the presence of large quantities of glucose. Arch. Biochem. 11, 269-278

27. Brown, R. G., Button, G. M. and Smith, J. T. (1965): A rapid method for the assay of small quantities of sulfomucopolysaccharides. Anal. Biochem. 12, 195-198

28. Buffington, L. A., Pysh, E. S., Chakrabarti, B. (1977): Far-ultraviolet circular dichoism of N-acetylglucosamine, glucuronic acid and hyaluronic acid. J. Am. Chem. Soc. 99, 1730-1734

29. Capelletti, R., Rosso, M. D. and Chiarugi, V. P. (1979): A new electrophoretic method for the complete separation of all known animal glycosaminoglycans in a monodimensional run. Anal. Biochem. 99, 311-315

30. Carroll, S. F. and Nelson, D. R. (1979): Fluorometric quantitation of amino sugars in the picomole range. Anal. Biochem. 98, 190-197

31. Castor, C. W. and Dorstewitz, E. L. (1964): Identification of acid mucopolysaccharides by paper chromatography. J. Chromatogr. 13, 157-165

32. Chakrabarti, B. and Balazs, E. A. (1973): Optical properties of hyaluronic acid. Ultraviolet circular dichroism and optical rotatory dispersion. J. Mol. Biol. 78, 135-141

33. Cifonelli, J. A. (1976): Nitrous acid depolymerization of glycosaminoglycans. Methods Carbohydr. Chem. 7, 139-141

34. Constantopoulos, G., Dekaban, A. S. and Carroll, W. R. (1969): Determination of molecular weight distribution of acid mucopolysaccharides by Sephadex gel filtration. Anal. Biochem. 31, 59-70

35. Coppa, G., Sani, S., Paris, D., Gabrielli, O. and Maiorana, A. (1979): Isoelectric focusing of serum lysosomal enzymes and characterization of urinary and tissual glycosaminoglycans in "I-cell" disease (mucolipidoses II). In: B. Berra, C. Balduini, S. di Donato and G. Tettamanti (Eds.). Glycoproteins and glycosaminoglycans in health and disease, Vol. 2 of Perspective in Inherited Metabolic Diseases, Edi Ermes, Milan, 149-164

36. Davidson, E. A. and Meyer, K. (1954): Chondroitin, a new mucopolysaccharide. J. Biol. Chem. 211, 6o5-611

37. Dietrich, C. P. and Dietrich, S. M. (1972): Simple micro method for identification of heparin and other acidic mucopolysaccharides from mammalian tissues. Anal. Biochem. 46, 209-215

38. Dietrich, C. P., McDuffie, N. M. and Sampaio, L. O. (1977): Identification of acidic mucopolysaccharides by agarose gel electrophoresis. J. Chromatogr. 130, 299-304

39. Di Ferrante, N. (1954): Quantitative colorimetric assay of acid mucopolysaccharides. J. Biol. Chem. 209, 579-583

40. Di Ferrante, N., Donnelly, P. V. and Berglund, R. K. (1971): Colorimetric measurement of dermatan suphate. Biochem. J. 124, 549-553

41. Dische, Z. (1947): A new specific color reaction of hexuronic acids. J. Biol. Chem. 167, 189-198

42. Dische, Z. (1955): New color reactions for determination of sugars in polysaccharides. Methods Biochem. Anal. 2, 313-358

43. Dische, Z. and Borenfreund, E. (1950): A spectrophotometric method for the microdetermination of hexosamines. J. Biol. Chem. 184, 517-522

44. Dodgson, K. S. and Price, R. G. (1962): A note on the determination of the ester sulfate content of sulfated polysaccharides. Biochem. J. 84, 106-110

45. Dorfman, A. and Matalon, R. (1972): The mucopolysaccharidoses. In: J. B. Stanbury, J. B. Wyngaarden and D. S. Fredrickson (Eds.). The Metabolic Basis of Inherited Disease, 3rd McGraw-Hill Book Co., 1218-1272

46. Dorner, R. W., Antonopoulos, C. A. and Gardell, S. (1968): Application of cetylpyridinium-complex elution methods to analysis of glycosaminoglycans of tendon. Biochem. Biophys. Acta 158, 336-343

47. Downs, F. and Pigman, W. (1976): Determination of hexosamines and hexosaminitols using the amino acid analyzer. Methods Carbohydr. Chem. 7, 244-248

48. Dutton, G. G. S. (1973, 1974): Application of gas-liquid chromatography to carbohydrates. Advan. Carbohydr. Chem. Biochem. 28, 11-160; 30, 9-110

49. Dziewiatkowski, D. (1962): Separation of hexuronic acids by ion-exchange process. Biochem. Biophys. Acta 56, 167-169

50. Elliott, R. J. and Gardner, D. L. (1978): The O-, 4- and 6-sulfated disaccharides of chondroitin sulfates: their electrophoretic separation and detection with p-anisidine. Biochem. Soc. Trans. 6, 768-769

51. Endo, M. and Yosizawa, Z. (1975): A method for quantitative recovery of acidic glycosaminoglycans after electrophoresis on cellulose acetate strip. Anal. Biochem. 65, 537-539

52. Exley, D. (1957): Determination of 10-100 mr quantities of hexosamine. Biochem. J. 67, 52-60

53. Fessler, J. H. and Fessler, L. I. (1966): Electron microscopic visualization of the polysaccharide hyaluronic acid. Proc. Natl. Acad. Sci. U.S.A. 56, 141-147

54. Franek, M. D. and Dunstone, J. R. (1966): Density-gradient centrifugation in the isolation of polysaccharide-protein complexes from aortic tissue. Biochim. Biophys. Acta 127, 213-222

55. Gaal, O., Vereczkey, L. and Medgyesi, G. (1980): Electrophoresis in the separation of Biological Macromolecules. John Wiley and Sons. New York. Chapter 4, 375-379

56. Galambos, J. T. (1967). The reaciton of carbazole with carbohydrates. I. Effects of borate and sulfamate on the carbazole color of sugars. Anal. Biochem. 19, 119-132

57. Galambos, J. T. (1967 a): The reaction of carbazole with carbohydrates. II. Effect of borate and sulfamate on the ultraviolet absorption of sugars. Anal. Biochem. 19, 133-143 (Ger)

58. Gardell, S. (1957): Separation of mucopolysaccharides on a cellulose column. Acta Chem. Scand. 11, 668-675

59. Gardell, S. (1961): The analysis of mucopolysaccharides. In: F. Clark and J. K. Grant (Eds.). The Biochemistry of Mucopolysaccharides of Connective Tissue. Biochemical Society Symp. No. 20, Cambridge University Press, 39-50

60. Gardell, S. (1965): Fractionation on cellulose columns. Methods Carbohydr. Chem. 5, 9-14

61. Gardell, S., Gordon, A. H. and Aquist, S. (1950): Electrophoresis of mucopolysaccharides in a slab of Hyflo Super-Gel. Acta Chem. Scand. 4, 907-913

62. Ginsberg, L. C. and Di Ferrante, N. (1977): Sensitive methods for the determination of ester sulfate in biological systems. Biochem. Med. 17, 80-86

63. Gold, E. W. (1979): A simple spectrophotometric method for estimating glycosaminoglycan concentrations. Anal. Biochem. 99, 183-188

64. Good, T. A. (1967): Ascending paper chromatography of sulfated acid mucopolysaccharides. Anal. Biochem. 19, 109-118

65. Greiling, H. (1974): Hyaluronic acid. In: H. U. Bergmeyer (Ed.): Methods of Enzymatic Analysis. Vol. 3, 2nd English Ed. Academic Press, New York, 1158-1164

66. Greiling, H. (1974 a): Glycosaminoglycans. In: H. Ch. Curtius and M. Roth (Eds.) Clinical Biochemistry: Principles and Methods, Vol. 2, Walter de Gruyter, Berlin, 941-971

66. Greiling, H. and Eberhard, A. (1974): Chondroitin 4-sulfate, chondroitin 6-sulfate and dermatan sulfate. In: H. U. Bergmeyer (Ed.) Methods in Enzymology, Vol. 3, Verlag Chemie, New York, 1210-1216

67. Haruki, F. and Kirk, J. E. (1967): A method for separation of chondroitin sulfate B from isomers A and C by paper electrophoresis with zinc sulphate or zinc acetate solution as electrolyte medium. Biochim. Biophys. Acta 136, 391-393

68. Hata, R. and Nagai, Y. (1973): A micro colorimetric determination of acidic glycosaminoglycans by two dimensional electrophoresis on a cellulose acetate strip. Anal. Biochem. 52, 652-656

69. Hatae, Y. and Makita, A. (1975): Colorimetric determination of hyaluronate degraded by Streptomyces hyaluronidase. Anal. Biochem. 64, 30-36

70. Havass, Z. and Szabo, L. (1972): Thin-layer chromatographic separation of glycosaminoglycans. J. Chromatogr. 71, 580-584

71. Hillborn, J. C. and Anastassiadis, P. A. (1969): Acrylamide gel electrophoresis of acidic mucopolysaccharides. Anal. Biochem. 31, 51-58

72. Hillborn, J. C. and Anastassiadis, P. A. (1971): Estimation of the molecular weights of acidic mucopolysaccharides by polyacrylamide gel electrophoresis. Anal. Biochem. 39, 88-92

73. Hirano, S. (1972): Molecular structure of heparin as examined by nuclear magnetic resonance, optical rotatory dispersion, and electron microscopy. Int. J. Biochem. 3, 677-683

74. Hoffman, P., Linker, A. and Meyer, K. (1956): Uronic acid of chondroitin sulfate B. Science 124, 1252

75. Horner, A. A. (1967): Electrophoresis of acidic mucopolysaccharides in agarose gel. Canad. J. Biochem. 45, 1009-1113

76. Hsu, D., Hoffman, P. and Mashburn, T. A. Jr. (1972): Micro colorimetric method of dye stainig and its application to cellulose polyacetate strip electrophoresis for the determination of acid glycosaminoglycans. Anal. Biochem. 46, 156-163

77. Hsu, D., Hoffman, P. and Mashburn,T. A. Jr. (1973): Acrylamide gel electrophoresis of acid glycosaminoglycans for the study of molecular size distribution. Anal. Biochem. 52, 382-394

78. Humbel, R. (1972): Determination of dermatane-sulfate in a mixture of glycosaminoglycans. J. Pharm. Belg. 26, 321-334

79. Humbel, R. (1974): Identification and quantitation of keratan sulfate in urine. Clin. Chim. Acta 52, 173-177

80. Ibrahim, A. N. and Streitfeld, M. M. (1973): The microassay of hyaluronic acid concentration and hyaluronidase activity by capillary turbidity (CT) and capillary turbidity reduction (CTR) tests. Anal. Biochem. 56, 428-434

81. Inou, S. and Miyaki, M. (1975): Quantitative analysis of iduronic acid and glucuronic acid in sulfated galactosaminoglycuronans by gas chromatography. Anal. Biochem. 65, 164-174

82. Iverius, P. H. (1971): Coupling of glycosaminoglycans to agarose beads (Sepharose 4B). Biochem. J. 124, 677-783

83. Jourdian, G. W., Wolfman, M., Sarber, R. and Distler, J. (1979): A specific sensitive method for the determination of hyaluronate. Anal. Biochem. 96, 474-480

84. Keleti, G. and Lederer, W. H. (1974): Handbook of Micromethods for the Biological Sciences. Van Nostrand Reinhold, New York

85. Kennedy, J. F. (1974): Methodology and instrumentation in the chromatography of carbohydrates on ion-exchange resins. Biochem. Soc. Trans. 2, 54-64

86. Khan, S. U., Morris, G. F. and Hidiroglou, M. (1979): Microdetermination of total sulfur and inorganic sulfate in biological materials. Microchem. J. 24, 291-297

87. Khym, X. J. and Doherty, D. G. (1952): The analysis and separation of glucuronic and galacturonic acids by ion exchange. J. Amer. Chem. Soc. 74, 3199-3200

88. Kiimura, A. and Tsurumi, K. (1969): Electrophoretic separation of acid mucopolysaccharides on cellulose acetate sheets. J. Biochem. (Tokyo) 65, 303-304

89. Kinoshita, T., Kanada, K. and Tsuji, A. (1975): A simple chemical method for the determination of dermatan sulfate in the presence of chondroitin 4-sulfate and chondroitin 6-sulfate. J. Biochem. (Tokyo) 77, 399-403

90. Kochetov, N. K. and Chizhov, O. S. (1972): Mass spectrometry of Carbohydrates. Methods carbohydr. Chem. 6, 540-554

91. Korn, E. D. (1959): The synthesis of heparin by slices of mouse cell tumor. J. Biol. Chem. 234, 1321-1324

92. Kosaki, M. and Yosizawa, Z. (1975): A rapid method for separation and identification of several hexuronic acids and hexuronic acid-containing oligosaccharides. Anal. Biochem. 69, 415-419

93. Kosheleva, G. I., Il'chenko, G. Y and Glebco, L. I. (1976): Uronic acid content in polysaccharides: estimation involving the use of gas-chromatography. Anal. Biochem. 73, 115-119

94. Lagunoff, D. and Warren, G. (1962): Determination of 2-deoxy-2-sulaminohexose content of mucopolysaccharides. Arch. Biochem. Biophys. 99, 396-400

95. Lakatos, M. and Di Ferrante, N. (1974): Simple measurement of urinary glycosaminoglycans and degradation products. Biochem. Med. 9, 256-260

96. Laurent, T. C., Ryan, M. and Pietruszkiewicz, A. (1960): Fractionation of hyaluronic acid. The polydispersity of hyaluronic acid from the bovine vitreous body. Biochim. Biophys. Acta 42, 476-485

97. Lee, G. J.-L., Evans, J. E. and Tieckelmann, H. (1978): Rapid and sensitive determination of enzymic degradation products of isomeric chondroitin sulfates by high-performance liquid chromatography. J. Chromatogr. 146, 365-380

98. Lee, G. J.-L., Evans, J. E. and Tieckelmann, H. (1978 a): Rapid and sensitive determination of enzymic degradation products of isomeric chondroitin sulfates by high-performance liquid chromatography. J. Chromatogr. 146, 439-448

99. Lee, Y. C. and Montgomery, R. (1965): Separations with molecular sieves. Methods Carbohydr. Chem. 5, 28-34

100. Lewandowski, J. (1973): Simple determination of hexosamines in column effluents. Anal. Biochem. 54, 325-332

101. Lippiello, L. and Mankin, H. J. (1971): Thin-layer chromatographic separation of the isomeric chondroitin sulfates, dermatan sulfate, and keratan sulfate. Anal. Biochem. 39, 54-58

102. Lloyd, P. F., Roberts, G. P. and Lloyd, K. O. (1960): Isolation of kerato-sulphate from bovine pulposus and cornea. Biochem. J. 75, 14p

103. Lönngren, J. and Svennsson, S. (1974): Mass spectrometry in structural analysis of natural carbohydrates. Methods Carbohydr. Chem. Biochem. 29, 41-106

104. Ludowieg, J. and Benmaman, J. D. (1967): Colorimetric differentiation of hexosamines. Anal. Biochem. 19, 80-88

105. Maeda, M., Kinoshita, T. and Tsuji, A. (1970): A novel fluorophotometric method for determination of hexosamines using pyridoxal and zinc (II) ion. Anal. Biochem. 38, 121-129

106. Makharadze, R. V., Chumburidze, B. I., Mshvidobadze, A. E., Sardzhveladze, O. V., Dzhordzhikiya, M. A. and Sepiashvili, V. N. (1977): Fractionation and purification of some biopolymers by chromatographic methods. Khromatogr. Metody Farm. 74-78

107. Margolis, R. U. and Margolis, R. K. (1972): Sulfate turnover in mucopolysaccharides and glycoproteins of brain. Biochim. Biophys. Acta 264, 426-431

108. Marsh, C. A. (1966): In: G. H. Dutton (Ed.). Glucuronic Acid: Free and Combined. Chemistry, Biochemistry, Pharmacology and Medicine. Academic Press, New York, 3

109. Marzullo, G. and Lash, J. W. (1967): Separation of glycosaminoglycans on thin layers of silica gel. Anal. Biochem. 18, 575-578

110. Masamune, H. and Satake, M. (1959): On paper chromatography of hexuronic acids. Tohoku J. Exptl. Med. 69, 167

111. Mashburn, T. A., Jr., Hoffman, P. and Hsu, D. S. (1974): The effect of cations in equilibrium gradient centrifugation of mucopolysaccharides. Biochim. Biophys. Acta 362, 366-374

112. Mathews, M. B. (1956): The molecular weight of sodium chondroitin sulfate by light scattering. Arch. Biochem. Biophys. 61, 367-377

113. Mathews, M. B. (1958): Isomeric chondroitin sulfates. Nature 181, 421-422

114. Mathews, M. B. (1976): Determination of the molecular weight of connective tissue glycosaminoglycans (acid mucopolysaccharides) by gel electrophoresis. Methods Carbohydr. Chem. 7, 116-119

115. Mathews, M. B. and Decker, L. (1971): Determination of molecular weight of acid mucopolysaccharides by gel electrophoresis. Biochim. Biophys. Acta 224, 30-34

116. Mathews, M. B. and Decker, L. (1977): Conformation of hyaluronate in neutral and alkaline solutions. Biochim. Biophys. Acta 498, 259-263

117. McDuffie, N. M., Dietrich, C. P. and Nader, H. B. (1975): Electrofocusing of heparin: fractionation of heparin into 21 components distinguishable from other acidic mucopolysaccharides. Biopolymers 14, 1473-1486

118. Mitra, S. K. and Blau, K. (1978): An improved determination of total glycosaminoglycans in body fluids by formation of complexes with quinacrine: changes in amniotic fluid total glycosaminoglycans during normal pregnancies and in pregnancies at risk for mucopolysaccharides. Clin. Chim. Acta 89, 127-134

119. Mukerjee, H. and SriRam, J. (1964): Paper chromatographic separation of glucosamine and galactosamine. Anal. Biochem. 8, 393-394

120. Nader, H. B. and Dietrich, C. P. (1978): Determination of sulfate by chromatography. In: Jaques, L. B. (1978): 234-235

121. Nader, H. B., Dietrich, C. P. and McDuffie, N. M. (1978): Isoelectric focusing. In: Jaques (1978), 304-305

122. Nalefski, L. A. and Takano, F. (1950): A photonephelometric method for the determination of sulfates in biologic fluids. J. Lab. Clin. Med. 36, 468-470

123. Neukom, H. and Kuendig, W. (1965): Fractionation on diethylaminoethylcellulose column. Fractionation of neutral and acidic polysaccharides by ion-exchange column chromatography on diethylaminoethyl (DEAE)-cellulose. Methods Carbohydr. Chem. 5, 14-17

124. Newton, D. J., Scott, J. E. and Whiteman, P. (1974): The estimation of acid glycosaminoglycan-Alcian blue complexes eluted from electrophoretic strips. Anal. Biochem. 62, 268-273

125. Niedermeier, W. (1971): Gas chromatography of neutral and amino sugars in glycoproteins. Anal. Biochem. 40, 465-475

126. Nikata, Y., Hata, R. and Nagai, H. (1971): Simple microanalysis of acid mucopolysaccharide. Jap. J. Clin. Pathol. 19, 349 (Jap.)

127. Nir, I. (1964): Determination of glucuronic acid by naphthoresorcinol. Anal. Biochem. 8, 20-23

128. Oates, M. D. and Schrager, J. (1967): The determination of sugars and amino sugars in the hydrolysates of mucopolysaccharides by gas-liquid chromatography. J. Chromatogr. 28, 232-245

129. Oguchi, M. and Oguchi, M. S. (1979): Tetraborate concentration on Morgan-Elson reaction and an improved method for hexosamine determination. Anal. Biochem. 98, 433-437

130. Ohya, T. and Kaneko, Y. (1970): Novel hyaluronidase from Streptomyces. Biochim. Biophys. Acta 198, 607-609

131. Orr, S. F. D. (1954): Infra-red spectroscopic studies of some polysaccharides. Biochim. Biophys. Acta 14, 173-181

132. Orr, S. F. D., Harris, R. J. C. and Sylven, B. (1952): Evidence from infra-red spectroscopy for the composition of certain polysaccharides. Nature 169, 544-545

133. Ototani, N. and Yosizawa, Z. (1978): Interaction of mucopolysaccharidases with glycosaminoglycans on glycosaminoglycan-bound AH-Sepharose. J. Biochem. (Tokyo) 84, 1005-1008

134. Pearce, R. H., Mathieson, J. M. and Grimmer, B. J. (1968): Fractionation of anionic glycosaminoglycans by ion-exchange chromatography. Anal. Biochem. 24, 141-156

135. Pearson, C. H. (1963): The rapid determination of small amounts of glucosamine and galactosamine in protein preparations with special reference to skin tissue. Biochem. J. 88, 540-545

136. Perlin, A. S. (1975): Proton and carbon-13 NMR studies on carbohydrate polymers. Structure of mucopolysaccharides. In: E. B. Mano (Ed.). Proceedings of International Symposium of Macromolecules. Elsevier, Amsterdam, Netherlands, 337-348

137. Perlin, A. S. (1976): Analysis and characterization of glycosaminoglycans by nuclear magnetic resonance spectroscopy. Methods Carbohydr. Chem. 7, 94-100

138. Perlin, A. S. (1978): Nuclear magnetic resonance spectroscopy. In: Jaques (1978), 226-229

139. Perlin, A. S., Casu, B., Sanderson, G. R. and Johnson, L. F. (1970): 220 MHz spectra of heparin, chondroitins and other mucopolysaccharides. Can. J. Chem. 48, 2260-2268

140. Preston, B. N., Davies, M. and Ogston, A. G. (1965): The composition and physiochemical properties of hyaluronic acids prepared from ox synovial fluid and from a case of mesothelioma. Biochem. J. 96, 449-471

141. Radhakrishnamurthy, B. and Berenson, G. S. (1963): Identification of uronic acids in mucopolysaccharides. Arch. Biochem. Biophys. 101, 360-362

142. Radhakrishnamurthy, B., Dalferes, E. R., Jr. and Berenson, G. S. (1968): Analyses of glycosaminoglycans by gas-liquid chromatography and the nature of hexuronic acids in heparin. Anal. Biochem. 24, 397-408

143. Radhakrishnamurthy, B., Dalferes, E. R., Jr. and Berenson, G. S. (1968 a): Determination of N-acetyl groups in glycosaminoglycans by gas-liquid chromatography. Anal. Biochem. 26, 61-67

144. Radhakrishnamurthy, B., Dalferes, E. R., Jr. and Berenson, G. S. (1976): Determination of hexuronic acids in glycosaminoglycans by automated ion-exchange chromatography. Anal. Biochem. 75, 160-167

145. Reissig, J. L., Strominger, J. L. and Leloir, L. F. (1955): A modified colorimetric method for the determination of N-acetylamino sugars. J. Biol. Chem. 217, 959-966

146. Ringertz, N. R. and Reichard, P. (1960): Chromatography on ECTEOLA of sulfate-containing mucopolysaccharides. Acta Chem. Scand. 14, 303-311

147. Rivkina, V. B., Tsimarkina, G. E., Oyachkova, A., Kasavina, B. S., Chukaeva, D. T. and Vinetskaya, M. I. (1974): Electrophoretic analysis of glycosaminoglycans in a polyacrylamine gel. Lab. Delo. (10), 597-600

148. Roberts, G. P. (1966): Thin-layer gel filtration of proteins and mucopolysaccharides. J. Chromatogr. 22, 90-94

149. Robertson, W. V. B. and Harvey, J. (1972): The determination of galactose in urinary acidic glycosaminoglycans as a measure of keratan sulfate-like substances in urine. Biochem. Med. 6, 246-256

150. Rosenquist, T. H. and Rosenquist, J. W. (1974): Histochemical changes in the affinity of Alcian blue 8GX for aging monkey kidneys, quantified by atomic absorption spectrophotometry. J. Gerontol. 29, 149-152

151. Saito, H., Yamagata, T. and Suzuki, S. (1968): Enzymatic method for the determination of small quantities of isomeric chondroitin sulfates. J. Biol. Chem. 243, 1536-1542

152. Sakamoto, S., Sakamoto, M., Goldhaber, P. and Glimcher, M. J. (1978): The purification of mouse bone collagenase by heparin-substituted Sepharose 4B affinity chromatography. Trans. Annu. Meet. Orthop. Res. Soc. 3, 61

153. Sato, C. S. and Gyorkey (1976): Zinc binding by glycosaminoglycans. J. Biochem. (Tokyo) 80, 883-886

154. Sato, C. S. and Gyorkey, F. (1977): Differential precipitation of glycosaminoglycans at pH 7.0. Current Sci. 46, 415-416

155. Scher, I. and Hamerman, D. (1971): Isolation of human synovial-fluid hyaluronate by density-gradient ultracentrifugation and evaluation of its protein content. Biochem. J. 126, 1073-1080

156. Schiller, S., Slover, G. A. and Dorfman, A. (1961): A method for the separation of acid mucopolysaccharides. Application to the isolation of heparin from the skin of rats. J. Biol. Chem. 236, 983-987

157. Schmidt, M. (1962): Fractionation of acid mucopolysaccharides on DEAE-Sephadex anion-exchanger. Biochim. Biophys. Acta 63, 346-348

158. Scott, J. E. (1960): Aliphatic ammonium salts in the assay of acidic polysaccharides from tissues. Methods Biochem. Anal. 8, 145-197

159. Scott, J. E. (1965): Fractionation by precipitation with quaternary ammonium salts. Methods Carbohydr. Chem. 5, 38-44

160. Scott, J. E. (1968): Periodate oxidation, pKa and conformation of hexuronic acids in polyuronides and mucopolysaccharides. Biochim. Biophys. Acta 170, 471-473

161. Scott, J. E. (1970): Critical electrolyte concentration (CEC) effects in interactions between acid glycosaminoglycans and organic cations and polycations. In: E. A. Balazs (Ed.). Chemistry and Molecular Biology of Intercellular Matrix. Academic Press, New York, Vol. 2, 1105-1119

162. Shively, J. E., Taylor, R. L. and Conrad, H. E. (1976): The estimation of uronic acids by radiochemical methods. In: D. A. Hall (Ed.). Methodology for Connective Tissue Research. Johnson-Vruvvers Ltd., Oxford, 257-262

163. Silpanata, P., Dunstone, J. R. and Ogston, A. G. (1968): Fractionation of a hyaluronic acid in a density gradient. Biochem. J. 109, 43-50

164. Smith, I. (1976): Chromatographic and Electrophoretic Techniques, 4th Ed., Vol. 1 and 2, Yearbook Medical Publications, Chicago

165. Smith, R. L. and Gilkerson, E. (1979): Quantitation of glycosaminoglycan hexosamine using 3-methyl-2-benzothiazolone hydrazone hydrochloride. Anal. Biochem. 98, 478-480

166. Spencer, B. (1960): The ultramicro determination of inorganic sulfate. Biochem. J. 75, 435-440

167. Spolter, L. and Marx, W. (1959): Paper chromatography of heparin and related sulfated mucopolysaccharides. Biochim. Biophys. Acta 38, 123-129

168. Srinivasan, S. R., Radhakrishnamurthy, B., Dalferes, E. R. and Berenson, G. S. (1970): Determination of sulfate in glycosaminoglycans by gas-liquid chromatography. Anal. Biochem. 35, 398-404

169. St. Cyr, M. J. (1970): Electrophoresis of hexuronic acids. J. Chromatogr. 47, 284-286

170. Stary, Z., Wardi, A. and Turner, D. (1964): Galacturonic acid in hydrolyzates of defatted human brain. Biochim. Biophys. Acta 83, 242-244

171. Staub, A. M. (1965): Removal of proteins, Sevag method. Methods Carbohydr. Chem. 5, 5-6

172. Stevenson, F. K., Kent, P. W. and Fisher, D. (1967): Quantitative determination of xylose in mucopolysaccharides derivatives. Chem. Industry 17, 703-704

173. Stimson, W. H. (1971): Gas-liquid chromatography of hexosamines. Fed. Eur. Biochem. Soc. Lett. 13, 17-20

174. Stone, A. L. (1965): Optical rotatory dispersion of mucopolysaccharides and mucopolysaccharide-dye complexes. Biopolymers 3, 617-624

175. Stone, A. L. (1976): Circular dichroism in polysaccharide structural analysis: intrinsic and extrinsic cotton effects. Methods Carbohydr. Chem. 7, 120-138

176. Stone, A. L. and Bradley, D. F. (1967): Aggregation of cationic dyes on acid polysaccharides. I. Spectrometric titration with acridine orange and other metachromatic dyes. Biochim. Biophys. Acta 148, 172-192

177. Suschke, J. (1973): Acid mucopolysaccharides in the urine, amniotic fluid and serum. Fortschr. Med. 91, 485-487 (Ger)

178. Svejcar, J. and Robertson, W. v. B. (1967): Micro separation and determination of mammalian acidic glycosaminoglycans (mucopolysaccharides). Anal. Biochem. 18, 333-350

179. Swann, D. and Balazs, E. A. (1966): Determination of the hexosamine content of macromolecules with manual and automated techniques using the dimethylaminobenzaldehyde reaction. Biochim. Biophys. Acta 130, 112-129

180. Szirmai, J. A., Van Boven-DeTyssonsk, E. and Gardell, S. (1967): Microchemical analysis of glycosaminoglycans, collagen, total protein and water in histological layers of nasal septum cartilage. Biochim. Biophys. Acta 136, 331-350

181. Taniguchi, N. (1970): Thin-layer gel filtration of urinary acid mucopolysaccharides in Hunter's syndrome and in normal individuals. Clin. Chim. Acta 30, 801-808

182. Taniguchi, N. (1976): Separation and identification of glycosaminoglycans. Separation and Purification Methods 5, 247-299

183. Thunell, S., Antonopoulos, C. A. and Gardell, S. (1967): Analysis of aortic glycosaminoglycans from various animal species by CPC-cellulose column procedures. J. Atheroscler. Res. 7, 283-294

184. Tira, M. E., Calatroni, A., Balduini, C., Torre, G., Moretti, R. and Casu, B. (1979): Structure of urinary glycosaminoglycans. Chemical and proton NMR characterization of heparan sulfates and dermatan sulfates from mucopolysaccharidoses type II, III, VI. In: B. Berra, C. Balduini, S. DiDonoto and G. Tettamanti (Eds.). Perspectives in Metabolic Disease. Vol. 2 (Glycoproteins and glycosaminoglycans in Health and Disease), 165-185

185. Tortolani, G. and Rogmanoli, E. (1975): Estimation of molecular weight of acid mucopolysaccharides by thin-layer electrophoresis on Sephadex-cellulose. Experientia 31, 389-390

186. Tracey, M. V. (1952): The determination of glucosamine by alkaline decomposition. Biochem. J. 52, 265-267

187. Tsignos, C. P. and Muir, H. (1966): Estimation of pentoses and methyl-pentoses in biopolymers, in particular of fucose and xylose. Anal. Biochem. 17, 495-501

188. Valle, U. D. and Shively, J. E. (1979): Two-column system for determination of glucosamine, galactosamine, and amino acids on a Beckman 121 MB Amino Acid Analyzer: separation of the anomers of glucosamine and galactosamine. Anal. Biochem. 96, 77-83

189. Varma, R. and Varma R. S. (1976): Simultaneous determination of neutral sugars and hexosamines in glycoproteins and acid mucopolysaccharides (glycosaminoglycans) by gas-liquid chromatography. J. Chromatogr. 128, 45-52

190. Varma, R. and Varma, R. S. (1977): A simple procedure for combined gas chromatographic analysis of neutral sugars, hexosamines and alditols. Determination of degree of polymerization of oligo- and polysaccharides and chain weights of glycosaminoglycans. J. Chromatogr. 139, 303-310

191. Varma, R., Vercellotti, J. R. and Varma, R. S. (1977): On arabinose as a component of brain hyaluronate. Confirmation by chromatographic, enzymatic and chemical ionization-mass spectrometric analyses. Biochim. Biophys. Acta 497, 608-614

192. Varma, R. S., Varma, R., Allen, W. S. and Wardi, A. H. (1974): A gas chromatographic method for determination of hexosamines in glycoproteins and acid mucopolysaccharides. J. Chromatogr. 93, 221-228

193. Vilter, V., Da Lage, C., Gaudin-Harding, F. et al. (1974): Spectrophotometric identification in vitro of glycosaminoglycan sulfates by the metachromatic bathochrome properties of pseudoisocyanine. C. R. Acad. Sci. (D) (Paris) 279, 815-818 (Fr)

194. Wagner, W. D. (1979): A more sensitive assay discriminating galactosamine and glucosamine in mixtures. Anal. Biochem. 94, 394-396

195. Walborg, E. F., Jr., Cobb, B. F., Adams-Mayne, M. and Ward, D.N. (1963): Semiautomatic analysis of glucosamine and galactosamine in protein hydrolyzates. Anal. Biochem. 6, 367-373

196. Wardi, A. H. and Michos, G. A. (1973): Spectrophotometric determination of acid mucopolysaccharides. Anal. Biochem. 51, 274-279

197. Wardi, A. H., Allen, W. S. and Varma, R. (1974): Spectrophotometric determination of glycosaminoglycans as their scandium complexes with xylenol orange. Anal. Chem. 46, 919-920

198. Wardi, A. H., Allen, W. S. and Varma, R. (1974 a): A simple method for the detection and quantitative determination of hexuronic acids and pentoses. Anal. Biochem. 57, 268-273

199. Wasserman, L., Ber, A. and Allalouf, D. (1977): Use of thin-layer chromatography in the separation of disaccharides resulting from digestion of chondroitin sulfates with chondroitinases. J. Chromatogr. 136, 342-347

200. Wasteson, A. and Wessler, E. (1971): Molecular size and xylose content of urinary glycosaminoglycans. Biochim. Biophys. Acta 252, 13-17

201. Wessler, E. (1968): Analytical and preparative separation of acidic glycosaminoglycans by electrophoresis in barium acetate. Anal. Biochem. 26, 439-444

202. Wessler, E. (1971): Electrophoresis of acidic glycosaminoglycans in hydrochloric acid: micro method for sulfate determination. Anal. Biochem. 41, 67-69

203. Whistler, R. L., Ed. (1964): Methods Carbohydr. Chem. Vol. 4, Academic Press, New York

204. Whiteman, P. (1973): The quantitative measurement of Alcian blue-glycosaminoglycan complexes. Biochem. J. 131, 343-350

205. Whiteman, P. (1973 a): The quantitative determination of glycosaminoglycans in urine with Alcian blue 8GX. Biochem. J. 131, 351-357

206. Wollin, A. (1978): Determination of uronic acid from carbon dioxide liberation. In: Jaques (1978), 235-237

207. Wollin, A. (1978 a): Determination of sulfate by atomic absorption spectrophotometry. In: Jaques, L. B. (1978), 232-234

208. Wortman, B. (1962): Quantitation of keratan sulfate in the presence of other corneal glycosaminoglycans. Anal. Biochem. 4, 10-16

209. Yamaguchi, H. and Makino, K. (1977): Studies on a micromethod for determination of the degree of polymerization of neutral oligo- and polysaccharides. II. Application to oligosaccharides and glucans. J. Biochem. (Tokyo) 81, 563-569

210. Yosizawa, Z. (1960): A note for identification of uronic acids in mucopolysaccharides. Tohoku J. Exp. Med. 72, 140-143

211. Yuki, H. and Fishman, W. H. (1963): A carbazole method for the differential analysis of glucuronate, glucosiduronate and hyaluronate. Biochim. Biophys. Acta 69, 576-578

212. Hjerpe, A., Antonopoulos, C. A. and Engfeldt, B. (1979): Determination of sulfated disaccharides from chondroitin sulfates by high performance liquid chromatography. J. Chromatogr. 171, 339-344

213. Murata, K. (1980): Enzymatic analysis of acid glycosaminoglycans. Methods in Carbohydrate Chemistry 8, 81-88

214. Dietrich, C. P., Michelacci, Y. M. C. and Nader, H. B. (1980): Mucopolysaccharidases from Flavobacterium heparinum: isolation, characterization, and use for structural analysis of chondroitin sulfates, heparin and heparan sulfates. In: J. J. Marshall (Ed.) Mechanism of Saccharide Polymerization and Depolymerization (Proc. 1978 sympos.), Academic Press, New York, pp. 317-329

Chapter 3

Biological Functions and Therapeutic Uses of Glycosaminoglycans

3.1 Introduction

The acidic glycosaminoglycans (mucopolysaccharides) form an important constituent of the connective tissue and have the general role in supporting and binding together the cells to form tissues, the tissues to form organs and the further organization of the organs into the systems of the body. Although all the connective tissues of the body consist of the cells, the extracellular fibers and the amorphous ground substance containing the glycosaminoglycans, yet the distribution of the relative quantities of these three components is different throughout the body providing modification of a particular connective tissue for adaptation to a specific biological function. Alterations in physiological processes with aging are accompanied by changes in the relative proportions and physico-chemical nature of the three components of the connective tissue (Vogel, 1973; Hall, 1976).

The key location of the acidic glycosaminoglycans, as a constituent of the amorphous ground substance, or matrix interspersed between the cells and extracellular fibers, makes these macromolecules very important in the physiology and pathology of the individual cells and tissues. All the nutrients going into the cells and the metabolites coming out of the cells have to pass through the ground substance. Therefore, any variations in the chemical composition and the physico-chemical properties of the glycosaminoglycans in the connective tissue ground substance influence the physiological and pathological processes (Otaka, 1974; Fricke and Hartman, 1974) in the surrounding cells and tissues.

Dorfman (1958) suggested the role of acidic glycosaminoglycans in biological processes, including lubrication of joints, wound healing, calcification, control of water and electrolytes in extracellular fluid, blood coagulation, blood lipid-clearing activity, and the maintenance of the stable transport and refractive medium in the eye. With the advancements made in the field and the realization that glycosaminoglycans occur in the tissues as the biologically important protein-glycosaminoglycan complexes (proteoglycans), several other biological functions have come to light. All these functions of the glycosaminoglycans and proteoglycans are attributable directly or indirectly to the polyanionic character due to the presence of the sulfate and carboxyl groups on these macromolecules. The various roles are described in the next sections.

3.2 Control of collagen formation and organization

The size and orientation of the collagen fibers and the glycosaminoglycan composition in various anatomical sites of the body is variable as an adaptation to the function performed by a tissue or organ. Thus, random meshwork arrangement of collagen fibers and elastin fibers is essential for elasticity of skin; for transparency of cornea, collagen fibers are present as fine fibrils, and to provide strength in binding muscles and bones collagen fibers are arranged as thick parallel bundles. The average collagen fiber diameter in a tissue is controlled by the mucopolysaccharide of that tissue (Jackson and Bentley, 1968). The evidence for this came from electronmicroscopic studies on in vitro formation of collagen fibrils from collagen solution in the presence of different glycosaminoglycan fractions (Keech, 1961; Wood, 1964). The collagen fiber formation consists of two stages: nucleation to form fibrils and fibril growth and stabilization. The chondroitin 4- and 6-sulfates and keratan sulfate affect the nucleation stage by a large number of nuclei formation, while hyaluronic acid and dermatan sulfate produce little or no nuclei. The greater the number of nuclei from a given collagen pool, the finer and thinner would be the collagen fibrils. For this reason, the collagen fibers formed in the presence of dermatan sulfate or hyaluronic acid are thicker. It is believed that the differences in glycosaminoglycan composition similarly affect the fiber size in various tissues. A theory of collagen fiber formation and stabilization (Jackson and Bentley, 1968) suggested that the smallest, initially formed collagen fibrils (proto-fibrils) were self-aggregating units stabilized by covalent cross linkages, the aggregation of several of these protofibrils was stabilized with a mucoprotein and still a higher aggregation into larger fibers was stabilized by mucoprotein plus mucopolysaccharide-protein complexes (proteoglycans). The most serious objection to the interpretation of the in vitro collagen aggregation experiments is that the aggregation observed in the test tube is the result of the interaction of some cationic groups on collagen with the anionic groups on acidic mucopolysaccharides. In the tissues, the acidic mucopolysaccharides (glycosaminoglycans) are already present as mucopolysaccharide-protein complexes called proteoglycans, which may interact with collagen in a manner different from that observed with pure mucopolysaccharides (Katzman and Jeanloz, 1971). The possible role of mucopolysaccharides in the distribution of proteins in the extracellular space of muscles has been described (Bozler and Kamishima, 1969).

The mucopolysaccharide production has been suggested to have a feedback control mechanism for regulating the biosynthesis of collagen in fibroblasts (Rokosova-Cmuchalova and Bentley, 1968). This relationship between the biosyntheses of chondroitin sulfate and collagen was observed by using epiphyseal cartilage slices incubated with ^{14}C-glucose, ^{3}H-proline and sodium salicylate. The decrease in collagen synthesis caused by salicylate was accompanied by a decreased incoporation of radioactive glucose into chondroitin sulfate.

3.3 Control of water in extracellular tissues

The control of the fluid dynamics of the tissues and tissue fluids is specifically related to the hyaluronic acid content of the connective tissue. The integrity of the connective tissue may be regarded as depending on the location of the interstitial water in sets of sponges on the molecular level (Fessler, 1957, 1960 a). Hyaluronic acid is noted for its ability to bind water because of its large size and its large hydrodynamic specific volume, which give rise to a high entropy of dilution permitting binding of up to 500 ml of water per gram of hyaluronic acid (Roger, 1961; Ogston, 1966). Also, the osmotic pressure of the two phases consisting of hyaluronic acid and plasma proteins in the intercellular ground substance was more than the algebraic sum of their individual osmotic pressures. At conditions normally present in the interstitium of the skin and muscle (0.6 % mucopolysaccharides and 2 % protein, osmotic pressure of 10 mm Hg) the partition of the fluid between gel-like mucopolysaccharide compartment and the free-fluid containing proteins is about 50 %. The volume exclusion effect of the interstitial mucopolysaccharides is of a significance in osmotic buffering mechanism which helps in the partition of the fluid between the circulation and the interstitial fluid (Wiederhielm et al., 1976).

The effects of the gel nature of the interstitium on tissue fluid have been described in detail elsewhere (Guyton et al., 1975). The coiled springlike glycosaminoglycans with gel-like characteristics offer tremendous resistance to the flow of water and water can hardly diffuse through this gel. However in some physiological states water retention and rigidity of a connective tissue like skin is increased and this is related to an increased concentration of hyaluronic acid. Such is the case in localized myxedema in which skin hyaluronic acid increases 6-16 times (see also sec. 18.3) and in the sexual skin of monkeys during the natural estrous cycle or artificially induced by injection of estrogens (Ogston, 1966 a; Bentley, 1970). During the subsiding phase occurring naturally or produced by progesterone injection, hyaluronic acid decrease correlates with loss of water from the swelling. The function of the swelling is to aid intromission during the mounting behavior (Washburn and DeVore, 1961; also see sec. 13.6).

The size of hyaluronic acid molecules is an important factor in the water-binding mechanism of the connective tissue. The molecular weight of the acidic glycosaminoglycans isolated from the animal skin decreased with age due to depolymerization and this was responsible for decreased water content and lowering of metachromasia in the aged skin (Hvidberg and Jensen, 1959). Similarly, severe dehydration produced a decrease in the molecular weight (Hvidberg, 1960) and so did the cortisone treatment. Estrogens, on the other hand, increased the polymerization of the skin glycosaminoglycans and its water content (Hvidberg and Jensen, 1959). Daily injections of saline increased the hexosamine contents of skin locally but the molecular weights of the glycosaminoglycans were decreased and restored during rehydration (Hvidberg, 1960). Besides the aforementioned conditions, in all conditions of inflammation, edema and suppuration, hyaluronic acid contents of the tissues involved are increased and these are responsible for higher water contents in those tissues (Ogston, 1966 a).

3.4 Role in urine concentrating process of kidney

A possible role of acidic glycosaminoglycans in urine concentration in the kidney has been proposed (Pinter, 1967). The renal medulla is rich in acidic mucopolysaccharides and these substances are continously synthesized in the interstitium. Upon release from the intracellular sites of synthesis, these substances become effective for their water-binding properties. Laurent and Ogston (1963) have reported that the mucopolysaccharide networks bind water by osmotic forces and the osmotic pressures generated show an extreme non-ideality increasing rapidly with increasing concentration. It has been proposed (Pinter, 1967) that these anomalous osmotic pressure properties of acidic glycosaminoglycans help to exert an effect on urine concentration in the kidney. Supporting this is the histochemical study of sections of renal medullae from animals in varying degrees of hydration (Vanhegan, 1968) which showed the presence of intraluminal acid mucopolysaccharide synthesized by the collecting tubule cells and less acidic mucopolysaccharides in the interstitium. During diuresis the interstitial mucopolysaccharides had open fibrillar appearance which became dense during water deprivation. The tubules also became very closely packed upon injection of antidiuretic hormone.

The reabsorption of water in the renal tubules has previously been shown to involve the hyaluronic acid-hyaluronidase system (Ginetsinskii and Ivanova, 1958). Hyaluronidase is a constituent of secretion from the collecting duct cells. Under normal conditions of diuresis at 1.2 ml/min, the secreted enzyme rapidly depolymerizes the mucopolysaccharide complexes of the intercellular cement and the basement membrane of the tubules. This makes the structures separating the renal tube lumen from the interstitial tissue permeable to water. The hypotonic fluid in the distal part of the nephron then undergoes facultative absorption by following an osmotic gradient. The role of hyaluronidase in the reabsorption of water has been suggested to be the mechanism for the action of antidiuretic hormone produced from the posterior pituitary gland or neurohypophysis (Ginetsinskii, 1958). Dehydration greatly increases the excretion of antidiuretic hormone which acts by secretion of hyaluronidase for making the renal canal membranes more permeable to water for absorption. The role of the GAG in the ground substance of the kidney medulla is further confirmed by the observations made in 14 dogs made hydropenic by protein deprivation, when one kidney was injected with hyaluronidase and the other with distilled water (Rivera et al., 1971). Kidneys injected with hyaluronidase formed hypotonic urine and had decreased metachromasia with Alcian blue.

3.5 Role in immunity, infection and inflammation

The hyaluronic acid in the ground substance has been postulated to provide a resistance to bacterial penetration, thus providing a barrier to bacteria and immunity to the body (Duran-Reynals, 1942). The bacteria are virulent by producing hyaluronidase which hydrolyzes hyaluronic acid (Day, 1949) and allows entry inside. However, studies with several microorganisms, including pneumococci, hemolytic streptococci have shown no correlation between virulence and hyaluronidase production. The relationship of hyaluronidase-hyaluronic acid system to the natural resistance has been reviewed (Zablocki, 1960).

Inflammation is one of the fundamental biological processes in which glycosaminoglycans, glycoproteins and collagen in the interstitial matrix of the connective tissue are involved. The glycosaminoglycans changes and their roles in inflammation have been reviewed (Delaunay and Bazin, 1964; Velican, 1966; Willoughby, 1970). During inflammation glycosaminoglycans are elevated in the inflammatory areas and these levels return to normal values when the cells have started secreting collagen fibers. Histamine released during the earliest changes produces vascular permeability and the appearance of polymorphonuclear leukocytes (Willoughby, 1970). Hyaluronic acid is the first uronic acid-containing mucopolysaccharide to increase especially in non-allergic type of inflammatory process (Berenson and Dalferes, 1960). This is followed by an increase of chondroitin sulfate with a slow increase but increasing to several fold after a few days of inflammation. After that the chondroitin sulfate levels decrease and dermatan sulfate begins to appear (Delaunay and Bazin, 1964; Jackson et al., 1960). When the collagen fibers are being laid the content of dermatan sulfate continues to rise suggesting a role in the cross linking of collagen fibers. Granulation tissue formation in response to an irritating substance has been used as a model for the study of inflammation. Histochemical changes resulting from glycosaminoglycan pattern changes have been observed in such studies (Menkin, 1950; Velican, 1966). The acidic glycosaminoglycans from oil turpentine inflammatory tissue were separated by zone electrophoresis and other methods into hyaluronic acid, heparan sulfate, chondroitin 4-sulfate and dermatan sulfate (Berenson and Dalferes, 1960). In experimental inflammation induced by subcutaneous injection of croton oil in dogs, fibrobalsts increased in number after 24 hours and reached maximum at 78 hours. The glycosaminoglycan changes paralleled changes in the number of fibroblasts (Feher and Jakab, 1972).

3.6 Role in wound healing

The glycosaminoglycans are involved in wound healing by acting in several ways including antiinflammatory action and a decrease of granulation tissue, effect on cell growth and control of collagen fiber morphology (Bentley, 1969). The mechanisms involved in these processes are not clear. During wound healing and tissue repair the glycosaminoglycans are synthesized at different rates in different repair tissues and the types of GAG found in the repair tissues are also variable. The action of GAG in wound affects the size and orientation of collagen fibers or the degree of differentiation of fibroblasts present in the wound (Bentley, 1969). Biochemical and biophysical studies during fracture repair in relation to glycosaminoglycans have been reviewed (Solheim, 1966). The wound healing process involves activation of fibroblasts first in the wound margins, then in the granulation tissue formed between these margins. This produces large amounts of matrix components, GAG and collagen fibers to turn the wounded tissue into a normal repaired tissue. Overproduction of the matrix components produces scars and keloids (Bazin et al., 1973).

Morphological changes during bone fracture repair and regeneration in rabbits correlated with GAG contents of bone fragments and bone callus in fractures. The GAG contents reached a maximum on the 7th day after fracture and the GAG contents reached values close to those of the mature bone when bone formation was completed on the 55th day (Kasavina and Zenkevich, 1961). A comparison of the GAG contents in standard and accelerated wound healing revealed that the accelerated healing was accompanied by a high GAG content (Weeks and Bryant, 1967).

It has been repeatedly demonstrated that the application of cartilage powder or crude cartilage extracts to wounds accelerate the rate of healing (Prudden et al., 1957; Prudden, 1964; Prudden and Allen, 1965). Similarly, the cartilage-bone marrow extract "Rumalon" increased the rates of healing of skin wounds and also antagonized the inhibitory effect of papain administration (Cunitz, 1967). The effect of Rumalon on increased rate of healing is due to the presence of glycosaminoglycans in this preparation. The chondroitin 4-sulfate has been discovered as a potent chemical agent in increasing the rate of healing wounds in experimental animals and human patients (Prudden et al., 1970).

A comparative study of ^{35}S-sulfate and total hexosamine in the skin wounds of normal, thyroidectomized and thyroid-treated guinea pigs revealed a more rapid synthesis of sulfated mucopolysaccharides in thyroidectomized animals with no change in total GAG content of the wound tissue (Moltke and Lorenzen, 1960). The formation of granulation tissue and synthesis of sulfated mucopolysaccharides was markedly reduced in cotton pellet-induced granulomas in alloxan diabetic rats (Wagner et al., 1970). Both estradiol and progesterone enhanced skin wound healing by an increase of tensible strength, accumulation of fibroblasts and glycosaminoglycans and production of procollagen and collagen fibers. Estradiol was more effective than progesterone in healing wounds of the female genital tract than its healing effect on skin wounds.

3.7 Role in eye

The glycosaminoglycans of aqueous and vitreous humors are important for providing a suitable refractive index for the optical system. Hyaluronic acid in the vitreous humor maintains the turgor of this organ and serves as a shock absorber. The GAG of the fluids are also essential for regulation of intraocular pressure (Francois, 1975). Disturbances in the GAG content of the aqueous humor lead to elevation of intraocular pressure in glaucoma. In hypothyroidism increased GAG content of the vitreous humor leads to protrusion (proptosis) of the eye. The transparency of the cornea is related to its hydration, which varies directly with its GAG content both in the live and dead subjects (Tsunenari and Kanda, 1978; Harris, 1957). Corneal clouding is one of the ocular manifestations of the disorders of GAG metabolism called the mucopolysaccharidoses excluding the Hunter's syndrome (Francois, 1974; McKusick, 1974). The corneal clouding in these metabolic disorders results from higher GAG contents (Danes, 1973) leading to an increased hydration and swelling of the ground substance, which in turn disturb the orderly spatial arrangement of the collagen fibrils in the corneal stroma (Harris, 1957). Hyaluronic acid present in the subretinal fluid has been implicated in retinal detachment resulting from local ischaemia and retinal tear (Godtfredsen, 1950; Berman, 1969). Also see subretinal fluid (section 7.5, chapter 7).

3.8 Role in central nervous system

The role of GAG in the central nervous system (CNS) is not clear but several pieces of evidence suggest an essential role for these macromolecules. Metabolic disorders of glycosaminoglycans seen in the mucopolysaccharidoses are accompanied by CNS dysfunction and mental retardation. Effects of intrathecal injections of hyaluronidase on behavior of human subjects and experimental animals suggest the role of GAG as a component of the blood-brain barrier (Cameron et al., 1957). Hyaluronic acid has been demonstrated in both exoplasm and neurilemma sheath of frog and bovine sciatic nerves. Since this macromolecule acts as a cationic exchange resin and does not seem to bind K^+ in preference to Na^+, it may play an important role in nerve conduction (Abood and Abul-Haj, 1956). By virtue of the polyanionic nature and high negative charges the glycosaminoglycan molecules can bind biogenic amines. The unmyelinated nerve fibers in the neurohypophysis and sympathetic nerve fibers contain mucopolysaccharides between the nerve swellings and the blood. A role in the storage and release of transmitter substances and hormones has been suggested (Barer, 1967). Other workers have provided evidence for the involvement of mucopolysaccharides in the storage of catecholamines in the central nervous system (Pycock et al., 1975; Mizokami et al., 1976; Peterson and Loh, 1972). Glycosaminoglycans have been identified in the intercellular substance of rat, cerebral cortex (Bondareff, 1967), subcellular particles from other animals (Blaschke et al., 1976), the nerve cell cytoplasm of rodents (Castejon, 1970; Margolis et al., 1979), and in hypothalmic neurons of rat (Martinez-Rodriguez et al., 1976). Histamine and 5-hydroxytryptamine have been shown to bind with mucopolysaccharides in in vitro and in vivo uptake studies (Parrot and Mordelet-Dambrine, 1973). Evidence has been provided for storage of 5-hydroxytryptamine bound to carboxyl groups of heparin proteoglycan in rat and mast cells (Bergendorf and Uvnas, 1972). On the other hand, DaPrada et al. (1972) found that the hexuronic acid/monoamines and hexosamines/monoamines molar ratios were less than unity in isolated 5-hydroxytryptamine organelles of platelets from rabbit, guinea pigs, rats as well as in chromaffin granules of bovine adrenal medulla. From this the authors concluded that the storage of the major part of the endogenous and exogenous monoamines in the isolated amine storage organelles cannot be explained by their direct binding to anionic groups of glycosaminoglycans.

3.9 Role in biological lubrication and biomechanics

Some biological fluids are jelly-like, sticky or slippery and these have this property due to the presence of glycosaminoglycans, particularly hyaluronic acid, in them. This helps these fluids to act as lubricants. Examples of such fluids are: serous fluids in the pleural and pericardial cavities and synovial fluid in joint movements. Both the articular cartilage and synovial fluid are important in joint lubrication, and any pathological change in either of these two leads to disruption of the lubrication system (Davies, 1966; Dowson and Wright, 1976). The viscosity and thixotropy of synovial fluid are due to its hyaluronic acid content (Sundblad, 1953). However, the differences in viscosity and rheology of synovial fluids are not due to differences in hyaluronic acid in normal and pathological states. In traumatic arthritis the hyaluronate content was 0.10-0.17 g/dl, while in rheumatoid arthritis it was 0.15 g/dl (Sundblad, 1954). In pathological synovial fluid a breakdown of intermolecular bonding of hyaluronic acid molecules and aggregates results in a decrease of viscosity and a change from thixotropic to Newtonian type thus impairing lubrication. Other studies (Castor et al., 1966) have shown that with advancing age and in synovitis there is a decrease in hyaluronic acid concentration and a multimodal distribution of molecular weights.

Synovial hyaluronate is a boundary lubricant in joints helping the sliding of one joint suface over the other. In pericardial and pleural cavities, the mucins in the corresponding serous fluids form swollen films of adsorbed mucin molecules which prevent adhesion between the two membranes of the cavity and help movement of the heart and the lungs, respectively. In the synovial space, the viscous hyaluronate lubricates the opposing joint surfaces by virtue of the physicochemical properties associated with its large size and conformation in solution (Roger, 1961). Also see Chapter 12.

3.10 Role in stone formation (lithiasis)

The polyanionic character of the glycosaminoglycans implicates these macromolecules in the etiology of stone formation by interaction with and precipitation of calcium cabonates, calcium oxalates and bile salts (Matsushiro and Nemoto, 1968; Fawzi et al., 1966). These stones are usually formed in the biliary tract or gallbladder (cholecystolithiasis, Nakamura, 1967) in kidneys (renal lithiasis) or in the urinary tract (urolithiasis). The physicochemical nature and the epidemiological surveys of the various human stones have been reviewed (Lonsdale, 1968). Calcium carbonate is rarely a constituent of urinary calculi though it occurs frequently in gallstones (Matsushiro and Nemoto, 1968).

The presence of mucoproteins in urine was thought to act as a protective colloid in preventing the formation of urinary calculi from precipitation of calcium urates, phosphates, carbonates etc. (Maclagan and Anderson, 1958). Several previous reports (for review see Butt, 1955) claimed that repeated subcutaneous injections of hyaluronidase increased mucosubstances in urine and stopped the formation of urinary concrements in persons predisposed to urinary lithiasis. On the contrary, Boyce et al. (1955) found no change in total urinary colloids or the precipitability of the urinary GAG upon subcutaneous administration of hyaluronidase to normal subjects or patients with urinary calculi. The urinary colloids measured by gravimetry were higher in patients with urinary calculi and these, probably, played a similar role in the formation of urinary calculi as in the calcification of bone and cartilage. The hexuronic acid, hexosamine, mucoproteins and other carbohydrate components in 16 kidney stones have been analyzed and related with physical characteristics such as hardness and with respect to the role of carbohydrates in renal lithiasis (Szederkenyi and Jozsa, 1968). In calcium oxalate stone formation induced in rats by feeding a pyridoxine-deficient diet, an increase in the total glycosaminoglycan complex was observed as compared to the control rats (Varalakshimi and Anandam, 1979). Cellulose column chromatography revealed that the elevation in glycosaminoglycan complex was due to an increase in heparan sulfate fraction with a simultaneous drop in dermatan sulfate. These changes may be related to urolithiasis. However, two recent studies confirm the concept originated by Butt (1955) that the glycosaminoglycans in urine acted as natural inhibitors of urolithiasis. In dilute normal urine, chondroitin 4-sulfate and chondroitin 6-sulfate were the major inhibitors of calcium oxalate crystal growth and aggregation (Bowyer et al., 1979). Several fractions obtained by electrophoresis of normal urine glycosaminoglycans on Vinylite support medium inhibited the formation of calcium oxalate crystals in vitro (Sallis and Lumley, 1979). This suggested the possible role of urinary glycosaminoglycans such as hyaluronic acid and chondroitin sulfate and an unknown macromolecule as natural inhibitors of calcium oxalate stone formation. Similar roles of mucopolysaccharides (glycosaminoglycans) in stone formation (Fawzi et al., 1966) and in gallstone formation (Nakamura, 1967) have also been discussed. The involvement of glycosaminoglycans in the formation of gallstones is further described in section 8.5.1.

3.11 Role in hair growth

Glycosaminoglycans, more specifically the sulfated ones, have been suggested to be involved in the growth of hair (Meyer et al., 1961). In experiments carried out in the shaved skin of pigmented rabbits, repeated intradermal injections of the sulfated glycosaminoglycans, heparan sulfate or dermatan sulfate, initiated a growth cycle. The injections of these mucopolysaccharides as their protamine complexes were equally effective in initiating hair growth, but heparan sulfate was more effective than dermatan sulfate. The correlation between metachromasia and the hair growth rate was noted and this suggested the essential role of these sulfated glycosaminoglycans in the normal hair growth cycle. In other studies (Maekawa, 1976, 1979) a strong parallelism between the various stages of hair growth cycle and the glycosaminoglycans contents of skin at those stages of the hair growth lends further support to the role of the sulfated glycosaminoglycans in normal hair growth. Plucking of the hair from the backs of rats reinitiated their hair cycle with the GAG content of the skin reaching maximum on the 6th day of the growth stage and decreasing in the degenerating stage to reach the minimum in the resting stage. Hyaluronic acid and mast cells were abundant in the skin during the growth stage and the least in the resting stage (Maekawa, 1976). Similar parallelism between the mast cells around the hair follicles, hyaluronic acid and total glycosaminoglycan contents was noted in the anagen, telogen and catagen stages of hair growth in rats (Maekawa, 1979).

Further evidence to the role of the chondroitin 4-sulfate in hair growth was provided by the observation that rabbit skin grafts soaked in solutions of chondroitin sulfate were stimulated to be accepted better when grafted and these also produced more and better pigmented hair as compared to the untreated skin grafts (Meyer et al., 1961). Also, in human patients receiving chondroitin 4-sulfate orally for treatment of coronary heart disease, rejuvenation of hair growth in terms of amount of hair, areas of growth and color from grey to dark was noted as a side effect, This effect due to oral ingestion was a nuisance in female patients and a useful side effect providing evidence for entry of orally ingested GAG into circulation in male patients (Morrison and Schjeide, 1974).

Histochemical staining with Alcian blue in various $MgCl_2$ concentrations revealed the presence of sulfated GAG in the basement membrane of hair follicles of dog skin (Butler, 1975). The active hair matrix contained GAG in amounts similar to that of the epidermis. The GAG content of the dog hair follicles increased with age and varied with the breed. Based on the aforementioned work in experimental animals and human subjects, it is tempting to speculate that the thick hair growth commonly seen in patients with mucopolysaccharidoses may be due to the stimulation of hair growth brought about by the storage of heparan sulfate and dermatan sulfate in tissues of these patients.

3.12 Role as ion-exchangers

Because of the similarities of physicochemical properties between the synthetic polyelectrolytes like polyacrylic and polyvinylsulfonic acid and the connective tissue polyelectrolytes, acidic glycosaminoglycans have been suggested to have a similar ion-exchanger role and have been studied extensively for their possible biological role in cation binding.

There is evidence that connective tissue (Headings et al., 1960; Barrett, 1968) and the polyanionic glycosaminoglycans, because of their carboxyl (Gilbert and Myers, 1960) and sulfate groups may act as anion-exchange resins in regulating the connective tissue electrolytes (Manery, 1966). The various individual glycosaminoglycans per se or because of their specific concentration in a particular connective tissue affect different ions. Calcium is bound tightly and selectively by hyaluronic acid, which does not significantly bind monovalent cations. This ability makes hyaluronic acid specifically suited to the role of intracellular Ca-cement important for cell-adhesion, excitability in nerve and muscle, physical and chemical reactions at cell surfaces, etc., (Manery, 1966 a). The calcium ions function as an adhesive between cells by chemically binding membrane components of one cell to another and in the absence of calcium a decrease in adhesiveness between cells occurs and in case of toad bladder epithelial cells, which are tightly joined to each other and to the basement membrane, the components appear to separate with wide aqueous channels between them under the electron microscope (Hays et al., 1965).

In addition to binding calcium ions, the synovial fluid hyaluronic acid carries Na^+ counterions for its negative charges.

The cation-exchange property of the various types of cartilages is due to the presence of the polyelectrolyte chondroitin sulfate and its carboxyl and sulfate ester groups (Barret, 1968). Each repeating unit of chondroitin sulfate at physiological pH carries two negative sites and this is responsible for binding of inorganic and organic (cationic dyes) ions. When powdered demineralized calf cartilage was used as an ion-exchange resin, its cation-binding capacity was found related to its sulfate content (Boyd and Neuman, 1951; Dunstone, 1959). Methylation of the carboxyl group and desulfation resulted in loss of the binding capacity. Since 25 % of the bovine nasal septum consists of chondroitin sulfate and also since it is not normally a calcifiable tissue, it has been used in several cation-binding studies (Dunstone, 1959, 1962; Farber et al., 1957; Farber, 1960). The chondroitin sulfate component is solely responsible for the cation-binding reactions of the ion-exchange type and different cations have been found to exhibit varying degrees of affinity for the cartilage (Dunstone, 1959) and for different isomers of chondroitin sulfate (Dunstone, 1962). These differences may be attributable to the structural differences of their conformations in solutions. These investigations may have some possible significance to the processes involved in calcification (Dunstone, 1962; Barrett, 1968). Dunstone (1962) considered the interactions of cations with aqueous solutions of sodium salts of glycosaminoglycans equivalent to ion-exchange reactions. The orders of increasing cation affinities measured and recorded in the form of exchange parameters were: K^+, Na^+, Mg^{2+}, Ca^{2+}, Sr^{2+}, Ba^{2+} for chondroitin 4-sulfate; K^+, Na^+, $Mg^{2+} \cong Ca^{2+} \approx Sr^{2+} \cong Ba^{2+}$ for dermatan sulfate; Na^+, K^+, Mg^{2+}, Sr^{2+}, Ba^{2+}, Ca^{2+} for heparin; Na^+, Mg^{2+}, $Ca^{2+} \cong Sr^{2+}$, Ba^{2+} for heparan sulfate.

The acidic glycosaminoglycans extracted from cartilage by mild procedures bound cations more strongly than the preparations obtained by harsh procedures, suggesting that higher degree of polymerization promotes cation binding (Dunstone, 1962). The affinity of acidic glycosaminoglycans for Ca^{2+} under physiological conditons was found to decrease in the order: Heparan > dermatan sulfate > chondroitin 4-sulfate > hyaluronic acid (Buddecke and

Drzenick, 1962). Mathews (1964) has pointed out that the cation-binding is not 100 % and the exchange constants vary up to 30-fold with change in ionic strength. Also, since at constant ionic strength, dermatan sulfate has a greater affinity for Ca^{2+} and Mg^{2+} but not for Sr^{2+} and Ba^{2+}, this selective affinity for hydrated cations may be due to some differences in minor ionic properties other than electrostatic charge on the glycosaminoglycans.

The physiological significance of cations associations with acidic glycosaminoglycans is two fold. Firstly, such an interaction between the cationic groups of proteins like collagen and glycosaminoglycans helps in the structural organization of the connective tissue. Secondly, it is implicated in homeostatic functions like regulation of water and electrolytes in the extracellular fluid, ionic changes during nerve and muscle activity, the process of calcification inactivation, etc.

3.13 Role in morphogenesis

Evidence indicates that the glycosaminoglycans (GAG) act as signalling agents for differentiation. Hyaluronic acid accumulated in the early phase of regenerating newt limb. Hyaluronidase then appeared and hyaluronic acid was replaced with chondroitin sulfate (Toole and Gross, 1971). The role of glycosaminoglycans in brain and other tissues has been reviewed (Slavkin et al., 1975; Toole, 1976).

3.14 Role in cell proliferation

The polyanionic glycosaminoglycans have been suggested to have a role in regulation of cell division. The growth regulatory role of GAG and their role in cell proliferation have been reviewed (Regelson, 1968; Long and Williamson, 1979). This makes the GAG significant in carcinogenesis. This may be related especially to the role of sulfated GAG, which are also important for cell recognition and cell adhesiveness (Dietrich et al., 1977).

3.15 Role in calcification

Chondroitin sulfate prevents precipitation of calcium and for this reason cartilage, where chondroitin sulfate concentration is high, remains uncalcified. In the neighborhood of calcifying areas, there is a rapid synthesis and a high concentration of proteoglycans of sulfated GAG. The role of the GAG in the calcification process has been reviewed (Bowness, 1968; Toole and Linsenmayer, 1977). Calcification is associated with a decrease of GAG at onset and a decrease of non-collageneous, N-containing material near its completion (Pugliarello et al., 1970).

3.16 Therapeutic uses of glycosaminoglycans

Heparin is used as an anticoagulant and it acts by preventing the conversion of prothrombin to thrombin (Douglas, 1956). Sulfated glycosaminoglycans have been obtained by chemical modifications which have increased anticoagulant activity with undesirable properties of heparin eliminated in the synthetic polymer (Horton and Usui, 1978). Heparin and heparinoids have plasma lipid-clearing and antiplatelet aggregation activities (Houck, 1957; Jorpes, 1971; Marini et al., 1974). The glycosaminoglycans increase fibrinolytic activity and this activity varies among the various polyanionic mucopolysaccharides depending upon the structure and degree of sulfation (Németh-Csóka, 1974).

Glycosaminoglycans prevent atherogenesis by activation of the enzyme lipoprotein lipase (Gero, 1966). The fibrinolysin action of heparin and heparan sulfate is caused by activation of plasminogen (Astrup et al., 1966). Sulfated GAG have been used in therapy of atherosclerosis in geriatrics (Bompani and Violi, 1974). Chondroitin 4-sulfate has been successfully used in prevention and therapeutic management of arteriosclerosis and coronary artery disease (Morrison and Schjeide, 1974).

Heparin and some other sulfated glycosaminoglycans have the ability of inhibiting the action of aldosterone. In hypertensive patients N-formylchitosan polysulfuric acid produced a fall in the secretion of aldosterone and its probable immediate precursor 18-hydroxycorticosterone (Abbott et al., 1966).

The glycosaminoglycans possess antiviral and antibacterial effects, which provide resistance to infection and allergy and immune reactions, probably mediated indirectly through endocrine mechanisms (Meier, 1958). Sulfated GAG have been used for therapy of viral diseases in experimental animals (Museteanu et al., 1975) and as antitumor agents (Regelson, 1968 a).

Sulfated polysaccharides such as pentosan polysulfuric, cellulose sulfuric and dextran sulfuric esters injected intraperitoneally inhibit local edema (Rocha e Silva et al., 1969). A preparation from bone marrow and cartilage called "Rumalon" and a glucuronyl polysulfated mucopolysaccharide "Arteparon" have been used in therapy of rheumatoid arthritis patients.

Hyaluronic acid has been used in therapeutic replacement of liquid vitreous in retinal detachment (Balazs and Sweeney, 1968). Similarly, viscous jellies of hyaluronic acid have been used for replacement of the aqueous humor and were well-tolerated as a mechanical buffer and in preventing formation of excessive scar following eye surgery (Shafer, 1965; Balazs et al., 1972). Intraarticularly implanted hyaluronic acid jellies aided the healing of cartilage wounds. Hyaluronic acid jellies injected into equine and human joints afflicted with chronic osteoarthritis and traumatic arthritis improved the healing process (Rydell et al., 1970; Balazs et al., 1972).

Mucopolysaccharide preparations have been used to speed up healing of superficial burns (Acharya, 1973) and as regeneration stimulants (Kasavina and Zenkevich, 1966).

Hyaluronic acid or chondroitin sulfate (1-5 parts) have been used in skin cosmetics or scalp lotions, respectively (Lubowe, 1972, 1975). Because of the role of sulfated GAG in hair growth, a cosmetic lotion containing these macromolecules has been patented (Schrek-Purola and Setala, 1974) to aid preservation and renewal of hair. A toothpaste containing 0.2 parts of mucopolysaccharide has been patented (Hashimoto and Aimoto, 1974). A cosmetic for reducing wrinkling of the skin contained 0.6 g carotenoids (from carrots) conjugated with 0.9 g of hexosaminoglycans (Rovesti and Curri, 1978).

3.17 Bibliography

1. Abbott, E. C., Gornall, A. G., Sutherland, D. J. A., Stiefel, M. and Laidlaw, J. C. (1966): The influence of a heparin-like compound on hypertension, electrolytes and aldosterone in man. Can. Med. Assoc. J. 94, 1155-1164

2. Abood, L. G. and Abul-Haj, S. K. (1956): Histochemistry and characterization of hyaluronic acid in axons of peripheral nerve. J. Neurochem. 1, 119-125

3. Acharya, B. S. (1973): Treatment of superficial burns in the accident department. Practitioner 211, 371-372

4. Astrup, T., Nissen, U. and Rasmussen, R. (1966): Effects of preparations of heparin and heparitin sulfate on fibrinolysis. Arch. Biochem. Biophys. 113, 643-640

5. Balazs, E. A., Freeman, M. I., Kloti, R., Meyer-Schwickerath, G., Regnault, F. and Sweeney, D. B. (1972): Hyaluronic acid and replacement of vitreous and aqueous humor. Mod. Probl. Ophthal. (s. Karger, Basel) 10, 3-21

6. Balazs, E. A. and Sweeney, D. B. (1968): The injection of hyaluronic acid and reconstituted vitreous into the vitreous cavity. In: A. McPherson (Ed.). New and Controversial Aspects of Retinal Detachment. Harper and Row, New York, 371-375

7. Barer, R. (1967): Speculations on the storage and release of hormones and transmitter substances. Bibl. Anat. 8, 72-75

8. Barrett, A. J. (1968): Cartilage. In: M. Florkin and E. H. Stotz (Eds.). Comprehensive Biochemistry, Vol. 26B. Elsevier, Amsterdam, Chap. VI, 425-474

9. Bazin, S., Nicoletis, C. and Delaunay, A. (1973): Intercellular matrix of hypertropic scars and keloids. In: E. Kulonen and J. Pikkarainen (Eds.). Biology of Fibroblast. Academic Press, New York, 571-578

10. Bentley, P. J. (1969): Mucopolysaccharide synthesis in healing wounds. In: J. E. Dunphy (Ed.). Repair and Regeneration, Centen. Symp. 1968. McGraw Hill, New York, 151-160

11. Bentley, J. P. (1970): The biological role of the ground substance mucopolysaccharides. In: W. Montagna, J. P. Bentley and R. L. Dobson (Eds.). The Dermis. Adv. Biol. of Skin, Vol. X, Appleton-Century-Crofts, New York, 103-121

12. Berenson, G. S. and Dalferes, E. R., Jr. (1960): Identification of acid mucpolysaccharides from granulation tissue in rats. Br. J. Exp. Pathol. 41, 422-429

13. Bergendorf, A. and Uvnas, B. (1972): Storage of 5-hydroxytryptamine in rat mast cells. Evidence for an ionic binding to carboxyl groups in a granule heparin-protein complex. Acta Physiol. Scand. 84, 320-331

14. Berman, E. R. (1969): Mucopolysaccharides (glycosaminoglycans) of the retina: Identification, distribution and possible role. Modern Problems of Ophthalmology, S. Karger, Basel, Vol. 8, 5-31

15. Blaschke, E., Bergovist, U. and Üvnas, B. (1976): Identification of the mucopolysaccharides in catecholamine-containing subcellular particle fractions from various rat, cat and ox tissues. Acta. Physiol. Scand. 97, 110-120

16. Bompani, R. and Violi, G. (1974): Glucuronyl-glucosaminoglycan sulfate in therapy of atheroarteriosclerotic syndromes in geriatrics. Minerva Med. 65, 4315-4329 (Ita)

17. Bondareff, W. (1967): An intercellular substance in rat cerebral cortex submicroscopic distribution of ruthenium red. Anat. Rec. 157, 527-535

18. Bowness, J. M. (1968): Present concept of the role of ground substance in calcification. Clin. Orthop. 59, 233-247

19. Bowyer, R. C., Brockis, J. G. and McCulloch, R. K. (1979): Glycosaminoglycans as inhibitors of calcium oxalate crystal growth and aggregation. Clin. Chim. Acta 95, 23-28

20. Boyce, W. H., Garvey, F. K. and Norfleet, C. M., Jr. (1955): Metal chelate compounds of urine; their relation to initiation and growth of calculi. Am. J. Med. 19, 87-95

21. Boyd, E. S. and Neuman (1951): The surface chemistry of bone V. The ion-binding properties of cartilage. J. Biol. Chem. 193, 243-251

22. Bozler, E. and Kamishima, Y (1969): Distribution of proteins in the extracellular space of muscles and the possible role of mucopolysaccharides. Experientia, 25, 910-911

23. Buddecke, E. and Drzenick, R. (1962): Stability constants of the calcium complexes of acid mucopolysaccharides. Hoppe-Seyler's Z. Physiol. Chem. 327, 49-64

24. Butler, W. F. (1975): Glycosaminoglycans of hair follicles of dog skin. Histochem. J. 7, 67-75

25. Butt, A. J. (1955): Pathogenesis and treatment of renal lithiasis - newer concepts. Adv. Int. Med. 7, 11-32

26. Cameron, D. E., Levy, L. and Hunzinger, W. (1957): Intrathecal administration of hyaluronidase. Effects upon the behavior of patients suffering from senile and arteriosclerotic behavior disorders. Amer. J. Psychiat. 113, 893-900

27. Catejon, H. V. (1970): Histochemical demonstration of acid glycosaminoglycans in the nerve cell cytoplasm of mouse central nervous system. Acta Histochem. (Jena) 35, 161-172

28. Castor, C. W., Prince, R. K. and Hazelton, M. J. (1966): Hyaluronic acid in human synovial effusions-sensitive indicator of altered connective tissue cell function during inflammation. Arthritis Rheum. 9, 783-794

29. Cunitz, G. (1967): Wound healing and mucopolysaccharide metabolism. Med. Pharmacol. Exp. 16, 49-56

30. Danes, B. S. (1973): Corneal clouding in the genetic mucopolysaccharidoses: A cell culture study. Clin Genet. 4, 1-7

31. DaPrada, M., von Berlepsch, K. and Pletscher, A. (1972): Storage of biogenic amines in blood platelets and adrenal medulla. Lack of evidence for direct involvement of glycosaminoglycans. Naun-Schmiedeberg's Arch. Pharmacol. 275, 315-322

32. Davies, D. V. (1966): Synovial fluid as a lubricant. Fed. Proc. 25, 1069-1076

33. Day, T. D. (1949): The nature and significance of the cementing substance in intersitial connective tissue. J. Path. Bact. 59, 567-573

34. Delaunay, A. and Bazin, S. (1964): Mucopolysaccharides, collagen and non-fibrillar proteins in inflammation. Int. Rev. Connect. Tissue Res. 2, 301-325

35. Dietrich, C. P., Sampaio, L. O., Toledo, O. M. et al. (1977): Cell recognition and adhesiveness: a possible biological role of the sulfated mucopolysaccharides. Biochem. Biophys. Res. Commun. 75, 329-336

36. Dorfman, A. (1958): Studies on the biochemistry of connective tissue. Pediatrics 22, 576-589

37. Douglas, A. S. (1956): The action of heparin in the prevention of prothrombin conversion. J. Clin. Invest. 35, 533-536

38. Dowson, D. and Wright, V. (Eds.) (1976): An Introduction to Bio-Mechanics of Joints and Joint Replacements. Elsevier, Amsterdam

39. Dunstone, J. R. (1959): Some cation-binding properties of cartilage. Biochem. J. 72, 465-473

40. Dunstone, J. R. (1962): Ion-exchange reactions between acid mucopolysaccharides and various cations. Biochem. J. 85, 336-351

41. Duran-Reynals, F. (1942): Tissue permeability and the spreading factors in infection. Bact. Rev. 6, 197-252

42. Farber, S. J. (1960): Mucopolysaccharides and sodium metabolism. Circulation 21, 941-947

43. Farber, S. J., Schubert, M. and Schuster, N. (1957): Binding of cations by chondroitin sulfate. J. Clin. Invest. 36, 1715-1722

44. Fawzi, R., El-Ridi, M. S., Abdel-Hay, A. et al. (1966): Mucopolysaccharides in stone formation. J. Egypt Med. Ass. 49, 639-645

45. Feher, J. and Jakab, L. (1972): The role of connective tissue cells in the metabolic changes of glycosaminoglycans and glycoprotein in experimental inflammation. Acta Morphol. Acad. Sci. Hung. 20, 27-37

46. Fessler, J. H. (1957): Water and mucpolysaccharide as structural components of connective tissue. Nature 179, 426-427

47. Fessler, J. H. (1960): Mode of action of testicular hyaluronidase. Biochem. J. 76, 132-135

48. Fessler, J. H. (1960 a): A structural function of mucopolysaccharide in connective tissue. Biochem. J. 76, 124-132

49. Francois, J. (1974): Ocular manifestations of the mucopolysaccharidoses. Opthalmologica 169, 345-361

50. Francois, J. (1975): The importance of the mucopolysaccharides in intraocular pressure regulation. Invest. Ophtamol. 14, 173-176

51. Fricke, R. and Hartman, F. (Eds.) (1974): Connective Tissues. Biochemistry and Pathophysiology. Springer Verlag, New York

52. Gero, S. (1966): Investigations on the mechanism of atherogenesis Effect of some mucopolysaccharides on the lipoprotein lipase). Rev. Atheroscler. Suppl. Arch. Mal. Coeur. Vaisseaux. 8, 152-158

53. Gilbert, I. G. F. and Myers, N. A. (1960): Metal binding properties of chondroitin sulfate. Biochim. Biophys. Acta 42, 469-475

54. Ginetsinskii, A. G. (1958): Role of hyaluronidase in reabsorption of water in renal tubules: mechanism of action of diuretic hormone. Nature 182, 1218-1219

55. Ginetsinskii, A. G. and Ivanova, L. N. (1958): Role of the system of hyaluronic acid-hyaluronidase in the reabsorption of water in kidney canals. Doklady Akad. Nauk S.S.S.R. 119, 1043-1045

56. Godtfredsen, E. (1950): Further investigations into hyaluronic acid and hyaluronidase in the subretinal fluid in retinal detachment. Acta XVI Concilium Opthalmol. Britania 2, 1162-1168

57. Guyton, A. C., Taylor, A. E. and Granger, H. J. (1975): Dynamics and Control of Body Fluids. Circulatory Physiology II. W. B. Saunders, Philadelphia. Chap. 1 and 5

58. Hall, D. A. (1976): The Aging Connective Tissue. Academic Press, Press, New York, 1976

59. Harris, J. E. (1957): The physiologic control of corneal hydration. Am. J. Opthalmol. 44, 262-280

60. Hashimoto, S. and Aimoto, K. (1974): Schizophyllans in toothpaste. Japan, Kokai 74 81, 554 Aug. 6, 1974. Appl. 72, 125,004. Dec. 13, 1972

61. Hays, R. M., Singer, B. and Malamed, S. (1965): Effect of calcium withdrawal on the structure and function of the toad bladder. J. Cell. Biol. 25, Pt. 2, 195-208

62. Headings, V. E., Rondell, P. A. and Bohr, D. F. (1960): Bound sodium in artery wall. Am. J. Physiol. 199, 783-787

63. Horton, D. and Usui, T. (1978): Sulfated glycosaminoglycans obtained by chemical modification of polysaccharides. In: R. G. Schweiger (Ed.). Carbohydrate sulfates. ACS Symp. Ser. 77, ACS, Washington, D. C., 95-112

64. Houck, J. C. (1957): Anticoagulant, lipemia clearing, and other effects of anionic polysaccharides extracted from seaweed. Proc. Soc. Exptl. Biol. Med. 96, 528-530

65. Hvidberg, E. (1960): Water binding by connective tissue and the acid mucopolysaccharides of the ground substance. Acta Pharmacol. Toxicol. 17, 267-276

66. Hvidberg, E. and Jensen, C. E. (1959): Changes in molecular weight of acid mucopolysaccharides in connective tissue due to hormone treatment, dehydration and age. Acta Chem. Scand. 13, 2047-2056

67. Jackson, D. S. and Bentley, J. P. (1968): Collagen-glycosaminoglycan interactions. In: B. S. Gould (Ed.). Treatise on Glycogen 11A, Academic Press, New York, 189-214

68. Jackson, D. S., Flickinger, D. B. and Dunphy, J. E. (1960): Biochemical studies of connective tissue repair. Ann. N. Y. Acad. Sci. 86, 943-947

69. Jorpes, J. E. (1971): Heparin. In: F. Markwardt (Ed.). Handbook of Experimental Pharmacology, F. Springer, Berlin, Vol. 27 143-191

70. Kasavina, B. S. and Zenkevich, G. D. (1961): Mucopolysaccharides of the bone tissue under normal and pathological conditions. Clin. Chim. Acta 6, 874-882

71. Kasavina, B. S. and Zenkevich, G. D. (1966): Theoretical (experimental) substantiation of the clinical use of mucopolysaccharide and mucopolysaccharidase preparations as regeneration stumulants. Eksp. Khir. Anest. 11, 62-63 (Rus)

72. Katzman, R. L. and Jeanloz, R. W. (1971): Are acidic polysaccharides involved in collagen fibril formation or stabilization? Biochem. Biophys. Acta 229, 516-521

73. Keech, M. K. (1961): The formation of fibrils from collagen solutions. IV. J. Biophys. Biochem. Cytol. 9, 193-209

74. Laurent, T. C. and Ogston, A. G. (1963): The interaction between polysaccharides and other macromolecules. 4. The osmotic pressure of mixtures of serum albumin and hyaluronic acid. Biochem. J. 89, 249-253

75. Long, W. F. and Williamson, F. B. (1979): Glycosaminoglycans, calcium ions and the control of cell proliferation. IRCS Med. Sci. Libr. Compend. 7, 429-434

76. Lonsdale, K. (1968): Human Stones. Science 159, 1199-1207

77. Lubowe, I. I. (1972): Medicinal compositions for application to the skin and/or hair. British 1, 283, 892 Aug. 2, 1972. US App. 44601 June 28, 1970, 5 pp

78. Lubowe, I. I. (1975): Medicinal compositions for application to the skin and/or hair. British 1, 283, 892 March 13, 1975. US App. 44610 June 8, 1970, 4 pp

79. Maclagan, N. F. and Anderson, A. J. (1958): Colloidal properties of urinary mucopolysaccharides. In: G. E. Wolstenholme and M. O'Connor (Eds.). Ciba Found. Symp. on Chemistry and Biology of Mucopolysaccharides. Little, Brown and Company, Boston, 268-282

80. Maekawa, Y. (1976): Behavior of dermal glycosaminoglycan during the hair growth cycle of rats. Nippon Hifuka Gakkai Zasshi 86, 917-927 (Japan)

81. Maekawa, Y (1979): Dermal glycosaminoglycan concentration throughout hair cycle of rats. J. Dermatol. 6, 191-195

82. Manery, J. F. (1966): Connective tissue electrolytes. Fed. Proc. 25, 1799-1803

83. Manery, J. F. (1966 a): Effects of Ca ions on membranes. Fed. Proc. 25, 1804-1810

84. Margolis, R. K., Thomas, M. D., Crockett, C. P. and Margolis, R. U. (1979): Presence of chondroitin sulfate in the neuronal cytoplasm. Proc. Natl. Acad. Sci. U. S. A. 76, 1711-1715

85. Marini, U., Bastoni, F. and Gibelli, A. (1974): Clearing and anti-platelet-aggregation effects of a heparinoid of natural origin. Clin. Ter. 71, 13-31

86. Martinez-Rodriguez, R., Toledano, A., Garcia-Segura, L. M. et al. (1976): Mucopolysaccharides in hypothalamic neurons of the rat. J. Anat. 121, 231-239

87. Mathews, M. B. (1964): Structural factors in cation binding to anionic polysaccharides of connective tissue. Arch. Biochem. Biophys. 104, 394-404

88. Matsushiro, T. and Nemoto, T. (1968): Biochemical studies on muco-polysaccharides in calcium carbonate stones. Tohoku J. Exp. Med. 94, 397-406

89. McKusick, V. A. (1974): Discussion of classification. Birth Defects 10, 63-65

90. Meier, R. (1958): The pharmacological effects of polysaccharides. In: G. E. W. Wolstenholme and M. O'Connor (Eds.) Ciba Found. Symp. on the Chemistry and Biology of Mucopolysaccharides. Little, Brown and Company, Boston, 157-186

91. Menkin, V. (1950): Newer concepts of inflammation. C. C. Thomas, Springfield

92. Meyer, K., Kaplan, D. and Steigleder, G. K. (1961): Effect of acid mucopolysaccharides on hair growth in the rabbit. Proc. Soc. Exptl. Biol. Med. 108, 59-63

93. Mizokami, K., Simoyama, T. and Tajino, M. (1976): Glycosamino-glycans in the central nervous system. Acta Soc. Ophthalmol. Jpn. 80, 108-112

94. Moltke, E. and Lorenzen, I. (1960): Effect of thyroidectomy and thyroxine in the mucopolysaccharides of wounds and skin. Acta Endocrinol. 34, 407-410

95. Morrison, L. M. and Schjeide, O. A. (1974): Coronary Heart Disease and the Mucopolysaccharides (Glycosaminoglycans). Charles C. Thomas, Springfield

96. Museteanu, C., Voss, H., Haase, J. and Ahlert, M. (1975): Sulfated glycosaminoglycans as virus inhibitors. 3. Therapy of viral diseases by glycosaminoglycan polysulfates in laboratory animals. Zentralbl. Bakterial. Parasitenkd. Infektionskr. Hyg. Abt. 1. Orig. Reihe A. 231, 375-388 (Ger)

97. Nakamura, N. (1967): Mucopolysaccharides and their significance in gallstone formation. Tohoku J. Exp. Med. 93, 235-247

98. Németh-Csóka, M. (1974): Relation between chemical structure and fibro genic character of glycosaminoglycans. In: R. Fricke and F. Hartmann (Eds.). Connective Tissues: Biochemistry and Pathophysiology. Springer, Berlin, Germany, 61-63

99. Ogston, A. G. (1966): Macromolecular hydration in solution. Fed. Proc. 25, 993

100. Ogston, A. G. (1966 a): On water binding. Fed. Proc. 25, 986-988

101. Otaka, Y. (Ed.) (1974): Biochemistry and Pathology of Connective Tissue. Igaku Shoin, Tokyo

102. Parrot, J. L. and Mordelet-Dambrine, M. (1973): Histamine and 5-hydroxytryptamine uptake by some mucopolysaccharides. Agents and Actions 3, 292-296

103. Peterson, R. P. and Loh, Y. P. (1972): The role of macromolecules in neuronal function in aplysia. In: G. A. Kerkut and J. W. Phillis (Eds.). Progress in Neurobiology, Vol. 2, Pergamon Press, New York, 177-203

104. Pinter, G. G. (1967): A possible role of acid mucopolysaccharides in the urine concentrating process. Experientia 23, 100-101

105. Prudden, J. F. (1964): Wound healing produced by cartilage preparation. Arch. Surg. 89, 1046-1059

106. Prudden, J. F. and Allen, J. (1965): The clinical acceleration of healing with a cartilage preparation. J. Am. Med. Assoc. 192, 352-356

107. Prudden, J. F., Migel, P., Hanson, P., et al. (1970): The discovery of a potent pure chemical wound-healing accelerator. Am. J. Surg. 119, 560-564

108. Prudden, J. F., Nishihara, G. and Baker, L. (1957): The acceleration of wound healing with cartilage. Surg. Gyn. Obs. 105, 283-286

109. Pugliarello, M. C., Vittur, F., De Bernard, B., Bonucci, E. and Ascenzi, A. (1970): Chemical modifications in osteones during calcification. Calcif. Tissue Res. 5, 108-114

110. Pycock, C. J., Blaschke, E., Bergquist, U., et al. (1975): On the possible involvement of sulfomucopolysaccharides in the storage of catecholamines within the central nervous system. Acta. Physiol. Scnad. 95, 373-382

111. Regelson, W. (1968): The growth regulating activity of polyanions: A theoretical discussion of their place in intercellular environment and their role in cell physiology. Adv. Cancer Res. 11, 223-304

112. Regelson, W. (1968 a): The antimitotic activity of polyanions (antitumor, antiviral and antibacterial action of heparin, heparinoids, anionic dyes and synthetic polymers). Adv. Chemotherapy 3, 303-370

113. Rivera, E. S., Cristostomo, C. P., Juliano, S. J. and Alfonso, R. L. (1971): Ground substance of the kidney medulla. Its role in concentrating urine. J. Philipp. Med. Assoc. 47, 187-202

114. Rocha e Silva, Cavalcanti, R. Q. and Reis, M. L. (1969): Anti-inflammatory action of sulfated polysaccharides. Biochem. Pharmacol. 18, 1285-1295

115. Rogers, H. J. (1961): The structure and function of hyaluronate. In: F. Clark and J. K. Grant (Eds.). The Biochemistry of Mucopolysaccharides of Connective Tissue. Cambridge University, Cambridge, 51-79

116. Rokosova-Cmuchalova, B. and Bentley, J. P. (1968): Relation of collagen synthesis to chondroitin sulfate synthesis in cartilage. Biochem. Pharmacol. Suppl., 315-328

117. Rovesti, P. and Curri, S. B. (1978): Cutaneous bioactivating effect of carotenoids conjugated with permeable molecules. Riv. Ital. Easenze, Profumi, Piante Off. Aromat. Syndets, Saponi Cosmet. Aerosols 60, 360-364

118. Rydell, N. W., Butler, J. and Balazs, E. A. (1970): Hyaluronic acid in synovial fluid. VI. Effect of intra-articular injection of hyaluronic acid on the clinical symptoms of arthritis in track horses. Acta Vet. Scand. 11, 139-155

119. Sallis, J. D. and Lumley, M. F. (1979): On the possible role of glycosaminoglycans as natural inhibitors of calcium oxalate stones. Invest. Urol. 16, 296-299

120. Schreck-Purola, I. and Setala, K. E. M. (1974): Composition for preserving and renewing the hair. Finn. 48,532 July 31, 1974. Appl. 880172, March 29, 1972, 25 pp

121. Shafer (1965): Intraocular injections as adjuncts to other retinal detachment procedures. In: C. L. Schepens and C. D. J. Regan (Eds.). Controversial Aspects of the Management of Retinal Detachment. Little, Brown and Company, Boston, 186-203

122. Slavkin, H. C., Croissant, R. D. and Guenther, H. (1975): Role of extracellular matrix macromolecules upon cell differentiation. Protides Biol. Fluids, Proc. Colloq. 22, 23-32

123. Solheim, K. (1966): Biophysical and biochemical studies of healing fractures. With special reference to the glycosaminoglycans. Acta Univ. Upsaliensis, Abst. Uppsala Dissertation Med. (28) 30 pp

124. Sundblad, L. (1953): Studies on hyaluronic acid in synovial fluid. Acta Soc. Med. Upsalien. 58, 113-238

125. Sundblad, L. (1954): Determination of anomalous viscosity in pathological joint fluids. Scand. J. Clin. Lab. Invest. 6, 295-302

126. Szederkenyi, Gy and Jozsa, L. (1968): Carbohydrate and mucoproteins in kidney stones. Z. Urol. Nephrol. 61, 289-294 (Ger)

127. Toole, B. P. (1976): Morphogenetic role of glycosaminoglycans (acid mucopolysaccharides) in brain and other tissues. In: S. Barondes (Ed.). Neural Recognition. Plenum Press, New York, 275-329

128. Toole, B. P. and Gross, J. (1971): The extracellular matrix of the regenerating newt limb: Synthesis and removal of hyaluronate prior to differentiation. Dev. Biol. 25, 57-77

129. Toole, B. P. and Linsenmayer, T. F. (1977): New knowledge of skeletogenesis: Macromolecular transitions in the extracellular matrix. Clin. Orthop. Relat. Res. 129, 258-278

130. Tsunenari, S. and Kanda, M. (1978): The roles of corneal mucopolysaccharides and water contents in post-mortem corneal clouding. Forensic Sci. 11, 87-92

131. Vanhegan, R. I. (1968): A method for characterization of acid mucopolysaccharides in histological sections of renal medullae from animals in varying states of hydration. J. Physiol. 192, 16P-18P

132. Varalakshmi, P. and Anandam, B. (1979): Alterations in kidney tissue glycosaminoglycans in experimental urolithiasis. Indian J. Med. Res. 69, 1019-1022

133. Velican, C. (1966): Histochemical investigation of the role of carbohydrates in the pathogenesis of inflammation. Studii Cercetari Med. Insterna 7, 357-376

134. Vogel, H. G. (Ed.) (1973): Connective Tissue and Aging. Excerpta Medica, Amsterdam

135. Wagner, H., Junge-Huelsing, G., Otto, H., Alder, J., Hauss, W. H. (1970): Animal experiments concerning wound healing in diabetes mellitus. Diabetologia 6, 612-613

136. Washburn, S. L. and DeVore, I. (1961): The social life of baboons. Scientific American 204, 62-71

137. Weeks, P. M. and Bryant, W. M. (1967): Mucopolysaccharides in secondary wounds exhibiting accelerated versus standard healing. Surg. Forum. 18, 502-504

138. Wiederhielm, C. A., Fox, J. R. and Lee, D. R. (1976): Ground substance mucopolysaccharides and plasma proteins: Their role in capillary water balance. Am. J. Physiol. 230, 1121-1125

139. Willoughby, D. A. (1970): Some views on the pathogenesis of inflammation. In: W. Montagna, J. P. Bentley and Dobson, R. L. (Eds.) The Dermis, Vol. X of Advances in Biology of Skin. Appleton-Century-Crofts, New York, 221-230

140. Wood, G. C. (1964): The precipitation of collagen fibers from solution. Intern. Rev. Connect. Tissue Res. 2, 1-3

141. Zablocki, B. (1960): Role of hyaluronidase-hyaluronic acid system in natural resistant. Postepy Hig. i. Med. Doswiadczalnej 14, 463-474

Chapter 4

Gastrointestinal Tract Fluids

4.1 Introduction

The gastrointestinal secretions contain mucins (consisting of neutral and sulfated glycoproteins and glycosaminoglycans) secreted by specialized cells located throughout the gastrointestinal tract. The important mucous-secreting glands are located in the esophagus, cardiac and pyloric parts of the stomach, Brunner's glands and goblet cells in the small intestine, and goblet cells in large intestine. In addition to these, superficial mucous glands found throughout the gastrointestinal tract also contribute. The mucus of the GI tract not only provides protection against stresses and auto-digestion but also it is important in healing of the mucosa in disease states. The sulfated glycoproteins in the mucus secretion and the mucosa protect against ulceration and action of digestive juices by inhibiting the proteolytic enzymes (Lambert et al., 1968). The sulfated polyanions (sulfated glycoproteins and sulfated glycosaminoglycans) have been considered as regulators of cell division and therefore may be of significance in the formation of carcinomas in the GI tract (Lippman, 1965).

In general, the gastrointestinal mucins contain neutral glycoproteins and glycoproteins of low sialic acid contents > sulfated glycoproteins > sulfated glycosaminoglycans. The fuco-, sialo- and sulfated glycoproteins present in the gastric secretions are modified by the action of the gastric juice; the hydrochloric acid (pH 1-2) in the juice liberates the terminal fucose and sialic acid residues, while pepsin removes some peptide fragments from the protein portions of these glycoproteins. The secretory functions of the gastro intestinal tract with reference to mucus synthesis, cellular specialization and other physiological aspects have been reviewed (Hendrix, 1974; Jacobson and Shanbour, 1974; Kent, 1975).

The glycoproteins (Horowitz, 1977; Allen, 1978) and mucopolysaccharides of the gastrointestinal tract and the viscous gel-forming and biological activities of these components have been reviewed (Glass, 1964; Horowitz, 1967). A study of the chemical nature of the gastrointestinal mucus revealed that the glycoproteins present contained a high (75- 85 %) carbohydrate content, had blood group specificity and the protein-carbohydrate linkages were the alkali-labile O-glycosidic type. The intestinal glycoprotein had significant contents of sialic acid and varying threonine/serine and galactose/N-acetylglucosamine/N-acetylgalactosamine ratios as compared with gastric glycoproteins (Schrager, 1974).

4.2 Methods of analysis and fractionation

All the methods described in chapter 2 could be used for isolation, fractionation and analysis of the glycoproteins and glycosaminoglycans of the fluids of the gastrointestinal tract. The technics which have been commonly used for this purpose include, precipitation of the GAG with cetylpyridinium salts from a dialyzed fluid, fractionation of the isolated material using ion-exchange chromatography on Amberlite IRC-50, AG 1 X-2, Dowex DEAE- and ECTEOLA-cellulose, preparative electrophoresis on starch gel or Pevikon blocks, gel filtration on Sephadex columns, ultrafiltration and ultracentrifugation, etc. The methods used for gastric juice have been reviewed (Glass, 1964). Gas chromatographic methods and other methods for analysis of the bound carbohydrates in the fluids are available (Masevich and Grach, 1972; Oette, and Gheorghiu, 1975). Methods for analysis of the total mucopolysaccharide contents of the gastric juice have been described (Ono and Eto, 1967; Masevich and Grach, 1972). Gastric mucoproteins have been characterized by equilibrium density-gradient centrifugation in cesium chloride (Starkey et al., 1974).

4.3 Gastric juice

This secretion is produced by 10-30 million gastric glands, which are localized principally in the body and the fundus of the stomach. These glands contain three different types of cells, each of which produces secretions of different chemical nature. The hydrochloric acid, which gives a pH of 1-2 to the gastric juice is secreted by parietal cells found scattered between the chief cells and absent from the cardiac and pyloric portions of the stomach. The *chief* cells located in the body of the gland produce and secrete pepsinogen. The mucous cells in the neck of the gastric glands contain mucinogenous granules and secrete a thick, viscous fluid which is very rich in mucoproteins. Besides these mucous cells, a complex mixture of mucus is produced by the surface epithelial cells of the gastric mucosa, the chief cells of the necks of the fundic glands, and cells of the pyloric and cardiac glands. The gastric juice mucus contains two types of mucosubstances. Those produced by the surface epithelium are present as insoluble, stringy masses, while the ones produced by the mucous cells of the gastric glands are present as viscous solutions. These two mucuses provide double-layered mucous barrier against irritation and autodigestion of the mucosa by the gastric secretions (Hollander, 1954). One way the mucins provide mechanical protection is by forming a viscous solution (Florey, 1955). For this function, the distribution of the mucosubstances is different in different parts of the GI tract.

The gastric mucins consist of neutral glycoproteins (blood-group substances, intrinsic factor of M_r 60,000), sulfated glycoproteins and acidic glycosaminoglycans. These macromolecules also possess biological activities, which have been reviewed (Glass, 1964 a).

4.3.1 Human gastric mucosubstances

The gastric tissue and gastric secretions from human subjects contain neutral (fuco-), sialo- and sulfated glycoproteins and acidic glycosaminoglycans. A few histochemical studies on human gastric tissue have been reported (Lev, 1965; Lisochkin, 1966; Molinari and Marcrina, 1977). The epithelium of fundus and pylorus was rich in glycoproteins. The epithelial cells of the pyloric glands were rich in acidic glycosaminoglycans, while peptic cells of the fundus contained no GAG and oxynctic cells lacked both glycoproteins and glycosaminoglycans (Lisochkin, 1966). Although the sulfated glycoproteins could be detected biochemically, these were not detected histochemically except in a few cases in mucous neck cells or gastric pits in some cases. It has been, therefore, suggested that the sulfated glycoproteins isolated biochemically from human gastric tissue may not originate in gastric mucous cells.

Analyses using specific enzymes, streptomyces hyaluronidase and chondroitinases, revealed that human gastric acidic glycosaminoglycans (GAG) consisted of, in order of decreasing amounts, heparan sulfate, dermatan sulfate, hyaluronic acid, chondroitin 4-sulfate, chondroitin 6-sulfate, and oversulfated chondroitin sulfate (Sekino et al., 1977). The acidic glycosaminoglycans were separated in 3 buffer systems and the enzymic analyses were carried out after fractionation on an anion-exchange resin column. The purification and fractionation of the GAG was carried out by Pevikon-C 870 block electrophoresis and Dowex 1 X2 column chromatography of human gastric juice. The fractions eluting with NaCl concentrations from 0.5-2.0 M were identified as hyaluronic acid, heparan sulfate and chondroitin 4- and 6-sulfates and dermatan sulfate (Masuda et al., 1977). An investigation of the distribution of the GAG content in human gastric wall showed that the total GAG and hyaluronic acid contents were the same in the mucosa and muscular layers but heparan sulfate was higher in the muscular layer (including serosa) than in the mucosa, whereas the chondroitin sulfate was higher in the mucosa than in the serosa (Masuda et al., 1978).

An earlier report on isolation of human gastric juice has described the presence of hyaluronic acid, dermatan sulfate and a new acid mucopolysaccharide (Terho et al., 1966). This newly isolated GAG was identified by its infrared absorption bands at 820 and 775 cm^{-1} to be similar to chondroitin 6-sulfate. Martin et al. (1967) reported isolation of 250 mg of a sulfated mucopolysaccharide from 1500 ml of human gastric juice. This sulfated mucopolysaccharide had no uronic acid and contained fucose 17 %, hexosamine >

20 %, hexoses 34 %, sialic acid 2.5 % and sulfate 5.1 %. It was actually a sulfated glycoprotein having blood group activity. This material was degraded by papain but not by pepsin (Martin et al., 1968).

The glycosaminoglycans measured as hexuronic acid represented only a small fraction (less than 0.4 mg/100 ml human gastric juice) of the total bound carbohydrates present in gastric juice (Glass et al., 1958). The effluents from Amberlite IRC-150 column (citrate buffer 0.2 M, pH 3.2-3.5) contained less than 1 % of the total carbohydrate as uronic acid (Glass, 1963 a). The remaining carbohydrates were those present as neutral and sulfated glycoproteins. The gastric glycoproteins in normal and pathological conditions have been reviewed (Horowitz, 1977; Pallavicini and Sinigaglia, 1979). Neutral glycoproteins in the form of blood group substances containing galactosamine, galactose, glucose, mannose and sialic acid and sulfoglycoproteins containing galactosamine/sulfate ratios of 1.0-3.2 were present. The sulfated glycoproteins comprised 60-80 % of the total carbohydrate content and contained glucosamine, galactose, fucose and sulfate (Schrager, 1964). Gas-liquid chromatography of the neutral sugars in the total mucosubstances of gastric juice has been reported (Oates and Schrager, 1965). Some of the carbohydrates in gastric juice may be derived from swallowed saliva.

The glycoproteins form a three-dimensional gel in the form of a thin, continous film covering the gastric mucosa, which protects the underlying mucosa from chemical and mechanical injury. The mucin in this gel represents two-thirds by weight of the nondialyzable material in normal human gastric juice. The gel mucin was separated from the extraneous glycoproteins and proteins by filtration when it was retained on a ultrafine glass filter (Waldron-Edward and Skoryna, 1964). It formed a tenaceous gum at a concentration of 7 % or more in water, saline or dilute acid and at lower concentrations up to 1 % it excluded large molecules like proteins. A major glycoprotein component of the gastric gel mucin from gastric juice aspirated from fasting human volunteers has been isolated (Waldron-Edward and Skoryna, 1970). It had a molecular weight of over $2 \cdot 10^6$ and contained galactose, fucose, N-acetylglucosamine, N-acetylgalactosamine and N-acetylneuraminic acid. Its ability to make tenaceous gum may be important for its protection of the mucosal surface. This glycoprotein had blood group activity characteristic of the donor.

Glycopeptides formed by trypsin digestion of the gastric mucosa have been isolated from six human autopsies. The fractions obtained by Pevikon C-870 block electrophoresis showed the presence of uronic acid and sulfate in three of the five fractions (Masuda et al., 1977). Fractionation of the trypsin digest of human gastric mucosae on DEAE-cellulose gave two glycopeptide fractions with 0.1 M and 0.05 M NaCl elutions, while the mucosae from the bovine fourth and porcine stomachs behaved differently and yielded five fractions at 0.01-0.45 M NaCl. These glycopeptides contained hexose, hexosamine, sialic acid and sulfate; those eluting with higher salt molarity were richer in sulfate and sialic acid (Masuda et al., 1977 a). Goso and Hotta (1977) have reported isolation of a glycopeptide of molecular weight 9,600, which contained 69 % carbohydrate linked mainly O-glycosidically to threonine but also had some N-glycosidic linkages. The carbohydrates present were N-acetylgalactosamine, N-acetylglucosamine, galactose, mannose, fucose and sialic acid in the proportions of 2:10:7:4:12:1.

4.3.2 Gastric glycosaminoglycans in disease states

The gastrointestinal mucus in various diseases (Goodman, 1975), including peptic ulcer (McGuigan, 1976) has been reviewed. Variations in GAG composition of the gastric mucosa have been observed by histochemical (Molinari, 1977) and biochemical analysis (Velev et al., 1973) in gastritis.

Although the etiology of the peptic ulceration is not clear, it has been considered a gastrointestinal disease of adaptation (Selye, 1951) and the role played by stress factors has been evidenced by increases in acidity, peptic activity, and gastric histamine and depletion of gastric mucopolysaccharides (Duttachoudhary et al., 1975). Similarly, the stress of immobilization in guinea pigs produced degranulation of the mast cells of the gastric mucosa and other cytological changes (Ludwig and Lipkin, 1969). This restraint in rats reduced the incorporation of radioactive sulfate to 54-77 % of the control values before the appearance of ulcers (Lambert et al., 1969). The ulcerogenesis, therefore, in these rats resulted perhaps from decreased formation of the sulfated glycoproteins for protection against acid and pepsin. Along with this, increased secretion of pepsinogen and its increased secretion (Mirsky et al., 1950) in urine as uropepsinogen have been noted in stressed human subjects. A part of these changes may be mediated by stimulation of the vagus nerve, since vagotomy resulted in modifications of mucopolysaccharide in gastric mucosa (Cordiano et al., 1968).

Evidence for the decrease in mucopolysaccharides in the gastric mucosa and mucin from patients with gastric ulcers has been provided by biochemical analysis of the gastric wall and gastric juice of the patients (Sekino et al., 1976; Umehara et al., 1976). Also, those substances that enhance synthesis of mucopolysaccharides in the gastric mucosa prevent gastric ulcers. This has been found the case with p-chlorocarbobenzoxy-L-lysine-O-Me-Hcl, also called KL-11 (Ezer and Szporny, 1970), pilocarpine (Gerald, 1965), antihistaminic drugs (Lagrton et al., 1965), all of which increased synthesis of sulfated mucopolysaccharides in gastric mucosa. Physical exercise has been shown to be antiulcerogenic due to antiserotonin and antihistamine effects produced (Frenkl, 1976). Conversely, the steroidal (cortisone, hydrocortisone) and non-steroidal (salicylate, phenylbutazone) antiinflammatory drugs inhibit sulfate incorporation into mucopolysaccharides and are ulcerogenic (Pallavicini et al., 1971).

According to a recent hypothesis, the gastric cancer occurs in two stages which involve piercing of gastric mucosal protective barrier due to interaction between mucoproteins and mucopolysaccharides of the barrier and the injested food starches, which then houses gastric flora that produce carcinogenic nitrosamines and/or nitrosamides (Lilienfeld and Garagliano, 1979). Other aspects of the etiology and prevention of the gastric cancer have been reviewed (Weisburger et al., 1976). Histochemical studies showed that the acidic glycosaminoglycan fraction was quantitatively similar in adenomatous polyps (Lisochkin, 1967) as compared with normal mucosa, but it was increased in carcinomas (Magri, 1968). The sulfate-containing gastric mucus components are related to peptic ulcer (Familiar, 1969) and to its progress to carcinoma (Lilienfeld and Garagliano, 1976). A glycopeptide was isolated from 20 human gastric carcinoma tissues (Masuda et al., 1978 a). This glycopeptide had an electrophoretic mobility slower than hyaluronic acid and was both PAS and Alcian blue positive. It contained galactose, fucose, glucosamine, galactosamine, sialic acid and threonine. Serine and proline were the predominant amino acids.

4.3.3 Relation between gastric mucopolysaccharides and blood lipid patterns

The gastric mucopolysaccharides seem to have some kind of regulatory control of serum lipids. Correlations have been observed between serum lipid levels and the gastric juice mucopolysaccharides in normal subjects. In 150 normal subjects without any coronary, gastrointestinal or metabolic diseases, the decreasing hexosamine concentration in gastric juice was correlated with serum cholesterol levels in the same subjects. Similarly, significant correlations were found between the sialic acid and mucoid contents of the gastric juice and the serum cholesterol levels (Gero et al., 1965, 1968). The changes in mucopolysaccharide components of gastric juice correlated directly with the serum mucopolysaccharide changes. Although the sialic acid, Winzler's gastromucoid and hexosamine contents of gastric juice and serum cholesterol levels were similar in quantitative correlations in normal subjects and hypercholesterolemic patients, yet the absolute values of the hexosamine, protein-bound hexose, Winzler's gastromucoid and sialic acid were lower in the patients. But the fucose and hexuronic acid contents were not different. The lowest gastric mucopolysaccharides were found in atherosclerosis but the distributions of the various carbohydrate components were similar to those of the controls (Szekely et al., 1967).

There was no correlation between gastric juice acidity and mucopolysaccharide composition of the fluid. The patterns of distribution of the mucopolysaccharides in the gastric wall of control and atherosclerotic patients were similar although out of the diverse abnormalities of lipid metabolism, the concentration of mucopolysaccharides in the gastric juice of atherosclerotic patients was the lowest. The relation between the individual mucopolysaccharide components in gastric juice and the serum of the same subject was such that the hexosamine concentration correlated linearly in the controls as well as in hypercholesterolemia patients, but the curve was parabolic in the control group (Szekely et al., 1968). From the aforementioned findings it appears that the gastric mucopolysaccharides may have a role in lowering the blood lipid levels. A similar antidyslipemic action of sulfomucopolysaccharide of pancreatic origin has been described (Svetoni et al., 1974). Evidence for the lipid clearing effect of the gastric juice mucopolysaccharides has been previously provided by plasma lipid clearance in rat and guinea pigs upon oral administration of mucin prepared from hog gastric mucosa. Similar clearing was seen when a glycosaminoglycan related to hyaluronic acid was fed to rabbits (Capraro et al., 1956, 1957).

4.3.4 Gastric secretion in experimental animals

The presence of glycosaminoglycans (GAG) in the gastric mucins has been known for a long time. Levene and Lopez-Suarez (1916, 1918) isolated a glycosaminoglycan from pig gastric mucin, which contained an equimolar proportion of N-acetyl glucosamine, glucuronic acid, ester sulfate but the N/S ratio was 4.0. Those investigators considered this sulfated polysaccharide isomeric with chondroitin sulfate in which galactosamine was replaced by glucosamine and named it "mucoitin sulfate". A similar material was present at a concentration of 22-46 mg per liter in dog gastric juice collected through a fistula (Komarov, 1938). Meyer et al. (1937) isolated a mucoitin sulfuric acid and a neutral polysaccharide from pig gastric mucin. The mucoitin sulfate obtained had a $\alpha_D = -22°$, 3.0-4.0 % nitrogen, 3.1-4.7 % sulfur and 30-40 % hexuronic acid. Using the procedure of Meyer et al. (1937) and purification by filtration through activated charcoal and Super-Cel followed by ethanolic precipitation, Wolfrom et al. (1943) isolated 1.6 g of purified mucoitin sulfate from 40 g of commercial pig mucin. Electrophoresis of this preparation in phosphate buffer revealed three components: 40 % stationary component and two components 41 % and 19 % with mobilities less than that of heparin (Wolfrom and Rice, 1947). The hexuronic acid present was identified as D-glucuronic acid (Wolfrom and Rice, 1947 a). Werner (1953) found that the acidic GAG consisted of about 22 % of the total carbohydrate in dialyzed digests of mucous scrapings of the stomach. The mucoitin sulfuric acid isolated from pig gastric mucosa contained 33 % hexosamine, 26 % glucuronic acid, 1 % fucose, 4 % nitrogen and 11 % sulfate. Sato (1952) reported isolation of a new galacturonic acid-containing acid mucopolysaccharide named gastroitin sulfuric acid. This GAG came from pig gastric mucosa and contained N-acetylglucosamine, galacturonic acid and sulfate.

Smith et al. (1953) found no mucoitin sulfate in pig gastric mucin and showed that the mucoitin sulfate reported in previous literature was actually a mixture of heparin and chondroitin 4-/6-sulfate and dermatan sulfate. Meyer (1959) also did not find any mucoitin sulfuric acid and reported isolation of heparan sulfate from commercial pig gastric mucin, which he tentatively identified since it did not show any negative rotation.

A systematic study of the glycosaminoglycans from commercial pig gastric mucin was reported by Kimura et al. (1964, 1965). Following pronase digestion of 150 g pig gastric mucin dissolved in 500 ml water, Sevag's deproteinization and dialysis, the GAG were recovered from the retentate in dialysis tubing by precipitation with four volumes of ethanol. The precipitate obtained upon centrifugation was dissolved in 5 % sodium acetate -0.5 N acetic acid solution, was subjected to fractional precipitation with ethanol and the fractions obtained were pooled as fraction A obtained at ethanol concentration below 40 % and fraction B obtained as the supernatant after this precipitation. Further fractionation of fraction A on Dowex 1-X8 (chloride form, 100-200 mesh) gave fractions which were identified by infrared spectrum, polyacrylamide gel electrophoresis, hyaluronidase digestion and chemical analysis. Dermatan sulfate was the major glycosaminoglycan with heparan sulfate and chondroitin 4- or 6-sulfate was the minor components (Kimura et al., 1964). A similar fractionation of fraction B on Dowex 1-X2 (chloride form, 200-400 mesh) gave hyaluronic acid, heparan sulfate, chondroitin 4- and 6-sulfates in yields of 0.25 g, 0.20 g, 0.05 g and 0.10 g from 50 g of pig stomach mucin, respectively (Kimura et al., 1965).

Fractionation of the tryptic digests of bovine and porcine gastric mucosa gave 5 fractions eluting at 0.01-0.45 M NaCl and these fractions had increasing contents of sialic acid and sulfate (Masuda et al., 1977 a, 1977 c). The glycoprotein contents of gastric mucosa from 11 mammalian species were higher than those of the muscular-serosa layer, especially in the pig and the fourth stomach of ruminants (Masuda et al., 1977 d). The glycosaminoglycans isolated from equine gastric mucosa by Dowes 1 X-2 column chromatography contained hyaluronic acid, chondroitin 4-sulfate, dermatan sulfate and chondroitin 6-sulfate (Masuda et al., 1979).

The sulfated glycoproteins isolated from hog gastric mucosa contained (in residue per 2 residues of N-acetylgalactosamine) fucose 1.9-4.1, galactose 5.9-7.9 and N-acetylglucosamine 3.7-4.0. Sulfate was present as N-acetylglucosamine 6-sulfate and the hexosamine/sulfate molar ratios varied from 2.4-5.4. The carbohydrate-protein linkage was of O-glycosidic type between N-acetylgalactosamine and serine and threonine (Slomiany and Meyer, 1972). The sulfated glycoproteins purified from papain-digested mucosal scrapings of the dog fundic stomach followed by digestion with nuclease and hyaluronidase were present as 3 fractions with equimolar ratios of hexosamine and galactose. The hyaluronidase-resistant glycosaminoglycans were identified as dermatan sulfate and heparan sulfate, (which were in addition to) hyaluronic acid, chondroitin 4- and 6-sulfates present in the extracts of the mucosal scrapings (Pamer et al., 1968). Kupchella and Steggerda (1972) reported heparin in addition to the GAG fractions identified by Pamer et al. (1968) in the canine mucosa.

4.3.5 Effects of drugs and hormones on gastric mucosubstances and ulcerogenesis

Gastrin, acetylcholine, histamine and calcium ions are the stimulants of gastric secretion. Being natively present, these interact and potentiate one another's stimulant effects (Konturek, 1979). The mechanisms for mucus secretion are poorly understood. In a healthy man mucus is secreted continuously even in the resting state. However, the composition of the gastric juice and of the mucosubstances elaborated under a stimulus differs from that of the resting state.

Reserpine is known to increase the irritation of the stomach and produces ulcers probably through increased acid secretion. Reserpine blocks the uptake of 5HT so that the latter cannot increase mucus in the stomach (Kaulbersz and Bugajki, 1967). Noradrenaline and isoprenaline increased the concentrations of mucopolysaccharides and the pH in the gastric juice of rats, but decreased the volume. On the other hand, reserpine and methyl tyrosine (an inhibitor of catecholamine synthesis) decreased the pH and the concentration of mucopolysaccharides in the gastric juice (Julkunen et al., 1970 a). This suggested that the catecholamine depletion may play a major role in gastric effects of reserpine. The gastric effects of reserpine could be antagonized by isoprenaline, whereas histamine antagonized the catecholamine effects (Julkunen et al., 1970).

The effects of the anti-inflammatory agents have already been described in section 4.3.2. Cinchophen is ulcerogenic in dogs. Following 6 days of cinchophen feeding, dermatan sulfate and 3 sulfomucopolysaccharides increased and hyaluronic acid decreased, while after 14 days of feeding dermatan sulfate and an epithelial sulfopolysaccharide decreased throughout the gastric and duodenal wall prior to ulcer formation (Hakkinen and Hartiala, 1966). Zinc deficiency increased the non-dialyzable hexosamine and sialic acid to nearly double the values for controls (Quarterman et al., 1970). In the athyroid rats the hyaluronic acid was increased by 12 % whereas dermatan sulfate decreased by 34 % in the gastric mucosa (Kofoed and Bacelo, 1978) Aspirin and indomethacin induced gastric ulceration in rat (Szollosi et al., 1976). A short term treatment with ACTH or cortisone produced an increase in high molecular weight mucopolysaccharide-glycoprotein complexes in the pyloric and fundic mucosa with distinct changes in ratios of hexosamines, hexuronic acids and sulfated mucopolysaccharides (Kozlowska, 1973). The effect of whale gastrone on experimental peptic ulcers has been reported (Chang et al., 1972).

4.4 Intestinal fluids

The duodenum, small intestine and large intestine down to the terminal part of the rectum are all associated with mucus. However, due to difficulties of collection of the secretions containing the mucus, the majority of the information has been obtained from human intestinal wall extracts or from experimental animals. A study of the duodenal juice in chronic pancreatitis (Takeda, 1975) and in cystic fibrosis patients (Dische et al., 1959) have been reported. The mucus in the intestinal tract is produced by the goblet cells present throughout the intestinal epithelium. In samples obtained from biopsy of human duodenal mucosa, histochemical analysis showed the presence of neutral polysaccharides in the brush borders, Brunner's glands and Paneth cells, while the goblet cells contained sulfated and carboxyl group-containing mucopolysaccharides and fuco- and sialo-mucins (Sinitsina, 1966). The duodenum secretion in young cattle contained a hyaluronidase-resistant chondroitin 4-sulfate type macromolecule (Koshevaya, 1971). Radioactive glucose was incorporated into GAG precipitable with CPC, chiefly in the 1.2 M NaCl anion exchange fraction (Gans, 1972). Abnormalities of mucopolysaccharides in duodenal ulcerations have been reported (Donaldson et al., 1974, 1976), which have been evaluated by monosaccharide analysis of hydrolyzed gastric secretion by gas chromatography. In addition to the monosaccharides, the free and total sialic acid contents were also abnormal. The quantity of mucopolysaccharides in 12 patients with duodenitis and ulcers was significantly reduced as compared to 36 normal subject biopsies (Cheli and Zaccheo, 1971).

The small intestinal mucosa of rabbit contained 2 sulfated glycopeptides and a highly sulfated uronic acid-containing glycosaminoglycan in the protease digested mucosal scrapings. The small intestinal mucosa contained more highly sulfated fractions than did the mucosa of colon. The sulfated GAG consisted of uronic acid, fucose, galactose, sulfate and glucosamine/galactosamine ratio of 0.5 (Nemoto and Yosizawa, 1969). About 25 % of the acidic mucopolysaccharides isolated from pig intestinal mucosa scrapings were testicular hyaluronidase-susceptible and contained hyaluronic acid 23.1 %, heparan sulfate 33 %, chondroitin sulfate 26.4 % and heparin 17.5 % (Srinivas et al., 1971). In addition to uronic acid and hexosamine, mannose, glucose, galactose, arabinose and xylose were also present. The mucin from rat small intestine contained a sulfated glycoprotein with sialic acid > fucose > glucosamine > galactose > galactosamine with O-glycosidic linkages to serine and threonine (Bella and Kim, 1972). The characterization of sulfomucins and sialomucins from pig small intestine has been reported (Degand et al., 1973).

A glycoprotein of molecular weight $15 \cdot 10^6$ consisted of 80 % of the weight of water-soluble pig colonic mucus glycoprotein. It contained fucose 10.4 %, glucosamine 23.9 %, galactosamine 8.3 %, sialic acid 9.9 %, galactose 20.8 %, sulfate 3.0 %, protein 13.3 % (Marshall and Allen, 1978).

The neutral and acidic mucopolysaccharides in the mucous membrane of the appendix have been investigated (Sll'chenko, 1966). The terminal part of the rat rectum contained neutral and acidic mucopolysaccharides synthesized by the goblet cells (Czabak, 1963). A significant increase in fucose content of 11 cystic fibrosis specimens of rectal mucus was observed by gas chromatography and the rectal mucus from these patients showed a significantly different L-fucose/N-acetylneuraminic acid ratio (Roelfs et al., 1967). The sulfated sialo-glycoproteins in the upper and lower halves of colon were different (Reid et al., 1975). The synthesis of mucopolysaccharides in colon homogenates was impaired in vitamin A deficiency (Moretti and Wolf, 1961). In experimentally induced peritonitis increase in the GAG was noted in colon, small intestine and duodenum (Szulc et al., 1967). Histochemical alterations occurred in the secretory elements of vermiform appendix following surgery (Platash, 1976), in known specific ulcerative colitis (Levin and Zabelina, 1966) and in intestinal coliform bacterial infection (Tikhonova, 1964). Increases in the neutral and acidic mucopolysaccharides occurred in epithelial secretions from patients with adenomatous polyps and primitive adenocarcinomas of the rectal mucosa (Magri, 1968 a).

The meconium from the neonates contained mucus type and mannose containing blood group active glycoproteins (Fraser and Clamp, 1975). Profoundly different patterns of 8 proteins and 6 mucoproteins were noted in meconium ileus (Knauff and Adams, 1968). The mucosubstances of meconium have been reviewed (Deroches et al., 1974; Antonowicz and Schwachman, 1979). In 5 cystic fibrosis newborns the alpha-glycosidase was elevated in the meconium extracts, which also had very low albumin contents (Di Matteo et al., 1979). The enzyme lysozyme (a glycoprotein) originating from lysed pus cells appeared in feces of ulcerative coilitis patients (Sammons, 1950/1951).

The sulfated macromolecules of the exocrine pancreas and ducts of cystic fibrosis patients contained blood group and virus hemagglutinating activities (Michalsen et al., 1979). The bovine pancreatic juice contained 8 anionic protein components, 6 of which were proteolytic enzymes (Keller et al., 1958). The Golgi complex in the pancreatic acinar cells is involved in the synthesis of sulfated mucopolysaccharides (Berg and Young, 1971). The mucopolysaccharide content of blood vessels in the pancreas and small intestine increased above normal in experimental chronic pancreatitis (Kiseleva et al., 1978). Three glycopeptides and a GAG were isolated by DEAE-cellulose chromatography of a protease-digested human pancreatic cyst.

Uronic acid was found in the 0.45 M and other fractions (Masuda et al., 1977 b).

4.5 Aging-related changes in gastrointestinal fluids

The gastrointestinal functions are altered in aging and these have been reviewed (Bhanthumnavin and Schuster, 1977). The gastric changes include atrophic gastritis, changes in mucoprotein composition of gastric juice (Peleshchuk and Voloshchenko, 1971) and replacement of glandular structures by adipose tissue and increased mucopolysaccharide content of parietal cells (Smolyanskii and Zhukova, 1967). After the age of 50, PAS reaction decreased and acidic GAG almost completely disappeared from the human anal glands (Yakhnitsa and Shaposhnikov, 1972). Incidence of diverticular disease of the colon increased with age.

4.6 Gastrointesinal changes in cystic fibrosis

Alterations of the GI tract in cystic fibrosis have been reviewed (Schwachman, 1975; Barbero et al., 1976; Nadler et al., 1978). An obstruction of the small intestine by a sticky meconium in utero (meconium ileus) and in children and young adults (fecal impaction syndrome) has been reported (Boat and Cheng, 1976). Meconium is rich in blood group substance and mucoproteins of fetal intestinal origin (For a review, see Antonowicz and Schwachman, 1979). The mucoprotein isolated by starch gel electrophoresis from cystic fibrosis newborns differed in carbohydrate composition of the mucoprotein from normal neonates (Schachter and Dixon, 1965). In normal neonates, the purified meconium mucoprotein contained (gm/100 gm dry weight) hexosamines 29.7, fucose 9.1, galactose 27.0, sialic acid 9.2 and total carbohydrates 67.8, whereas the corresponding values in meconium ileus cases were 27.2, 8.1, 23.3, 6.5 and 58.8. The glucosamine/galactosamine ratios in normal cases were 1.91 as compared to 1.77 in meconium ileus patients. Since the secretions of the GI tract are of the exocrine type, all the fluids are affected with an increase in mucus content. Thickened mucosal folds, mucosal smudging and nodules appear in the duodenum in 70 % of the older patients. However, the incidence of peptic ulcers is not increased. A mucoid not present in controls appeared in the duodenal fluids of patients with cystic fibrosis (Dische et al., 1959).

4.7 Bibliography

1. Allen, A. (1978): Structure of gastrointestinal mucus glycoproteins and the viscous and gel-forming properties of mucus. Br. Med. Bull. 34, 28-33

2. Antonowicz, I. and Schwachmann, H. (1979): Meconium in health and in disease. Adv. Pediatrics 26, 275-309

3. Barbero, G. J., Schwachman, H., Grand, R. and Woodruff, C. (1976): Gastrointestinal and nutritional manifestations of cystic fibrosis. In: J. A. Mangos and R. C. Talamo (Eds.). Cystic Fibrosis: Projections into the Future. Stratton Intercontinental Med. Book Corp New York, 83-111

4. Bella, A., Jr. and Kim, Y. S. (1972): Rat small intestinal mucin: Isolation and characterization of a water-soluble mucin fraction. Arch. Biochem. Biophys. 150, 679-689

5. Berg, N. B. and Young, R. W. (1971): Sulfate metabolism in pancreatic acinar cells. J. Cell. Biol. 50, 469-483

6. Bhanthumnavin, K. and Schuster, M. M. (1977): Aging and gastrointest inal function. In: C. E. Finch and L. Hayflick (Eds.). Handbook of Aging. Van Nostrand Reinhold, New York, 709-723

7. Boat, T. F. and Cheng, P. W. (1976): Mucous Glycoproteins. In: J. A. Mangos and R. C. Talamo (Eds.). Cystic Fibrosis: Projections into the Future. Stratton International Med. Book Corp., New York, 165- 177

8. Capraro, V., Cantone, A. and Cresseri, A. (1956): Über die Anwesenheit einer resorbierbaren mit "clearing"-Vermögen versehenen Substanz im normalen Magensaft. Naturwissenschaften 43, 36-37

9. Capraro, V., Cresseri, A. and Cantone, A. (1957): Some biological properties of gastric mucin preparations. In: Vitamin B_{12} and Intrinsic Factor. Ist Europ. Sympos. Hamburg, 1956: Enke, Stuttgart, 283-291

10. Chang, F. M., Yii, J. Y. and Sung, J. L. (1972): Effect of whale gastrone preparation on the experimental peptic ulcers. J. Formosan Med. Assoc. 71, 160-169

11. Cheli, R. and Zaccheo, D. (1971): Histochemical aspects of the duodenal mucosa obtained from healthy and sick human subjects by biopsy: mucopolysaccharides. In: Ber Ges. Inn. Med. DDR 1970 7, 37-39 (Ger). Pub. in Z. Inn. Med. Ihre Grenzgeb. 26

12. Cordiano, C., Ottolenghi, A., Perrino, G. et al. (1968): On the modifications of some mucopolysaccharide fraction in the gastric mucosa of the rat after vagotomy. Acta Chir. Ital. 24, 577-580 (Ita).

13. Czabak, O. (1963): Distribution and nature of mucopolysaccharides in the differentiating terminal part of the rectum in the rat embryo. Folia Histochem. Cytochem. (Cracow) 1, 347-359

14. Degand, P., Gaveriaux, M. and Haver, R. (1973): Biochemical characterization of the sulfomucins and sialomucins from pig small intestine. C. R. Soc. Biol. 166, 622-627 (Fr)

15. Deroches, A., Galerne, D., Baudon, J. and Dastugue, G. (1974): The biochemical compositions of meconium (glucides, lipids, proteins and related substances). J. Gynecol. Ostet. Biol. Reprod. 3, 543-552 (Fr).

16. Di Matteo, G., Romano, M., Mastella, G., Castellani, E., Rutigliano, B. and Casola, L. (1979): Chemical comparison of normal and cystic fibrosis meconium: a quantitative and qualitative analysis of carbohydrate splitting enzymes. Mongr. Pediatr. 10 (Recent Adv. Cystic Fibrosis), 19-26

17. Dische, Z., di Sant'Agnese, P., Pallavincini, C. and Youlos, J. (1959): Composition of mucoproteins fractions from duodenal fluid of patients with cystic fibrosis of the pancreas and from controls. Pedriatics 24, 74-91

18. Donaldson, J. D., MacRae, K. D. and Parks, T. G. (1976): Proceedings: The study of gastric mucopolysaccharides in patients with duodenal ulceration and X-ray negative by discriminant function analysis. Gut 17, 401-402

19. Donaldson, J. D., MacRae, K. D. and Rodgers, H. W. (1974): Abnormalities of mucopolysaccharides in duodenal ulceration. Proc. R. Soc. Med. 67, 11-12

20. Duttachoudhury, M., Rai, V., Singh, R. H. and Udupa, K. N. (1975): Role of stress factors in causation of peptic ulcer. A study of local factors. Q. J. Surg. Sci. 11, 68-72

21. Ezer, E. and Szporny, L. (1970): Prevention of experimental gastric ulcer in rats by a substance which increases biosynthesis of acid mucopolysaccharides. J. Pharm. Pharmac. 22, 143-144

22. Familiar, R. G. (1969): Sulfate-containing gastric mucus components and peptic ulcer; possible relation. J. Amer. Geriat. Soc. 17, 680-684

23. Florey, H. (1955): Mucin and protection of the body. Proc. Roy. Soc. Biol. 143, 147-159

24. Fraser, D. and Clamp, J. R. (1975): Glycoprotein content of meconium. Clin. Chim. Acta 59, 301-307

25. Frenkl, R. (1976): The antiulcerogenic mechanism of physical exercise. In: G. Mozsik and T. Javor (Eds.). Progress in Peptic Ulcer, Akad. Kiado, Budapest, 407-422

26. Gans, J. H. (1972): In vitro incorporation of carbon-14 from 2^{-14}C-glucose into rabbit kidney medullary and duodenal glycosaminoglycans. Proc. Soc. Exp. Biol. Med. 140, 1077-1081

27. Gerald, A. (1965): Histochemical modification of the fundic mucous membrane, induced by pilocarpine, histamine, and colchicines in dogs. Compt. Rend. Soc. Biol. 159, 1245-1249 (Fr).

28. Gero, S., Bihari-Varga, M. and Szekely, J. (1968): Gastric mucosubstances and serum lipid pattern. Progr. Biochem. Pharmacol. 4, 183-190

29. Gero, S., Bihari-Varga, M., Szekely, J. and Steingaszner, O. (1965): Interrelations between the concentration of gastric mucopolysaccharides and that of the serum mucopolysaccharide and lipid constituents. Eur. Clin. Sect. Int. Ass. Gerontol., Congr., 4th, Sanremo, Italy, 607-612

30. Glass, G. B. J. (1963): Mucosubstances of gastric secretion in man and their biological activity. Ann. N. Y. Acad. Sci. 106, 775-793

31. Glass, G. B. J. (1964): Fractionation of macromolecular components of human gastric juice by electrophoresis. Chromatography, and other physicochemical methods. Adv. Clin. Chem. 7, 373-479

32. Glass, G. B. J. (1964 a): Proteins, mucosubstances and biologically active components of gastric secretion. Adv. Clin. Chem. 7, 235-372

33. Glass, G. B. J., Rich, M. and Stephanson, L. (1958): Comparative study of mucoproteins of human gastric juice and serum. Gastroenterology 34, 598-615

34. Goodman, M. S. (1975): Gastrointestinal mucous in relation to disease. In: S. C. Truelove and M. J. Goodman (Eds.). Topics in Gastroenterology, Vol. 3, Blackwell Scientific Publications, Oxford

35. Goso, K. and Hotta, K. (1977): A glycoprotein isolated from human gastric juice. Biochem. J. 163, 160-172

36. Hakkinen, I. and Hartiala, K. (1966): The effect of cinchophen on the acid polysaccharides of the gastric and duodenal wall in dog. Acta Physiol. Scand. 66, 326-332

37. Hendrix, T. R. (1974): The secretory function of the alimentary canal. In: V. B. Mountcastle (Ed.). Medical Physiology, C. V. Mosby Company, St. Louis, 1178

38. Hollander, F. (1954): Two component mucousbarrier; its activity in protecting the gastro-duodenal mucosa against peptic ulceration. Arch. Intern. Med. 93, 107-120

39. Horowitz, M. I. (1967): Mucopolysaccharides and glycoproteins of the alimentary canal. In: C. F. Cole (Ed.). Handbook of Physiology. Alimentary Canal II. William and Wilkins Co., Baltimore, 1063-1085

40. Horowitz, M. I. (1977): Gastrointestinal glycoproteins. In: M. I. Horowitz and W. Pigman (Eds.). The Glycoconjugates, Vol. I. Academic Press, New York, 189-213

41. Jacobson, E. D. and Shanbour, L. L. (Eds.) (1974): MTP International Review of Sciences: Physiology Series One, Vol. 4: Gastrointesinal Physiology. University Park Press, Baltimore, -385

42. Julkunen, R. J. K., Airaksinen, M. M. and Puro, K. (1970): Effects of reserpine and 5-hydroxytryptamine on the secretion of mucopolysaccharides in gastric juice of rats. Ann. Med. Exp. Fenn. 48, 149-154

43. Julkunen, R. J. K., Airaksinen, M. M. and Puro, K. (1970 a): Possible role of catecholamines in mucopolysaccharides secretion into gastric juice of rats. Ann. Med. Exp. Biol. Fenn. 48, 155-159

44. Kaulbersz, J. and Bugajki, J. (1967): Studies on mechanism of action of reserpine on gastric secretion. Acta Physiol. Pol. 18, 503-516

45. Keller, P. J., Cohen, E. and Neurath, H. (1958): The proteins of bovine pancreatic juice. Pancreatic Proteins. 233, 344-349

46. Kent, P. W. (1975): Mucus synthesis and cellular specialization in gastro-intestinal epithelium. In: S. C. Truelove and J. Trowell, (Eds.). Topics in Gastroenterology, Vol. 2, Blackwell Scientific Publications, Oxford

47. Kimura, A., Watanabe, T., Nagai, Y. and Tsurumi, K. (1965): Acid mucopolysaccharides of hog gastric mucin. II. Seikagaku 37, 419-424 (Japan)

48. Kimura, A., Watanabe, T. and Yashushi, N. (1964): Studies on acid mucopolysaccharides of pig gastric mucin. I. Fukushima J. Med. Sci. 11, 71-83

49. Kiseleva, A. F., Shatrova, K. M. and Keisevich, L. K. (1978): Morphological changes in the small intestine in experimental chronic pancreatitis. Vrach. Delo (6), 86-90 (Russ)

50. Knauff, R. E. and Adams, J. A. (1968): Meconium proteins and mucoproteins in meconium ileus. Proc. Soc. Exp. Biol. Med. 127, 801-804

51. Kofoed, J. A. and Barcelo, A. C. (1978): The acidic glycosaminoglycans in gastric mucosa of athyroid rats: The effects of L-triiodothyronine. Experientia 34, 808-809

52. Komarov, S. A. (1938): The partition of nitrogen in canine gastric juice. J. Lab. Clin. Med. 23, 822-832

53. Konturek, S. J. (1979): Gastric secretion: physiological aspects. In: H. L. Duthie and K. G. Wormsley (Eds.). Scientific Basis of Gastroenterology. Churchill Livingstone, London, Chap. 6, 133-162

54. Koshevaya, G. I. (1971): Mucopolysaccharides of the secretory elements of the duodenum in young cattle. Tr. Khar'kov. Med. Inst., 260-263

55. Kozlowska, K. (1973): The effect of ACTH and cortisone on gastric acid mucosubstances in rats. Digestion 9, 241-247

56. Kupchella, C. E. and Steggerda, F. R. (1972): Acid mucopolysaccharide distribution in the canine gastrointestinal mucosa. Trans. N. Y. Acad. Sci 34, 351-360

57. Lagrton, A., Dorcos, V. and Wegmann, R. (1965): Effect of histamine and antihistaminic compounds on the gastroduodenal mucosa of normal rats. Ann. Histochem. 10, 289-295 (Fr.).

58. Lambert, R., Andre, C., Martin, F. and Martin, M. S. (1969): Radiosulfate incorpotation in rat gastric mucosa; effect on the length of the restraint. Digestion 2, 187-195

59. Lambert, R., Martin, F., Andre, C., Descos, L. and Vouillon, G. (1968): Apical localization of glycoproteins in gastric mucous cells. Am. J. Dig. Dis. 13, 941-951

60. Lev, R. (1965): The mucin histochemistry of normal and neoplastic gastric mucosa. Lab. Invest. 14, 2080-2100

61. Levene, P. A. and Lopez-Suarez, J. (1916): The conjugated sulfuric acid of the mucin of pig's stomach (mucoitin sulfuric acid). J. Biol. Chem. 25, 511-516

62. Levene, P. A. and Lopez-Suarez, J. (1918): Mucin and mucoids. J. Biol. Chem. 36, 105-126

63. Levin, V. N. and Zabelina, T. N. (1966): Some histochemical peculiarities of the colon epithelium in non-specific ulcerative colitis. Vestn. Akad. Med. Nauk SSSR 21, 79-86 (Russ)

64. Lilienfeld, D. E. and Garagliano, C. F. (1976): Gastric cancer. J. Natl. Cancer Inst. 57, 9

65. Lilienfeld, D. E. and Garagliano, C. F. (1979): Gastric Cancer Etiology: A biochemical hypothesis. Med. Hypotheses 5, 145-151

66. Lippman, M. (1965): A proposed role for mucopolysaccharides in the initiation and control of cell division. Trans. N. Y. Acad. Sci. 27, 342-360

67. Lisochkin, B. G. (1966): Some histochemical peculiarities of human gastric mucosa. Arkh. Anat. Gistol. i Embriol. 51, 109-111 (Russ)

68. Lisochkin, B. G. (1967): Histochemical characteristics of simple, proliferating, and malignized adenomatous polyps of the stomach. Arkh. patol. 29, 36-40 (Russ)

69. Ludwig, W. M. and Lipkin, M. (1969): Biochemical and cytological alterations in gastric mucosa of guinea pigs under restraint stress. Gastroenterology 56, 895-902

70. Magri, P. (1968): Histochemical study of adenomatous polypes and primitive carcinomas of the gastric mucosa. Atti Accad. Med. Lomb. 23, 619-623 (Ital)

71. Magri, P. (1968 a): Histochemical aspects of adenomatous polypes and primitive adenocarcinomas of the rectal mucosa. Atti Accad. Med. Lomb. 23, 624-628 (Ital)

72. Marshall, T. and Allen, A. (1978): The isolation and characterization of the high molecular weight glycoproteins from pig colonic mucus. Biochem. J. 173, 569-578

73. Martin, F., Berard, A., Jouvenceaux, A. and Lambert, R. (1967): Isolation of a sulfated mucopolysaccharide from human gastric juice. C. R. Soc. Biol. 161, 848-851 (Fr)

74. Martin, F., Lambert, R., Berard, A. and Martin, M. (1967): Effect of pepsin and papain on a mucopolysaccharide sulfate from gastric juice. C. R. Soc. Biol. 161, 1727-1730 (Fr)

75. Masevich, Ts. G. and Grach, A. G. (1972): Determination of mucopolysaccharides in stomach contents. Lab. Delo (7), 443-444 (Russ)

76. Masuda, H., Shichijo, S. and Takeuchi, M. (1977): Electrophoresis of glycosaminoglycans of the human gastric mucosa. Seibutsu Butsuri Kagaku 21, 215-220 (Japan)

77. Masuda, H., Shichijo, S. and Takeuchi, M. (1977 a): The glycopeptides from the human, bovine and porcine stomachs. Int. J. Biochem. 8, 165-169

78. Masuda, H., Shichijo, S. and Takeuchi, M. (1977 b): Isolation and characterization of the glycopeptide from a pancreatic cyst. Int. J. Biochem. 8, 627-631

79. Masuda, H., Shichijo, S. and Takeuchi, M. (1977 c): Further studies on the glycopeptides from the human, bovine and porcine stomachs Int. J. Biochem. 8, 623-626

80. Masuda, H., Shichijo, S. and Takeuchi, M. (1977 d): Comparative studies on the gastric glycopeptide in eleven animal species. Comp. Biochem. Physiol. 58B, 163-165

81. Masuda, H., Shichijo, S. and Takeuchi, M. (1978): Distribution of glycosaminoglycans in the human gastric wall. Seibutsu Butsuri Kagaku 22, 29-32 (Japan)

82. Masuda, H., Shichijo, S. and Takeuchi, M. (1978 a): A glycopeptide from gastric cancer tissue. Biochem. Med. 20, 305-314

83. Masuda, H., Shichijo, S., Sasaki, Y and Kurakazu, T. (1979): Electrophoresis of glycosaminoglycans from equine gastric mucosa. Seibutsu Butsuri Kagaku 23, 15-19 (Japan)

84. McGuigan, J. E. (1976): Peptic ulcer disease. In: J. M. Dietschy (Ed.) Scientific Practice in Clinical Medicine. Grune and Stratton, New York, Vol. 1, 88-102

85. Meyer, K. (1959): Chondroitin sulfates. Conf. on Polysaccharides in Biol., Trans. of 4th Conf. 1958, The Josaih Macy Jr. Foundation, New York, 11-56

86. Meyer, K., Smyth, E. M. and Palmer, J. W. (1937): On glycoproteins. III. The polysaccharides from pig gastric mucosa. J. Biol. Chem. 119, 73-84

87. Michalsen, H., Roella, G., Melsen, B. and Soenju, T. (1979): Sulfated glycoproteins in pancreas. Monogr. Paediatr. 10 (Recent Adv. Cystic Fibrosis) 148-150

88. Mirsky, A., Kaplan, S. and Broh-Kahn, R. H. (1950): Pepsinogen excretion (uropepsin) as an index of the influence of various life situations on gastric secretion. Life Stress and Bodily Disease (Vol. XXIX of the 1949 proceedings Ass. for Res. in Nervous Men. Dis.) Chap. XLII, 628-646

89. Molinari, F. (1977): Histochemical studies on the mucopolysaccharides of the fundal mucosa in chronic aspecific gastritis (proceedings) Minerva Gastroenterol 23, 43-44 (Ital)

90. Molinari, F. and Marcrina, G. (1977): Histochemical studies on the mucopolysaccharides of the normal gastric mucosa (proceedings) Minerva Gastroenterol 23, 45 (Ital)

91. Moretti, A. and Wolf, G. (1961): Vitamin A and mucopolysaccharide biosynthesis in colon homogenates. Biochim. Biophys. Acta 53, 263-272

92. Nadler, H. L., Rao, G. J. S. and Taussig, L. M. (1978): Cystic Fibrosis. In: J. B. Stanbury, J. B. Wyngaarden and D. S. Fredrickson (Eds.). The Metabolic Basis of Inherited Disease. McGraw Hill, 4th Ed. Chap. 73, 1683-1710

93. Nemoto, T. and Yosizawa, Z. (1969): Sulfated glycopeptides and glycosaminoglycan peptides isolated from intestinal mucosae of rabbit. Biochim. Biophys. Acta 192, 37-48

94. Oates, M. D. and Schrager, J. (1965): The use of gas-liquid chromatography in the analysis of neutral monosaccharides in hydrolysates of gastric mucopolysaccharides. Biochem. J. 97, 697-700

95. Oette, K. and Gheorghiu, T. (1975): Hydrolysis of mucosubstances and gas-liquid chromatography of bound sugars in gastric juice analysis. In: T. Gheorghiu (Ed.). Experimental Ulcer, Gerhard Witzstrock, Publ. Baden-Baden, Germany, 259-270

96. Ono, T. and Eto, K. (1967): Determination of acid mucopolysaccharide in the gastric juice. Jap. J. Clin. Path. 15, 645-648

97. Pallavicini, G., Cetta, G., Quaroni, A. and Castellani, A. (1971): Glycopeptides from normal and ulcered gastric mucosae of pig. Experientia 27, 895-897

98. Pallavicini, G. and Sinigaglia, F. (1979): Gastric glycoproteins in normal and pathological conditions. In: B. Berra, C. Balduini, S. Di Donato and G. Tettamanti (Eds.). Glycoproteins and glycosaminoglycans in Health and Disease, Perspectives in Inherited Metabolic Diseases, Vol. 2, Edi Ermes, Milan, 91-100

99. Pamer, T., Glass, G. B. and Horowitz, M. I. (1968): Purification and characterization of sulfated glycoproteins and hyaluronidase-resistant mucopolysaccharides from dog gastric mucosa. Biochemistry (Wash) 7, 3821-3829

100. Peleshchuk, A. P. and Voloshchenko, I. I. (1971): Age and gastric secretion. Fiziol. Zh. (Kiev) 17, 357-363

101. Platash, V. I. (1976): Histochemical characteristics of the secretory elements of the vermiform appendix following surgical trauma, circulatory hypoxia and immunization. Biull. Eksp. Biol. Med. 81, 91-94 (Russ)

102. Quarterman, J., Humphries, W. R. and Florence, E. (1970): Changes in appetite and alimentary mucosubstances in zinc deficiency. In: C. F. Mills (Ed.). Trace Elem. Metab. Anim., Proc. W A A P (World Ass. Anim. Prod.)/IBP (Int. Biol. Progr.) Int. Symp. 1969. Livingstone, London, 167-170

103. Reid, P. E. Culling, C. F. A., Dunn, W. L., Ramey, C. W. and Clay, M. G. (1975): Differences in chemical composition between the epithelial glycoproteins of the upper and lower halves of rat colon. Can. J. Biochem. 53, 1328-1332

104. Roelfs, R. E., Gibbs, G. E. and Griffin, G. D. (1967): The composition of rectal mucus in cystic fibrosis. Amer. J. Dis. Child. 113, 419-421

105. Sammons, H. G. (1950/1951): Mucinases in ulcerative colitis. Gastroenterologia 76, 346-349

106. Sato, R. (1952): Biochemical studies on carbohydrates. CLVIII. Karl Meyer's mucoitinsulfuric and hyaluronosulfuric acids. Tohoku J. Exptl. Med. 56, 387-397

107. Schachter, H. and Dixon, G. H. (1965): A comparative study of the protein in normal meconium and in meconium ileus patients. Can. J. Biochem. 48, 381-397

108. Schrager, J. (1964): Mucopolysaccharides of the gastric secretion. Nature 201, 1220-1222

109. Schrager, J. (1974): The chemical nature of gastric and small intestinal mucus. Top. Gastroenterol. 2, 101-111

110. Schwachman, H. (1975): Gastrointestinal manifestation of cystic fibrosis. Pediat. Clin. N. Amer. 22, 787-805

111. Sekino, T., Murata, K., Saito, Y., Ishikawa, T. and Tsubura, K. (1976): Mucosubstances of gastric juice and the gastric wall in patients with gastric ulcer. Prog. Peptic Ulcer, (Proc. Conf. Exp. Ulcer), Akad. Kaido, Budapest, 299-314

112. Sekino, T., Murata, K., Saito, Y and Tsubura, K. (1977): A study of acidic glycosaminoglycans in human gastric tissue. Digestion 16, 28-39

113. Selye, H. (1951): The general-adaptation-syndrome and the gastrointestinal diseases of adaptation. Amer. J. Proctology 2, 167-184

114. Sinitisina, V. F. (1966): Mucopolysaccharides of human duodenal mucosa Arkj. Anat. Gistol. Embriol. 51, 111-114 (Russ)

115. Sll'chenko, V. P. (1966): Acid and neutral carbohydrates in the mucous membrane of the appendix. Vrach. Delo 3, 25-28 (Russ)

116. Slomiany, B. L. and Meyer, K. (1972): Isolation and structural studies of sulfated glycoproteins of hog gastric mucosa. J. Biol. Chem. 247, 5062-5070

117. Smith, H., Zwartouw, H. T., Gallop, R. C. and Harris-Smith, P. W. (1953): Virulence-enhancing factor of mucins. VII. Remaining components of the "third factor" involved in virulence enhancement. Biochem. J. 53, 673-678

118. Smolyanskii, B. L. and Zhukova, N. M. (1967): The state of the stomach mucous membrane in aged persons according to gastrobiopsy data. Klin. Med. (Moscow) 45, 92-96 (Russ)

119. Srinivas, H., Leela, R. and Rama Rao, P. B. (1971): Studies on tissue acid mucopolysaccharides: Part I. Pig intestinal mucosa. Indian J. Biochem. Biophys. 8, 289-294

120. Starkey, B. J., Snary, D. and Allen, A. (1974): Characterization of gastric mucoproteins isolated by equilibrium density-gradient centrifugation in caesium chloride. Biochem. J. 141, 633-639

121. Svetoni, N., D'Alonzo, D. and Rechichi, A. (1974): Clinical evaluation of the antidyslipemic action of a sulfomucopolysaccharide of pancreatic origin. Clin. Ter. 71, 33-48 (Ital)

122. Szekely, J., Bihari-Varga, M. and Gero, S. (1967): The role of mucopolysaccharides of gastric juice in the alterations of the blood lipid pattern. Orv. Hetil. 108, 927-930 (Hung)

123. Szekely, J., Bihari-Varga, M. and Gero, S. (1968): Relationship between the mucopolysaccharide content of gastric juice and the composition of serum lipids. Acta Med. Acad. Sci. Hung. 25, 141-151

124. Szollosi, G., Udvardi, J., Nafradi, J., Balint, G. A., Karacsony, G. and Varro, V. (1976): Indomethacin-induced gastric ulcer in the rat. In: G. Mozsik and T. Javor (Eds.). Prog. Peptic Ulcer, Akad. Kiado: Budapest, 529-542

125. Szulc, H., Majewski, C., Kadlubowska, M., Wirga, Z., and Wolowicka, L. (1967): Reactions for acid mucopolysaccharides in different parts of the alimentary tract in acute experimental peritonitis. Patol. Pol. 18, 535-542 (Pol)

126. Takeda, Y (1975): Clinical study of mucopolysaccharide of the duodenal juice in chronic pancreatitis. Jpn. J. Gastroenterol. 72, 491-500 (Jpn)

127. Terho, T., Hartiala, K. and Hakkinen, I. (1966): Infrared spectroscopic investigations of a new acid mucopolysaccharides isolated from human gastric juice. Nature 211, 198-199

128. Tikhonova, N. M. (1964): Histochemical data and characteristics of intestinal coliform bacterial infection. Arkh. Patol. 26, 53-59 (Russ)

129. Umehara, S., Yukawa, Y., Ishimori, N., Nabeshima, Y., Hayashi, T. and Ito, H. (1976): The acid mucopolysaccharides in the mucosa and submucosa of stomach with ulcer and submucosal inflammation. In: G. Mozsik and T. Javor (Eds.). Prog.Peptic Ulcer Akad. Kiado: Budapest, 315-323

130. Velev, G., Brailski, Kh. and Pacheva, G. (1973): Mucopolysaccharides in the gastric mucosa in chronic gastritis. Vutr. Boles 12, 78-84 (Bul)

131. Waldron-Edward, D. and Skoryna, S. C. (1964): Properties of the gel mucin of human gastric juice. Proc. Soc. Exp. Biol. Med. 116, 794-799

132. Waldron-Edward, D. and Skoryna, S. M. (1970): Studies on human gastric gel mucin. Isolation and characterization of a major glycoprotein component. Gastroenterology 59, 671-682

133. Weisburger, J. H., Fiala, E., Narisawa, T. and Reddy, B. S. (1976): On the etiology and prevention of cancer in the gastrointestinal tract. In: P. N. Magee, S. Takayama and T. Sugimara (Eds.). Fundamentals of Cancer Prevention. Univ. Tokyo Press, Tokyo, 121-142

134. Werner, I (1953): Studies on glycoproteins form mucous epithelium and epithelial secretion. Acta Soc. Med. Upsalien 58, 1-55

135. Wolfrom, M. L. and Rice, F. A. H. (1947): Electrophoretic resolution of heparin and related polysaccharides. J. Amer. Chem. Soc. 69, 2918-2919

136. Wolfrom, M. L. and Rice, F. A. H. (1947 a): The uronic acid component of mucoitin-sulfuric acid. J. Am. Chem. Soc. 69, 1833

137. Wolfrom, M. L., Weisblat, D. I., McNeeley, W. H., McLean, J. and Karabinos, J. V. (1943): Chemical studies on crystalline barium acid heparinate. J. Amer. Chem. Soc. 65, 2077-2085

138. Yakhnitsa, A. G. and Shaposhnikov, V. M. (1972): Morphological and histochemical age characteristics of human anal glands. Tr. Kar'kov. Med. Inst. 1971, (93) 253-259 (Russ). From Ref. Zh. Biol. Khim. Abstr. No. 22F1773

Chapter 5

Glycosaminoglycans in Fluids and Fluid-Associated Tissues of the Oral Cavity

5.1 Introduction

The oral cavity under normal physiological or pathological states contains the following fluids:

- Saliva
- Gingival fluid
- Oral cyst fluid
- Sputum.

Sputum is primarily a bronchial secretion to which secretions from the upper respiratory tract and saliva from the oral cavity are added upon expectoration. For these reasons, sputum has not been included here and will be described in the next chapter on fluids of the respiratory tract. The other three fluids of the oral cavity and the tissues associated with these fluids will be described with respect to the glycosaminoglycan composition and the enzymes hydrolyzing these macromolecules.

5.2 Salivary apparatus and salivary secretion

Saliva is a mixture of secretions from a number of sources in the oral cavity. The major contribution is from secretions of three pairs of salivary glands: parotid, submaxillary (or submandibular) and sublingual (Fig. 5.1) with a minor contribution from numerous small mucous glands and ectopic sebaceous glands located in the oral cavity. The mucus secretion from the epithelium of the salivary ducts and the oral mucosa also contribute to some extent. The sublingual glands in rodents (rat) are small and are closely associated with the submaxillary gland.

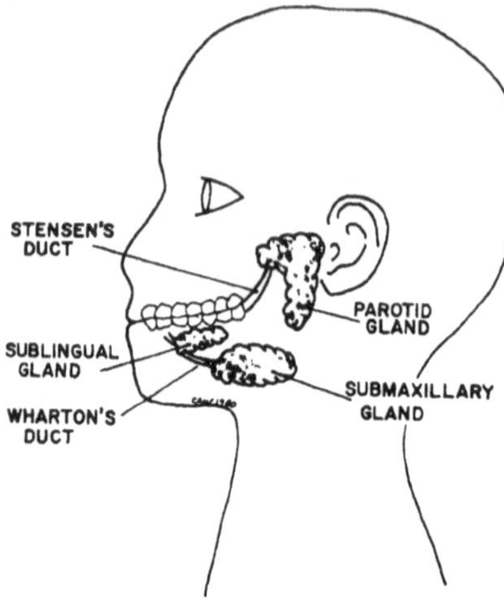

Fig. 5-1. The major salivary glands

The salivary glands are exocrine glands and drain their secretion through a duct; the parotid duct opens on the buccal surface adjacent to the second upper molar tooth, the Wharton's duct from the both sides of submaxillary gland opens into the floor of the mouth and the sublingual empties its secretion through many small ducts and a large duct. The rates of secretions of these glands are variable depending upon stimulation by psychic and mechanical factors. The secretions of these glands are different and each has a unique composition. The parotid gland lacks mucus cells and produces serous secretions low in mucosubstances. The sublingual and the submaxillary glands have a large number of mucus-producing cells. Their secretions are viscous and contain larger amounts of mucosubstances. Saliva is produced from the glands under the effect of a stimulus. The different types and doses of stimuli influence the composition of the salivary secretion. Human submaxillary saliva is produced only when a stimulus like chewing or a gustatory agent is involved. The sialic acid and fucose ratio changes with the intensity of stimulation in human secretors of blood group substances in saliva. Pilocarpine stimulates submaxillary saliva secretion; when given in high doses the secretion produced has a decreased sialic acid/fucose ratio in dog. The sulfate content and the ratios of sulfate to sialic acid and fucose in the submaxillary secretion of dog vary with different intensities of electrical stimulation (Pigman, 1977). Acetylcholine produces a secretion of a higher sialic acid/fucose ratio (Dische et al., 1962). Sympathetic stimulation produces a highly viscous saliva of a high content of mucosubstances.

The physiological aspects of the salivary apparatus (Diamant, 1976), and salivary secretion (Dawes, 1976), the chemistry of saliva (Shannon, 1972; Dawes, 1978) and the methods of saliva collection (Dawes, 1976; Wotman and Mandel, 1976) have been reviewed. Circadian rhythms in flow rate and in composition of unstimulated and stimulated saliva have been noted (Dawes, 1974). Saliva composition varies with the flow rate, nature and duration of stimulus, effect of previous stimulation, plasma composition and time of the day for sampling (due to circadian rhythm).

The ultrastructural organization of the embryonic salivary tissue of mouse submandibular gland has been studied (Banerjee et al., 1976; Cohn et al., 1976). The basal lamina was a highly ordered arrangement of the epithelium and contained equal amounts of nonsulfated GAG (hyaluronic acid with a small amount of chondroitin) and sulfated GAG (chondroitin 4-sulfate was twice chondroitin 6-sulfate). The supramolecular complexes of hyaluronic acid and proteoglycans were organized into an extracellular scaffolding which stabilized the morphology of the epithelium (Cohn et al., 1976). This role in maintaining the structural morphology of the epithelium and lobules is evident from the changes including rounding up of the cells, loss of cellular cohesion, appearance of microvilli and changes in organization of cytoskeleton structures produced upon brief treatment with testicular hyaluronidase in the absence of Ca^{2+} and their reversal within 2 ½ hr by epithelial synthesis of GAG for lamina organization (Banerjee et al., 1976).

5.2.1 Functions of saliva in oral health

The glycoproteins and mucoids present in saliva give this fluid characteristics making it capable of performing a number of functions such as:
1. Lubrication and digestion: It helps in deglutination of the food down the esophagus into the stomach by forming a slippery coating around the food bolus. This lubrication is due to highly viscous secretions of the sublingual gland and the minor mucous glands.
2. Protection: The submaxillary secretion inhibits the hemagglutination of red blood cells by binding to the virus and blocking the viral neuraminidase. Saliva attenuates dental caries and maintains the integrity of the teeth by virtue of its food debris' washing, buffering and antibacterial (lysozyme, lactoperoxidase) action. An inverse relation between salivary flow and a higher buffering action of bicarbonate and phosphate in saliva to the incidence of dental caries has been noted (Newburn, 1978). Dry mouth or xerostomia is followed by an increased incidence of caries and proliferation of bacteria, resulting in soreness of oral mucosa, gingivitis, cracking of lips, salivary duct infection and salivary calculi (Mandel, 1974; Dawes, 1978). A reduction in salivary flow during sleep also greatly reduces the protective effects of saliva.

5.3 Mucosubstances in saliva and salivary glands

The carbohydrates comprise about 20 % of the nondializable substances which themselves are present at a concentration of 50-300 mg/100 ml saliva. The carbohydrate-containing substances in saliva could be classified into the following three types of macromolecules:

Fucomucoids: These contain fucose, hexosamine, galactose and a small amount of mannose. These are associated with the blood group activity of saliva.

Sialomucoids: These are rich in sialic acid and acetylgalactosamine and are predominant in the submaxillary gland secretion and have the property of inhibiting the hemagglutination produced by influenza virus

Acidic glycosaminoglycans: These form a minor component of saliva and information on these is still fragmentary.

The majority of the available information on salivary mucosubstances is on fucomucoids and sialomucoids and the presence of hexuronic acid and sulfate was not detected in previous investigations. The dried, pooled saliva contained 2.1% sialic acid, 3.3% fucose, 6% hexosamine and 8% hexose; no hexuronic was present (Werner and Odin, 1965): The parotid gland contains a very low (0.1%) sialic acid content, whereas in the sublingual and submaxillary gland sialic acid contents are much higher than in pooled saliva. The glycoproteins in saliva and salivary glands have been reviewed extensively (Werner and Odin, 1965; Pigman, 1977). Chemical changes in carbohydrate components of saliva (McCoombe et al., 1961) occur during the menstrual cycle (Mancuso et al., 1978) but the components studied were confined only to the hexosamine and neutral sugars. There is no clear-cut report for the presence of glycosaminoglycans and their characterization from saliva. Indirect evidence for the presence of these macromolecules has been only from the histochemical tests on saliva and their isolation from salivary glands (Mogil'naia, 1966; Siervo et al., 1965; Spicer and Duvenci, 1964).

Using histochemical and biochemical techniques the mucous cells of the labial and palatine glands of the monkey were shown to elaborate sulfated mucosubstances (Hensten-Pettersen, 1976). In rat, the glycosaminoglycan concentration was the highest in the retrolingual, intermediate in the parotid and the lowest in the submandibular gland. Hyaluronic acid, heparan sulfate, chondroitin 4-sulfate and chondroitin 6-sulfate were present in all the three salivary glands. However, hyaluronic acid was the major uronic acid fraction in the parotid and submandibular glands (Kofoed et al., 1969). Shiozaki (1973) confirmed that glycosaminoglycans were highest in the sublingual, intermediate in the parotid and the lowest in the submandibular gland. However, this investigator reported hyaluronic acid, heparan sulfate and dermatan sulfate as the major components, while chondroitin 4-sulfate and heparin were the minor components in the salivary gland. Also, the major component in the submandibular, parotid and sublingual glands was heparan sulfate, hyaluronic acid and dermatan sulfate, respectively. Histochemical study of the parotid glands during their development revealed that during the embryonic and the postembryonic periods the acini elaborated coarse granules of a secretion containing glycosaminoglycans and proteins (Shubnikova and Chunaeva, 1966).

The pig and bovine salivary glands contained the same glycosaminoglycan components as the rat but differed in quantities. The glycosaminoglycans were characterized by cellulose acetate electrophoresis, CPC (cetylpyridinium chloride) fractionation, hexosamine and chondroitin sulfate isomer analysis (Nagura, 1976). In rabbit salivary glands, hyaluronic acid predominated followed by dermatan sulfate, heparan sulfate, chondroitin 4-sulfate and chondroitin 6-sulfate in the decreasing order. The parotid gland had a higher glycosaminoglycan content as compared to the submaxillary gland (Kim, 1978).

In dog, the glycosaminoglycans (GAG) contents varied per dry weight as sublingual > submandibular > parotid gland. The GAG components present in all the glands were the same. Hyaluronic acid, dermatan sulfate and heparan sulfate constituted the major components and chondroitin 4- and 6-sulfates were the minor components. The relative proportion of these components varied in the three salivary gland types. From the aforementioned information it is clear that the glycosaminoglycan fractions were the same in dog, bovine, rat and porcine salivary glands but these differed quantitatively.

Two glycosaminoglycan-degrading enzymes, β-glucuronidase and hyaluronidase were present in saliva. Salivary hyaluronidase was present in two forms (isoenzymes) and the values of both of these varied during the menstrual cycle (Hopps, 1978). The β-glucuronidase was present in the submandibular gland and its level was related to androgenic state (Hosoi et al., 1978). An inhibitor of the β-glucuronidase, which was glycoprotein in nature, was isolated from saliva (Sakamoto et al., 1974). This activity was related to the AB0 blood group and secretory status (Sakamoto et al., 1976).

5.3.1 Factors affecting glycosaminoglycans composition of saliva and salivary glands

The glycosaminoglycans in the connective tissue of the salivary glands and the oral mucosa are affected by several hormones (Baker et al., 1964). Insulin increased the in vitro incorporation of glucose in sublingual gland (Szymczyk, 1971) and the submaxillary gland (Ines Masi and Szymczyk,

1966) of rat. In alloxan diabetic rabbits, the glycosaminoglycan content of the submaxillary gland was decreased by 35 % of the control level. In the parotid gland chondroitin 4-sulfate was increased by 66 % and chondroitin 6-sulfate was decreased by 37 %. In the submaxillary gland of the alloxan diabetic rabbit, a decrease was observed in hyaluronic acid, heparan sulfate, dermatan sulfate and chondroitin 6-sulfate, while chondroitin 4-sulfate showed an increase (Kim, 1978).

Thyroid influences the size and structure of the submandibular gland. Atrophy of the granular tubules and the submandibular gland occurs in hypothyroidism and these changes are reversed by administration of thyroid hormone (Muhler et al., 1956). Thyroidectomy has also been reported to produce a significant atrophy of the parotid glands (Baker et al., 1964). Radiothyroidectomy in male rats increased the hexuronic acid content only in the hyaluronic acid fraction of only the submandibular gland and thyroid administration reversed these changes (Kofoed et al., 1973).

The effect of glucocorticoids on the glycosaminoglycans of salivary glands is opposite to their inhibitory effect on biosynthesis of these macromolecules in skin, cartilage and bone. Adrenalectomy in rats reduced the hexuronic acid contents of submandibular and retrolingual glands (without affecting the parotid glands). Adrenalectomy decreased glycoprotein, hyaluronic acid and dermatan sulfate fractions and this decrease was not reversed by cortisol administration in these adrenalectomized rats. On the other hand, cortisol acetate increased concentration of hexuronic acid reflected in the increase of hyaluronic acid, chondroitin 4-sulfate and chondroitin 6-sulfate fractions in submandibular and retrolingual glands. No changes in hexuronic acid occurred in the parotid glands (Kofoed et al., 1973 a). Another investigation reported changes in the parotid gland under the influence of hydrocortisone (Groza et al., 1966).

Hormonal changes occurring during puberty, pregnancy and menopause influence the salivary glands. Testosterone propionate increased and castration decreased the glycosaminoglycan of the rat submandibular gland, but did not affect those of parotid and retrolingual glands (Kofoed et al., 1971 a). Adrenal activity has been shown to be necessary for the maintenance of normal structure of the striated tubules in the submandibular glands in mice and rats (Raynaud, 1964). Androgens have been reported to regulate the N-acetylglucosaminidase in the submandibular gland. This activity depended upon the androgenic state; testosterone propionate increased this activity in castrated males and females, while estrogen and progesterone had no effect in males (Hosoi et al., 1979). Androgens also influence the β-glucuronidase activity in the submandibular gland of the mouse. The enzyme activity was the same in males and females before but increased following puberty to twice the activity in females. Castration returned this activity to the female level. Estrogens have effect on gingiva and salivary glands. Estradiol caused normal differentiation of epithelium and increased the survival of the isolated rat salivary glands in tissue culture (Rybakova, 1975). Progesterone decreased the hexuronic acid content of submaxillary gland in hyaluronic acid, heparin and glycoprotein fractions. 17-α-Hydroxyprogesterone produced decreases in glycoprotein and hyaluronic acid in castrated female rats (Kofoed et al., 1977). Changes in carbohydrate composition (Mancuso et al., 1978) and in the two hyaluronidases in saliva (Hopps, 1978) have been noted during the menstrual cycle.

The glycosaminoglycan composition of the salivary glands may also vary as a result of nervous factors like those in incisor amputations (Kofoed et al., 1971), isoproterenol administration (Kofoed et al., 1970) and certain other factors (Tomanczyk and Pankiewicz, 1972). Presence of cysts (see section 5.4) in the oral cavity will modify the glycosaminoglycan composition of saliva, as would the neoplasms of the oral cavity (see section 5.2.5).

5.3.2 Pathology of saliva and salivary gland

Several authors (Farb, 1970; Mason and Chisholm, 1976; Wotman and Mandel, 1976; Sweeney, 1978) have reviewed the abnormal salivary function encountered as decreased salivary flow in xerostomia and Sjögren's syndrome, increased salivary secretion (Ptyalism or Sialorrhea), inflammation of major glands by bacterial infection (Sialoadenitis) or viral infection (mumps or parotitis), salivary calculi (sialolithiasis), mucous cysts (mucocele), salivary gland tumors (neoplasia) and cystic fibrosis. The GAG composition of saliva may vary in pathological states affecting the parenchymal cells of the salivary glands or in those affecting the salivary flow.

Inflammatory enlargement of the salivary glands occurs due to infections by bacteria, virus, granulomatous or mycotic agents. The gland most commonly affected is the parotid gland. Infection of the gland occurs by entrance of the *Staphylococcus aureus, Streptococcus viridans* and *Pneumococci* through the salivary duct producing firm, nodular swelling of the gland and erythema of the skin overlying the gland. The swelling may be due to the bacterial infection of the gland or the duct and the secretion is a purulent discharge. The swelling of the gland also occurs in *Mycobacterium tuberculosis* and *sarcoidosis*.

The relationship of GAG changes in saliva to salivary gland inflammation is suggested by the fact that Sjögren syndrome (Triad of Keratoconjunctivitis sicca, rheumatoid arthritis and xerostomia) occurs in connective tissue disorders. Lymphocytic infiltration of the salivary glands causes atrophy of the glands' acini and narrows the lumen of the duct.

5.3.3 Glycosaminoglycans in salivary gland tumors

In addition to benign and malignant tumors of the major salivary glands, tumors may also occur in a minor salivary gland, being malignant more often than benign. The tumors of the minor glands may be located anywhere in the oral cavity, pharynx etc. (Mason and Chisholm, 1976; Lucas, 1976). These tumors are associated with mucosubstances and have been classified into various types (Farb, 1970; Lucas, 1976). Tumors are ten times more in the parotid gland than in other salivary glands. There is no association between neoplasia and secretor and nonsecretor status of blood groups but an association does exist with breast cancer. Quantitative ultrastructural comparison of the pleomorphic adenomas and normal salivary glands showed a higher proportion by volume of cells of duct region in pleomorphic adenoma. Both epithelial and mesenchymal mucins could be distinguished histochemically in the stroma of the pleomorphic adenoma (Mason and Chisholm, 1976).

Hyaluronidase-sensitive glycosaminoglycans have been shown to be present in tumors of mucous tissue or myxomatous mucin (Azzopardi and Smith, 1959) by histochemical methods. Histochemical analysis of mucoid substances in recurrent mixed tumors of the parotid gland showed the presence of acidic glycosaminoglycans (Musy and Gotzos, 1973). The orientation of these macromolecules in the tumors was demonstrated by methods using a polarizing microscope.

Histochemical and chemical analyses of glycosaminoglycans from mixed salivary tumors revealed that these tumors contained hyaluronic acid, dermatan sulfate and chondroitin 6-sulfate. Chondroitin 6-sulfate of varying degrees of sulfation and hyaluronic acid predominated in the myxomatous areas, whereas dermatan sulfate was associated with the mature collagen of the capsule and the septa (Lovell et al., 1966). The isolation of glycosaminoglycans from the surgically removed tumors involved drying with acetone and 1:1 chloroform methanol, papain digestion, deproteinization with 7.5 % trichloroacetic acid and precipitation of GAG from the supernatant with three volumes of ethanol containing sodium acetate. The components in the isolated material were fractionated and identified by elution from a cellulose column as GAG-CPC complexes, digestion with testicular and bacterial hyaluronidases and electrophoresis.

5.3.4 Saliva in cystic fibrosis

Since cystic fibrosis (mucoviscidosis) is a hereditary generalized abnormality of the exocrine glands, salivary gland and salivary composition are also affected. Since parotid gland is a serous type gland, only an insignificant change in protein-bound carbohydrates occurs in parotid saliva (Mandel et al., 1965). Sublingual and submaxillary (submandibular) glands being mucous glands are affected and produce thick viscous secretion which leads to stasis, duct dilatation, fibrosis and finally acinar atrophy (Mangos and Talamo, 1973, 1976; Nadler et al., 1978). The child with cystic fibrosis produces more saliva per minute per gram of the gland tissue than a normal child, since the submaxillary gland is enlarged 2-6 times. The submaxillary saliva of cystic fibrosis patients is turbid due to the presence of large amounts of mucosubstances and protein. These changes are evident from pathological changes observed in sublingual and submaxillary glands by histochemical reactions for carbohydrates (Anderson, 1962; Shackleford, and Bentley, 1964) and by changes in molar ratios of carbohydrate constituents of various fractions of these glands (Dische, 1964). Histochemical evidence has been provided for qualitative and quantitative alterations in carbohydrates in the small intercalated or small intralobular ducts which are connected directly with the gland. These intercalated ducts are modified from their normal natural form into mucus-secreting structures which react strongly with Alcian blue similar to the secretions of the mucous acini (Shackleford and Bentley, 1964 a).

5.4 Glycosaminoglycans of gingival fluid and gingiva

The gingival fluid has been mentioned in literature by different names like gingival pocket fluid, periodontal pocket exudate, gingival exudate, crevicular fluid. Because of the relationship of this fluid to gingival health, this fluid has been studied well and has been the subject of several reviews (Brill, 1962; Eastoe et al., 1971) and a monograph (Cimasoni, 1975).

Greeves (1977) from an investigation in 71 dental students observed a positive relationship of the width of the zone of keratinized gingiva and pocket depth for all teeth except the third molars. In gingivitis, a positive correlation between the zone of keratinized gingiva and plaque within the dentition was also noted.

5.4.1 Glycosaminoglycans in gingival fluid

The amount of gingival exudate collected on preweighed filter papers (1.5 x 10 mm) inserted for 30 sec. into pockets of teeth till resistance was felt, is a measure of gingival inflammation (Loe et al., 1965).

In 16 human subjects with a gingival index of one and in 13 with a gingival index of about two an average of 0.15 mg exudate and 0.50 mg per filter strip, respectively, were obtained (Hara and Loe, 1969). Thus the amount of exudate increased with the severity of inflammation.

The presence of mucoproteins and glycoproteins in gingival exudate could be demonstrated by histochemical staining of the exudate absorbed on filter paper strips (Sueda et al., 1966). The gingival tissue contains glycosaminoglycans in its connective tissue intercellular matrix, which contain approximately equal amounts of hexuronic acid and hexosamine (Schultz-Haudt, 1958). Hyaluronic acid and chondroitin sulfates associated with protein have been reported in gingival tissue prepared by different extraction procedures and these macromolecules have been identified by electrophoresis and histochemical staining with Alcian blue, toluidine blue and colloidal iron. Testicular hyaluronidase digestion has been used to differentiate between hyaluronidase-sensitive chondroitin 4- and 6-sulfates from the hyaluronidase-resistant dermatan sulfate in gingival tissue (Schultz-Haudt, 1965).

The microorganisms present in the oral cavity are known to elaborate hyaluronidase, chondroitinase and protease in pure and mixed cultures (Schultz-Haudt and Scherp, 1955, 1956; Schultz-Haudt and Lundquist, 1962). Increased activity of these oral microflora brings about breakdown of the glycosaminoglycans and glycoproteins and increased permeability (Quintarelli, 1960) and this may be responsible for the pathogenesis of inflammation seen in periodontitis (Schultz-Haudt, 1965). Furthermore, an increase in metachromasia is encountered proportional to the intensity of inflammation (Quintarelli, 1960 a) and in chronic inflammation increased amounts of glycosaminoglycans may be present (Engel et al., 1960).

In agreement with the aforementioned concepts of the gingival inflammation is the report of the presence of 1.13-7.09 μg glucose, 0.6 mg hexosamine and 50-60 μg hexuronic acid per mg weight of the gingival exudate collected on preweighed filter paper strips (1.5 x 10 mm) from 29 otherwise healthy human subjects but with some degree of gingival inflammation (Hara and Loe, 1969). However, it is puzzling why hexuronic acid was 100 times the hexosamine in the gingival exudate, when the two components were present in equal amounts in the gingival connective tissue (Schultz-Haudt, 1958). This discrepancy may be due to an unknown differential pattern of dissolution of the gingival ground substance and release of the products of degradation into the gingival pocket. Also, the concentrations of these have been reported to be influenced by the local microflora since these microorganisms in the oral cavity contain hexosamines and uronic acids (Cimasoni, 1975). For these reasons, determination of hydroxyproline coming from gingival collagen in the gingival exudate may be a better measure of connective tissue breakdown. However, results of analysis of exudates from four patients revealed that the hydroxyproline-containing components present in exudates from one single sulcus were on the average only 1.62 µg/strip and hexosamine was 22.8 µg/strip and 50 % of the hydroxyproline-containing components were dialyzable (Paunio, 1971). The two fractions obtained by Sephadex G-200 gel chromatography of the pool of gingival exudate extracted from the paper strips placed in single sulci gave two fractions resembling those from plasma. Thus this material in the exudate could have originated as an ultrafiltrate of plasma.

5.4.2 Glycosaminoglycans in gingiva

The dermis of the gingiva is composed of the collagen protein and glycosaminoglycans. The presence of glycosaminoglycans in the intercellular matrix of the gingival connective tissue has been demonstrated by metachromatic staining with basic dyes like Alcian blue and toluidine blue or colloidal iron (Engel et al., 1960; Quintarelli, 1960, 1960 a). Digestion with hyaluronidase hydrolyzes these water-soluble macromolecules and results in loss of this metachromasia of the gingiva in tissue sections. The glycosaminoglycan and collagen composition of gingiva have been studied in animals and human subjects in normal and pathological states. The glycosaminoglycans in gingiva from human subjects and various species of animals are qualitatively similar and differ only in various proportions of the component glycosaminoglycans. The major components found in all gingivae are chondroitin 4-sulfate, dermatan sulfate, hyaluronic acid and heparan sulfate.

The glycosaminoglycans of the gingiva in human subjects have been investigated because of the relationship of the lesions in periodontium to the periodontal disease, the most common type of which is periodontitis and involves inflammation of the gums (gingivitis). Schultz-Haudt (1957) examined acidic glycosaminoglycan contents of human gingiva without reference to the degree of periodontal involvement and found that the inflammed gingiva contained less hexuronic acid and hexosamines than normal gingiva. The glycosaminoglycans isolated from the acetone-defatted normal gingival tissue contained 0.32 % of the dry weight and paper chromatography of its acid hydrolysate showed the presence of glucuronic acid, glucosamine and galactosamine (Schultz-Haudt, 1958). Analysis of a 5 % aqueous trichloroacetic acid extract of the dried-defatted tissue contained two electrophoretic fractions. The slow moving fraction stained for protein and glycoprotein and upon paper chromatography of its hydrolyzate, it showed the presence of fucose, mannose and galactose. The fast electrophoretic band was stained by toluidine blue and colloidal iron, was hyaluronidase-sensitive and its hydrolyzate contained only glucosamine (Schultz-Haudt et al., 1961). The slow and the fast fractions were identified as glycoprotein and hyaluronic acid, respectively. The glycosaminoglycans isolated from the residue left after a 20-minute trichloroacetic acid extraction contained, besides the glycoprotein and hyaluronic acid, a fastest moving anodic band of chondroitin sulfate (Schultz-Haudt et al., 1964). The method for isolation of glycosaminoglycans used included proteolytic digestion and cetylpyridinium chloride (CPC) precipitation.

In a comparative study of gingival tissue from 30 normal subjects and 90 patients with periodontal disease (with gingival crevicular depth of more than 5 mm) the concentration of chondroitin sulfate was lower ($p < 0.01$) in periodontal gingival tissue (Ciancio and Mather, 1971). Only chondroitin 4- and 6-sulfates were present in measurable amounts. The hexuronic acid and hexosamine in normal gingival tissue were 1.96 ± 0.57 and 2.01 ± 0.55 mg/g dry wt, respectively. In periodontal disease the corresponding values were reduced to 1.47 ± 0.22 for hexuronic acid and 1.32 ± 0.31 for hexosamine. The glycosaminoglycans were isolated from gingival tissue obtained by employing reverse bevel gingivectomy, defatting with acetone, papain-trypsin digestion, deproteinization with trichloroacetic acid, CPC precipitation and elution of GAG from this precipitate with 1.2 M sodium chloride.

In contrast with glycosaminoglycan decrease in gingival tissue, the carbohydrate macromolecules in other periodontial tissues respond differently in periodontal disease. The perchloric acid-soluble but phosphotungstic acid-insoluble material isolated from the water extract of the acetone-defatted, dried inflammed gingiva contained higher orcinol-sulfuric acid positive carbohydrates than those from normal gingival tissue (Winzler et al., 1948). Heavy accumulations of Alcian blue-positive material and increased metachromasia were noted near the alveolar bone; necrotic areas in chronic gingivitis showed no metachromasia (Quintarelli, 1961). Munemato et al. (1970) found that twice as much chondroitin sulfate was present in human periodontal ligament as compared to its hyaluronic acid. No difference in the total glycosaminoglycan content was observed in normal and diseased states due to an apparent increase in chondroitin sulfate masked by a corresponding decrease in hyaluronic acid (Paunio, 1969).

The acidic glycosaminoglycans in the gingiva of patients with periodontal disease were studied by two-dimensional electrophoresis on cellulose acetate before and after digestion with hyaluronidase and chondroitinase ABC. The gingiva of periodontal disease patients contained hyaluronic acid, chondroitin 4-sulfate, chondroitin 6-sulfate, dermatan sulfate, heparan sulfate and heparin (Takao et al., 1977). In another investigation quantitative analysis gave values of 32.3 % hyaluronic acid, 22.7 % chondroitin 4-sulfate, 40.1 % dermatan sulfate, 4.9 % heparan sulfate and 0 % heparin in periodontal disease (Miki et al., 1977). Free gingiva, considered as a measure of periodontal pocket and the degree of bone loss in periodontal disease, contained higher concentrations of hyaluronic acid and lower dermatan sulfate than the attached gingiva present in healthy state (Miki et al., 1977).

Histochemical examination of the distribution of glycosaminoglycans in human flabby gum revealed larger quantities of hyaluronic acid, chondroitin 4-sulfate and chondroitin 6-sulfate than in normal alveolar mucosa (Kuwajima et al., 1977). The susceptibilities to enzyme digestion with testicular and bacterial hyaluronidases, chondroitinases and neuraminidase were used for identification of these macromolecules.

In the gingiva of cattle, chondroitin 4-sulfate, dermatan sulfate, hyaluronic acid and heparan sulfate contents accounted for 34.7, 36.3, 25.8 and 3.8 %, respectively, of the total glycosaminoglycans present. In rabbit the corresponding percentages were 33.2, 32.6, 29.3 and 4.9 %, respectively (Sakamoto, 1977). The aforementioned values are for full-grown animals and show changes at various ages. In the gingiva of two weeks to ten months old cattle, the glycosaminoglycans consisted of 32-40 % chondroitin 4-sulfate, 33-37 % dermatan sulfate, 17-27 % hyaluronic acid and 3-8 % heparan sulfate. During the first three years of age, chondroitin 4-sulfate and dermatan sulfate increased markedly and were unchanged thereafter. The hyaluronic acid content slightly decreased with increasing age. Heparan sulfate content significantly decreased during the first year and thereafter decreased slightly with increasing age (Sakamoto, 1977 a). The rabbit gingiva showed age-dependent decreases in the contents of chondroitin 4-sulfate, dermatan sulfate and hyaluronic acid, whereas heparan sulfate did not change with age.

In dog gingiva, the glycosaminoglycans constituted about 0.3 % of the defatted dry tissue. The isolated material consisted of the same glycosaminoglycans fractions as in the bovine and human gingiva. In dog, dermatan sulfate plus the chondroitin 4-sulfate were 70 % and hyaluronic acid was 30 % of the total glycosaminoglycans; a trace of heparan sulfate was also present (Tawa et al., 1976).

The glycosaminoglycans isolated from porcine gingiva upon fractionation by anion-exchange chromatography, ethanol precipitation and electrophoresis were found to contain 31.1 % chondroitin 4-sulfate, 29.9 % dermatan sulfate, 15.7 % hyaluronic acid, 15.3 % heparan sulfate and 8 % chondroitin 6-sulfate (Hiramatsu et al., 1976). The purified dermatan sulfate (DS), chondroitin 6-sulfate (CS) and hyaluronic acid (HA) were isolated from the chondroitin sulfate-rich fraction of the porcine gingiva as single electrophoretic bands. The molecular weights of these were in the decreasing order as HA > CS > DS (Hiramatsu et al., 1977). The dermatan sulfate so isolated was a hybrid containing iduronic acid as the major component with a small content of glucuronic acid (Hiramatsu et al., 1977 b).

Insulin affects the composition of the ground substance of the gingiva. This was evident from the chemical changes seen in gingiva from rabbits experimentally made diabetic with alloxan or streptozotocin. In tissue culture the gingiva from these rabbits had a markedly enhanced collagenolytic activity and in situ it contained collagen breakdown products not found in gingiva from control rabbits (Golub et al., 1978). From this it appeared that gingival collagenase activity/synthesis was stimulated in diabetes both in vitro and in vivo. This may be responsible for the changes in glycosaminoglycan and collagen observed by other workers, since other lysosomal enzymes (β-glucuronidase, N-acetyl-β-glucosaminidase, β-glycosidase and β-galactosidase) are unchanged in gingiva from alloxan diabetic rabbits (Hiramatsu et al., 1976 a). However, although the salt soluble, acid soluble and insoluble collagen contents were decreased at all stages of the diabetes, the aforementioned mucosaccharases tended to decrease at 3 and 6 weeks (Hiramatsu et al., 1977 a). This fluctuation in the enzyme activities may be responsible for the various reports of decreases (Kuwajima, 1978; Hiramatsu et al., 1977 a) or increases (Hiramatsu et al., 1976 a) and variations of glycosaminoglycans' contents of gingivae at various stages of the diabetes.

In experimental dystrophy produced by experimental damage of the trigeminal nerve of rat and in spontaneous periodontosis, depolymerization of glycosaminoglycans and lowered resistance of collagen against lysosomal enzymes was observed (Bromberg et al., 1971). Histochemical changes resembling those in gingivitis have been experimentally induced in monkeys by injecting testicular hyaluronidase in gingivae (Aisenberg and Aisenberg, 1951). Similarly, experimental gingivitis has been produced in human subjects (Loe et al., 1965).

5.5 Glycosaminoglycans in oral cyst fluid

A cyst is a pathological cavity containing a fluid or semifluid material and is usually (though not necessarily) lined.

Oral cysts consist of mucus-containing spaces lined by connective or granulation tissues. Small cysts arising in connection with the minor salivary glands are very common and many of these may be present anywhere in the oral submucosa. However, the most favorite sites for these are the lower lip, the cheeks, the under surface of the tip of the tongue and the floor of the mouth. The upper lip is rarely affected (Lucas, 1976). Although the exact etiology of these mucous cysts is obscure, duct obstruction and trauma do contribute to these pathological lesions.

Depending upon the tissue involved and the position of the cyst, oral cysts have been classified into: primordial cyst, gingival cyst, lateral cyst, periodon tal cyst, dentigerous cyst, eruption cyst, nasopalatine duct cyst, median cyst, globulomaxilliary cyst, nasolabial cyst, radicular cyst, surgical ciliated cyst, solitary bone cyst and aneury small bone cyst (Shear, 1976). Each of these cyst fluids may be associated with glycosaminoglycans which are responsible for a high viscosity and a high osmotic pressure of these fluids as compared to those of patients' blood sera. Digestion with testicular hyaluronidase reduces the viscosity of the cyst fluid due to depolymerization of the GAG (Skaug and Hofstad, 1972). The origin of GAG in the cyst fluid may vary depending on the nature of the cyst. In those cysts lined by a cyst wall consisting of the granulation or the connective tissues the GAG may be produced by the cells in the cyst wall. In those cysts such as maxillary cysts the GAG may originate by breakdown (osteolysis) of the bony structure (Skaug and Hofstad, 1972). Other workers have previously considered the oral cystic fluid as a plasma dialyzate because of similarities in protein electrophoresis patterns (Toller, 1970) and immunoelectrophoresis patterns (Toller and Holbrow, 1969).

The GAG composition of the fluid collected from five maxillary jaw cysts upon cellulose acetate electrophoresis followed by staining with 0.5 % aqueous Alcian blue solution of pH 4 revealed two electrophoretic bands (Skaug and Hofstad, 1972). The band with faster mobility disappeared upon digestion with testicular hyaluronidase and corresponded to hyaluronic acid. The slow migrating band was hyaluronidase-resistant and was not identified. From our own experience this latter band could be an acid glycoprotein or hyaluronic acid associated with proteins. The two possibilities can be resolved by proteolytic digestion of the cyst fluid with the enzyme pronase prior to electrophoresis. The methods used by Skaug and Hofstad (1972) consisted of preservation of the centrifuged cyst fluid with merthiolate or by freezing at -20 °C, electrophoresis in barbiturate buffer of pH 8.6, ionic strength 0.06 and current 0.5 mA/cm strip width. One ml of the cyst fluid was digested with 0.2 ml of hyaluronidase solution containing 3 mg enzyme/1.2 ml of 0.15 M NaCl by incubation at 37 °C for two hours.

5.6 Relationship between dental plaque, periodontal disease and salivary flow

Periodontal disease affects the supporting structures of the teeth or the "periodontium" which includes the gums, bone, the outer-layer of the root of the tooth (cementum) and fibers that connect the bone and the cementum. Depending upon the type of the periodontal disease, various parts of the periodontium may be involved, e. g., infection of the gums in gingivitis, destruction of supporting bone and fibers in periodontitis or pyorrhea, necrotizing ulcerative gingivitis in trench mouth. All these types of periodontal diseases affect the connective tissue components of the periodontium. The symptoms include bleeding gums, bad breath, looseness of the teeth, pain, decreased chewing ability and poor aesthetics. Periodontitis (pyorrhea) ends up in the loss of teeth and the bone that supports them (alveolar bone loss).

The primary cause of the dental caries and the periodontal disease has been related to the dental plaque accumulation. The dental plaque is a colorless sticky growth consisting of apparently intact micro-organisms embedded in a matrix of polysaccharide and protein (Schroeder and DeBoever, 1970) and clinging to the tooth. The protein forms a network of chains in which different kinds of bacteria are entrapped (Kleinburg, 1970; Cohn and Kramer, 1976). After being removed it reforms within 24-36 hours in the absence of oral hygiene and is therefore present in all people in different degrees. The maintenance of a plaque-free mouth through the use of repeated, good oral hygiene measures is the only way to prevent dental caries and periodontal disease (Loe et al., 1965; Eastoe et al., 1971; Williams and Howell, 1978).

Electron microscopic studies have shown the presence of bacteria adhering to the gingival sulcular epithelium and in widened inter-epithelial spaces in the sulcular epithelium (Freedman et al., 1968). These bacteria elaborate hydrolytic enzymes: proteinases, glycosidases and acid phosphatase. The proteinases bring about degradation of the collagen (Beutner et al., 1966; Jenkins, 1971; Gross, 1974; Barbanell et al., 1976). The hyaluronidase and β-glucuronidase (Schultz-Haudt and Scherp, 1955; Ginwalla et al., 1972) and chondroitinase (Schultz-Haudt and Scherp, 1956; Schultz-Haudt, 1965; Eastoe et al., 1971) bring about degradation of the glycosaminoglycans in the ground substance of the gingiva and other periodontal tissue. These changes are responsible for gingival inflammation (gingivitis) and osseous resorption leading to alveolar bone loss and falling of the tooth.

The proteolytic activity (Söder et al., 1966; Söder, 1967) and hyaluronidase activity have been measured in the dental plaque material by spectrophotometric and viscosity methods (Söder and NOrd, 1969). In eight patients, the supernatent from the homogenate in 0.1 M succinic acid-NaOH buffer, pH 5.9 contained 785-2045 HVh/mg protein in the supernatent using hyaluronic acid as the substrate. The enzyme seemed to be labile upon storage; even one day of storage at -20 °C reduced the activity markedly. By gel filtration on Sephadex G-200, at least two different enzymes with pH optima of 5.5 and 6.5 were recovered (Söder and Nord, 1969).

Because of the implication of dental plaque as the major etiological factor for initiation of dental caries and periodontal disease, the protein-polysaccharide matrix and the bacteria associated with the plaque have been investigated (Schroeder and DeBoever, 1970; Kleinburg, 1970; Critchley, 1976). Analysis of the saline extracts of human dental plaques by UV-spectroscopy, immunoelectrophoresis, gel filtration on Sephadex G-10 and isoelectric focusing revealed that the proteins present are serum-like and these might have originated from the gingival fluid (Levine et al. 1976). The glycosaminoglycans isolated by papain digestion and cetylpyridinium chloride precipitation from the dental plaque material collected from different individuals contained uronic acid, sulfate, hexosamine, hexoses and sialic acid. The amounts of these carbohydrate components and of proteins were, however, less than those in oral calculus and parotid saliva (Paunio and Paunio, 1976).

The presence of mucopolysaccharide-protein matrix in oral calculus was suggested by histochemical tests on subgingival calculus and by chemical and histochemical methods in supragingival calculus (Mandel and Levy, 1957). Glycosaminoglycans have been isolated from oral calculus and its components determined by chemical analysis and electrophoresis (Paunio and Paunio, 1970). The GAG present in supra- and subgingival calculi have been isolated by decalcification in EDTA, solubilization of the organic matrix by autoclaving, proteolytic digestion, ethanol and cetylpyridinium chloride precipitation (Osuoji and Rowless, 1972). The GAG fraction contained 38.4 % hexosamine, 25.6 % hexuronic acid and sulfate equal to 10.92 % of the dry weight of the starting material. By electrophoresis the GAG components were chondroitin sulfate, hyaluronic acid and heparan sulfate. The GAG, particularly chondroitin sulfates, have been suggested to have a role in the mineralization (calcification) of dental calculus as has been postulated for other calcified tissues (Sobel and Burger, 1954; Mandel and Levy, 1957; Bowness, 1968; Eastoe, 1968; Osuoji and Rowless, 1972; Irving, 1978). The extracellular polysaccharides extracted from dental plaques in monkeys and humans are heteroglycans composed of glucose, galactose, mannose, N-acetylhexosamine and hexuronic acid. Also, the cultures of plaque streptococci, grown on a simple medium produce similar heteroglycans (Critchley, 1976).

Human saliva contains a β-glucuronidase inhibitor which is glycoprotein in nature and is composed of sialic acid, hexose, fucose, hexosamine and protein (Sakamoto et al., 1974). The activity of this inhibitor was found to be related to the AB0 blood group and the secretor status but was not related to decayed, missing and filled (DMF) teeth scores. In non-secretors, there was a significant positive correlation between DMF scores and the inhibitor activity. The inhibitor activity measured in 0.1 ml saliva and expressed as percentage inhibition was significantly ($p < 0.005$) greater in non-secretors than in secretors (Sakamoto et al., 1976). The physiological siginificance of the glycoprotein inhibitor and these findings is not clear.

5.7 Age-related changes in polysaccharides of salivary glands and oral mucosa

Significant age-related changes occur in the oral mucosa, which include: loss of elasticity, keratosis tendency, microscopic degeneration of the underlying connective tissue, increased Fordyce spot (sebaceous glands) in cheeks, decreased rate of cementum formation, reduced keratinization of gingiva, increased prevalence of gingivitis and other periodontal diseases, increased susceptibility to atrophy of alveolar bone causing recession of gingiva and subcrevicular inflammation (Tonna, 1977). All these changes are related to the alterations in the connective tissue of the oral cavity. The mucopolysaccharides of the human oral mucosa increased with advancing age due to degenerative alterations of the mucosa (pleomorphic adenomas) and a benign tumor in the vicinity of the parotid gland (Cystadenoma lymphomatosum or Warthin's tumor) in males in the fifth and sixth decades.

There is a diminished salivary secretion with advancing age (Dawes, 1976). Lymphocytic infiltration and obstruction of the gland lumina of the salivary glands have been observed in the aging rats. In human subjects fatty and fibrotic changes are commonly seen in the parotid glands in the advanced ages. Alterations in the polysaccharides of the salivary glands have been noted in experimental animals. In aging guinea pigs the hexose/protein ratio of the glycoprotein from the submaxillary glands was decreased (Burzynski, 1971).

5.8 Bibliography

1. Aisenberg, M. S. and Aisenberg, A. D. (1951): Hyaluronidase in periodontal disease. Oral Surg. 4, 317-320

2. Anderson, D. H. (1962): Pathology of cystic fibrosis. Ann. N. Y. Acad. Sci. 93, 500-517

3. Azzopardi, J. G. and Smith, O. D. (1959): Salivary gland tumors and their mucins. J. Path. Bact. 77, 131-140

4. Baker, B. L., Clapp, H. W. and Light, J. A. (1964): Hormonal influences on the cytology and physiology of salivary glands. In: L. M. Sreebny and J. Meyer (Eds.). Salivary Glands and their secretions. Pergamon Press, Oxford, 63-80.

5. Banerjee, S. D., Cohn, R. H. and Bernfield, M. R. (1976): Basal lamina of embryonic salivary epithelia. Production by the epithelium and role in maintaining lobular morphology. J. Cell. Biol. 73, 445-463

6. Barbanell, R. L., Lian, J. B. and Keith, D. A. (1976): A review of collagen degradation with particular emphasis upon that which occurs during inflammatory periodontal disease. J. West. Soc. Periodont. 24, 43-55

7. Beutner, E. H., Triftshauser, C. and Hasen, S. P. (1966): Collagenase activity of gingival tissue from patients with periodontal disease. Proc. Soc. Exp. Biol. Med. 121, 1082-1085

8. Bowness, J. M. (1968): Present concepts of the role of ground substance in calcification. Clin. Orthop. 59, 233-249

9. Brill, N. (1962): The gingival procket fluid. Studies of its occurrence, composition and effect. Acta Odont. Scand. 20 (Suppl. 32), 1-115

10. Bromberg, E. D. and Musienko, N. A. (1971): Mucopolysaccharides of rat periodontium during experimental neurodystrophy and spontaneous periodontosis. Ter. Ortop. Stomatol. 1, 3-7 (Russ.)

11. Burzynski, N. J. (1971): Aging in the guinea pig salivary glands. J. Geront. 26, 204-207

12. Ciancio, S. G. and Mather, M. L. (1971): Acid mucopolysaccharides in gingivitis and periodontitis. J. Periodont. Res. 6, 188-193

13. Cimasoni, G. (1975): The Crevicular Fluid. Monographs in Oral Science Vol. 3, S. Karger, Basel, 1-122

14. Cohn, R. H., Banerjee, S. D. and Bernfield, M. R. (1976): Basal lamina of embryonic salivary epithelia. Nature of glycosaminoglycan and organization of extracellular materials. J. Cell. Biol. 73, 464-478

15. Cohn, B. and Kramer, I. R. H. (Eds.) (1976): Scientific Foundations of Dentistry. Year Book Medical Publications, Chicago, Chap. 31

16. Critchley, P. (1976): The Chemistry of Carbohydrates. In: B. Cohn and I. R. H. Kramer (Eds.). Scientific Foundations of Dentistry. Year Book Medical Publications, Chicago, Chap. 34, 410-415

17. Dawes, C. (1974): Rhythms in salivary flow rate and composition. Int. J. Chronobiol. 2, 253-279

18. Dawes, C. (1976): Salivary secretion. In: B. Cohen and I. R. H. Kramer (Eds.). Scientific Foundation of Dentistry. Year Book Medical Publications, Chicago, 516-527

19. Dawes, C. (1978): The chemistry and physiology of saliva. In: J. H. Shaw, E. A. Sweeney, C. C. Cappuccino and S. M. Meller (Eds.): Textbook of Oral Biology. W. B. Saunders, Philadelphia, 593-616

20. Diamant, H. (1976): The salivary apparatus. In: R. Hinchcliffe (Ed.). Scientific Foundations of Otolaryngology. Year Book Medical Publications, Chicago, 460-467

21. Dische, Z. (1964): Structural factors in mucous glycoproteins related to cystic fibrosis. In. P. A. Di Sant'Agnese (Ed.). Research on Pathogenesis of Cystic Fibrosis. Natl. Inst. Arthr. Metab. Dis. NIH, Bethesda, 229-242

22. Dische, Z., Pallavicini, C., Kawasaki, H., Smirnow, N., Cizek, L. J. and Chien, S. (1962): Influence of the nature of the secretary stimulus on the composition of the carbohydrate moiety of glycoproteins of the submaxillary saliva. Arch. Biochem. Biophys. 97, 459-469

23. Eastoe, J. E. (1968): Chemical aspects of the matrix concept in calcified tissue organization. Calc. Tiss. Res. 2, 1-19

24. Eastoe, J. E., Picton, D. C. A. and Alexander, A. G. (Eds.) (1971): The Prevention of Periodontal Disaese, Kimpton, London

25. Engel, M. B., Joseph, N. R., Laskin, D. M. and Catchpole, H. R. (1960): A theory of connective tissue behavior: Its implications in periodontal disease. Ann. N. Y. Acad. Sci. 85, 399-420

26. Farb, S. N. (1970): Otolaryngology. Medical Exam. Publ. Co., Flushing, Chap. 10, 245-256

27. Freedman, H., Listgarten, M. and Taichman, N. (1968): IADR Abstr., 166

28. Ginwalla, T. M. S., Nayak, R. P., D'Silva, I. R. and Pradhan, A. G. (1972): Beta-glucuronidase and periodontal destruction. J. Dent. Res. 51, 345-348

29. Golub, L. M., Schneir, M. and Ramamurthy, N. S. (1978): Enhanced Collagenase activity in diabetic rat gingiva: In vitro and in vivo evidence. J. Dent. Res. 57, 520-525

30. Greeves, A. (1977): Width of attached gingiva and vestibular depth in relation to gingival health. Thesis University of Amsterdam, Netherlands. Reviewed in: Netherland Dent. J. 85 (1978), 54-55

31. Gross, J. (1974): Collagen biology: Structure, degradation and disease. Harvey Lect. 68, 351-432

32. Groza, P., Ionescu, S., Corneanu, M., Buzoianu, V. and Busneag, C. (1966): Changes in parotid saliva under the influence of hydrocortisone. Stud. Cercet. Fiziol. 11, 5-8 (Rum)

33. Hara, K. and Loe, H. (1969): Carbohydrates components of the gingival exudate. J. Periodontal Res. 4, 202-207

34. Hensten-Pettersen, A. (1976): Histochemical and biochemical properties of the labial and palatine glands and their secretions in the Macaca irus and Cercopithecus aethiops monkeys. Acta Odontal. Scand. 34, 79-88

35. Hiramatsu, M., Abe, I. and Minami, M. (1976): Acid mucopolysaccharides in porcine gingival tissue. Josai Shika Daigaku Kiyo 5, 75-84 (Japan)

36. Hiramatsu, M., Abe, I. and Minami, M. (1976 a): Acid mucopolysaccharides and collagen contents in gingival tissue of diabetic rabbits. Josai Shika Daigaku Kiyo 5, 85-92 (Japan)

37. Hiramatsu, M., Hatakeyama, K., Abe, I., Atsumi, T. and Minami, M. (1977): Acid mucopolysaccharides in porcine gingival tissue. Josai Shika Daigaku Kiyo 6, 109-115 (Japan)

38. Hiramatsu, M., Hatakeyama, K., Abe, I. and Minami, N. (1977 a): Changes of acid mucopolysaccharides and collagen contents and some gingiva lysosomal enzyme activities of alloxan diabetic rabbits. Nippon Shishubyo. Gakkai Kaishi 19, 358-363 (Japan)

39. Hiramatsu, M., Hatakeyama, K. and Minami, N. (1977 b): Glycosaminoglycans in porcine gingiva. Ketsugo Soshiki 9, 13-24 (Japan)

40. Hopps, R. M. (1978): Salivary hyaluronidase changes and oral changes during the menstrual cycle. Acta Med. Rom. 16, 371-376

41. Hosoi, K., Kobayashi, S., Hiramatsu, M., Minami, N. and Ueha, T. (1979): Androgenic regulation of N-acetyl-β-glucosaminidase activity in the submandibular glands of mice. J. Biochem. (Tokyo) 85, 1483-1488

42. Hosoi, K., Kobayashi, S. and Ueha, T. (1978): Influence of androgens on β-glucuronidase activity in liver, kidney and submandibular salivary gland of the mouse. Arch. Oral Biol. 23, 905-909

43. Ines Masi, F. P. and Szymczyk, T. (1966): Glucose metabolism in vitro in rat submaxillary glands; the effect of insulin. Ann. Ist. Super Sanita 2, 287-295 (Ital.)

44. Irving, J. T. (1978): Theories of mineralization of bone and teeth. In: J. H. Shaw, E. A. Sweeney, C. C. Cappucino and S. M. Meller (Eds.). Text Book of Oral Biology, W. B. Saunders, Philadelphia, Chap. 14, 470-483

45. Jenkins, G. N. (1971): Plaque formation and metabolism. In: J. E. Eastoe, D.C.A. Picton and A. G. Alexander (Eds.). The Prevention of Periodontal Disease. Kimpton, London, 34-45

46. Kim, Y. B. (1978): Acid mucopolysaccharides of the salivary glands in normal and alloxan diabetic rabbits. Shika Igaku 41, 627-643 (Japan)

47. Kleinburg, I. (1970): Biochemistry of the dental plaque. Adv. Oral Biol. 4, 43-90

48. Kofoed, J. A., Curbelo, H. M. and Houssay, A. B. (1969): Acid glycosaminoglycans and sialic content in salivary glands of rats. J. Dent. Res. 48, 555-558

49. Kofoed, J. A., Houssay, A. B., Curbelo, H. M. and Tocci, A. A. (1970): Effect of isoproterenol on the glycosaminoglycans in the salivary glands of the rat. Arch. Oral. Biol. 15, 335-341

50. Kofoed, J. A., Houssay, A. B., Curbelo, H. M. and Tocci, A. A. (1971): Effects of incisor amputations upon the glycosaminoglycans in salivary glands in the rat. Arch. Oral. Biol. 16, 87-93

51. Kofoed, J. A., Houssay, A. B., Curbelo, H. M., Tocci, A. A., Gamper, H. G. and Barcelo, A. C. (1973): Influence of the thyroid upon the glycosaminoglycans in the salivary glands in the rat. Arch. Oral. Biol. 18, 481-485

52. Kofoed, J. A., Houssay, A. B., Curbelo, H. M., Tocci, A. A. and Gamper, C. H. (1973 a): Effects of glucocorticoids upon the glycosaminoglycans in salivary glands in the rat. Arch. Oral. Biol. 18, 681-688

53. Kofoed, J. A., Houssay, A. B., Tocci, A. A., Curbelo, H. M. and Barcelo, C. (1977): Effects of acetoxyprogesterone upon glycosaminoglycans in the uterus and salivary glands of rats. Horm. Metab. Res. 9, 85-89

54. Kofoed, J. A., Houssau, A. B., Tocci, A. A., Curbello, H. M. and Gamper, C. H. (1971 a): Effects of testosterone on glycosaminoglycans in the prostate, seminal vesicles and salivary glands of the rat. J. Endocrin. 51, 465-471

55. Kuwajima, S. (1978): Experimental study on changes of acid mucopolysaccharides in the gingiva of alloxan diabetic rabbits. Shika Igaku 41, 267-286 (Japan)

56. Kuwajima, S., Matsumoto, M., Miki, M., Ohkusa, W., Kamemizu, T., Taguchi, T., and Nishiura, M. (1977): Histochemical studies on acid mucopolysaccharides of human flabby gum. Shika Igaku 40, 279-285 (Japan)

57. Levine, M., Beeley, J. A. and Beeley, J. G. (1976): Comparative biochemical studies on the chemical composition of saline extracts of human dental plaque, saliva and serum. Arch. Oral. Biol. 21, 741-747

58. Loe, H. and Theilade, E. and Jensen, S. B. (1965): Experimental gingivitis in man. J. Periodont. 36, 177-187

59. Lovell, D., Briggs, J. C. and Schorah, C. J. (1966): Chemical analysis of acid mucopolysaccharides of mixed salivary tumors. Brit. J. Cancer 20, 463-468

60. Lucas, R. B. (1976): Pathology of Tumors of the Oral Tissues, 3rd Ed., Churchill Livingstone, New York, Chap VII (salivary gland tumors). 298-350, Chap. VIII (cysts of the oral tissue), 352-387

61. Mancuso, S., Bellante, F. P., Marana, R. and Angelini, A. (1978): Chemical changes in the saliva during the menstrual cycle. Acta Med. Rom. 16, 387-403

62. Mandel, I. D. (1974): Relation of saliva and plaque to caries. J. Dent. Res. 53, 246-266

63. Mandel, I. D. and Levy, B. M. (1957): Studies on salivary calculus. I. Histochemical and chemical investigations of supra and sub-gingival calculus. Oral. Surg. 10, 874-884

64. Mandel, I. D., Thompson, R. H., Wotman, S., Taubman, M., Kutscher, A. H., Zegarelli, E. V., Denning, C. R., Botwick, J. T. and Fahn, B. S. (1965): Parotid saliva in cystic fibrosis. Amer. J. Dis. Child. 110, 646-651

65. Mangos, J. A. and Talamo, R. C. (Eds.). (1973): Fundamental Problems of Cystic Fibrosis and Related Diseases. Year Book Medical Publishers, Chicago, 1-424

66. Mangos, J. A. and Talamo, R. C. (Eds.) (1976): Cystic Fibrosis: Projections into the Future. Year Book Medical Publishers, Chicago, 1-400

67. Mason, D. K. and Chisholm, D. M. (1976): Diseases of the salivary glands. In: B. Cohen and I. R. H. Kramer (Eds.). Scientific Foundation of Dentistry. Year Book Medical Publications, Chicago, Chap. 44, 528-539

68. McCoombe, G., Knox, K. W. and Sullivan, H. R. (1961): Carbohydrate constitutents of human saliva. Arch. Oral. Biol. 3, 171-175

69. Miki, T., Katsutani, Y., Kawai, J., Iwayama, Y., Nomura, Y., Ishida, H., Yamagami, H. and Kaya, H. (1977): Studies on gingival connective tissue components of patients with periodontal diseases. 4. Quantitative determination of acid mucopolysaccharides. Nippon Shishubyo Gakkai Kaishi 19, 364-368 (Japan)

70. Mogil'naia, G. M. (1966): On the histochemistry of human salivary glands. Stomatologia (Moskva) 45, 17-21 (Russ)

71. Muhler, J. C., Bixler, D. and Shafer, W. G. (1956): Effect of replacement therapy on dental caries experience of radiothyroidectomized rats. Proc. Soc. Exp. Biol. N. Y. 93, 328-330

72. Munemato, K., Iwayama, Y., Yoshida, M., Seva, M., Ano, M. and Yokomizo, I. (1970): Isolation and characterization of acid mucopolysaccharides of bovine periodontal membrane. Arch. Oral. Biol. 369-38

73. Musy, J. P. and Gotzos, V. (1973): Histochemical study in recurrent mixed tumor of the parotid gland. Ann. Anat. Pathol. 18, 347-356 (Fr)

74. Nadler, H. L., Rao, G. J. S. and Taussig, L. M. (1978): Cystic fibrosis. In: J. B. Stanbury, J. B. Wyngaarden and D. S. Fredrickson (Eds.). The Metabolic Basis of Inherited Disease, 4th Ed., McGrawHill, New York, Chap. 73, 1683-1710

75. Nagura, Y. (1976): Acid mucopolysaccharides in the salivary glands of the pig. Shika Igaku 39, 461-474 (Japan)

76. Newburn, E. (1978): Cariology, Williams and Wilkins, Baltimore, Chap. 2, 15-43

77. Osuoji, C. I. and Rowless, S. L. (1972): Isolation and identification of acid glycosaminoglycans in oral calculus. Arch. Oral. Biol. 17, 211-214

78. Paunio, K. (1969): Connective tissue, biochemical studies of disease in man. Suom. Hamaslaak. Toim. 65, 279-290

79. Paunio, K. (1971): On the hydroxyproline-containing components in the gingival exudate. J. Periodont. Res. 6, 115-117

80. Paunio, K. U. and Paunio, I. K. (1970): Glycosaminoglycan-like components in oral calculus, parotid saliva and dental plaque. A comparative study. Scand. J. Dent. Res. 78, 447-451

81. Pigman, W. (1977): Submandibular and sublingual glycoproteins. In: M. I. Horowitz and W. Pigman (Eds.). The Glycoconjugates, Vol. I., Academic Press, New York, 137-152

82. Quintarelli, G. (1960): Histochemistry of the gingiva. III. The distribution of amino-peptidase in normal and inflammatory conditions. Arch. Oral. Biol. 2, 271-276

83. Quintarelli, G. (1960 a): Histochemistry of gingiva. IV. Preliminary investigations of the mucopolysaccharides of connective tissue. Arch. Oral Biol. 2, 277-284

84. Quintarelli, G. (1961): Histochemical studies on human mucus-secreting salivary glands. Acta Histochem. 12, 1-11

85. Raynaud, J. (1964): The action of the thyroid gland and adrenal glands on the submaxillary gland of mice. In: L. M. Sreebny and J. Meyer (Eds.). Salivary glands and Their Secretions, Pergamon Press, Oxford, 47-62

86. Rybakova, M. G. (1975): Effect of estrogens on the growth and differentiation of the salivary glands of rats in organ cultures. Biull. Eksp. Biol. Med. 79, 97-100 (Russ)

87. Sakamoto, N. (1977): Biochemical studies on acidic glycosaminoglycans in gingiva and dental pulp. 1. A microquantitative determination of acidic glycosaminoglycans in gingiva and dental pulp. Hiroshima Daigaku Zasshi 9, 169-177 (Japan)

88. Sakamoto, N. (1977 a): Biochemical studies on acidic glycosaminoglycans in gingiva and dental pulp. 2. Changes in acidic glycosaminoglycans in gingiva and dental pulp with aging. Hiroshima Daigaku Shigaku Zasshi 9, 178-186 (Japan)

89. Sakamoto, W., Nishikaze, O. and Sugimara, T. (1974): Isolation of β-glucuronidase inhibitor from human saliva. Arch. Oral Biol. 19, 703-707

90. Sakamoto, W., Nishikaze, O., Sugimara, T. and Ishibashi, H. (1976): β-glucuronidase inhibitor activity in human saliva and its relation to dental caries, secretor status, and blood group. Arch. Oral Biol. 21, 761-763

91. Schroeder, H. E. and DeBoever, J. (1970): The structure of microbial dental plaque. In: W. D. McHugh (Ed.). Dental Plaque, Livingstone, Edinburgh, 49-72

92. Schultz-Haudt, S. D. (1957): Separation of acid mucopolysaccharides of human gingiva. Acta Chem. Scand. 11, 1070-1071

93. Schultz-Haudt, S. D. (1958): Observations on the acid mucopolysaccharides of human gingiva. Odontol. Tidskr. 66, 1-98

94. Schultz-Haudt, S. D. (1965): Connective tissue and periodontal disease Int. Rev. Conenct. Tissue Res. 3, 77-89

95. Schultz-Haudt, S. D., From, S. H. and Nordbo, H. (1964): Histochemical staining properties of isolated polysaccharide components of human gingiva. Arch. Oral. Biol. 9, 17-26

96. Schultz-Haudt, S. D. and Lundquist, C. (1962): The biochemistry of periodontal disease. Intern. Dental J. 12, 180-193

97. Schultz-Haudt, S. D., Paus, S. and Assex, S. (1961): Periodic acid-schiff reactive components of human gingiva. J. Dent. Res. 40, 141-147

98. Schultz-Haudt, S. D. and Scherp, H. W. (1955): Production of hyaluronidase and β-glucuronidase by viridans Streptococci isolated from gingival crevices. J. Dent. Res. 34, 924-929

99. Schultz-Haudt, S. D. and Scherp, H. W. (1956): The production of chondrosulfates by micro-organisms isolated from human gingival crevices. J. Dent. Res. 35, 299-307

100. Shackleford, J. M. and Bentley, Jr., H. P. (1964): Changes in carbohydrate histochemistry of salivary glands and pancreas in cystic fibrosis. In: P. A. Di Sant'Agnese (Ed.). Research on Pathogenesis of Cystic Fibrosis of the Pancreas. Natl. Inst. Arthr. Metab. Dis. Bethesda, 51-56

101. Shackleford, J. M. and Bentley, Jr. H. P. (1964 a): Carbohydrate histochemistry of the salivary glands and pancreas in cystic fibrosis. J. Histochem. Cytochem. 12, 512-521

102. Shannon, I. L. (1972): The biochemistry of human saliva in health and disease. In: N. H. Rowe (Ed.). Salivary Glands and Their Secretion. University of Michigan, Ann Arbor, 92

103. Shear, M. (1976): Cysts. In: B. Cohen and I. R. H. Kramer (Eds.). Scientific Foundations of Dentistry, Year Book Medical Publications Chicago, Chap. 21, 251-266

104. Shiozaki, M. (1973): Acid mucopolysaccharides in the salivary glands of rat. J. Osaka Dent. Univ. 7, 85-96

105. Shklar, G. (1966): The effects of aging upon oral mucosa. J. Invest. Dermatol. 47, 115-120

106. Shubnikova, E. A. and Chunaeva, M. Z. (1966): Histochemical study of the parotid glands during their development in rats. Arkh. Anat. Gisol. i. Embriol. 51, 18-23 (Russ)

107. Siervo, R., Galluzzo, F. and Ravetto, G. (1965): Mucopolysaccharides staining with the Alcian blue-Alcian yellow method in the salivary glands. Riv. Ital. Stomat. 20, 630-641 (Ital)

108. Skaug, N. and Hofstad, T. (1972): Demonstration of glycosaminoglycans in fluid from jaw cysts. Acta Pathol. Microbiol. Scand. Section A. 80, 285-286

109. Sobel, A. E. and Burger, M. (1954): Investigation of the role of chondroitin sulfates in the calcifying mechanism. Proc. Soc. Exp. Biol. (N. Y.) 87, 7-13

110. Söder, P. O. (1967): Proteolytic activity of dental plaque material. Part II. Action on gelatin of proteolytic enzymes from dental plaque material Odontol. Tidskr. 75, 50-61

111. Söder, P. O., Lundblad, G. and Lindqvist, L. (1966): Proteolytic activity of dental plaque material. Part V. Purification of proteinases from dental plaque material by gel filtration through Sephadex columns. Acta. Chem. Scand. 20, 1504-1512

112. Söder, P. O. and Nord, C. E. (1969): Determination of hyaluronidase activity in dental plaque material. J. Periodont. Res. 4, 208-214

113. Spicer, S. S. and Duvenci, J. (1964): Histochemical characteristics of mucopolysaccharides in salivary and exorbital lacrimal glands. Anat. Rec. 149, 333-357

114. Sueda, T., Cimasoni, G. and Held, A. J. (1966): Histochemical study of human gingival fluid. Paradontologie 20, 141-147

115. Sweeney, E. A. (1978): Salivary gland abnormalities. In: J. H. Shaw, E. A. Sweeney, C. C. Cappuccino and S. M. Meller (Eds.). Textbook of Oral Biology, W. B. Saunders, Philadelphia, Chap. 27, 897-915

116. Szymczyk, T. (1971): Metabolism of glucose in the sublingual gland of the rat, and the effect of insulin in vitro. Acta Biochim. Pol. 18, 171-176

117. Takao, M., Katsutani, Y., Iwayama, Y., Nomura, Y., Ishida, H., Fukuhara, H., Kaye, H. Yokomizo, I., Adachi, K. (1977): Studies on gingival connective tissue components of patients with periodontal disease. 3. Qualitative determination of acidic glycosaminoglycans. Nippon Shishubyo Gakkai Kaishi 19, 13-17 (Japan)

118. Tawa, T., Honda, T., Fujita, A., Miyai, Y., Kim, Y. B., Chung, T. Y., Kuwajima, S., Matsumoto, M. and Miki, M. (1976): Studies on acid mucopolysaccharide in the gingiva. II. The acid mucopolysaccharide components of normal dog gingiva. Shika Igaku 39, 799-811 (Japan)

119. Toller, P. A. (1970): Protein substances in odontogenic cyst fluids. Brit. Dent. J. 128, 317-322

120. Toller, P. A. and Holbrow, E. J. (1969): Immunoglobulins and immunoglobulin-containing cells in cysts of the jaws. Lancet 2, 178-181

121. Tomanczyk, W. and Pankiewicz, Z. (1972): The effects of certain factors on the changes of glycosaminoglucuronoglycans in the organism with reference to the hard tissues of the teeth. Czas. Stomatol. 25, 121-126

122. Tonna, E. A. (1977): Aging of skeletal-dental systems and supporting tissues. In: C. E. Finch and L. Hayflick (Eds.). Handbook of the Biology of Aging, Van Nostrand Reinhold, New York, 470-495

123. Werner, I. and Odin, L. (1965): Glycoprotein in salivary glands, saliva and sputum. In: E. A. Balazs and R. W. Jeanloz (Eds.). The Amino Sugars, Vol. II A, Academic Press, New York, Chap. 23, 141-148

124. Williams, R. C. and Howell, T. H.: Definition, epidemiology and etiology of periodontal disease. In: J. H. Shaw, E. A. Sweeney, C. C. Cappuccino and S. M. Meller (Eds.). Textbook of Oral Biology, W. B. Saunders, 1019-1031

125. Winzler, R. J., Devor, A. W., Mehl, J. W. and Smyth, I. M. (1948): Studies on mucoproteins of human plasma: determination and isolation. J. Clin. Invest. 27, 609-616

126. Wotman, S. and Mandel, I. D. (1976): The salivary secretions in health and disease. In: R. M. Rankow and I. M. Polayes (Eds.). Diseases of the Salivary Glands. W. B. Saunders Company, Philadelphia, Chap. 3, 32-53

Chapter 6

Mucopolysaccharides in Fluids and Fluid-Associated Membranes of Respiratory Tract

6.1 Introduction

The mucopolysaccharides are associated with the mucus secreted throughout the respiratory tract from the posterior part of the nose down to the pre-terminal bronchioles. For the sake of convenience we can divide this into nasal mucus and the bronchial mucus. Sputum, as we will discuss later, is a mixture of several secretions with bronchial secretion predominating in it. For this reason, sputum has been considered a representation of tracheo-bronchial mucus. Based on the fact that it would reflect the physicochemical changes in the bronchial secretions, sputum from patients with obstructive pulmonary conditions has been used for rheological and chemical structural studies. We will discuss these studies in nasal secretion, bronchial secretion and sputum. All these studies so far reported have been on neutral and sulfated glycoproteins, although the mucosae that elaborate these secretions do contain acidic glycosaminoglycans. Recently, the presence of the latter macromolecules has been reported by use of better methods of analysis made available. We have surveyed the current knowledge of both the mucus glycoproteins and the acidic glycosaminoglycans of the mucosa of the respiratory tract in the next sections. The pleural fluids present in the pathological state do contain acidic glycosaminoglycans and these studies have been described in Chapter 13 (section 13.8.3).

6.2 Mucus production and mucociliary function

Mucus is produced in the respiratory tract from the posterior two-third part of the nasal passage down to the bronchioles. It is derived from goblet cells on the surface of epithelium and the mixed bronchial glands beneath the epithelial basement membrane (Spicer et al., 1973). Direct transudation from serum also contributes to some glycoprotein components in mucus. This transudation is very small normally but may increase in pathological states. The goblet cells produce primarily the acidic glycoproteins. The mixed bronchial glands are complex submucosal glands and contain two types of cells: mucous and seromucoid cells. The mucous cells produce a whole spectrum of neutral to very acidic glycoproteins, while the seromucous cells produce serum type glycoproteins like those appearing in mucus as transudates of serum. The complex submucosal glands outnumber goblet cells 40 : 1 and the former may be the major source of mucus.

Besides the mucus cells the tracheobronchial tree also contains ciliary cells, which by rhythmic beating of their 200 hair-like 5-6 μm long projections (cilia) on each cell, transport the mucus to the top of the trachea into the esophagus where it is swallowed. The mucus in the nasal passage is likewise pushed down into the pharynx and is swallowed. The bronchial glands and the ciliated cells in the larger bronchi are present in a ratio of 1 : 5 per square millimeter. The terminal bronchioles have only a lining of a single layer of epithelial cells and lack goblet cells and bronchial glands. The mucus and ciliary cells are formed by differentiation of the basal cells in the respiratory epithelium. The surface fluid of the mucous is 5 μm thick and consists of a *sol* phase, probably a plasma transudate covered by a 1-2 μm thick visco-elastic layer of mucus gel which prevents the dehydration of the lower sol phase (Kilburn, 1968). The cilia beat rhythmically in the lower sol phase with their tips touching the upper viscous layer. It has been suggested that the serous sol phase is present throughout the lumen of the tracheo-bronchial pathway but the visco-elastic layer of the mucus may be present only at certain spots as patches (Irvani and van As, 1972).

The extrusion and upward transport of the mucus is at a rate of 1-3.5 cm/min in the larger bronchi and is only 1/10th of it in the small bronchi. This muco-ciliary system protects the epithelium of the respiratory system from harmful pollutants, noxious aerosols and airborne bacteria. Particles larger than 5 μm are retained by filtration in the nasal passages. Those smaller than 5 μm enter the tracheo-bronchial system; of < 3 μm size are engulfed by the phagocytes in the alveoli and others are handled by the mucociliary apparatus and cleared. The mucus also keeps the epithelium moist for its proper function (Dulfano, 1973) and contains immunoglobulins for defense against infection (Tomasi and Bienenstoc, 1968). Despite these defensive mechanisms, metaphasia of the mucous cell (Reid, 1963) in the respiratory bronchioles occurs in chronic bronchitis and the mucus epithelium becomes prone to infection with bacteria and virus due to a loss in the normal sterility of the bronchial passages by inhalation of cigarette smoke, alcohol and noxious aerosols.

In the normal state the rate of formation of mucus in the respiratory tract matches the rate of its elimination into the gastrointestinal tract so that only a small amount of mucus is present in the respiratory tract. In chronic bronchitis and cystic fibrosis far larger amount of mucus is produced than is removed so that an accumulation of this occurs in the tracheobronchial system. These patients eliminate 150 ml or more of sputum per day.

6.3 Mucopolysaccharides in nasal secretion

The nasal mucous membrane is covered by a protective blanket of mucus. This mucus is produced by mucous glands located in the posterior two-thirds of the nose. For physicochemical and structural studies human nasal mucus has been obtained by flushing the nose with isotonic saline (Alford, 1968). The presence of acidic glycosaminoglycans in the paranasal and nasal mucous membranes has been demonstrated in the biopsies taken during otolaryngological work (Michaels, 1976) and also in the peripheral smell organs of mammals and fishes (Herberhold, 1968).

The nasal secretion can be separated into the *sol* and *gel* types of mucus by using technics like centrifugation of saline flushings of the nose. The two types of the mucus have been reported to differ in carbohydrate composition (Odin, 1958; Werner, 1953). The *sol* type nasal mucus contained 11.9 % nitrogen, 6,8 % hexosamine, 4.2 % galactose, 2.0 % fucose and 3.1 % sialic acid. On the other hand, the *gel* type of the nasal mucus contained 11.6 % nitrogen, 6.9 % hexosamine, 4.0 % galactose, 2.2 % fucose and 3.3 % sialic acid. Both types contained traces of mannose (Werner, 1953).

The nonpurulent nasal mucus from patients with rhinitis contained the same carbohydrate residues. Paper chromatography detected the presence of hexosamine, galactose, fucose, sialic acid and a trace of mannose. The molar ratios of the major carbohydrate components were fucose/hexosamine = 0.3, hexose/hexosamine = 0.8 and sialic acid/fucose = 0.8 (Werner, 1953). In atrophic rhinitis, there are changes in the connective tissue of the capillaries and blood vessels of the nasal mucosa. In clinical conditions like atrophic rhinitis, estrogens have been found to reverse these changes (Soskin, 1939) and Bernheimer, 1939).

In normal adults the sialic acid content of the nasal secretion was 43 mg/dl. In the watery nasal secretion of patients with acute rhinitis it decreased to 20 mg/dl but was raised to 70 mg/dl in purulent secretion and to 90 mg/dl in slimy secretion (Kawamoto, 1969). The sialomucin has been shown to protect the nasal epithelium by adsorbing the myxovirus and its hemagglutination inhibiton activity is directly related to its sialic acid content (Mizukoshi and Saito, 1974).

6.4 Factors affecting nature and quantity of nasal secretion

The amount of nasal secretion increases with viral infection of the nasal mucous membrane. Depending upon the nature of infection, the secretion may be watery in acute rhinitis or it may be a purulent or a slimy nasal secretion. The differences in the physical properties of these nasal secretions are due to the contents of sialomucin mentioned above.

The nasal secretion is increased in conditions involving inflammation of the nasal mucous membrane and paranasal sinus membranes, such as respiratory allergy, antrochoanal polyps, exposure to noxious agents. Some drugs used locally for nasal disorders (decongestants like corticosteroids, naphazoline, ephedrine), those applied to the nasal mucosa for their systemic effects (like snuff and pituitary snuff) and those taken orally for disorders somewhere in the body (antihypertensive and anticonvulsant drugs) have been reported to have rhinotoxicity (Capel, 1976). The decongestants and topical corticosteroids (hydrocortisone and dexamethasone) decrease the nasal secretion by their vasoconstriction action on the capillaries, arterioles and venous sinuses in the nasal mucosa. The contraceptive pill, the alphaadrenergic blocking agents like guanethidine and bretylium tosylate and the antihypertensives such as reserpine and methyl DOPA increase nasal secretion and produce nasal stuffiness.

6.5 Nasal secretion in cystic fibrosis

In this generalized abnormality of the exocrine glands, the upper respiratory tract is also affected. By histochemical technics the mucous glands appear to be actively secreting an abundance of mucus, which may distend the glands and ducts (Anderson, 1962). Chronic sinusitis is a usual finding and is associated with hypersecretion of the nasopharyngeal mucous glands (Pennington, 1956) and in addtion to hyperactive mucus secreting cells the mucous membrane shows hypertrophy and edema (Gharib et al., 1964). Purulent sinusitis is an unusual finding in cystic fibrosis (Anderson, 1962). The histological changes in nasal mucous membrane are similar to those in nasal sinuses and respiratory allergy may coexist. Some (5-20 %) of the children of over two years of age suffer from recurrent nasal polyps (Shwachman et al., 1962; Taylor et al., 1974),. which are presumably the result of chronic rhinitis and sinusitis in these patients (Pennington, 1956). Both allergic rhinitis and bronchial asthma may be present in cystic fibrosis patients.

The incorporation of ^{3}H-labeled 6-D-glucosamine and $^{35}SO_4$ into glycoproteins was studied in cultured nasal polyp epithelium removed from patients with allergic rhinitis and cystic fibrosis (Boat et al., 1974). Similar patterns of secretions of lysozyme, sialic acid and fucose were observed in patients and controls. However, the partially purified mucus glycoproteins had higher $^{35}SO_4$ contents in patients. The higher content of sulfated polysaccharides may be related to the rheological properties of mucus from cystic fibrosis patients.

6.6 Glycosaminoglycans of nasal and paranasal mucous membranes

The acidic glycosaminoglycans present in the intercellular matrix of the connective tissue are altered during the course of inflammation (Delaunay and Bazin, 1964). The acidic glycosaminoglycans of human and animal nasal mucosa have been determined (Jackson and Arihood, 1971). The nasal mucous membrane and the paranasal sinuses mucous membranes are inflammed in conditions like chronic rhinitis, nasal polyposis and sinusitis. These inflammatory states are accompanied by histochemical (Michaels, 1976) and chemical changes of acidic glycosaminoglycans in these mucous membranes. From the nature of the changes in glycosaminoglycans, Saito and Nakamura (1977) have suggested that the edematous and suppurative types of chronic sinusitis and nasal polyps are related to the relative early stage of inflammation, while the fibrous type of sinusitis is related to the late stage.

Nasal polyps or polypoid swellings of the nasal mucosa are produced in several benign and malignant conditions of the nose. The benign nasal polyps are present in allergy (Michaels, 1976), may be present in asthma patients (Jones, 1976) and are seen in 5-20 % of the cystic fibrosis patients (Shwachman et al., 1962; Taylor et al., 1974). The polypoid swelling is due to edema of the connective tissue and dilatations of the lymphatic vessels. Antrochoanal polyps show chronic inflammation. Histochemically, intense hyperplasia of goblet cells, hyperactive mucinous glands and marked thickening of the epithelial basement membrane are noted. The acidic glycosaminoglycans (AGAG) of the nasal mucous membrane are increased to over twice the values in normal mucous membrane.

The total amount of AGAG measured as glucuronic acid in human nasal polyps extracted with NaCl reported as mean ± S.E. was 0.29 ± 0.03 % (Jackson and Arihood, 1971) and in acetone-defatted, pronase-digested material it was 0.32 ± 0.04 % dry weight. Other workers (Sato et al., 1977) have reported a value of 4.06 mg uronic acid per gm dry weight. When specific AGAG components were compared, chondroitin sulfate was the major component with small amounts of heparin (Jackson and Arihood, 1971).

Saito and Nakamura (1977) could not detect heparin in nasal polyps but identified the presence of hyaluronic acid, dermatan sulfate and chondroitin 4- and 6-sulfate. Significant increase in total AGAG occurred in nasal polyps which was actually a resultant of significant increase ($P < 0.001$) in chondroitin 4- and 6-sulfate to 26 ± 2.4 % with a concomitant significant ($P < 0.005$) decrease of dermatan sulfate to 28 ± 2.8 %. Although the change in hyaluronic acid level in nasal polyp (47 ± 4.3 %) was not significant, the ratio of hyaluronic acid to dermatan sulfate was increased from 1.07 ± 0.07 S.E. to 1.64 ± 0.13 due to the aforementioned decrease in dermatan sulfate (Saito and Nakamura, 1977). Changes in collagen content, activities of some mucopolysaccharidases and other lysosomal enzymes have also been reported in nasal polyps (Sato et al., 1977).

6.7 Acidic glycosaminoglycans in paranasal sinus mucosa

The acidic glycosaminoglycans in the ground substance of the mucous membrane lining the maxillary sinus have been investigated in detail in both health and disease (Sato, 1966; Saito and Nakamura, 1976, 1977; Sato et al., 1977, 1977 a). The normal maxillary sinus mucosa contained 0.15 ± 0.03 (S.E.) % total glucuronic acid per dry weight which was distributed as 7 ± 2.5 % chondroitin 4- and 6-sulfate, 45 ± 2.3 % dermatan sulfate and 48 ± 4.3 % hyaluronic acid (Saito and Nakamura, 1977). In chronic inflammation of the paranasal sinal mucosa there is an increase in the AGAG content (Sato, 1966; Saito and Nakamura, 1976; Sato et al., 1977). The total amounts of chondroitin and dermatan sulfates were increased several fold in chronic sinusitis of edematous, suppurative types and also these components showed differential changes in the three types of sinusitis (Mizukoshi and Saito, 1974). This differential variation in the three chondroitin sulfate isomers was found by electrophoresis, enzyme analysis and $^{35}SO_4$ incorporation studies to be due to differences in the activities of the enzymes involved in the synthesis of *active sulfate*, 3-phospho-adenosine-5'-phophosulfate (PAPS) (Sato, 1966). Whereas the chondroitin sulfotransferase, involved in the transfer of sulfate from PAPS to chondroitin, was not significantly different in normal and pathological maxillary sinus membranes, the PAPS-synthesizing enzyme activity was increased in the three types of sinusitis in the order of suppurative > fibrous > edematous sinusitis as compared to that in the normal. These differences correlated with the changes in the chondroitin sulfate isomers in the three types of sinusitis (Sato, 1966; Mizukoshi and Saito, 1974).

A later investigation (Sato et al., 1977 a) reported an increased amount of glucosamine-containing glycosaminoglycans (heparan sulfate and hyaluronic acid) in the paranasal mucous membranes from patients with chronic sinusitis. In these patients the acidic glycosaminoglycans consisted of 53.3 % heparan sulfate, 21 % hyaluronic acid, 10.9 % dermatan sulfate, 8.1 % chondroitin 6-sulfate and 2.9 % chondroitin 4-sulfate.

A detailed investigation of the variations in different acidic glycosaminoglycan fractions in edematous, suppurative and fibrous maxillary sinusitis did not detect the presence of heparan sulfate in either normal or pathological paranasal membranes (Saito and Nakamura, 1977). The total AGAG measured as mean ± S.E. glucuronic acid and expressed as per cent dry weight increased from the normal mucosal value of 0.15 ± 0.03 to 0.32 ± 0.05 % in edematous, 0.31 ± 0.08 in suppurative and 0.24 ± 0.04 in fibrous maxillary sinusitis. Hyaluronic acid content of 48 ± 4.3 % in the normal sinus mucosa did not change significantly. Dermatan sulfate decreased from the normal value of 45 ± 2.3 % in the three types of sinusitis, while chondroitin 4- and 6-sulfate increased in the three sinusitises in the order of fibrous > suppurative > edematous over the normal value of 7 ± 2.5 % of dry weight (Saito and Nakamura, 1977).

6.8 Mucopolysaccharides in nasal and paranasal carcinoma

The nasal and paranasal mucous membranes of the sinuses are subject to a wide variety of carcinomas. The most common of these are the ones in which the sero-mucinous glands of the nose and paranasal sinuses produce tumors of the salivary gland type. These include mixed tumor (pleomorphic adenoma) muco-epidermoid tumor and adenocystic carcinoma (Michaels, 1976). The muco-epidermoid tumor forms a polypoid mass in the nose and affects both the columner mucus secreting cells and squamous cells producing a large quantitiy of mucus in the nose. Squamous carcinomas of the upper respiratory tract are also encountered. Such tumors are particularly common in the nasal pharynx.

Galactosamine-containing acidic glycosaminoglycans have been isolated from a human nasopharyngeal chordoma. The two macromolecules present in this isolated material were identified by their optical rotation as chondroitin 4-sulfate and chondroitin 6-sulfate (Meyer et al., 1956). The acidic glycosaminoglycans in nasal polyps carcinoma of the paranasal sinus mucosa were analyzed (Saito and Nakamura, 1977). Both the normal and the cancerous maxillary sinus mucosae contained hyaluronic acid, dermatan sulfate and chondroitin 4-/6-sulfate. However, in the maxillary sinus cancer the total uronic acid content was increased from 0.15 ± 0.03 (S.E.) to 0.23 ± 0.05 % per dry weight and the hyaluronic acid increased from 48 % to 56 %. The dermatan sulfate, on the other hand, was significantly decreased while the chondroitin 4-/6-sulfates showed significant increase.

6.9 Mucopolysaccharides of sputum

6.9.1 Source of sputum

Sputum is a mixture of varying proportions of bronchial secretion, saliva, upper respiratory tract secretion and serum proteins originating as a serum transudate or those synthesized by the seromucous cells of the complex submucosal glands in the respiratory tract. Bronchial secretion is the major contributor to sputum and to the carbohydrate residues. During expectoration the secretions from the upper respiratory tract and saliva are added to the tracheobronchial secretion to constitute sputum. The contribution of saliva varies and since the glycoprotein content of saliva is only one-tenth of that in bronchial secretion, a sputum containing a 10 % saliva volume accounts for only 1 % of the total carbohydrate residues in sputum (Keal, 1971).

6.9.2 Collection of sputum

Sputum from normal subjects free of respiratory diseases is difficult to obtain and is of a very small quantity. For this reason the physico-chemical and structural studies have been carried out primarily on sputum collected from patients with pathological states like cystic fibrosis, chronic bronchitis, allergic asthma etc., in which cases large quantities of bronchial secretions accumulate in the tracheobronchial system and these are eliminated as large quantities of sputum during cough. Sputum from patients with chronic bronchitis or cystic fibrosis may be of the order of 150 ml or more a day. Since bronchial secretion is the major contributor to the sputum volume and its carbohydrate composition, the sputum from these patients has been considered and studied as a *pathological bronchial secretion*.

6.9.3 Methods for analysis of sputum and bronchial secretion

Since sputum is in a way a pathological bronchial secretion, the methods for analysis of the sputum and the bronchial secretion are the same. Histochemical methods for mucosubstance in the human and animal respiratory tract have been described (Anderson, 1962; Pozkladka and Andreichenko, 1968; Andreichenko et al., 1969; Spicer et al., 1971). Rheology using magnetic microrheometer has been discussed (Litt et al., 1976).

The methods for isolation and characterisation of sputum glycoproteins (Roberts, 1974) and for analyses of their chemical structures (Clamp et al., 1978; Litt et al., 1976) have been reviewed. The protein content of glycoproteins of bronchial origin may be carried out by direct analysis of the washed and dried sputum gel phase. Isolation of the bronchial glycoproteins from this water-insoluble gel, however, requires its solubilization prior to fractionation. This solubilization has been achieved by ultrasonication, treatment with hydrogen-bonding breaking agents like buffered 6 M urea, guanidine hydrochloride or by treatment with disulfide bond cleavage reagents like mercaptoethanol, dithiothreitol, dithioerythritol or by digestion with proteolytic enzymes. The solutions so obtained may be used, directly or after initial precipitation, for fractionation according to the molecular size by gel filtration on agarose, polyacrylamide, dextran or according to charge by ion-exchange chromatography on DEAE-cellulose or AG 1 X-2 resin and gel electrophoresis. The acidic components may also be separated from the total mixture by precipitation with cetylpyridinium salts or rivanol. Density gradient ultracentrifugation may be used for isolation of bronchial glycoprotein in an undegraded form from serum glycoproteins present in sputum (Creeth et al., 1977).

6.9.4 Gross composition in relation to function

The sputum and the bronchial secretion contain two phases which may be separated by centrifugation. The fluid phase is the serous phase and contains soluble glycoproteins of salivary and bronchial origin. The gel phase contains the fibrillar mucous and cell debris mainly of bronchial origin and under the microscope appears as a network of glycoprotein fibrils (Havez et al., 1967; Burgi, 1973). The relative proportions of the two phases vary among the various sputum samples. The nondialyzable material in sputum is also variable but on an average it is 1 % with a carbohydrate content of 20-50 % in this nondialyzable material (Werner and Odin, 1965).

Both the sputum and the bronchial secretion contain neutral and acidic glycoproteins and deoxyribonucleic acid (DNA). The DNA is probably derived from the nuclei of degenerated lung or white blood cells and its amount is directly related to the extent of bronchial infection, being very high in purulent sputum. The viscoelastic properties of the bronchial mucus are important for the mucociliary function for elimination of noxious substances from the respiratory tract. The viscoelastic properties are attributable to the presence of sulfate and sialic acid-containing acidic glycoproteins. A single sample from a patient may contain a large number of glycoproteins varying in degrees of acidity (Havez et al., 1967). The relative proportions of the glycoproteins of bronchial origin vary in different disease states. In pulmonary destructive diseases (West, 1977) an increase occurs in the acidic glycoproteins, particularly the sulfated glycoproteins, which increases the viscosity of the bronchial secretions. Increased production of mucus and its higher viscosity result in failure of the bronchial mucociliary transport system, which is the main mechanism of defense in the respiratory tract. The relationship between mucus structure, rheology and biological transport function has been reviewed (Meyer, 1976; Litt et al., 1976; Havez and Roussel, 1976; Reid, 1978).

6.9.5 Mucopolysaccharides of sputum

Most of the studies on mucopolysaccharides of sputum (Burgi, 1965) have been carried out as a measure of human bronchial secretion in chronic obstruction lung diseases (Reid, 1968; Dulfano, 1973; Kasama, 1974) notably cystic fibrosis (Potter et al., 1963; Anderson, 1962; Reid, 1968 a; Roussel et al., 1975; Lamblin et al., 1977; Sturgess, 1977; Boat et al., 1977; Lopez-Vidriero and Reid, 1977), asthma, chronic bronchitis and bronchiectasis (Burgi and Regli, 1967; Charman and Reid, 1972; Lamblin et al., 1973; Boat et al., 1976; Barton et al., 1977; Lamblin et al., 1977). In all these conditions hypersecretion of respiratory tract mucus and increased sputum production occur. The inabilitiy of the cilia of the mucociliary system to eliminate the excess mucus by their escalator-like action results in mucus accumulation in air passages thereby reducing airflow and making the tracheobronchial membranes susceptible to infection.

Several workers have isolated glycoproteins of differing carbohydrate composition. White et al. (1959) isolated a glycoprotein from the sputum of chronic bronchitis patients which contained 6.1 % hexosamine, 5.1 % fucose, 4.2 % sialic acid and 19.5 % hexose. A similar glycoprotein isolated by Atassi et al. (1959) contained 2.0 % galactosamine, 12.8 % glucosamine, 5.4 % sialic acid and 13.4 % neutral sugars comprised of equivalent amounts of fucose, mannose and galactose. This glycoprotein was prepared by isoelectric precipitation of a dilute water extract of freeze dried sputum. It was not homogenous by electrophoretic criteria like the glycoprotein isolated by White et al. (1959).

Starting with acetone-dried pooled sputum from chronic bronchitis patients, Brogan (1959) isolated two mucopolysaccharides and an electrophoretically homogenous mucoprotein. The procedure used involved the steps: Pepsin digestion at pH 2.3 for 40 hours at 37 °C, ethanol and ether precipitation and centrifugation. The insoluble material in the pellet contained a mucoprotein, which was similar to the one isolated from urine and contained fucose, mannose, galactose, N-acetylglucosamine and N-acetylgalactosamine. It was composed of 14.3 % hexosamine expressed as N-acetylhexosamine, 26.3 % reducing sugars and 8.7 % nitrogen. Two mucopolysaccharides were isolated from the supernatent by precipitation with ethanol containing sodium acetate, dialysis, reprecipitation with ethanol and fractionation into a soluble and insoluble fraction by dissolving in 90 % phenol water (w/v). The two fractions contained 5.2 and 5.7 % N, respectively. These mucopolysaccharides contained fucose, galactose, N-acetylgalactosamine and N-acetylglucosamine.

From the sputum gel solubilized by ultrasonic vibration, two heterogeneous fractions were obtained by isoelectric precipitation (Brogan, 1960). The supernatent left after isoelectric precipitation of a glycoprotein (Atassi et al., 1959) was used for recovering two other glycoproteins by anion exchange chromatography on DEAE-cellulose. The glycoproteins I and II contained sialic acid 2.2 and 0 %, neutral sugars (fucose, mannose and galactose) 17.2 % and 12.3 %, glucosamine 2.1 % and 2.4 %, galactosamine 13.3 % and 0.5 % (Atassi et al., 1962), respectively. A glycoprotein isolated by ethanolic precipitation of sputum previously solubilized in calcium chloride solution in 10 % ethanol contained 8.3 % sialic acid, 12.5 % fucose, 21.9 % hexoses measured as galactose, 19.1 % hexosamine and 6.3 % nitrogen (Gibbons, 1963).

A purified mucus glycoprotein fraction was isolated from the sputum of a patient with chronic bronchitis, who was a secretor of blood group A substance (Havez et al., 1969). This glycoprotein resembled glycoproteins from epithelial secretions in carbohydrate composition. Mannose was absent and it contained 14.8 % fucose, 18.8 % galactose, 15.1 % N-acetylglucosamine, 6.7 % N-acetylgalactosamine, 4 % sialic acid and 11.5 % protein. The presence of sulfate was not checked (Havez et al., 1969). A similar acidic glycoprotein isolated from the sputum of an 0-secretor chronic bronchitis patient contained a small amount (0.4 %) of mannose, 1.9 % sulfate and 9.6 % protein

(Holden and Griggs, 1977). Other carbohydrates present were: Fucose 9.1 %, galactose 18.7 %, sialic acid 8 %, N-acetylglucosamine 13.6 % and N-acetylgalactosamine 8.1 %.

Five different sulfated glycoproteins differing in sulfate contents were isolated from the sputum of a patient with chronic bronchitis (Degand, 1972). Fractions I, II a, II b; III and IV contained sulfate 0.9, 2.0, 2.5, 2.7 and 2.9 %; sialic acid 4.7, 9.0, 9.1, 4.8 and 3.1 %; fucose 18.7, 15.6, 13.6, 14.3 and 12.3 %; galactose 26.9, 22.3, 23.2, 23.3 and 25.8 %; hexosamine 26.5, 24.0, 23.2, 25.5 and 23.6 %, respectively.

Lamblin et al. (1973) isolated a neutral mucin from the fibrillar portion of the sputum of a patient with chronic bronchitis by reduction with mercaptoethanol, chromatography on Sepharose 2B and 6B and CM-cellulose. This mucin contained antigenic determinants which were not destroyed by trypsin. The sputum from chronic bronchitis patients contained a higher total content of mucus glycoproteins than from normal subjects. Gel filtration of sputum solubilized with mercaptoethanol on Sepharose 4B gave a very high molecular weight fraction in the excluded volume and a small molecular weight fraction retarded on this column. The increase in sputum glycoprotein in chronic bronchitis was due to a large increase in this small molecular weight fraction (Barton et al., 1977).

The fibrillary mucus of sputum has been found to contain mucins as the essential constituent on which the fibril network is formed. Two groups of glycoproteins were isolated from this material. The group of neutral glycoproteins was composed of substances with blood group activity while the other group of acidic, polyanionic glycoproteins was rich in sialic acid and sulfate residues. These glycoproteins were of 600,000 molecular weight and were composed of a single peptide chain of 700-800 amino acids bearing 150-180 glycan chains linked by O-glycosidic bonds to serine and threonine in the peptide chain. Enzyme hydrolysis and ion-exchange chromatography of sputum from 12 patients with chronic bronchitis gave an average distribution of 99.7 mg neutral glycoprotein (fucomucin), 55.9 mg of sialic acid-rich acidic glycoprotein (sialomucin) and 9.75 mg of sulfated acid glycoprotein (sulfomucin) per 100 ml of expectoration (Degand et al., 1973). Another batch of 12 sputum samples from patients with diagnosed chronic bronchitis gave values of protein 41.3 % (average of five samples only), fucose 5.8 %, mannose 0.6 %, galactose 11.7 %, N-acetylglucosamine 9.9 %, N-acetylgalactosamine 5 %, sialic acid 4 % and sulfate 1.7 % (Holden and Griggs, 1977).

By density gradient ultracentrifugation, bronchial glycoproteins were separated from serum protein components in the sputum of three patients diagnosed as chronic bronchitis (Creeth et al., 1977). The three glycoproteins from these patients differed in sialic acid contents which were 12, 16 and 23 %, in mannose contents being 0.7, 1.3 and 0.6 and in glucose contents of 0.4, 0.7 and 0 %, respectively. All these glycoproteins contained fucose 8 %, galactose 21 %, N-acetylglucosamine 22-24 % and N-acetylgalactosamine 9 %.

Other workers have noted differences in relative distribution of fucomucin, sialomucin and sulfomucin in sputum and tracheobronchial secretions from patients with various hypersecretory diseases. Acidic mucins from sputum of three chronic bronchitis patients contained more sialomucin than sulfomucin and had shorter chain lengths (Lamblin et al., 1977 a). On the other hand, the acidic mucins isolated from sputum of three children with cystic fibrosis were more sulfated than sialylated (sulfomucin > sialomucin) and their average carbohydrate chain lengths were relatively shorter.

Whereas neutral glycoproteins with blood group activity have been reported previously (Degand et al., 1973), Boat et al. (1976) reported isolation of two major sulfated fractions with blood group activity by reduction, carboxymethylation and deoxyribonuclease digestion of purulent sputum from patients with bronchitis and cystic fibrosis. The first fraction contained 1.4-2.6 % sulfate and was not an inhibitor of virus hemagglutination while the second fraction contained 7.8 % sulfate and was a potent inhibitor of hemagglutination. Furthermore, the second fraction was more abundant in two sputum samples from cystic fibrosis patients than in purulent bronchitic secretions or in nonpurulent secretions. Roussel et al. (1975) have also reported isolation of heterogeneous sulfated glycoproteins with blood group H activity from sputum of a child suffering from cystic fibrosis, using reduction of fibrillar mucus, chromatography on ECTEOLA-cellulose and Sepharose 4B. The sialic acid oligosaccharide chains were shorter than the sulfated chains and among the latter the size increased with the degree of sulfation. The sulfate residues were present as esters of hydroxyl groups at C-6 of galactose.

The relative amounts of mucus glycoproteins and serum glycoproteins differed in mucoid and purulent sputum from patients with chronic bronchitis, asthma, cystic fibrosis or bronchiectasis. Whereas in infected sputum both mucus and serum glycoproteins contributed to the viscosity, in mucoid sputum only the serum glycoprotein contributed (Lopez-Vidriedo and Reid, 1978).

Rheological studies of tracheobronchial mucus in cystic fibrosis have been carried out largely on sputum. The mucus from these patients is more abundant and more viscous and sticky (Reid, 1968; Charman and Reid, 1972; Nadler et al., 1978). These disturbances of secretion in cystic fibrosis are due to an increase in the gland and cell size and increased formation of sulfated glycoproteins (sulfomucins). Histochemically hyperactivity and hyperplasia of the gland have been demonstrated (Reid, 1968, 1978). Increased accumulation of mucus in the respiratory tract results from mucociliary failure which in turn is responsible for impaired respiratory function and an increased infection in cystic fibrosis (Litt, 1977; Nadler et al., 1978; Boat and Cheng, 1978).

The sputum is more viscous in asthma and other diseases, although the viscosity varies over a wide range in asthma (Keal, 1971). This high viscosity is attributable to the higher concentrations of acidic glycoproteins in the sputum of the asthmatics. The contents of N-acetylneuraminic acid and fucose have been used as markers and their concentrations correlate well with viscosity in asthma and other diseases (Charman and Reid, 1972; Reid, 1973). The sputum viscosity measured on a Ferranti-Shirley Viscometer was in the order of asthma > chronic bronchitis > cystic fibrosis > bronchorrhoea. Also, the contribution of saliva in sputum was higher than normal (Keal, 1971).

Histochemical studies in asthma showed striking volumetric increases in muscle and mucous gland contents (Dunnill et al., 1969; Takazawa and Thurlbeck, 1971). The mucous gland mass was $12.7 \pm 2.74\,\%$ in 11 normal subjects, 23.4 ± 5.36 % in 12 status asthmaticus, 27.8 ± 8.67 in 19 chronic bronchitis and 19.6 ± 6.10 % in 13 patients with emphysema. The values for connective tissue correspondingly were 55.8 ± 5.18 % in normals, 42.6 ± 4.01 in status asthmaticus , 42.5 ± 8.44 % in chronic bronchitis and 50.6 ± 9.82 % in emphysema. Thus the percentage of connective tissue decreased slightly in all these patients (Dunnill et al., 1969). The increase in muscle was noted primarily in the smaller bronchi and bronchioles, and this could represent an attempt to clear luminal exudate by peristaltic action following the breakdown of the ciliary activity (Takazawa and Thurlbeck, 1971). The bronchial basement has also been reported to thicken in asthma (McCarter and Vazquez, 1966).

Various physiological and biochemical studies on asthma have been reviewed (Austen and Lichenstein, 1973; Jones, 1976; Havez and Roussel, 1976). Bronchial glycoproteins were isolated by density gradient ultracentrifugation from a patient diagnosed for intrinsic asthma associated with bronchorrhoea and another patient with extrinsic asthma (Creeth et al., 1977; Reid, 1978). The carbohydrate analyses of these glycoproteins gave values of fucose 14 and 7 %, galactose 26 and 20 %, N-acetylglucosamine 22 and 18 %, N-acetylgalactosamine 11 and 14 %, glucose 1.2 and 0.5 %, mannose 1.4 and 1.5 % and N-acetylneuraminic acid 1 and 11 %, respectively.

6.10 Mucopolysaccharides in bronchial secretion

As described in the section 6.1, the bronchial secretion is produced by the mucosal goblet cells and the bronchial mucous glands. Since the latter have 40-100 times the volume of the former the bronchial glands are considered to be the source of most of the intrabronchial mucus. In conditions like chronic bronchitis in which there is hypersecretion of bronchial mucous, these glands are enlarged correspondingly (Thurlbeck et al, 1963). The size of the gland expressed as *Reid-Index* (= Bronchial gland thickness/bronchial wall thickness) is increased from normal value of 0.43 to 0.53 in chronic bronchitis (Thurlbeck, 1975). The increased bronchial secretion is associated with chronic sputum production. For this reason, the majority of investigations on bronchial secretion have been on its indirect analysis in sputum. Only a limited studies on bronchial secretion per se have been carried out, since it is nearly impossible to obtain a sufficient amount of bronchial mucus. The very few human studies have been on the so-called normal secretions collected by aspiration during laryngectomy, tracheotectomy, surgical bronchogenic cyst removal or tracheal intubation during surgery. The average volume of bronchial secretions collected by these procedures is only 10 ml /24 hr. (Havez and Roussel, 1976). However, in pathological pulmonary states this volume increases several fold (Reid, 1973).

Investigations of the so-called normal bronchial secretion have been the subject of a few investigations (Potter et al., 1963; Thiemann and Wohlrab, 1968; Degand, 1976). Analysis of four samples of human bronchial fluid collected by mild suction through a tracheotomy performed on patients free of respiratory disease gave average values of protein 53.7 %, fucose 3.2 %, mannose 1.1 %, galactose 10 %, N-acetylglucosamine 7.7 %, N-acetylgalactosamine 2.1 %, sialic acid 5.4 %, and sulfate 1 %. It has been argued that these analyses do not represent normal values, since the trauma of cannulation or tracheal intubation may change the nature of secretion present under normal conditions. The mucosubstances in human and canine bronchial secretions have been studied by histochemical methods (Spicer et al., 1971). The effect of trauma is evident from differences in carbohydrate composition of bronchial secretions collected from the same beagle dogs using two different methods of collection. When 12 samples were obtained by mild suction through a flexible catheter introduced through a bronchoscope, the average results obtained were: Protein 45.4 %, fucose 3.8 %, mannose 0.8 %, galactose 6.6 %, N-acetylglucosamine 6 %, N-acetylgalactosamine 5.5 %, sialic acid 5.8 % and sulfate 1.8 %. The 12 samples collected from tracheal pouches in the same dogs gave different values as: Protein 57.1 %, fucose 3 %, mannose 1.6 %, galactose 6.5 %, N-acetylglucosamine 6.3 %, N-acetylgalactosamine 4.9 %, sialic acid 4.2 %, and sulfate 1.2 % (Wardell et al., 1970). Differences in tracheobronchial protein contents and electrophoretic profiles were observed in mucins isolated from tracheobronchial secretions collected from a canine tracheal pouch and those collected in intact dogs (Reasor et al., 1978).

The total carbohydrate contents of nonprocessed human pulmonary secretions in normal subjects and patients with cystic fibrosis and bronchiectasis were different and were 0.951, 1.050 and 1.137 % wet weight (Spicer, 1963). The overall carbohydrate composition of the nonprocessed normal human bronchial secretion was found to be 4.9 % fucose, 5.9 % hexose, 7.3 % N-acetylhexosamine and 4.8 % sialic acid (Spicer, 1963). Hyaluronic acid was isolated from pulmonary secretions of asthma patients. This was identified by electrophoretic mobility on cellulose acetate and disappearance of this band upon digestion with testicular hyaluronidase (Sahu and Lynn, 1978).

6.11 Factors affecting bronchial secretion

Bronchial secretion is increased by all the conditions which involve infection, inflammation or irritation of the bronchial mucous membrane. All these cause hypertrophy of the mucous glands in the bronchial wall. Ether has been used as an irritant to induce increased bronchial secretion during collection (Wardell et al., 1970) in beagle dogs. During various types of anaesthesias, the mucociliary clearance in dog's lungs were delayed by 2-6 hours (Forbes and Gamsu, 1979). Similarly, the relative mucociliary transport for dog tracheal mucus went down from 98 % before anesthesia to 79 % after phenobarbital anesthesia (King et al., 1979). Hypersecretion and increase of mucous cells and a decrease in goblet cells occurred by intratracheal instillation of elastase (EC 3.4.4.7) at 0.02 mg/kg body weight in hamsters (Hayes and Christensen, 1978). Breathing of air containing 0.01-0.06 ml SO_2/m^3 produced a reversible increase in tracheal and bronchial epithelium of rats (Zlateva and Vulcheva, 1973). Increased secretion in response to such noxious agents is an attempt to provide protection by diluting the offending agent.

The effects of chemical and nervous stimulations on tracheobronchial secretion have been reviewed (Baker and Chakrin, 1977; Widdicombe, 1978). Cholinergic drugs like acetylcholine and parasympathetic drugs increase the rate of mucus secretion, whereas the anticholinergic drugs like atropine delayed but did not prevent secretions. The adrenergic drugs are less consistent in their effect. Guanethidine in two adults decreased secretion but had no effect on the third person (Reid, 1968).

6.12 Bibliography

1. Alford, R. H. (1968): Effects of chronic bronchopulmonary disease and aging on human nasal secretion IgA concentrations. J. Immunol. 101, 984-988

2. Anderson, D. H. (1962): Pathology of cystic fibrosis. Ann. N. Y. Acad. Sci. 93, 500-517

3. Andreichenko VI, Roekladka, A. I. and Kalinóvskaia, L. P. (1969): Histochemical studies of tissue mucopolysaccharides and nonspecific phosphatases in the mucous membrane of the upper respiratory tract of albino rats under conditions of the inhalation effect of caprolactame. Zh. Ushn. Nos. Gorl. Bolez. 29, 86-91 (Rus).

4. Atassi, M. Z., Barker, S. A. and Stacey, M. (1959): Neuraminic acid and its relation to chronic bronchitis. III. Carbohydrate constituents of sputum. Clin. Chim. Acta 4, 823-827

5. Atassi, M. A., Barker, S. A. and Stacey, M. (1962): Neuraminic Acid and its relation to chronic bronchitis. IV. Isolation of homogeneous mucoproteins. Clin. Chim. Acta 7, 588-591

6. Austen, K. F. and Lichenstein, L. M. (Eds.) (1973): Asthma: Physiology, Immunopharmacology and Treatment. Academic Press, New York

7. Baker, A. P. and Chakrin, L. W. (1977): Glycosyltransferases and hypersecretion of respiratory mucus. Modern Problems in Paediatrics 19, 99-109

8. Barton, A. D., Weiss, S. G., Lourenco, R. V., Dralle, W. M. and Shamsuddin, M. (1977): Mucous glycoprotein content of chronic sputum. Proc. Soc. Exp. Biol. Med. 156, 8-13

9. Boat, T. F. and Cheng, P. W. (1978): Mucous glycoproteins in cystic fibrosis. In: E. F. Walborg, Jr. (Ed.). Glycoproteins and Glycolipids in Disease Processes. ACS Sympos. Series 80, Am. Chem. Soc., Washington, D. C., 198-121

10. Boat, T. F., Cheng, P. W., Iyer, R. N., Carlson, D. M. and Polony, I. (1976): Human respiratory tract secretions. Mucous glycoproteins of nonpurulent thracheobronchial secretions, and sputum of patients with bronchitis and cystic fibrosis. Arch. Biochem. Biophys. 177, 95-104

11. Boat, T. F., Cheng, P. W. and Wood, R. E. (1977): Tracheobronchial mucus secretion in vivo and in vitro by epithelial tissues from cystic fibrosis and control subjects. Modern Problems in Paediatrics 19, 141-152

12. Boat, T. F., Kleinerman, J. I., Carlson, D. M., Maloney, W. H. and Matthews, L. W. (1974): Human respiratory tract secretions. I. Mucous glycoproteins secreted by cultured nasal polyp epithelium from subjects with allergic rhinitis and with cystic fibrosis. Am. Rev. Resp. Dis. 110, 428-441

13. Brogan, T. D. (1959): Carbohydrate complexes of bronchial secretion. Biochem. J. 71, 125-131

14. Brogan, T. D. (1960): The high molecular-weight components of sputum Brit. J. Exp. Pathol. 41, 288-297

15. Burgi, H. (1965): Sputum mucopolysaccharides. Ther. Umsch. 22, 331-332

16. Burgi, H. (1973): Fibre systems in sputum. Bull. Physio-Pathol. Respir. 9, 191-196

17. Burgi, H. and Regli, J. (1967): Mucopolysaccharides in asthma and chronic bronchitits. Med. Thorac. 24, 78 (Ger)

18. Capel, L. H. (1976): Rhinotoxic drugs. In: R. Hinchcliffe and D. Harrison (Eds.): Scientific Foundations of Otolaryngology, Year Book Medical Publications, Chicago, 862-863

19. Charman, J. and Reid, L. (1972): Sputum viscosity in chronic bronchitis, bronchiectasis, asthma and cystic fibrosis. Biorheology 9, 185-199

20. Clamp, J. R., Allen, A., Gibbons, R. A. and Roberts, G. P. (1978): Chemical aspects of mucus. Br. Med. Bull. 34, 25-27

21. Creeth, J. H., Bhaskar, K. R., Horton, J., Das, I., Lopez-Vidriero, M. T. and Reid, L. (1977): The separation and characterization of bronchial glycoproteins by density gradient methods. Biochem. J. 167, 557-569

22. Degand, P. (1972): Classification et éléments de structure des mucines bronchiques. Thése de Sciences, Université de Lille

23. Degand, P. (1976): Biochemical study of bronchial secretions. Interest in the treatment of bronchial hypersecretion. In: C. Voisin (Ed.). Secretions, Excretions Bronchiques, Expans. Science Fr., Paris cedex, France 33-36 (Fr)

24. Degand, P., Lamblin, G., Durand, G. and Havez, R. (1973): Biochemical and rheological data in sputum. Biochemical definition of mucins in sputum. Bull. Physio-pathol. Resp. 9, 199-217 (Fr)

25. Delaunay, A. and Bazin, S. (1964): Mucopolysaccharides, collagen, and nonfibrillar protein in inflammation. Int. Rev. Connect. Tissue Res. 2, 301-325

26. Dulfano, M. J. (Ed.) (1973): Sputum, Fundamentals and Clinical Pathology. Thomas, Springfield

27. Dunnill, M. S., Massarella, G. R. and Anderson, J. A. (1969): A comparison of the quantitative anatomy of the bronchi in normal subjects, in status asthmaticus, in chronic bronchitis, and in emphysema. Thorax 24, 176-179

28. Forbes, A. R. and Gamsu, G. (1979): Mucociliary clearance in the canine lung during and after general anesthesia. Anesthesiology 50, 26-29

29. Gharib, R., Allen, R. P., Joos, H. A. and Brano, L. R. (1964): Paranasal sinuses in cystic fibrosis. Am. J. Dis. Child. 108, 499-502

30. Gibbons, R. A. (1963): The sensitivity of the neuraminosidic linkage in mucosubstances towards acid and towards neuraminidase. Biochem. J. 89, 380-391

31. Havez, R. and Roussel, P. (1976): Bronchial mucus: physical and biochemical features. In: E. B. Weiss and M. S. Segal (Eds.). Bronchial Asthma: Mechanism and Therapeutics. Little, Brown and Company, Boston, Chap. 25, 409-422

32. Havez, R., Roussel, P., Degand, P., Delmas-Marsalet, Y. and Biserte, G. (1969): Etude de substances de groups sanguin A isolées du mucus bronchique. Bull. Soc. Chim. Biol. 51, 245-259

33. Havez, R., Roussel, P., Degand, P., Randoux, A. and Biserte, G. (1967): Etude des structures fibrillaires de la sécrétion bronchique humaine. Clin. Chim. Acta 17, 281-295

34. Hayes, J. A. and Christensen, T. G. (1978): Bronchial mucus hypersecretion induced by elastase in hamsters. J. Pathol. 125, 25-31

35. Herberhold, C. (1968): Comparative histochemical studies on the peripheral smell organ in mammals and fishes. Arch. Klin. Exp. Ohren Nasen Kehlkopfheilkd. 190, 166-182 (Ger)

36. Holden, K. G. and Griggs, L. J. (1977): Respiratory tract. In: M. I. Horowitz and W. Pigman (Eds.). Glycoconjugates, Vol. I, Academic Press, New York, 215-237

37. Iravani, J. and van As, A. (1972): Mucus transport in the tracheobronchial tree of normal and bronchitic rats. J. Pathol. 106, 81-93

38. Jackson, R. T. and Arihood, S. A. (1971): The acid mucopolysaccharide and collagen content of human nasal polyps, human and animal nasal mucosa. Ann. Otol. Rhino. Laryngol. 80, 586-592

39. Jones, R. S. (1976): Asthma in Children. Publishing Sciences Group, Acton, Chap. 3, 26-38

40. Kasama, K. (1974): Studies on acid mucopolysaccharides of the sputum in chronic obstructive lung diseases. Jpn. J. Thorac. Dis. 12, 310-319 (Jpn)

41. Kawamoto, K. (1969): Puncture of the maxillary sinus. Otolaryngology (Tokyo) 41, 727-729 (Jpn)

42. Keal, E. E. (1971): Biochemistry and rheology of sputum in asthma. Postgrad. Med. J. 47, 171-177

43. Kilburn, K. H. (1968): A hypothesis for pulmonary clearance and its implication. Amer. Rev. Resp. Dis. 98, 449-463

44. King, M. Engel, L. A. and Macklem, P. T. (1979): Effect of pentobarbital anesthesia on rheology and transport of canine tracheal mucus. J. Appl. Physiol. Resp., Environ. Exercise Physiol 46, 504-509

45. Lamblin, G., Lafitte, J. J., L'hermitte, M., Degand, P. and Roussel, P. (1977): Mucins from cystic fibrosis sputum. Modern Problems in Paediatrics 19, 153-164

46. Lamblin, G., L'hermitte, J., Degand, P., Sergeant, Y. H. and Roussel, P (1973): Isolation and immunological properties of neutral mucins from a patient with chronic bronchitis. Biochim. Acta 322, 372-382

47. Lamblin, G., L'hermitte, M., Lafitte, J. J., Filliat, M., Degand P. and Roussel, P. (1977 a): Comparative study of bronchial mucins isolated from the sputum of patients with mucoviscidosis or other chronic bronchial diseases. Bull. Eur. Physiopathol. Respir. 13, 175-190 (Fr)

48. Litt, M. (1977): Mucus rheology and mucociliary clearance. Modern Problems in Paediatrics 19, 175-181

49. Litt, M., Khan, M. A. and Wolf, D. P. (1976): Mucus rheology: relation to structure and function. Biorheology 13, 37-48

50. Lopez-Vidriero, M. T. and Reid, L. (1977): Pathophysiology of mucus secretion in cystic fibrosis. Modern Problems in Paediatrics 19, 129-140

51. Lopez-Vidriedo, M. T. and Reid, L. (1978): Chemical markers of mucous and serum glycoproteins and their relation to viscosity in mucoid and purulent sputum from various hypersecretory diseases. Am. Rev. Resp. Dis. 117, 465-477

52. McCarter, J. H. and Vazquez, J. J. (1966): The bronchial basement membrane in asthma. Arch. Path. 82, 328-335

53. Meyer, F. F. (1976): Mucus structure: relation to biological transport function. Biorheology 13, 49-58

54. Meyer, K., Davidson, E. A., Linker, A. and Hoffman. P. (1956): The acid mucopolysaccharides of connective tissue. Biochim. Biophys. Acta 21, 506-518

55. Michaels, L. (1976): Histophathology of nose and throat. In: R. Hinchcliffe and D. Harrison (Eds.). Scientific Foundations of Otolaryngology, YearBook Medical Publications, Chicago, 667-700

56. Mizukoshi, O. and Saito, H. (1974): Mucopolysaccharides in the otolaryngological division. In: Y. Otaka (Ed.). Biochemistry and Pathology of Connective Tissue, Igaku Shoin, Tokyo, Chap. 7, 111-126

57. Nadler, H. L., Rao, G. J. S. and Taussig, L. M. (1978): Cystic fibrosis. In: J. B. Stanbury, J. B. Wyngaarden and D. S. Fredrickson (Eds.). The Metabolic Basis of Inherited Disease, McGrawHill, 4th Ed., chap. 73, 1683-1710

58. Odin, L. (1958): Mucopolysaccharides of epithelial mucus. In: G. E. W. Wolstenholme and M. O'Connor (Eds.). Ciba Foundation Symposium on the Chemistry and Molecular Biology of Mucopolysaccharides. Little, Brown and Company, Boston, 234-244

59. Pennington, C. L. (1956): Paranasal sinus changes in fibrocystic disease of the pancreas. A. M. A. Arch. Otolaryng. 63, 576-579

60. Potter, J. L., Matthews, L. W., Lemm, J. and Spector, S. (1963): Part VI. Studies on mucus in relation to human disease. Human pulmonary secretions in health and disease. Annals N. Y. Acad. Sci. 106, 692-697

61. Pozkladka, A. I. and Andreichenko, V. I. (1968): Histochemistry of cell mucopolysaccharides in the mucous membrane of respiratory tract in the dynamics of caprolactame effect. Zh. Ushn. Nos. Gorl. Bolez. 28, 76-82 (Ger)

62. Reasor, M. J., Adams, G. K. III., Proctor, D. F. and Rubin, R. J. (1978): Tracheobronchial secretions collected from intact dogs. I. Protein and mucous glycoprotein composition. J. Appl. Physiol. Respir. Environ.Exercise Physiol. 45, 182-189

63. Reid, L. (1963): Experimental study of hypersecretion of mucus in the bronchial tree. Brit. J. Exp. Path. 44, 437-445

64. Reid, L. (1968): Bronchial mucus production in health and disease. In: A. A. Liebow and D. E. Smith (Eds.). The Lung, International Academy of Pathology Monograph No. 8, Williams and Wilkins Co., Baltimore, Chap. 8, 87-108

65. Reid, L. (1968 a): Disturbances in the pattern of secretion of bronchial mucous glands. In: R. Porter and M. O'Connor (Eds.). Cystic Fibrosis. Ciba Found. Sympos. No. 32. Little, Brown and Company, Boston, 45-67

66. Reid, L. (1973): Bronchial secretion. General comments. Bull. Physiopathol. Resp. 9, 15-30

67. Reid, L. (1978): The cell biology of mucus secretion in the lung. In: W. M. Thurlbeck and M. R. Abell (Eds.). The Lung: Structure, Function and Disease, Williams and Wilkins Co., Baltimore

68. Roberts, G. P. (1974): Isolation and characterisation of glycoproteins from sputum. Eur. J. Biochem. 50, 265-280

69. Roussel, P., Lamblin, G. and Degand, P. (1975): Heterogeneity of the carbohydrate chains of sulfated bronchial glycoproteins isolated from a patient suffering from cystic fibrosis. J. Biol. Chem. 250, 2114-2122

70. Sahu, S. and Lynn, W. S. (1978): Hyaluronic acid in the pulmonary secretions of patients with asthma. Biochem. J. 173, 565-568

71. Saito, H. and Nakamura, A. (1976): Analysis of acid mucopolysaccharides in paranasal sinus diseases. J. Otolaryngol. 79, 450-458 (Jap)

72. Saito, H. and Nakamura, A. (1977): Acidic glycosaminoglycans in the ground substance in paranasal sinus disease. Acta Otolaryngol. 84, 434-439

73. Sato, Y. (1966): Biochemical studies on acid mucopolysaccharides in paranasal mucous membrane in chronic sinusitis. J. Otorhinolaryng Soc. Jap. 69, 1039-1064 (Jap)

74. Sato, Y., Hiramatsu, M., Hatakeyama, K., Minami, N. and Sasaki, Y. (1977): Ground substances and pathochemical features of the human nasal polyp. Josai Shika Daigaku Kujo 6, 381-386 (Jap)

75. Sato, Y., Hiramatsu, M., Hatakeyama, K., Minami, N. and Sasaki, Y. (1977 a): Acid mucopolysaccharides of paranasal sinus mucous membrane. Josai Shika Daigaku Kujo 6, 375-379 (Jap)

76. Shwachman, H., Kulczycki, L. L., Mueller, H. L. and Flake, C. G. (1962): Nasal polyposis in patients with cystic fibrosis. Pediatrics 30, 389-401

77. Soskin, S. and Bernheimer, L. B. (1939): Mechanism of the estrogen effect on the nasal mucosa in atrophic rhinitis. Proc. Soc. Exp. Biol. (N. Y.) 42, 223-224

78. Spicer, S. S. (1963): Histochemical differentiation of mammalian mucopolysaccharides. Ann. N. Y. Acad. Sci. 106, 379-388

79. Spicer, S. S., Chakrin, L. W., Wardell, J. R. and Kendrick, W. (1971): Histochemistry of mucosubstances in the canine and human respiratory tract. Lab. Invest. 25, 483-490

80. Spicer, S. S., Chakrin, L. W. and Wardell, J. R. (1973): In: M. J. Dulfano (Ed.). Sputum Fundamentals and Clinical Pathology. Thomas, Springfield, 22-68

81. Sturgess, J. F. (1977): Bronchial mucus secretion in cystic fibrosis. Modern Problems in Paediatrics 19, 129-140

82. Takazawa, T. and Thurlbeck, W. M. (1971): Muscle and mucus gland size in the major bronchi of patients with chronic bronchitis, asthma and asthmatic bronchitis. Am. Rev. Resp. Dis. 104, 331-336

83. Taylor, B., Evans, J. N. G. and Hope, G. A. (1974): Upper respiratory tract in cystic fibrosis. Arch. Dis. Child. 49, 133-136

84. Thiemann, H. H. and Wohlrab, L. (1968): Content of mucoid substances of bronchial secretions in terms of the bronchologic aspect. Broches 18, 91-96

85. Thurlbeck, W. M. (1975): Chronic obstructive lung disease. In: S. C. Sommers (Ed.). Pulmonary Pathology Decennial 1966-1975, Appleton-Century-Crofts, New York, 381-417

86. Thurlbeck, W. M., Angus, G. E. and Pare, J. A. P. (1963): Mucus gland hypertrophy in chronic bronchitis and its occurrences in smokers. Brit. J. Dis. Chest 57, 73-78

87. Tomasi, T. B. and Bienenstoc, J. (1968): Secretory immunoglobulins. Adv. Immunol. 9, 1-96

88. Wardell, J. R., Chakrin, L. W. and Payne, B. J. (1970): The canine tracheal pouch. A model for use in respiratory mucus research. Amer. Respir. Dis. 101, 741-754

89. Werner, I. (1953): Studies on glycoproteins form mucus epithelium and epithelial secretions. Acta Soc. Med. Upsalien. 58, 1-55

90. Werner, I. and Odin, L. (1965): Glycoproteins in salivary glands, saliva and sputum. In: E. A. Balazs and R. W. Jeanloz (Eds.). The Amino Sugars, Vol. II A, Academic Press, New York, Chap. 23, 141-148

91. West, J. B. (1977): Pulmonary Pathophysiology, the essentials. Williams and Wilkins, Baltimore, 1-227

92. White, J. C., Elmes, P. C. and Whitley, W. (1959): Mucoprotein of bronchial mucus gel. Nature 183, 1810-1811

93. Widdicombe, J. G. (1978): Control of secretion of tracheobronchial mucus. Br. Med. Bull. 34, 57-61

94. Zlateva, M. and Vulcheva, V. (1973): Changes in the mucus content of tracheal and bronchial epithelium of white rats placed in a working environment full of sulfur dioxide. Khig. Zdraveopazvane 16, 433-440 (Bulg)

Chapter 7

Ocular Fluids

7.1 Introduction

Four fluids are associated with the eye. One of these is the extraocular fluid called the tears or the lacrimal secretion. The remaining three fluids are intraocular and one of these (called subretinal fluid) is present only in pathological states such as the retinal detachment. Under normal, physiological state only two intraocular fluids, the aqueous and vitreous humors are present. The aqueous humor is clear and watery and is present in front and at the periphery of the eye lens in the anterior chamber of the eye. The vitreous humor is gel-like, transparent and fills the posterior chamber of the eye. The tears have the functions for protection of cornea against bacterial infections. The aqueous and vitreous humors provide proper refractive media of the optical system, and by pressure keep the eyeball distended to the dimensions necessary for focusing the visual image on the retina.

The physiological, pathological and biochemical aspects of the four aforementioned fluids in relation to their glycosaminoglycan composition will be described in this chapter. Mimura (1973) has reviewed the glycosaminoglycans of the eye system.

7.2 Lacrimal secretion or tears

The tear fluid is a combined secretion of the lacrimal glands and the accessory sebaceous glands called the glands of Zeis and the Meibomian glands. The lacrimal apparatus in human subjects is shown in Figure 7-1. The tear fluid is continously secreted and it forms a layer on the surface on the cornea. The presence of proteins in the tear fluid lowers its surface tension and enables the tears to wet cornea and the epithelial surfaces. The surface chemistry of tears has been reviewed (Iwata, 1976): Under mild stimulus the flow rate is slow and it evaporates to appear hypertonic. The flow rate is very much enhanced by the presence of foreign substances and then it is isotonic. The composition of lacrimal fluids and their importance during tearing in response to the foreign presence of contact lenses has been reviewed (Haberich, 1978). A slit fluorometer for clinical determination of tear turnover rate and volume without eye irritation has been described (Furukawa et al., 1976).

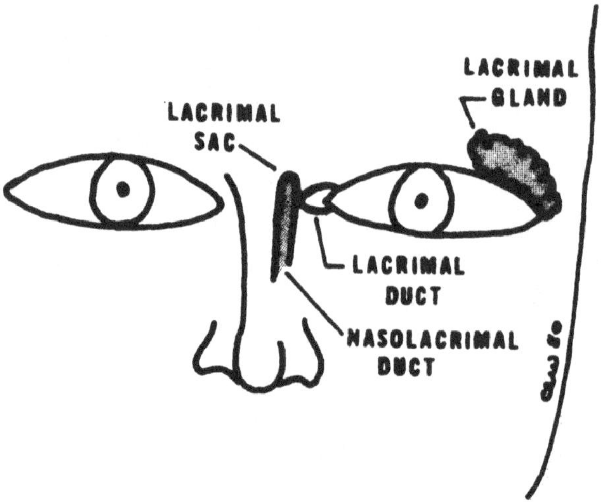

Fig. 7-1. The human lacrimal system

7.2.1 Chemical composition of tears

The fluid is an ultrafiltrate of plasma and is formed in the lacrimal glands. The fluid has normally a pH of 7-7.4 but corneal injuries induce alkaline tears. The composition of diffusible electrolytes and nitrogenous substances is similar to that of the plasma but it is much lower (0.2-0.8 g/dl) in proteins with an albumin/globulin ratio of 2. The enzyme lysozyme present in tears is a glycoprotein. It protects the eye from infections of the airborne cocci by hydrolyzing β-1,4-N-acetylglucosaminidic linkage of the repeating disaccharide units of the neuraminic acid-containing mucopeptide in the cell walls of these microorganisms.

The lacrimal system could be divided into the lacrimal secretory system and the lacrimal drainage system. The secretory portion consists of basic secretors, which are essential for the integrity of eye and consists of mucin, lacrimal fluid and oil which together form a protective film on the cornea. The reflex secretors pour tears (lacrimal fluid) in response to psychic stimuli and irritation. The mucin glands include the conjuctival goblet cells, the crypts of Henle behind the tarsi and the glands of Manz of the perilimbal bulbar conjuctiva. They pour their secretions directly into the conjunctival sac and provide the innermost film of polysaccharides for lubrication of the cornea (Jones, 1970).

The authors have not come across any report on the analysis of the polysaccharide of the tears except that of the lysozyme (a glycoprotein enzyme) mentioned above. However, the mucosubstances in the acini of the human lacrimal glands have been investigated by histochemical methods. These substances were strongly alcianophilic, PAS-positive but hyaluronidase resistant and neuraminidase sensitive. The presence of a high content of sialic acid and small amounts of weakly sulfated mucosubstances localized peripherally in the cytoplasm was suggested (Jensen et al., 1969). Normal acini contained mucosubstances identical to epithelial mucins consisting of sulfomucins and sialomucins. The mucopolysaccharides in the homogenate of human lacrimal glands in cold water were precipitated with 1 % cetylpyridinium chloride. Following dialysis against distilled water, the mucopolysaccharides obtained consisted of 13 % hexosamine, 20 % hexose, 7 % methylpentose, 40 % sialic acid and 46 % protein. The sialic acid may be a part of the tear liquid (Falbe-Hansen and Jensen, 1969).

Similar morphological and histochemical studies in the adenomeres of the lacrimal glands of horse, sheep, ox and dog have been reported in other studies (Filotto, 1954). The preterminal portion of the gland showed mostly mucus and mucoprotein substances. These animals had differences in the nature of mucopolysaccharides in their lacrimal glands. Another electron microscopic study also revealed distinct differences of mucosubstances in the glandular acini of the tear glands of adult rabbits and cats of both sexes (Kuehnel, 1968). In the lacrimal glands of goat, the glandular epithelium was seen to consist of light and dark cells. The dark cells were rich in acidic mucopolysaccharides. The gland cells and the ductural cells contained large granules containing acid mucopolysaccharides (Kuehnel, 1968 a).

A mucoprotein has been isolated from the Harder's gland and the ventral lacrimal gland of guinea pig following treatment with the exophthalmos producing substance of pituitary origin. It contained, in addition to hexuronic acid, 21.2 % glucosamine, 21.6 % sulfate and 4.04 % nitrogen. This macromolecule had a molecular weight of 10,200 (Brunish, 1962).

7.2.2 Tumors of the lacrimal gland

The mucosubstances of mixed tumors of the human lacrimal gland have been investigated. These mucosubstances were analyzed by histochemical methods and were found to be different than those in normal gland epithelium. The epithelium of mixed gland tumors contained mainly neutral mucosubstances in contrast to the abundance of sialomucins in normal gland epithelium (Jensen, 1970).

Like the mixed tumors of the lacrimal gland, tumors of the lacrimal sac are a rare entity in ophthalmology. Excision of the entire lacrimal sac (dacryocystectomy) is usually carried out as a treatment of the epithelial tumor of the lacrimal sac (Flanagan, 1979). However, no report on analysis of the mucopolysaccharides in the sac fluid or the sac is available.

7.2.3 Lacrimal secretion in disease states

In Sjogren's disease (Sicca syndrome) atrophy of the lacrimal glands occurs and leads to reduced or absence of lacrimation and trophic disorders of the epithelium of cornea and conjuctiva (Godtfredsen, 1954). Abnormalities of flow rates and elctrolytes of tears have been reported in patients with cystic fibrosis (Nadler et al., 1978).

7.2.4 Effects of hormones on the lacrimal apparatus

The effect of hormones has been studied extensively on Harderian gland and the ventral lacrimal gland in experimental animals. A peptide hormone from the anterior pituitary called *Exophthalmos producing substance* (EPS), which is distinct from the thyroid stimulating hormone (TSH), acted on the Harderian gland and ventral lacrimal gland of guinea pigs and stimulated sulfate incorporation (Wegelius et al., 1959). When radioactive sulfate ($^{35}SO_4$) was given to EPS-treated guinea pigs, over 50 % of the radioactivity was found associated with a mucoprotein in the ventral lacrimal gland (Brunish, 1962). It was suggested that this hexuronic acid-containing mucoprotein was the main sulfate receptor and the major product of EPS stimulation. Thyrotropin preparations contain EPS as a contaminant and therefore thyrotropin (TSH) produces effects similar to EPS. The sulfated macromolecules in the TSH-treated Harderian gland of mice were characterized by Sephadex G-25 and cellulose chromatography and radioactive sulfur was found incorporated in increased amounts in sulfated glycosaminoglycans, especially keratan sulfate and dermatan sulfate (Singh and McKenzie, 1970). In another study (Andrews and Dunn, 1973) TSH was found to increase hexosamine levels in hyaluronic acid, chondroitin sulfate and glycoprotein fractions of the Harderian gland of guinea pig and decreased hexosamine in the keratan sulfate fraction. These effects were in addition to increased sulfate incorporation. Another investigation identified hyaluronic acid as the major GAG of the Harderian gland. TSH or EPS isolated from TSH produced a 2-fold increase in the total GAG of this gland; 90 % of this increase being in hyaluronic acid and 10 % in the sulfated GAG (Winand and Kohn, 1973).

7.3 Aqueous humor

7.3.1 Location, formation and fate

This clear flowing fluid fills the anterior chamber of the eye, the space in front of and toward the periphery of the lens. It is maintained at a pressure of approximately 15 mm Hg for keeping the eyeball distended to the same dimensions for an exact focusing of the visual image on the retina. It also provides nourishment to the avascular cornea and the eye lens. In human eye it is present in a volume of only 0.125 ml and this volume is maintained by a balance between its formation at a rate of 2-5 mm^3/min in the ciliary body and its reabsorption into the canal of Schlemm. The volume of this fluid varies in the various animal species depending upon the size of the eyeball and the anterior chamber of the eye.

The ciliary body consists of a highly vascular layer covered by epithelial folds called ciliary processes followed by a layer of secretory epithelial cells. Following secretion by the ciliary body, this fluid circulates between the ligaments of the eye to the angle between the cornea and iris. The fluid flows through a lattice of traculae located in the angle into a modified vein called the canal of Schlemm. The canal of Schlemm empties through the aqueous veins into the blood veins of the Sclera (Cole, 1970). The biochemistry of aqueous humor formation has been reviewed (Maren, 1974; Guyton et al., 1975; Thill, 1976). The position of the humors of the eye is shown in Fig. 7-2.

7.3.2 Composition

The diffusible substances are similar as in plasma, except that the ascorbic acid concentration is twenty times in aqueous humor. Since the epithelium of the ciliary processes is almost impermeable to large macromolecules, the aqueous humor differs from plasma in the macromolecular composition. The protein content is negligible (0.025 g/dl) but the albumin/globulin ratio is the same as in plasma of the same subject. When aqueous humor is removed, it is rapidly replaced with a fluid which is called plasmoid aqueous humor and it contains larger quantities of protein. Repetition of the paracentesis fills the anterior chamber with a fluid resembling plasma. A small amount of glycosaminoglycans are present.

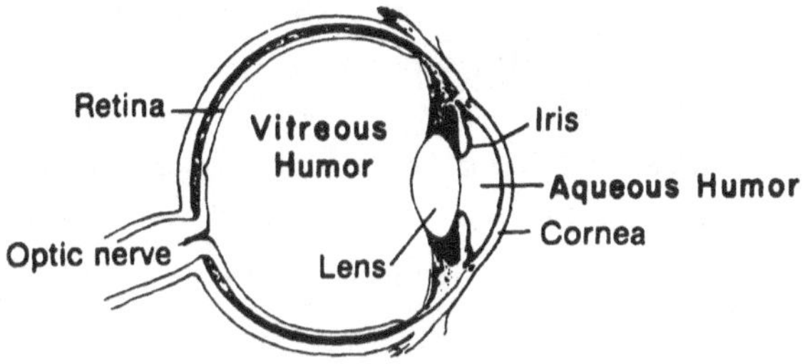

Fig. 7-2. The structure of the human eye showing the location of aqueous and vitreous humors

7.3.3 Glycosaminoglycans in aqueous humor

The presence of glycosaminoglycans in aqueous humor was first reported by Meyer and Palmer (1936) who isolated this material from bovine aqueous humor, which contained 23.5 % hexosamine. These investigators reported hexosamine contents of the aqueous humor from various animal species. The hexosamine contents, however, represent both glycoproteins and glycosaminoglycans. The hyaluronic acid content of bovine aqueous humor has been reported to be 45 µg/ml (Balazs, 1965). Human aqueous humor contained much lower amounts of GAG (17.2-31 µg hexosamine per ml). The literature on aqueous humor has been well covered in the review by Balazs (1965).

The hyaluronic acid from aqueous humor has not been studied to the extent as that from the vitreous humor has been. We prepared 848 mg of hyaluronate from the aqueous humor collected from 462 bovine eyes (Varma et al., 1975). The glycosaminoglycan-glycoprotein mixture from the pooled aqueous humor was precipitated as acetone powder which was then digested with papain-trypsin. Hyaluronate was isolated as the 0.5 M NaCl fraction by anion-exchange chromatography on an AG 1 X-2 (Cl^-) column. Following purification this hyaluronate contained 0.34 % protein, 39.1 % glucosamine, 0.2 % galactosamine and 42.13 % glucuronic acid. The hydrolyzate of this preparation showed the presence of neutral sugars by gas-liquid chromatography, distributed as 0.28 % fucose, 0.05 % arabinose and glucose 0.004 %. It had $[\alpha]_D^{32} = -62.8°$ in water, (6 g/l) viscosity (ml/g) 360 and the molecular weight calculated from viscosity was 164,900. Alkaline treatment and reduction of the hyaluronate in alkaline medium resulted in reduction in viscosity, optical rotation and molecular weight. A similar reduction in optical rotation was noted in urea for the unreduced hyaluronate but not with the reduced sample (Varma et al., 1975).

7.3.4 Sulfated glycosaminoglycans in aqueous humor

During the isolation of hyaluronic acid from the bovine aqueous humor, we observed the presence of sulfated GAG consisting of heparan sulfate and chondroitin sulfate which eluted from the anion-exchange column with higher than 0.5 M NaCl and showed the presence of sulfate in the infrared spectra and had electrophoretic mobility comparable to the standards. The total sulfated GAG in bovine aqueous humor amounted to approximately 3 % of the total GAG (unpublished work). The sulfated components can originate from the trabecular meshwork, since monolayer cultures derived from the isolated trabecular meshwork from the eyes of the monkeys incorporated ^{14}C-glucosamine into mainly 60-85 % hyaluronic acid (Schachtschabel et al., 1977). The GAG were also demonstrated within and around the endothelial cells of the corneoscleral and endothelial meshwork and those lining Schlemm's canal by using electron microscopic histochemistry (Tawara, 1978).

The injection of the enzyme hyaluronidase into the chamber of the eye of two patients with secondary glaucoma produced a temporary beneficial effect (Linn and Ozment, 1950). The temporary reduction in intraocular pressure resulted from gradual dissolution of the hyaluronidase-sensitive component of the trabecular barrier to the aqueous outflow. An immediate 5-fold increase in outflow with stabilization over three hours and subsequent perfusion occurred (Van Buskirk and Brett, 1978). However, hyaluronidase injection produced no effect on outflow in congenitally buphthalmic rabbits (Brown and Geeraets, 1973). The significance of glycosaminoglycans in relation to intraocular pressure regulation and glaucoma has been discussed by several investigators (Kashintseva and Plevinskis, 1973; Francois et al., 1974; Francois, 1975). The presence of the hyaluronidase-sensitive GAG on the surface of the trabecular meshwork cells and the endothelial cells lining the Schlemm's canal has been shown by electron microscopy (Grierson and Lee, 1975).

7.3.5 Physiological and pathological changes in aqueous outflow

7.3.5.1 Age-related changes

Normally a reduction in aqueous humor production is balanced by an increase in resistance to aqueous outflow through the trabeculum. In 1-2 % of persons over 40, this resistance rises to the point where the intraocular pressure rises to abnormal levels causing blindness due to open-angle glaucoma (Anderson, 1971).

7.3.5.2 Effects of drugs and hormones

Some anticholinergics have an adverse effect of producing elevated intraocular pressure (Nicholson, 1976). The changes in glycoproteins of aqueous humor occurred during the sexual cycle, which disappeared after castration. Estradiol or progesterone administration to castrated rats produced similar changes (Houssay and Blumenkrantz, 1964). Similarly, castration in male rats produced changes in aqueous humor glycoproteins, which were reversed by testosterone administration. In normal human volunteers topical application of 0.1 % β-methasone raised intraocular pressure and induced glaucoma due to its effect on the hyaluronidase-sensitive GAG (Perkins, 1965). Corticoids (Kasavina and Pantieleva, 1969; Kasavina et al., 1970) injected in the rabbits' conjuctiva produced intraocular hypertension, while ascorbic acid increased the outflow (Lieb and Staerk, 1966). These effects were attributable to GAG changes in the trabeculum of the chamber angle, which affected its permeability.

Qualitative and quantitative changes occurred in aqueous humor along with an amassment of GAG in the angle of the anterior chamber in response to epinephrine administration in rabbits (Pantieliva et al., 1971).

7.3.5.3 Glaucoma and osteogenesis imperfecta

Although the etiology of glaucoma is not known, the hexosamine content of the aqueous humor is increased (Kasavina and Vinetskaya, 1976) in this disorder. Whereas glaucoma was associated with higher intraocular pressure, the pressure was low in osteogenesis imperfecta due to a lower resistance to outflow (Berggren et al., 1969).

7.4 Vitreous humor

7.4.1 Nature of the vitreous humor

The vitreous humor is a gel-like mass in the vitreous cavity of the eye-ball and is also called the vitreous body. It is transparent (vitreous = glass-like), rigid and elastic and by virtue of these properties it allows the passage of light rays unimpeded from lens to retina, acts as a barrier to invading cells or large molecules and acts as a shock absorber to protect the surrounding structures, especially the delicate retina. Because of its transparency it has only a few cells, *hyalocytes*, (hyalos = glassy), which are confined to the periphery and overlie the retina as a fine layer. The major portion (99 %) of the vitreous is water and the remaining 1 % dry matter consists of a special form of collagen (vitrosin), glycoproteins and glycosaminoglycans, chiefly hyaluronic acid. The gel structure prevents the infiltration of blood vessels and thus maintains its transparency.

The vitreous is a collagen gel consisting of a special type of very fine, short protocollagen fibrils which make up the framework, reinforced and stabilized by spongy giant molecules of glycosaminoglycans, chiefly hyaluronic acid. Around the retina and behind the lens, the vitreous is always present as a solid state (semisolid gel) and is therefore called the cortical tissue-layer. The rest of the vitreous may be present as liquid in some animals like squid and as a gel in others. In human eye, the vitreous is present as a gel except in old age, when liquid pockets are present (Balazs, et al., 1959 a). The molecular biology of the vitreous with respect to contraction-expansion, excluded-volume and molecular sieve effects in retarding molecules larger than 150-200 Å diameter, the optical and other physicochemical properties (Balazs, 1958) have been discussed in excellent reviews (Balazs, 1965, 1968).

7.4.2 Glycosaminoglycans in vitreous humor

Both the nonsulfated and sulfated glycosaminoglycans are present. The nonsulfated GAG is hyaluronic acid, which forms almost all of the macromolecular polysaccharides of the vitreous humor, although it also contains some glycoproteins (for reviews and references, see Balazs, 1965; Allen and Wardi, 1973) and some sulfated glycosaminoglycans, which will be described later.

7.4.2.1 Vitreous humor hyaluronate

The bovine vitreous humor is the first source from which hyaluronic acid was isolated by Meyer and Palmer (1934, 1936 a). These investigators obtained the hyaluronate preparation by repeated water extractions of the acetone powder followed by its precipitation from the combined extract with alcohol acidified with acetic acid. They named it *hyaluronic acid*. Later this macromolecule was also isolated from porcine (pig) vitreous humor (Meyer and Palmer, 1936). The GAG contents of the vitreous from animal species vary and have been tabulated (Balazs et al., 1959; Balazs, 1965). In human subjects the glycosaminoglycan concentration calculated from hexuronic acid determination is 140 μg/ml for the age group 31-65 years and 338 μg/ml for the ages of 65-83 years.

This hyaluronate does not seem to be bound to protein, yet in spite of repeated fractionations it is always associated with a small amount of protein and contains neutral sugars. The nature of the association of this hyaluronate was the subject of a doctoral dissertation (Pahuja, 1977). Although hyaluronate is always associated with protein upon isolation by different methods, yet it may not be covalently linked to protein (Varma et al., 1974). In spite of any evidence for the presence of a covalent linkage like the one known for the sulfated glycosaminoglycans (Chapter 1), a small amount of neutral sugars were present in bovine vitreous humor preparations (Varma et al., 1973, 1974 a). A bovine vitreous humor hyaluronate preparation composed of 35 % glucosamine, 1.3 % galactosamine and 41.6 % glucuronic acid with a hexosamine/hexuronic acid molar ratio of 0.945 contained 1.85 mg L-arabinose per gram of the hyaluronate (Varma et al., 1973). A more purified preparation with galactosamine less than 0.23 %, 40.5 % glucosamine, 43.94 % glucuronic acid and molar ratio of hexosamine/hexuronic acid of 1.0 contained 0.34 % fucose, 0.2 % L-arabinose and another neutral sugar which was tentatively identified as the 7-deoxy-L-glycero-D-manno-heptose (Varma et al., 1974 a).

In view of the lack of evidence for a protein-carbohydrate covalent linkage, the presence of the small amount of neutral sugars and amino acids is puzzling. The fractional molar ratio of an amino acid to a neutral sugar should be approximately unity if these two types of residues are involved in a protein-carbohydrate linkage. However, working with highly purified bovine vitreous humor, the authors found that the fractional molar ratios of the three possible candidates, serine, theronine and aspartic acid to neutral sugars was not unity. Also, a 4-5 fold increase in the percentages of these amino acids and the absence of sugar alditols in hyaluronate reduced with sodium borohydride-palladium chloride after alkali treatment indicated the absence of a carbohydrate-protein linkage. Measurements of physicochemical characteristics before and following this reduction suggested that this reduction decreased the molecular weight to one-half, probably due to the two antiparallel chains coming apart. It was suggested that the bound silicon (109.2 ppm) present in this hyaluronate may bridge the two antiparallel chains through the neutral sugars and/or through the hydroxyl group of the hexuronic acid moiety (Varma et al., 1974).

The vitreous humor hyaluronate is one of the most studied hyaluronates. However, the physicochemical properties reported by different investigators vary due to the differences in protein contents and the varying degrees of degradation associated with different methods used for its isolation. Laurent (1955) prepared hyaluronate from the cattle eye which by light scattering appeared to be a random coil of an average molecular weight of 340,000-500,000. The large polysaccharide degradable with hyaluronidases and isolated from gel part of the vitreous had a M_r of 420,000 (Balazs, 1965). Balazs et al. (1959) reported changes during development and aging in the vitreous gel and its glycosaminoglycans. Berman (1962) isolated a hyaluronate preparation which was separated on DEAE-Sephadex (A-50, medium) into two fractions by elution with 0.94 and 0.96 M NaCl fractions. These fractions contained 4.2 % and 5 % protein and had limiting viscosity numbers of 980 and 1880 g/ml, respectively. Balazs (1961) has reviewed the molecular morphology of the vitreous body. A number of other studies have been reported and some of the details of these may be found elsewhere (Balazs, 1965, 1977).

7.4.2.2 Sulfated glycosaminoglycans in vitreous humor

Although hyaluronate was considered as the only GAG of the vitreous humor, the preparations obtained by various investigators were reported to contain galactosamine and sulfur, which are not components of hyaluronate. Laurent (1955) reported a sulfur content of 0.08 % of the dry weight and 1.7 % galactosamine in his preparation but the presence of the sulfated GAG was not suspected or mentioned. Balazs et al. (1959 a) noted sulfur values between 0.2-0.4 mg/dl of the vitreous humor along with a galactosamine content of 20-30 % in the embryos and calves and only 5 % in the adult steers. Since the molar concentration of ester sulfate seemed to be approximately equivalent to that of galactosamine, Balazs et al. (1959 a) suggested that galactsoamine-containing sulfated macro-molecules might be present in the bovine vitreous humor.

The vitreous body of tuna contained a glucosamine/galactosamine ratio of 2.0 (Kobets, 1977). A sulfated galactosamine-containing proteoglycan was isolated from the human vitreous humor by pronase digestion followed by alcohol precipitation and zone elctrophoresis (Breen et al., 1977). The sulfated glycosaminoglycans have been isolated and characterized from bovine vitreous humor in our laboratory (Allen et al., 1977). The sulfated GAG consisted of 3 % of the total glycosaminoglycans isolated from the protease-digested bovine vitreous body by stepwise fractionation of AG 1 X-2 (Cl^-) and by gel filtration on Biogel P-300 columns. The isolated sulfated GAG were identified by digestion with specific chondroitinases AC and ABC and were found to consist of two heparan sulfate and two chondroitin sulfate fractions. The two heparan sulfate fractions were undersulfated and contained the same N-sulfate/O-sulfate ratio of 0.5, although the total sulfate content of the two fractions differed by approximately 100 %. Digestion with chondroitinases ruled out the presence of any chondroitin 6-sulfate.

A more recent study on the embryonic chick vitreous body reported the presence of chondroitin sulfate as the major sulfated GAG (Smith and Newsome, 1978). Based on radio-incorporation studies the vitreous humor chondroitin sulfate was found to be initially synthesized by the neural retina and was then secreted into the vitreous space. Dabrowska (1975) reported that the vibrations affect the metabolism of sulfomucopolysaccharide. The incorporation of intraperitoneally injected $^{35}SO_4$ was reduced in proportion to the length of exposure to vibration. An injection of methylprednisone in rabbits increased the GAG content of the vitreous humor by producing an increase of hyaluronate (Kaplan and Fisher, 1964). On the other hand, some of the drugs have adverse effects on vitreous humor (Nicholson, 1976).

7.4.3 Changes in vitreous humor during development and aging

The newborn rabbit contained 2 mg nitrogen/ml of the vitreous humor, which decreased to the adult level of 0.2 mg/ml in 15-18 days (Bembridge and Antoinette, 1951). Balazs et al. (1959) studied changes in glycosaminoglycans during development of calves, cattle, etc.. In the homogenized vitreous of 4-9 months old bovine embryo, the concentration of hyaluronic acid was 50 µg/ml, which increased to 10 times in the three-year-old adults. The molecular weight of hyaluronate calculated from the limiting viscosity did not change during development of the vitreous. In human infants the hyaluronate of the vitreous is less than that of the adults. With aging the liquification of the vitreous gel (semisolid) occurred. The changes in the macromolecular components, including hyaluronic acid, collagen in human eye vitreous in senescence have been extensively reviewed (Anderson, 1971; Osterlin, 1975; Balazs, 1977). The gel structure breaks, loculi of fluid form in the cavity and the remaining gel condenses. These condensations and partial opacities in the vitreous produce shadows on the retina (muscae volitantes or vitreous floaters), which commonly annoy older people. Tractions of the vitreous gel on retina produce bright streaks of light in the temporal light (Moore's lightning streaks). A shower of vitreous floaters with such light streaks signals a retinal detachment (see below).

7.5 Subretinal fluid

This is a pathological fluid which accumulates between the pigmented epithelial layer and the layer of rods of the retina in detachment of retina. Since the retinal detachment could result from several different causes, the source of this fluid is not precisely known.

The retina is attached to the choroid by almost no tissue fibers. It is so loosely attached that even the slightest excess pressure will result in retinal detachment. Thus the retina is held to choroid by the pressure of the vitreous humor and this pressure is nicely regulated. Thus experimental retinal detachment may be produced in experimental animals by aspiration or by injection into vitreous of substances like concentrated sodium chloride solution, hyaluronic acid, sulfate or mercurial sublimate, which cause shrinkage of the vitreous or by experimental induction of inflammatory exudate from the choroid and by other means (Foulds, 1969). These serve as models for retinal detachments in human subjects.

Liquification of the vitreous such as occurring naturally or that induced experimentally by hyaluronidase injection in the vitreous can produce primary retinal detachments and formation of holes in retina (Aggarwal et al., 1969). Analysis of the glycosaminoglycans by DEAE-Sephadex (A-50) fractions revealed that whereas normal vitreous had 5 GAG fractions, only three fractions were present in the vitreous from rats with hyaluronidase injected vitreous. It was suggested that the low molecular weight hyaluronic acid products produced a reaction upon coming in contact with retina and produced retinal hole and detachment (Aggarwal et al., 1969). If this is the case, it will explain the vitreous origin of the subretinal fluid and also the presence of hyaluronic acid in the subretinal fluid reported in 15 cases of retinal detachment by Godtfredsen (1949) and by other workers. Other workers (Seimiya et al., 1974) also have used a similar approach for experimental retinal detachment induction by lytic technics.

A trauma that ruptures the retina will also cause a separation of the retina from the choroid and the accumulation of the subretinal fluid (Hudson, 1969). In 18 out of 21 cases of primary serous retinal detachments, the subretinal fluid contained hyaluronic acid but no hyaluronidase (Godtfredsen, 1950). It was suggested that local retinal ischemia caused cystic degeneration and formation of the retinal tear. Hyaluronidase was liberated during these retinal changes which broke down the vitreous body and the breakdown products flowed into the subretinal space as the subretinal fluid (Godtfredsen, 1950). When a visible hole was present the subretinal fluid in retinal detachment had higher concentrations of the protein and ascorbic acid (Heath et al., 1962).

Besides the aforementioned reports only limited analyses have been carried out on the subretinal fluid, since the cryosurgical and other treatments of the retinal detachments are carried generally without drainage of this fluid (Johnston et al., 1975). Analysis of a number of subretinal fluid revealed that 10-11 % of the total hexosamine present was galactosamine and also the amount of hexosamine present was more than that can be accounted for by the hyaluronate content of this fluid. Also, sialic acid (1.24 % average) was present. These observations indicate the presence of glycoproteins, in additio to the hyaluronic acid, which varied very widely from less than 0.1 mg/ml to over 3 mg/ml (Sweeney et al., 1962) depending upon the type and duration of the retinal detachment. Laszczyk (1969) has reported the results of the attempt in differentiation of the retinal detachment types on the basis of the cytological picture of the subretinal fluid.

The physiology of retina (Kaneko, 1979) and the biochemistry of retinal disorders (Reading, 1970, 1974) have been reviewed.

7.6 Bibliography

1. Aggarwal, L. P., Dhiri, S., Khosla, P. K. and Srivastava, O. P. (1969): Experimental retinal detachment. Mod. Probl. Ophtalmol. (S. Karger, Basel) 8, 106-110

2. Allen, W. S., Otterbein, E. C. and Wardi, A. H. (1977): Isolation and characterization of the sulfated glycosaminoglycans of the vitreous body. Biochim. Biophys. Acta 498, 167-175

3. Allen, W. S. and Wardi, A. H. (1973): Pentose-rich glycoprotein from bovine vitreous humor. Isolation and partial characterization. Biochim. Biophys. Acta 297, 1-10

4. Anderson, B. Jr. (1971): The aging eye. Postgraduate Med. 50, 235-239

5. Andrews, A. D. and Dunn, J. T. (1973): Effects of TSH on the chemical composition of orbital tissue in guinea pigs. Endocrinology 93, 527-531

6. Balazs, E. A. (1958): Physical chemistry of hyaluronic acid. Fed. Proc. 17, 1086-1093

7. Balazs, E. A. (1961): Molecular morphology of the vitreous body. In: G. K. Smelser (Ed.). The Structure of the Eye. Academic Press, New York, 293-310

8. Balazs, E. A. (1965): Amino sugar-containing macromolecules in the tissues of the eye and the ear. In: E. A. Balazs and R. W. Jeanloz (Eds.). Amino Sugars, Vol. II A. Academic Press, New York, Chap. 37, 401-460

9. Balazs, E. A. (1968): The molecular biology of the vitreous. In: A. McPherson (Ed.). New and Controversial Aspects of Retinal Detachment. Harper and Row, New York, 3-15

10. Balazs, E. A. (1977): Intercellular matrix of connective tissue. In: C. E. Finch and L. Hayflick (Eds.). Handbook of Biology of Aging. Van Nostrand Reinhold, New York, Chap. 9, 222-240

11. Balazs, E. A., Laurent, T. C. and Laurent, U. B. G. (1959): Studies on the structure of the vitreous body. VI. Biochemical changes during development. J. Biol. Chem. 234, 422-430

12. Balazs, E. A., Laurent, T. C., Laurent, U. B. G., DeRoche, M. H. and Bunney, D. M. (1959 a): Studies on the structure of the vitreous body. VIII. Comparative biochemistry. Arch. Biochem. Biophys. 81, 464-479

13. Bembridge, A. and Antoinette, P. (1951): Biochemical and histological changes in developing rabbit eyes. British J. Ophthalmol. 35, 784-789

14. Berggren, L., Wessler, E. and Wennerstrom, J. (1969): Intraocular pressure and excretion of mucopolysaccharides in osteogenesis imperfecta. Acta Ophthal. 47, 122-128

15. Berman, E. R. (1962): Isolation of bovine vitreous hyaluronic acid on diethylaminoethyl-Sephadex. Biochim. Biophys. Acta 58, 120-122

16. Breen, M., Bizell, J. W. and Weinstein, H. G. (1977): A galactosamine containing proteoglycan in human vitreous. Exp. Eye Res. 24, 409-412

17. Brown, J. L. and Geeraets, W. J. (1973): Effect of hyaluronidase on the facility of outflow in normal and buphthalmic rabbits. Acta Ophthalmol. 50, 486-494

18. Brunish, R. (1962): The nature of the lacrimal mucoprotein stimulated by exophthalmos producing substance. J. Biol. Chem. 237, 2816-2819

19. Cole, D. F. (1970): Aqueous and ciliary body. In: C. N. Graymore (Ed.): Biochemistry of the Eye. Academic Press, New York, Chap. 2, 105-181

20. Dabrowska, J. (1975): Effect of vibration on sulfomucopolysaccharide metabolism in the eye. Invest. Ophthalmol. 14, 250-252

21. Falbe-Hansen, I. and Jensen, O. A. (1969): Mucosubstances of the acini of the human lacrimal gland (orbital part). II. Biochemical identification. Acta Ophthalmol. 47, 620-624

22. Filotto, U. (1954): Morphological and histochemical research on lacrimal glands of some domestic mammals. Riv. istochim. norm. et patol. 1 125-142

23. Flanagan, J. C. (1979): In: A. Hornblass (Ed.). Tumors of the Ocular Adenoxa and Orbit. Mosby, St. Louis, 89-101

24. Foulds, W. S. (1969): Experimental detachment of the retina and its effect on the intraocular fluid dynamics. Mod. Probl. Ophthal. 8, 51-63

25. Francois, J. (1975): Importance of the mucopolysaccharides in intraocular pressure regulation. Invest. Ophthalmol. 14, 173-176

26. Francois, J., Victoria-Troncoso, V., Maudgal, P. C. et al. (1974): Mucopolysaccharides of the chamber angle and ocular pressure. Bull. Soc. Belge. Ophthalmol. 168, 765-776

27. Furukawa, R. E., Polse, K. A. and Emori, Y. (1976): Slit lamp flurophotometry. Opt. Eng. 15, 321-324

28. Godtfredsen, E. (1949): Investigations into hyaluronic acid and hyaluronidase in subretinal fluid in retinal detachment, partly due to ruptures and partly secondary to malignant choroidal melanoma Preliminary report suggesting new hypothesis concerning pathogenesis of retinal detachment. Brit. J. Ophth. 33, 721-732

29. Godtfredsen, E. (1950): Further investigations into hyaluronic acid and hyaluronidase in the subretinal fluid in retinal detachment. Acta XVI Concilium Ophthalmol. Brittania 2, 1162-1168

30. Godtfredsen, E. (1954): Mesenchymal aspects in ophthalmology. In: G. Asboe-Hansen (Ed.). Connective Tissue in Health and Disease. Munksgard, Copenhagen, 296-307

31. Grierson, I. and Lee, W. R. (1975): Acid mucopolysaccharides in the outflow. Exp. Eye Res. 21, 417-431

32. Guyton, A. C., Taylor, A. E. and Granger, H. J. (1975): Dynamics and Control of Body Fluids. Circulatory Physiology II., W. B. Saunders, Philadelphia, Chap. 15, 206-224

33. Haberich, F. J. (1978): The composition of lacrimal fluids and their importance during tearing with contact lenses. DOZ, Dtsch. Optikerztg. 1978, 243-249 (Ger)

34. Heath, H., Beck, T. C. and Foulds, W. C. (1962): Chemical composition of subretinal fluid. Br. J. Ophthalmol. 46, 385-396

35. Houssay, A. B., and Blumenkrantz, N. (1964): Changes in glycoproteins in serum, urine, and aqueous humor after gonadectomy or sex hormone injections in rats. Endocrinology 74, 825-832

36. Hudson, J. R. (1969): The role of trauma in retinal detachments. Modern Probl. Ophthal. 8, 235-241

37. Iwata, S. (1976): Surface Chemistry of Tears. Hyomen 14, 464-471 (Japan)

38. Jensen, O. A. (1970): Mucosubstances of mixed tumors of the human lacrimal gland. Histochemical and histogenic aspects. Acta Pathol. Microbiol. Scand., Sect. A 78, 110-126

39. Jensen. O. A., Falbe-Hansen. I., Jacobsen, T. and Michelsen, A. (1969): Mucosubstance of the acini of the lacrimal gland (orbital part). I. Histochemical identification. Acta Ophthalmol. 47, 605-619

40. Johnston, G. P., Okun, E., Boniuk, O. and Arribas, N. P. (1975): Drainage of Subretinal Fluid: Why, when, where and how. Mod. Probl. Ophthal. (S. Karger, Basel) 15, 197-206

41. Jones, L. T. (1970): The lacrimal apparatus. In: Ophthalmic Anatomy, a manual. Am. Acad. Ophthal. Otolaryngol., Rochester, Minnesota, USA

42. Kaneko, A. (1979): Physiology of retina. Ann. Rev. Neurosci. 2, 169-191

43. Kaplan, D. and Fisher, B. (1964): The effect of methylprednisolone on mucopolysaccharide of rabbit vitreous humor and coastal cartilage. Biochim. Biophys. Acta 83, 102-112

44. Kasavina, B. S. and Pantieleva, V. M. (1969): The effect of corticosteroids on mucopolysaccharides of the aqueous humor and the status of regulation of ophthalmotonus. Vestn. Oftal. 82, 7-11 (Russ)

45. Kasavina, B. S., Pantieleva, V. M. and Shapkina, A. M. (1970): Effect of glucocorticoids on ocular mucopolysaccharides in rabbits and the regulation of intraocular tension. Vestn. Oftal'mol., 13-17

46. Kasavina, B. S. and Vinetskaya, M. I. (1976): Characteristics in changes of mucopolysaccharides in various pathological processes of eye. Oftal'mol. Zh. 31, 346-349 (Russ)

47. Kashintseva, L. T. and Plevinskis, V. P. (1973): Change in the mucopolysaccharide content in the tissues of the ocular drainage system in primary glaucoma and its role in the development of the glaucomatous process (a histochmical study). Oftal'mol. Zh. 28, 57-62 (Russ)

48. Kobets, M. L. (1977): Chemical composition of the vitreous body of the tuna. Izv. Sib. Otd. Akad. Nauk SSSR, Ser. Biol. Nauk, 96-104

49. Kuehnel, W. (1968): Comparative histological, histochemical, and electron microscopic studies of tear glands. I. Rabbits and cats. Z. Zellforsch. Mikrosk. Anat. 85, 408-440

50. Kuehnel, W. (1968 a): Comparative histological, histochemical and electron microscopic studies of lacrimal glands. III. Goat. Z. Zellforsch. Mikrosk. Anat. 86, 430-443

51. Laszczyk, W. A. (1969): Trials on the differentiation of retinal detachment on the basis of cytological pictures of the subretinal fluid. Klin. Oczna 39, 227-230

52. Laurent, T. C. (1955): Studies on hyaluronic acid in the vitreous body. J. Biol. Chem. 216, 263-271

53. Lieb, W. A. and Staerk, N. (1966): Interrelation of ascorbic acid and facility of outlfow; steroid hormones as possible regulating mechanisms of intraocular pressure. Drug Mech. Glaucoma, 105-133

54. Linn, J. G., Jr. and Ozment, T. L. (1950): Some effects of injection of hyaluronidase into the anterior chamber. Am. J. Ophthal. 33, 33-44

55. Maren, T. H. (1974): The cerebrospinal fluid. With notes on aqueous humor and endolymph. In: V. B. Mountcastle (Ed.). Medical Physiology, 13th Ed., Mosby, St. Louis, Vol. 2, 1116-1141

56. Meyer, K. and Palmer, J. W. (1934): The polysaccharides of the vitreous humor. J. Biol. Chem. 107, 629-634

57. Meyer, K. and Palmer, J. W. (1936): The nature of the ocular fluids. Am. J. Ophth. 19, 859-865

58. Meyer, K. and Palmer, J. W. (1936 a): On glycoproteins. II. The polysaccharides of vitreous humor and of umbilical cord. J. Biol. Chem. 114, 689-703

59. Mimura, Y (1973): Eye system and glycosaminoglycan. Kagaku No Ryoiki, Zakan 99, 155-179

60. Nadler, H. L., Rao, G. J. S. and Taussig, L. M. (1978): Cystic fibrosis. In: J. B. Stanbury, J. B. Wyngaarden and D. S. Fredrickson (Eds.). The metabolic Basis of Inherited Disease. McGraw Hill, New York, Chap. 73, 1683-1710

61. Nicholson, D. H. (1976): Detection of Ocular Toxicity. In: Mod. Pharmacol.-Toxicol., 6 (Horiz. Clin. Pharmacol.) Chap. 9, 67-81

62. Osterlin, S. (1975): Changes in the vitreous with age. Trans. Ophthalmol. Soc. U. K. 95, Pt. 3, 372-377

63. Pahuja, S. L. (1977): Evidence for the association of protein with hyaluronic acid in vitreous body. Univ. Minnesota, Minneapolis, Minn. Diss. Abstr. Int. B. 37 (12, Pt. 1), 6102-6103

64. Pantieliva, V. M., Kasavina, B. S. and Shapkina, A. M. (1971): Changes in the tissue structure, mucopolysaccharides level, and hydrodynamics in the eyes of rabbits under the effect of epinephrine. Vestn. Oftal'mol., 39-42

65. Perkins, E. S. (1965): Steroid-induced glaucoma. Proc. Roy. Soc. Med. 58, 531-533

66. Reading, H. W. (1970): Biochemistry of Retinal Dystrophy. J. Med. Genet. 7, 277-284

67. Reading, H. (1974): Biochemistry of Retinal Degeneration. Biochem. Soc. Trans. 2, 1230-1233

68. Schachtschabel, D. O., Bigalke, B. and Rohen, J. W. (1977): Production of glycosaminoglycans by cell cultures of the trabecular meshwork of the primate eye. Exp. Eye Res. 24, 71-80

69. Seimilya, T., Takahashi, J., Hayasaka, S. et al. (1974): Experimental retinal detachment due to lysis of mucoprotein. Acta Soc. Ophthalmol. Jpn. 78, 808-818 (Japan)

70. Singh, S. P. and McKenzie, J. M. (1970): Influence of thyrotropin on harderian gland glycosaminoglycans of the mouse. Proc. Soc. Exp. Biol. Med. 135, 478-482

71. Smith, G. N., Jr. and Newsome, D. A. (1978): The nature and origin of the glycosaminoglycans of the embryonic chick vitreous body. Dev. Biol. 62, 65-77

72. Sweeney, D. B., Balazs, E. A. and Karlin, D. B. (1962): Invest. Ophthalmol. 1, 427, Quoted by Balazs (1965)

73. Tawara, A. (1978): Demonstration of acid mucopolysaccharides within and around endothelial cells of normal human trabecular meshwork. Nippon Ganka Gakkai Zasshi 82, 335-346

74. Thill, H. J. (1976): Biochemistry of aqueous humor formation. Buech. Augenartzes 1976, 69 (Fortschr. Diagn. Ther. Primärer Glaukomas, Hauptref. Essener Fortbild. Augenärzte, 10th), 28-43 (Ger)

75. Van Buskirk, E. M. and Brett, J. (1978): The canine eye: In vitro dissolution of the barriers to aqueous outflow. Invest. Ophthalmol. Visual Sci. 17, 258-263

76. Varma, R., Allen, W. S. and Wardi, A. H. (1973): The arabinose content of cerebral and vitreous humor hyaluronate-peptide. J. Neurochem. 20, 241-244

77. Varma, R., Varma, R. S., Allen, W. S. and Wardi, A. H. (1974): On the carbohydrate-protein linkage in vitreous humor hyaluronate. Biochim. Biophys. Acta 362, 584-588

78. Varma, R., Varma, R. S., Allen, W. S. and Wardi, A. H. (1967 a): Neutral sugars from the hyaluronate-peptide of vitreous humor. Carbohyd. Res. 32, 386-395

79. Varma, R., Varma, R. S., Allen, W. S. and Wardi, A. H. (1975): Hyaluronate from aqueous humor: Some physicochemical studies. Biochem. Exp. Biol. 11, 407-417

80. Wegelius, I., Naumann, J. and Brunish, R. (1959): Uptake of ^{35}S-labelled sulfate in the harderian and the ventral lacrimal glands of the guinea pig during stimulation with ophthalmotropic pituitary agents. Acta Endocr. (Kbh) 30, 53-60

81. Winand, R. J. and Kohn, L. D. (1973): Retrobulbar modifications in experimental exophthalmos. Effect of thyrotropin and an exophthalmos-producing substance derived from thyrotropin on the sulfate ion (sulfur-35) incorporation and glycosaminoglycan content of Harderian glands. Endocrinology 93, 670-680

Chapter 8

Bile

8.1 Physiological aspects

Bile is secreted by the cells of the liver at a rate of 500 ml/day and between the meals it is stored in the gallbladder. During this storage the mucosa of the gallbladder synthesizes and secretes the mucosubstances into the bile (Harding, 1931; Esterly and Spicer, 1968). The cystic duct from the gallbladder joins the hepatic duct coming out from the liver to form the common bile duct, which just before entering the duodenum joins the pancreatic duct.

In response to food in the mouth, the sphinctor around the orifice, (the sphinctor of Oddi) of the bile duct in the duodenum relaxes. When the gastric contents are emptied into the duodenum a gastrointestinal hormone *Cholecystokinin-Pancreozymin* (CCK-PZ) is secreted by the intestinal mucosa, which causes the contraction of the gallbladder and the flow of the bile into the duodenum, and also the secretion of enzyme-rich juice from the pancreas. Fatty acids of more than ten carbon atoms and some amino acids, particularly tryptophan and phenylalanine present in the contents entering the duodenum are the stimuli for the secretion of CCK-PZ. The bile and the pancreatic juice entering the duodenum in response to this hormone promote the digestion of fats and protein, respectively. The digestive products so formed further stimulate secretion of CCK-PZ in a positive feedback control manner.

8.2 Source and functions of bile mucins

Gallbladder is the source of mucosubstances in the bile. Ontogenetically the gallbladder develops from the primordial intestine before the 16th week of gestation (Bianchi, 1977). Histochemical studies have shown that the bile mucus is secreted by the epithelial cells and the true mucinous glands present predominantly in the neck region of the gallbladder (Harding, 1931; Hulten, 1968; Esterly and Spicer, 1968).

The bile mucus has been claimed to protect the epithelium from attack by digestive enzymes. A mucus may exist in the form of a gel or a solute. The interconversion of the two states can be brought about by a conformational change in the mucin glycoprotein. While investigating radioactive mucoproteins in pig gastric mucosa, Snary and Allen (1972) observed that sodium deoxycholate (0.2 %, w/v) solubilized 37.7 % of the water-soluble mucoprotein radioactivity. Thus the detergent action of the bile salts may convert the mucin gel into soluble form in vivo, probably, by a conformational change induced in the macromolecule. Consistent with this is the report of the decrease in viscosity and plasticity of the mucus due to a structural breakdown of the mucus affected by bile salts and synthetic dodecyl sulfate (Martin et al., 1978). In this respect it is necessary to allude to the protective effect of the sulfated acidic glycosaminoglycans and sulfated glycoproteins on hepatic damage. A 0.3 ml/kg body weight of carbon tetrachloride given through the gastric tube produced acute liver damage but no such liver damage occurred in rabbits receiving chondroitin sulfate at 5 mg/kg body weight along with the same quantity of carbon tetrachloride (Jui-Yun, 1960).

The undesirable role of the presence of mucus in bile is the connection of the sulfated glycoproteins and sulfated glycosaminoglycans with the pathogenesis of gallstones (see 8.5.1).

8.3 Mucosubstances in bile

Three types of mucosubstances have been identified in bile. These are: neutral glycoprotein, sulfated glycoproteins and acidic glycosaminoglycans. The mucus present in bile is a soluble type of mucus and was originally considered to be a fucomucin containing only hexosamine, galactose, fucose and sialic acid (Odin, 1957). With the advancement and sophistication of methods of analysis, sulfate and hexuronic acid have been identified as components of sulfated glycoproteins and acidic glycosaminoglycans.

The presence of carbohydrates in the protein material in bile isolated from various origins has been known since 1905 when Hammarstem first observed reducing substances in the hydrolyzates of protein isolated from bile. A glycoprotein having blood group activity was isolated from a patient with cancer of the common bile duct. This glycoprotein contained nitrogen (10 %), hexosamine (14 %) and 24 % hexose (Tiba, 1950). Werner (1953) obtained similar analytical results on a glycoprotein isolated from the white bile of a patient with hydrops vesical fellea. This glycoprotein contained 6.8 % nitrogen, 23.6 % hexosamine, 20.3 % hexose, 7.6 % L-fucose and 5.7 % sialic acid. A trichloroacetic acid-soluble glycoprotein fraction isolated from the bile of a patient with duct obstruction contained per 100 ml bile, 0.85 mg galactose, 0.4 mg fucose, 0.89 mg hexosamine and 4.6 mg protein (Dische, 1960). Maxfield and Wolins (1959) isolated from bile a homogeneous mucoprotein which resembled urinary mucoprotein of Tamm-Horsfall. These authors suggested that it may play a role in gallstone formation similar to that for urinary mucoprotein in urolithiasis. Serum and biliary mucoprotein levels were found to change in hepatobiliary damage (Jui-Yun, 1960). A preparation isolated from bile was a glycoprotein containing 5 % nitrogen, 35.7 % hexosamine, 16.2 % L-fucose and 5 % sialic acid. A decrease in hexosamine content of bile was reported in a patient with gallbladder stone (Bouchier et al., 1965). The hexosamine content in the patient's bile was 5.7 ± 4.9 mg/dl as compared to 8.3 ± 5.3 mg/dl in the normal gallbladder bile. This was related to the decreased viscosity of the patient's bile.

The presence of a sulfated glycoprotein in the epithelium of 38 surgically removed normal and diseased human gallbladders was reported by Hakkinen and Laitio (1970). This sulfated glycoprotein was present as a specific antigen in gallbladder mucosa extracts and it was detected by histochemical and immuno-fluorescence methods using specific rabbit antisera. Both neutral and sulfated glycoproteins were present in normal human gallbladder bile samples and accounted for 70-80 % of the non-dialyzable carbohydrate content of bile. The remainder of 10-30 % of the carbohydrates came from plasma glycoproteins present in bile. In some of these sulfated glycoproteins the hexosamine/sulfate molar ratio was unity (Schrager et al., 1972).

Matsushiro et al. (1970) separated glycoproteins in bile into water-soluble and water-insoluble fractions from the dialyzed retentate of the acid-soluble fraction obtained by acidification of diluted bile to 0.5 M HCl concentration. The water-insoluble fraction was virtually absent in normal control biles, while the water-soluble fraction consisting of glycoproteins was present in all normal and abnormal biles. After solubilization by papain digestion, the sulfated glycoproteins and glycosaminoglycans were present in the 1.0 and 1.25 M NaCl fractions from a Dowex 1-X2 column. Hybrids of a highly sulfated glycoprotein and sulfated glycosaminoglycans with electrophoretic mobilities similar to heparan sulfate and chondroitin sulfate were also present in calcium stones (Matsushiro and Nemoto, 1968).

The presence of the sulfated glycoproteins and sulfated glycosaminoglycans has been reported both in pathological human bile (Nagashima et al., 1974 a) and in normal human bile (Nagashima et al., 1974).

The mucopolysaccharides present in human gallbladder and bile have been isolated and characterized (Terho and Laito, 1974). In addition to a small amount of acidic glycosaminoglycans, neutral and acidic mucin glycoproteins from the major portion of the non-dialyzable hexosamine. The major portion of the 500 µg/g wet weight of the human gallbladder wall (Terho and Laito, 1974) and 70-80 % of the non-dialyzable carbohydrates of the human gallbladder bile (Schrager et al., 1972) comprised of the neutral and acidic glycoproteins. The polysaccharide parts of these so-called mucins were composed of glucosamine, galactosamine and galactose in the ratio of 3 : 1 : 4 or 9 : 1 : 10 and contained different amounts of fucose, sialic acid and sulfate (Terho and Laito, 1974). The polysaccharides from the gallbladder could initially be separated into fractions soluble and insoluble in 0.9 % NaCl; the latter supposedly originated from the structural or membrane-bound glycoproteins. In addition to large molecular weight fractions, which could be separated by anion-exchange chromatography on the basis of sulfate/sialic acid molar ratios of 0.1-0.2, large amounts of small molecular weight polysaccharides, consisting of varying amounts of glucosamine, mannose, galactose, fucose and sialic acid were also present (Terho and Laito, 1974).

Besides the human bile, mucins isolated from the bile of several animal species have been studied. Bile mucins were isolated from cattle, pig and human bile by dialysis against water and veronal buffer of pH 8.6, cetylpyridinium chloride precipitation of the supernatent, extractions with 20 % (w/v) $CaCl_2$, precipitation of the supernatent with 30-70 % ethanol. These mucins contained 20-30 % protein, sialic acid, fucose, galactose and hexosamine (Hashimoto et al., 1961). The fucose and sialic acid contents were inversely related. However, these mucins differed in sialic acid contents; the bovine product contained the highest, while human bile mucin contained the least sialic acid contents (Hashimoto et al., 1961).

In another investigation (Neiderhiser et al., 1971) two distinct mucins with blood group A and H activitiy were isolated from blood group A and H pigs' bile, respectively. These purified mucins constituted more than 85 % (80-90 %) of the total macromolecular carbohydrate content of glucosamine, galactosamine, galactose and fucose in pig bile. The pig bile contained 0.64-4.80 mg/ml of protein and was similar in size and composition to blood group glycoproteins isolated from gastric and intestinal mucosa and the ovarian cyst fluid. The relative proportion of glucosamine and galactosamine varied in the blood group A and H glycoproteins; the glucosamine/galactosamine ratio was unity in the former and three in the latter. The pig bile mucin preparations contained only a small amount of sialic acid and had an isoelectric pH of 2.1 and were not retained on a Dowex 1-X2 column (Neiderhiser et al., 1971). Although these authors did not analyze the mucins for sulfate, the physicochemical characteristics suggested the possible presence of at least a small sulfate content.

8.4 Hexuronic acid conjugates in bile

Hexuronic acid is present in bile in either the low molecular weight conjugate of the bile pigments (bilirubin and biliverdin diglucuronides) or as a component of the macromolecules. Besides the glycosaminoglycans, a macromolecular compound of bilirubin called *bilimucin* has been isolated from human bile. This macromolecule was isolated by ethanolic precipitation of the bile in the dark and Sephadex G-200 gel filtration of the precipitate dissolved in bicarbonate buffer. It has been considered to be a macromolecular complex in which bilirubin may be linked to acidic glycosaminoglycans (Gent et al., 1971). The bile-pigment diglucuronides are formed in the liver, while the acidic glycosaminoglycans of bile are synthesized by the epithelium of the gallbladder wall. The origin of the *bilimucin* is not known.

The first report of the presence of hexuronic acid associated with macromolecules in bile was, probably, that of Miyake and Katsuki (1956). Upon paper electrophoresis of bile dialyzed in phosphate buffer, these workers noticed a specific component (in addition to plasma glycoproteins) which upon acid hydrolysis contained hexuronic acid, glucose, galactose and hexosamine. The presence of sulfate esters in locally synthesized mucosubstances was detected histochemically in the mucosa of gallbladders from normal human subjects and from patients with cholelithiasis, cystic fibrosis and adenocarcinoma (Esterly and Spicer, 1968). The presence of mucopolysaccharides in bile and their significant role in gallstone formation was emphasized (Nakamura, 1967). The sulfated glycopeptides and sulfated glycosaminoglycans were present in 1.25 M and 1.5 M NaCl fractions from anion-exchange chromatography of the papain digests of calcium carbonate stones (Matsushiro and Nemoto, 1968). Sulfated glycoproteins along with sulfated glycosaminoglycans were absent from normal bile but were present only in the 1.0 and 1.25 M NaCl fractions from anion-exchange chromatography of the papain-digested water-insoluble fractions of the biles from patients with abnormal liver function or with gallstone (Matsushiro et al., 1970). These two fractions contained hexose, hexosamine, sulfate and peptide as the major constituents, with small amounts of L-fucose and uronic acid. Their electrophoretic mobilities were similar to those of heparan sulfate and chondroitin 4-sulfate.

Sulfated glycosaminoglycans along with sulfated glycoproteins, originally thought to be present only in bile in diseased state (Nagashima et al., 1974 a), have also been reported in normal human bile (Nagashima et al., 1974). These exert strong coagulating effects on calcium carbonate suspensions in water in vitro and may have a similar effect in producing gallstones However, it has been suggested that the presence of these sulfated macromolecules in normal bile might be a result of bacterial infection, although these samples were prepared by collecting bile under aseptic conditions and fractionation by centrifugation in presence of salts (Nagashima et al., 1974).

In human gallbladder wall, chondroitin sulfates formed the major acidic glycosaminoglycan, although hyaluronic acid, heparan sulfate, dermatan sulfate and chondroitin 4- and/or 6-sulfates were present in this tissue, primarily as proteoglycans. Most of the hyaluronic acid and proteoglycans of 4- and/or 6-sulfates were extractable with 0.9 % NaCl, whereas the dermatan sulfate proteoglycan was almost insoluble in 0.9 % NaCl (Terho and Laito, 1974). In human bile only a trace amount of acidic glycosaminoglycans were present and these consisted of hyaluronic acid and chondroitin 4- and 6-sulfates.

Hyaluronidase activity was present in the bile from dog. Administration of sodium glycocholate (20 mg/kg, i.v.) increased bile secretion and decreased hyaluronidase activity in the bile (Mel'nikov et al., 1972). Examination of the so affected dog's bile ducts showed depolymerization of the nonsulfated acidic glycosaminoglycans of the connective tissue and intercellular spaces.

Besides the human bile, hexuronic acid-containing materials have been detected in bile and gallbladders from several animal species. Pig bile contained only a trace amount of acidic glycosaminoglycans; less than 1 % of hexosamine in bile could be attributed to these macromolecules (Neiderhiser et al., 1971). Morphochemical analysis by histochemical examination revealed the presence of sulfated and carboxylated mucopolysaccharide-protein complexes in the epithelium of fowl gallbladder (Yamada and Hoshino, 1972). Similarly, sulfomucin-protein complexes were detected in striated border and in granules grouped in the apical region of columner epithelium of cat gallbladder (Varute and Nalavade, 1973). A series of acidic mucpolysaccharide-containing ultrastructures were observed upon electron microscopic examination of the rabbit gallbladder epithelium stained with dialyzed iron, colloidal Th and ruthenium red (Yamada, 1974).

8.5 Changes in physiological and pathological states

A large number of physiological and pathological states affect the functions of the liver, which in turn directly or indirectly affect the secretion of bile and the composition of mucosubstances in bile. Since the mucosubstances are secreted by the epithelium of the gallbladder, the conditions which involve the inflammation or hyperplasia of this mucous membrane affect its secretory activity and the rates of mucus addition to the bile. Obstruction of the cystic and bile ducts also affect the mucus composition of the bile.

8.5.1 Mucopolysaccharides of bile and gallstone in cholelithiasis

Formation of gallstone in the gallbladder (cholelithiasis) and bile passages of the liver occurs in some pathological states. The etiological factors include high serum cholesterol level, high bile cholesterol, stasis of bile, infection etc. Depending upon the etiological factor, different types of gallstones are formed. Calcium bilirubinate gallstones develop in 10 % of the patients with hemolytic jaundice and are present as small, multiple black stones. Calcium carbonate gallstones are grey, amorphous, solitary stones. Cholesterol gallstones develop as solitary, metabolic stones. Only in 10 % of the cases the gallstones are pure calcium carbonate, cholesterol or calcium bilirubinate stones. The gallstones formed in infection in 90 % of the cases are of the mixed, multifaceted type (Lonsdale, 1968).

Human gallstones contain glycoproteins. The changes in structure and physicochemical properties of these biliary mucosubstances have been suggested to be connected with the formation of gallstones (Nakamura, 1967). Womack et al. (1963) observed by histochemical staining technics that human gallstones have a carbohydrate-containing lattice or framework into which precipitation of calcium carbonate, cholesterol or calcium bilirubinate may occur. In the formation of human cholesterol gallstone, a mucin gel, normally present in bile, acts as a matrix in which precipitation and aggregation of cholesterol from bile takes place (Hulten, 1968). From these and other studies suggestions have been made that the bile mucins, particularly the sulfated glycoproteins and sulfated glycosaminoglycans of abnormal bile, are significant in cholelithiasis. Another active area of research is related to the role of bile acids (Van der Berg et al., 1977) and lecithin in the pathogenesis of gallstones (Neiderhiser et al., 1966; Thistle, 1973; Lukie and Dietschy, 1973; Matern et al., 1975). Lecithin per se is important for maintaining cholesterol in an aqueous phase. Cholesterol, the principal constituent of gallstones, is relatively insoluble in aqueous solutions and is solubilized by lecithin and by bile salts through the formation of mixed micelles (Neiderhiser et al., 1966; Lukie and Dietschy, 1973; Lukie, 1976). These and other aspects of gallstone disease have been reviewed (Soloway, 1974; Hammerman et al., 1977; Billing, 1978). In another study, cholesterol supersaturation was found to be the predominant driving force for cholesterol precipitation and cholesterol stone formation (Carey and Small, 1978).

The significance of mucopolysaccharides in gallstone formation is evident from the results of analysis of calcium carbonate gallstones and calculus-like concrements. Histochemical examination of thin serial sections of these stones revealed the presence of mucopolysaccharides which were evenly distributed and filled the spaces between the calcium carbonate crystals in a reticular fashion. The mucopolysaccharides present were mainly of the sulfated form with some neutral or weakly acidic mucopolysaccharides (Nakamura, 1966). Similar results were obtained by Matsushiro and Nemoto (1968) on analysis of a calcium carbonate stone and a calcium carbonate mantle from a stone with a cholesterol center. Analyses of the hydrolyzates of these stones showed the presence of hexose, hexosamine and peptide as the major components and L-fucose, uronic acid, sulfate and sialic acid as the minor components. Papain digestion and fractionation on a DEAE-Sephadex A-25 column with stepwise NaCl elution gave fractions of glycopeptides, sulfated glycopeptides and a small amount of acidic mucopolysaccharides (glycosaminoglycans) in the 0.5, 0.7, 1.25 and 1.5 M NaCl fractions (Matsushiro and Nemoto, 1968). Hyaluronic acid was associated with the 0.7 M fraction, while the sulfated glycosaminoglycans (chondroitin and heparan sulfates) were present as hybrids with sulfated glycoproteins in the latter fractions.

Further support to the role of sulfated glycoproteins and sulfated mucopolysaccharides in the pathogenesis of gallstones is provided by a comparative analysis of mucopolysaccharide contents of normal and lithogenic bile (Gupta and Udupa, 1974; Maki et al., 1971), histochemical demonstration of mucopolysaccharides in calcium bilirubinate stones (Maki and Nakamura, 1969) and by the in vitro coagulation of aqueous suspensions of calcium carbonate by sulfated glycoproteins or sulfated glycosaminoglycans isolated from normal human bile (Nagashima et al., 1974) and from pathological human bile (Nagashima et al., 1974 a) and also by various other mucosubstances (Yamashita, 1971). Histochemical studies have been reported on hyperplasia of the secretory mucosa of the gallbladder in cholelithiasis, which would cause increased secretion of mucus in bile (Esterly and Spicer, 1968; Elfving et al., 1969).

The mucoid substances of the bile in biliary tract diseases have been investigated (Sheleketina and Leliukhina, 1974). The effect of estrogens upon the stone-forming bile (Tritapepe et al., 1975) and the pathophysiological aspects of gallbladder stones and calculi in the biliary tract have been reviewed (Leuschner, 1977).

8.5.2 Mucopolysaccharides of white bile in hydrops of gallbladder

Normal bile has a golden yellow color due to the presence of the glucuronides of the bile pigments, bilirubin and biliverdin. In case of obstruction of the bile from the gallbladder without infection, the mucin plugs the ducts (cystic and bile ducts), the gall bladder becomes filled and distended with mucin. The mucin so accumulated is called the *white bile* and the distension of gallbladder is termed the *hydrops of the gallbladder.*

The glycosaminoglycans of the white bile from hydrops of the gallbladder were isolated by papain digestion and cetylpyridinium chloride precipitation. The cellulose acetate electrophoresis of the isolated material revealed three Alcian blue-positive components. The first component with a mobility slower than hyaluronic acid was the major component in all white bile samples. The second component with a mobility similar to hyaluronic acid was a minor component and was absent in half of the samples. The third component was also a minor component, it was present in only some of the samples and moved between hyaluronic acid and chondroitin sulfate (Imada, 1970). Despite having electrophoretic mobilities similar to those of glycosaminoglycans, these components lacked hexuronic acid in their hydrolyzates and contained galactose, fucose, glucosamine, galactosamine and sialic acid as the carbohydrate components. The component I was composed of 27.3 % neutral sugars, 22.9 % amino sugars, 4.7 % sialic acid and 2.7 % nitrogen. The galactosamine : galactose : fucose : glucosamine molar ratios were 1.0 : 8.9 : 3.9 : 8.2.

8.5.3 Bile in cystic fibrosis and chronic irritation

The primary abnormality in cystic fibrosis is an alteration of secretion in most of the exocrine glands. The gall bladder mucosa synthesizes the glycoproteins and GAG present in bile and since bile is an exocrine secretion, this fluid is also affected in cystic fibrosis (Esterly and Spicer, 1968). The secretion of the mucous glands of the gallbladders of these patients is very viscous due to increased quantities of the mucins in it and it is precipitated to form casts, which cause obstruction to the outflow of secretion (Andersen, 1962). The alterations in chemical composition of the bile mucins in these patients is suggested by the deviations from the fucose/hexose and (sialic acid + fucose)/hexose ratios of 0.4-0.52 and 0.57-0.71, respectively, in normal human subjects (Dische, 1964).

Chronic irritation can also increase the mucus secretion from the epithelial cells due to metaplasia. The glands present in the normal state increase in number and spread throughout the mucosa of the gallbladder. These glands formed histologically resemble the pyloric and Brunner's glands with islands similar to gastric epithelium. This has been called gastro-intestinal heterotropia and has been reported to play a role in neoplasmogenesis (Jarvi, 1961; Jarvi and Meurman, 1964).

8.5.4 In tumors of gallbladder

Histochemical changes have been noted in the sections of adenomatous polyps and adenocarcinomas (Esterly and Spicer, 1968; Magri, 1968). A quantitative fall in the Alcian blue-positive GAG and PAS-positive glycoproteins was seen for carcinomas as compared with normal gallbladder mucosa. Histochemical investigations have also been carried out in 180 cases of benign tumors and pseudotumors of the gallbladder (Christensen and Ishak, 1970). In these cases where the glandular mucosa is affected the bile shows qualitative and quantitative changes.

8.5.5 In hepatic disorders

Metabolic alterations in the collagen and glycosaminoglycans of the hepatic connective tissue are encountered in a number of disease states, including alcoholic hepatitis, fibrogenesis, hepatic carcinoma, mucopolysaccharidoses, amyloidosis, regenerating liver, autogenesis and hypervitaminosis A. The qualitative and quantitative changes occurring in bile in response to these hepatic changes have not been investigated, although a survey of some of these aspects has been presented in a monograph (Bianchi et al., 1977).

8.6 Bibliography

1. Andersen, D. H. (1962): Pathology of cystic fibrosis. Ann. N. Y. Acad. Sci. 93, 500-517

2. Bianchi, L., Gerok, W., Sickinger, K., Eds. (1977): Liver and Bile, Falk Symposium No. 23, University Park Press, Baltimore, 1-448

3. Billing, B. H. (1978): Bile acid metabolism in cholestasis. Ital. J. Gastroenterol. 10, 37-40

4. Bochier, I. A. D., Cooperband, S. R., and el-Kodsi, B. M. E. (1965): Mucous substances and viscosity of normal and pathological human bile. Gastroenterology 49, 343-353

5. Carey, M. C. and Small, D. M. (1978): The physical chemistry of cholesterol solubility in bile. Relationship to gallstone formation and dissolution in man. J. Clin. Invest. 61, 998-1026

6. Christensen, A. H. and Ishak, K. G. (1970): Benign tumors and pseudotumors of the gallbladder. Arch. Path. 90, 423-432

7. Dische, Z. (1960): Fucose and sialic acid in glycoprotein of the mucus of the digestive tract. Fed. Proc. 19, 904-910

8. Dische, Z. (1964): Structural factors in mucous glycoproteins related to cystic fibrosis. In: P. A. DiSant'Agnese (Ed.). Research on Pathogenesis of Cystic Fibrosis. Natl. Inst. Arth. Metab. Dis., Bethesda, 229-242

9. Elfving, G., Silvonen, E. and Teir, H. (1969): Mucosal hyperplasia of the gallbladder in cases of cholecystolithiasis. Acta Chir. Scand. 135, 519-522

10. Esterly, J. R. and Spicer, S. S. (1968): Mucin histochemistry of human gallbladder: Changes in adenocarcinoma, cystic fibrosis and cholecystitis. J. Natl. Cancer Inst. 40, 1-12

11. Gent, W. L. G., Haslewood, G. A. D. and Montesdeoca, G. (1971): Macromolecular compounds of bilirubin in human bile. Biochem. J. 122, 15P-16P

12. Gupta, S. and Udupa, K. N. (1974): Mucopolysaccharide content of normal and lithogenic bile. J. Indian Med. Assoc. 69, 306-307

13. Hakkinen and Laitio, M. (1970): Epithelial glycoproteins of human gallbladder. Immunological characterization. Arch. Pathol. (Chicago) 90, 137-143

14. Hammerman, H. J., Shatz, B. A. and Evens, R. G. (1977): Biliary matrix: Matrix stones as a cause of biliary obstruction. Radiology 124, 31-32

15. Harding, H. E. (1931): The secretion of mucus by the epithelial cells of the gallbladder and the experimental production of mucocele. Br. J. Surg. 22, 355-364

16. Hashimoto, Y., Tsuiki, S. and Pigman, W. (1961): Comparison of the composition of submaxillary and bile mucins from several animal species. Fed. Proc. Abstracts 20, 87

17. Hulten, O. (1968): Formation of gallstones II. Acta Chir. Scand. 134, 557-561

18. Imada, T. (1970): Origin of white bile mucopolysaccharide. Fukuoka-Igaku-Zasshi 61, 53-66 (Japan)

19. Jarvi, O. (1961): A review of the part played by gastro-intestinal heterotopias in neoplasmogenesis. Proc. Finn. Acad. Sci. Lett. 151-187

20. Jarvi, O. and Meurman, L. (1964): Heterotropic gastric mucosa and pancreas in gallbladder with reference to the question of heterotopias in general. Ann. Acad. Sci. Fenn. 106, 1-42

21. Jui-Yun Yu (1960): Mucoprotein in hepatobiliary diseases. II. Serum and liver mucoprotein in hepatobiliary damage. Tai-wan i-hsueh-hui tsa-Chih 59, 487-507

22. Kulcsar, A., Foldes, I. and Kulcsar-Gergely, J. (1969): Importance of mucopolysaccharides in the development of the protective hepatropic effect of hypothyrosis. Tohoku J. Exp. Med. 97, 373-384

23. Leuschner, U. (1977): Pathophysiological aspects of gallstone formation. In: W. Stille and R. S. Timmler (Eds.). Int. Arbeitstag. Gallenweg Infekt. 1976 (Publ. 1977), Steinkoff, Darmstadt, 24-29 (Ger)

24. Lonsdale, K. (1968): Human stones. Science 159, 1199-1207

25. Lukie, B. E. (1976): The pathophysiology of cholesterol gallstone formati In: J. M. Dietschy (Ed.). Scientific Practice in Clinical Medicine. Grune and Stratton, New York, Vol. I, 333-335

26. Lukie, B. E. and Dietschy, J. M. (1973): Cholesterol gallstone formation Recent advances in pathophysiology. Texas Med. 69, 90-100

27. Magri, P. (1968): Histochemical studies of adenomatous polyps and primitive adenocarcinomas of the gallbladder. Atti. Accad. Med. Lomb. 23, 628-632 (Ital)

28. Maki, T., Mathsushiro, T., Suzuki, N. et al. (1971): Role of sulfated glycoprotein in gallstone formation. Surg. Gynecol. Obstet. 132, 846-854

29. Maki, T., Matsushiro, T., Suzuki, N. et al. (1971): Role of sulfated saccharides in calcium bilirubinate gallstones. Tohoku J. Exp. Med. 98, 81-86

30. Martin, G. P., Marriott, C. and Kellaway, I. W. (1978): Direct effect of bile salts and phospholipids on the physical properties of mucus. Gut 19, 103-107

31. Matern, S., Hackenschmidt, J. and Back, P. (Eds.) (1975): Advances in Bile Acid Research. Bile Acid Meet. 3rd, 1974 (Publ. 1975), Stuttgart

32. Matsushiro, T. and Nemoto, T. (1968): Mucopolysaccharides in calcium stones (gallstones). Tohoku J. Exp. Med. 94, 397-406

33. Matsushiro, T., Nemoto, T., Masahiko, E. and Yosizawa, Z. (1970): Glycoproteins and sulfated glycoproteins isolated from human bile. Clin. Chim. Acta 30, 645-651

34. Maxfield, M. and Wolins, W. (1959): A periodic acid-Schiff-staining macromolecular component of human hepatic bile. Clin. Chim. Acta 4, 647-651

35. Mel'nikov, V. P., Ginovker, A. G., Goroshcheniya, Yu. B. and Solov'ev, G. S. (1972): Mechanism of bile-expelling effect of bile acids. Fiziol. Zh. SSSR im. I. M. Schenova 58, 458-462 (Russ)

36. Miyake, H. and Katsuki, T. (1956): Filter paper electrophoretic studies on bile proteins. Sogo Igaku 13, 905-911

37. Nagashima, H., Suzuki, N. and Yosizawa, Z. (1974): Sulfated glycoproteins in normal human bile. Tohoku J. Exp. Med. 113, 225-229

38. Nagashima, H., Suzuki, N. and Yosizawa, Z. (1974 a): Coagulating effect on calcium carbonate of sulfated glycoproteins isolated from pathological human bile. Tohoku J. Exp. Med. 113, 159-168

39. Nakamura, N. (1966): Resin-reinforcing technique for sectioning gallstones. Tohoku J. Exp. Med. 90, 347-354

40. Nakamura, N. (1967): Mucopolysaccharides and their significance in gallstone formation. Tohoku J. Exp. Med. 93, 235-247

41. Neiderhiser, D. H., Plantner, J. J. and Carlson, D. M. (1971): The purification and properties of the glycoproteins of pig gallbladder bile. Arch. Biochem. Biophys. 145, 155-163

42. Neiderhiser, D. H., Roth, H. P. and Webster, L. T. (1966): Studies on the importance of lecithin for cholesterol solubilization in bile. J. Lab. Clin. Med. 68, 90-102

43. Odin, L. (1957): Mucopolysaccharides of epithelial mucus. In: G. E. W. Wolstenholme and M. O'Connor (Eds.). Ciba Found. Sympos. on the Chemistry and Biology of Mucopolysaccharides. Little, Brown and Company, Boston, 234-244

44. Schrager, J., Oates, M. D. G. and Rosbottom, A. (1972): The isolation and partial characterization of the principal billiary glycoprotein. Digestion 6, 338-355

45. Sheleketina, II. and Leliukhina, E. V. (1974): Mucoid substances of the bile in biliary tract diseases. Vrach Delo (8), 39-43 (Rus)

46. Snary, D. and Allen, A. (1972): Studies on gastric mucoproteins. The production of radioactive mucoproteins by pig gastric mucosal scrapings in vitro. Biochem. J. 127, 577-587

47. Soloway, R. D. (1974): Gallstone disease. In: F. P. Brooks (Ed.). Gastrointestinal Pathophysiology. Oxford Univ. Press, New York, 172-191

48. Terho, T. and Laito, M. (1974): Polysaccharides in human gall bladder wall and bile. Biochim. Biophys. Acta 338, 135-150

49. Thistle, J. L. (1973): Gallstones: Pathophysiology and dissolution. Postgrad. Med. 53, 65-71

50. Tiba, H. (1950): Biochemical studies on carbohydrates CXXV. Glycidamins in human bile. Tohoku J. Exptl. Med. 52, 103-110

51. Tritapepe, R., DiPadova, C. and Podda, M. (1975): Estrogens and gallstones. Arch. Ital. Chir. 99, 267-274

52. Van der Berg, J. W. O., Frankel, M., Zivanovic, M. B., Svircevic, A. and Wilson, J. H. P. (1977): Bile salts in cholesterol gallstones. Med. Pregl. 30, 445-451

53. Varute, A. T. and Nalavade, M. N. (1973): Histochemical analysis of mucosubstances in the gall bladder epithelium of the domesticated cat. Folia Histochem. Cytochem. 11, 21-32

54. Werner, I. (1953): Studies on glycoproteins from mucous epithelial secretions. Acta. Med. Upsalien 58, 1-55

55. Womack, N. A., Zeppa, R. and Irvin, G. L. (1963): The anatomy of gallstones. Ann. Surg. 157, 670-686

56. Yamada, K. (1974): Acid-mucopolysaccharide containing structures in the gall bladder epithelium of the rabbit as seen with the electron microscope. Histochemistry 39, 351-360

57. Yamada, K. and Hoshino, M. (1972): Morphochemical analysis of epithelial mucosubstances in the gall bladder of the fowl (Gallus domesticus). Histochemie 29, 120-128

58. Yamashita, T. (1971): An experimental study of coagulating effect of various mucosubstances on calcium carbonate suspensions. Tohoku J. Exp. Med. 105, 75-85

Chapter 9

Fluids Associated with Embryonic Membranes

9.1 Introduction

The embryonic membranes are structures or tissues that develop from the fertilized egg yet do not form a part of the embryonic body. Placenta and umbilical cord though not strictly speaking membranes are included among the embryonic membranes by definition (Hamilton et al., 1962). These membranes are of a functional importance during embryonic life, being concerned with supply and storage of nutriment, respiratory exchange, excretion and mechanical protection of the embryo and these are shed at birth or absorbed at hatching. The embryonic membranes include the yolk sac, the chorion (or serosa), the amnion, the allantois, the trophoblast, placenta and umbilical cords. The GAG in fluids asscociated with some of these membranes, will be discussed in this section. The GAG of amniotic fluid will be described in another section of this monograph (Chapter 11).

9.2 Wharton's jelly of umbilical cord

The umbilical cord consists of primarily the extramesenchymal tissue of gelatinous consistency and soft pulpy nature called the Wharton's jelly with a small contribution of two arteries and a vein. During isolation of GAG these three blood vessels need to be carefully removed to prevent contamination of the cord. Embryologically, this tissue, unlike other connective tissues, is not derived from the embryonic mesenchyme but from the extra-embryonic mesenchyme. Wharton's jelly is a matrix of collagen, fibers, salts, water, proteins and GAG; the major portion of GAG and proteins being only loosely held. The two functions ascribed to this jelly and its gelatinous nature are attributable to the protein-polysaccharide or proteoglycans present in this tissue. The Wharton's jelly protects the three blood vessels in the cord by cushioning any compression during fetal life and birth. Based on studies of umbilical cord and of artificial mixture of collagen fibers and hyaluronic acid, Fessler (1960) suggested that the combined structure of hyaluronic acid, collagen and water in the Wharton's jelly provides a definite resistance to compression and thus prevents any contortion of the cord. Studies with deuterated and tritiated water have shown that the Wharton's jelly provides a diffusion medium for exchange between the fetus, amniotic fluid and maternal tissue, and also between this jelly and the blood vessels in the cord at a rate of 40-50 ml/min (Plentl, 1961). This exchange property is because of the proteoglycans in the jelly which bind water and generate osmotic difference (Fessler, 1960).

9.2.1 Glycosaminoglycan composition of Wharton's jelly

The Wharton's jelly from human umbilical cord contains both nonsulfated and sulfated GAG. A 100 g of this wet tissue contained 198 mg of hyaluronic acid, 106 mg of chondroitin 6-sulfate, 38 mg of dermatan sulfate and 17.6 mg of a glucosamine-containing sulfomucopolysaccharide (Danishefsky and Bella, 1965, 1966). All of the hyaluronic acid and over 60 % of the chondroitin 6-sulfate present could be extracted by homogenization with 0.2 % NaCl at 0-5 °C. The remaining sulfated GAG could be recovered from the precipitate left after NaCl extraction only after pepsin-trypsin digestion. This difference in extractability is probably related to different degrees of binding to protein.

The sulfated glucosamine-containing GAG present at a concentration of 17.6 mg/100 g of wet tissue reported by Danishefsky and Bella (1965, 1966) has been shown to consist of heparan sulfate at a concentration of 70 mg/100 g dry weight of the tissue (Cifonelli and King, 1970) with a very small amount of heparin (Moore and Schoenberg, 1957). The GAG isolated by proteolysis, deproteinization and ethanolic precipitation upon acid hydrolysis and ion-exchange chromatography on Dowex 50 contained 72.5 % glucosamine and 27.5 % galactosamine (Wosicki, 1970). The two hexosamines together constituted 1.92 % of the dry tissue weight.

Jeanloz and Forchielli (1950) using two different methods, the one involving pepsin-trypsin digestion followed by removal of residual proteins by Sevag's procedure and the other involving precipitation of the saline extract of the human umbilical cords with ammonium sulfate and pyridine followed by purification of hyaluronate by digestion with pepsin-trypsin and reprecipitation, obtained a yield of 5.6 % and 4.2 %, of pure hyaluronate, respectively. About 1 % of chondroitin sulfate was also present in umbilical cord.

With the development of better, improved and modernized methods of preparation and analysis it is now known that GAG represent nearly 10 % of the dry weight of the human umbilical cord tissue and this tissue is the major source of commercial hyaluronic acid. Besides the nonsulfated GAG (hyaluronic acid), sulfated GAG are also present to a lesser extent. All the sulfated GAG are present in the cord as polysaccharide-proteins or proteoglycans.

A procedure for extraction and analysis of the GAG components (Lee, 1973) consisted of extraction of the cords with 10 % calcium chloride, saturated calcium hydroxide and 6 M urea at 60 °C, deproteinization of the extracts and precipitation of GAG with cetylpyridinium chloride. Fractionation into GAG components was carried out on Dowex 1-X2 anion exchange resin column and eluted fractions were examined by cellulose acetate electrophoresis. By this procedure hyaluronic acid was the major component and at least two kinds of heparan sulfate were present.

Antonopoulos et al. (1974) have reported another procedure for isolation of proteoglycans from this tissue which involves extraction with 4 M guanidinium chloride, reextraction of the extract with 7 M urea and chromatography on DEAE-cellulose column packed in 7 M urea. Following the elution of proteins with urea in a weak salt solution, the proteoglycans were eluted with strong salt solutions.

9.2.2 Hyaluronic acid

Hyaluronic acid is the major GAG in the human umbilical cord and has been the primary source for studies on hyaluronic acid. Hyaluronic acid was the first identified glycosaminoglycan and it was originally extracted in a crude form from human umbilical cord as Wharton's jelly (Wharton, 1659). Meyer and Palmer (1936) isolated 2.1 g of hyaluronic acid by sodium hydroxide extraction of 100 g of acetone dried material, which contained no sulfate. Suzuki (1938) isolated from the human umbilical cord a sulfate free material with a chemical composition similar to that of hyaluronic acid and he named it *mucoitin*.

Later workers improved the yield of glycosaminoglycans by using proteolytic enzyme digestion of the acetone-dried material followed by precipitation with ethanol. Using pepsin-trypsin digestion of the acetone extracted and dried umbilical cords, Hadidian and Pirie (1948) obtained a yield of 2.5-3 % glucosamine which corresponded to 6-8 % of hyaluronic acid. They also obtained a series of fractions differing in viscosity. A simple procedure for preparation of hyaluronic acid from umbilical cords has been reported (Shermatov, 1974). The method consisted of defatting the cords with acetone-alcohol (1:2, v/v), drying at 90 °C, extraction with water containing a few drops of chloroform at 37 °C for 2-3 days, centrifugation and precipitation of hyaluronic acid in the supernatant with acetone-alcohol (1:2, v/v).

Some workers have suggested the possibility that some of the GAG isolated from the Wharton's jelly may be from contamination of the blood vessels in the umbilical cord (Danishefsky and Bella, 1966). In attempts to resolve this possibility, the GAG of arteries and veins of umbilical cord have been analyzed. Although the total GAG in the artery and vein were virtually identical (5.45 ± 0.70 and 5.38 ± 0.82, respectively), and both contained hyaluronic acid (21.2 ± 1.4 and 26.3 ± 2.0); heparan sulfate (21.0 ± 1.1 and 19.9 ± 1.1); dermatan sulfate (20.5 ± 1.3 and 21.9 ± 1.7); chondroitin 4- and/or 6- sulfate (37.3 ± 1.1 and 31.9 ± 1.8, respectively), hyaluronic acid and chondroitin 4- and/or 6- sulfate were higher in the vein (Stefanovich and Levental, 1972). Another report (Wosicki, 1975) contradicts this. Although hyaluronic acid was greater in the venous wall in the cord, chondroitin 6-sulfate was less than that in the arterial wall. Both the tissues contained hyaluronic acid, heparan sulfate, chondroitin 6-sulfate and dermatan sulfate.

The hyaluronic acid from this source has a high viscosity which varies with the method of preparation, which involve different degrees of depolymerization during isolation. The different values for physicochemical parameters for hyaluronic acid prepared by different procedures have been reported. Based on physicochemical parameters different shapes have been assigned to this macromolecule in solution. The various shapes assigned may be due to the effect of shearing force applied to the easily deformable asymmetric particle as a highly solvated coil structure (Laurent, 1957). The viscosity of this hyaluronate also varies with the ionic strength of the solvent. At ionic strengths less than 0.1 mutual repulsion between the ionized carboxyl groups ($-COO^-$) causes an expansion and stiffening of the coil leading to an increase in viscosity. At ionic strengths greater than 0.2 the mutual repulsion of $-COO^-$ groups is eliminated and is reflected in the loss of dependence of viscosity on ionic strength (Blumberg and Ogston, 1958).

Hyaluronate is the only GAG for which a linkage to protein has not been clearly established to classify it among the proteoglycans, although in literature some workers have used the term proteohyaluronic acid (Balazs, 1970). However, hyaluronate isolated from umbilical cord and other sources even after repeated proteolysis always contains a small amount of amino acids and neutral sugars. In view of the lack of evidence for a carbohydrate-protein linkage, the presence of these amino acids and the minority sugars and their contribution to the structural organization of the hyaluronate molecule is not clear (Varma et al., 1977). The neutral sugar content and amino acid composition of the purified human umbilical cord hyaluronate and their role in carbohydrate-protein linkage has been studied (Varma et al., 1975). The purified preparation contained 37.2 % glucosamine (had no galactosamine), 40.4 % glucuronic acid, 0.21 % fucose, 0.05 % arabinose and less than 0.02 % glucose. The molar ratios of serine, threonine and aspartic acid to neutral sugars were not unity suggesting the absence of linkage between these. The increases in the percentages of the amino acids and an absence of sugar alditols in the hyaluronate reduced with $NABH_4/PdCl_2$ in an alkaline medium further confirmed the absence of protein-carbohydrate linkage. However, this reduction decreased the specific rotation and intrinsic viscosity with a decrease in the molecular weight to one-half of the original molecular weight. Bound silicon was present in the umbilical cord hyaluronate. In view of these results and left handed, double-stranded helical structure of hyaluronate, Varma et al. (1975) suggested that the bound silicon may cross link the two chains of hyaluronate at interspersed intervals through uronic acid and/or neutral sugar moieties.

The earlier work on chemical composition and structure has been reported elsewhere (Meyer and Palmer, 1936; Kaye and Stacey, 1950; Jeanloz and Forchielli, 1950). Periodate consumption of 0.1-0.2 mole and liberation of 0.08 mole of formic acid was observed when the periodate oxidation of three different preparations of human umbilical cord hyaluronate having the same purity was carried out at pH 4.5 and 5 °C in dark (Jeanloz and Forchielli, 1951). From this these authors concluded that C-6 of N-acetylglucosamine or C-2 and C-4 of glucuronic acid are not involved in any linkage. This hyaluronic acid, like hyaluronic acid from other sources, consists of equimolar amounts of D-glucuronic acid and N-acetyl-D-glucosamine. The uronic acid and hexosamine are linked alternately to form an unbranched polymer, with a pyranose configuration and β-linkages (Jeanloz and Forchielli, 1951). Upon irradiation with an electric lamp depolymerization of umbilical cord hyaluronate and a change of metachromasia to orthochromasia was noted (Castellani and Torlone, 1954).

9.2.3 Sulfated GAG in Wharton's jelly

The GAG composition of Wharton's jelly resembles the GAG present in the fluid-filled cavities, e. g. aqueous and vitreous humors, synovial and labyrinthine fluids, in having mainly hyaluronic acid with sulfated components forming only a minor fraction of the GAG. The presence of these GAG was suggested in the earlier work by metachromatic staining with basic dyes. Levene and López-Suárez (1916, 1918) first isolated a glycosaminoglycan *mucoitin sulfuric acid* by NaOH extraction of human umbilical cords freed from blood vessels. This material contained equimolar amounts of glucosamine, glucuronic acid and acetyl group in addition to 4.0 % sulfur. The presence of sulfated GAG in addition to hyaluronic acid was first reported by Meyer and Palmer (1936) and later by others (Hadidian and Pirie, 1948; Jeanloz and Forchielli, 1950).

Umbilical cord contains 0.5-2 % sulfate which corresponds to 10-20 % of the total GAG present (Jeanloz and Forchielli, 1950; Weissmann et al., 1953). The presence of chondroitin 6-sulfate at a yield of 1.9 g/100 g of acetone dried tissue was reported and identified on the basis of solubility and optical rotation (Meyer and Rapport, 1951; Adams, 1959) and by infrared spectroscopy (Mathews, 1958). The presence of chondroitin 6-sulfate in the cord was also reported by Meyer et al. (1965). Mast cells are present in this tissue and heparin has been identified in the mast cells on the basis of histochemical studies (Moore and Schoenberg, 1957). Aqueous extracts of the umbilical cord have anticoagulant acitivity (Sundberg et al., 1954) indicating the presence of heparin-related compounds in this tissue although Cifonelli and King (1970) failed to detect any heparin and reported that if heparin was present it could be in a negligible amount as compared to heparan sulfate of which only 70 mg/100 g dry tissue were present.

Chondroitin 6-sulfate and dermatan sulfate are the principal sulfated GAG of umbilical cord. The dermatan sulfate present is about half of that of chondroitin 6-sulfate (Danishefski and Bella, 1966). The dermatan sulfate was prepared by papain digestion, cetylpyridinium chloride (CPC) precipitation, pronase digestion, CPC precipitation and finally precipitation with 25 % ethanol (Fransson, 1968). The dermatan sulfate so isolated was a hybrid containing both chondroitin 6-sulfate type disaccharide units and iduronic acid-containing disaccharide units. The hybrid structure contained 90 % D-glucuronic acid and 10 % L-iduronic acid and both 4- and 6-sulfated N-acetylgalactosamine residues. Analysis of small (di-, tetra- and hexa-) oligosaccharides and hybrid octasaccharides obtained by digestion with testicular hyaluronidase revealed that D-glucuronic acid-containing disaccharide units were arranged in clusters and these units were 6-sulfated. A suggestion was made (Fransson, 1968) that sulfation, which occurs after polymerization, at position 4 or 6 may be determined by the nature of uronic acid adjacent to the N-acetylgalactosamine acceptor.

Inoue and Iwasaki (1976) reported that the major galactosaminoglycans from human umbilical cord consisted of copolymeric structures of various proportions of disaccharide units, D-glucuronosyl-N-acetylgalactosamine, D-glucuronosyl-N-acetylgalactosamine 4-sulfate, D-glucuronosyl-N-acetylgalactosamine 6-sulfate, and L-iduronosyl-N-acetylgalactosamine 4-sulfate. A yield of 700 mg of dermatan sulfate-chondroitin sulfate polymers was isolated from 100 g of ethanol-dried tissue by papain (EC 3.4.22.2) digestion followed by precipitation with cetylpyridinium chloride. Electrophoresis on cellulose acetate in 0.15 M pyridine-formic acid, pH 3.0 and gel filtration on Sephadex G-200 showed polydispersity of the copolymers. By fractionation with varying percentages of ethanol the polymer was subfractionated into 4-5 fractions, each of which differed in relative contents of L-iduronic acid and D-glucuronic acid, but none of these contained solely one of these two hexuronic acids. The copolymeric nature of these subfractions was evident from the oligosaccharide fragments by digestion with chondroitinase-AC (EC 4.2.2.5).

Because of the recent interests and significance of proteoglycans rather than glycosaminoglycans in the tissues, intact proteoglycans have been isolated by using milder extraction procedures without proteolysis. Intact chondroitin 6-sulfate proteoglycans were isolated from human umbilical cord by repeated homogenization and extraction of the tissue with 0.2 % NaCl at 0 °C, followed by precipitation with 5 % cetylpyridinium chloride (Danishefsky and Abraham, 1971). The coprecipitated hyaluronic acid was recovered in supernatant obtained by centrifugation of the homogenate of the precipitate with 0.4 M NaCl. The chondroitin sulfate was extracted from the residual precipitate with 1.2 M NaCl and recovered by precipitation with ethanol. Further purification was carried out by dialysis, passage through Dowex 50 (H^+) and gradient NaCl elution from Dowex 1 X-2 (Cl^-). Two fractions I and II were eluted with 1.25 M and 1.5 M Nacl. Fraction I contained 24.8 % hexosamir (19.13 % galactosamine + 3.1 % glucosamine), 28.5 % uronic acid, 13.54 % sulfate and 3.63 % nitrogen. Fraction II was composed of 24.68 % hexosamine (23.8 % galactosamine and 1.38 % glucosamine), 30.0 % uronic acid, 13.96 % sulfate and 3.74 nitrogen.

9.2.4 Physiological and pathological variations of glycosaminoglycans in Wharton's Jelly

The umbilical cord actively synthesizes hexosamine-containing macromolecules. In vitro incorporation of U-^{14}C-glucose into GAG in the umbilical cord slices has been studied (Brachet, 1972). The GAG increase gradually during embryonic development as shown by histochemical methods. The GAG content of the umbilical cord, as shown by metachromasia, increases with a concomitant rise of β-glucuronidase (Szirmai, 1965). By means of luminiscence microscopy, GAG gave weak orange luminiscence in the cytoplasm as compared to green luminiscence of nucleic acids in human umbilical cord preparations stained with acridine orange. Investigation of GAG at various stages of embryonic development by this technic detected no GAG in 5-7 weeks old embryos; in 3-4 month old embryos GAG were present in the cytoplasm of fibroblasts and intermediary material and older embryos contained these macromolecules in subendothelial layers and muscles (Saliichuk, 1967).

The hexosamine composition of human umbilical cord (Wosicki, 1972) and human chorionic tissue (Wosicki, 1972 a) in the first trimester of fetal development have been investigated. In the first trimester of gestation, a total of 0.49 % of the dry mass consisted of hexosamine of which 76.8 % was glucosamine and 23.2 % galactosamine. The GAG comprised 27.7 % of the total hexosamine and 0.13 % of the dry mass of the tissue. This is different from hexosamines and GAG in human extraplacental fetal membranes which contained 0.91 % glucosamine and 0.21 % of galactosamine of the dry weight of the tissue and half of the total hexosamine present was bound as GAG (Wosicki, 1972 b). In full term human umbilical cord tissue the acid hydrolyzate contained 72.5 % glucosamine and 27.5 % galactosamine, with the total hexosamine making 1.92 % of the dry tissue weight (Wosicki, 1970).

The relative amounts of GAG in Wharton's jelly of the umbilical cord vary in different species of animals depending upon the relative amounts of avascular and vascular components of the connective tissue. In human subjects and animals like goat, horse, cow, sheep and monkey the surrounding connective tissue of the jelly is avascular, whereas in other animals like pig, dog, etc. it is vascular (Szirmai, 1965). Polysaccharides have been reported to vary in different stages of pregnancy in sheep (Kasamanly, 1970), with the highest level of GAG at 45 days and falling by 3-5 months.

In vitro incorporation of radioactive glucose into GAG in the umbilical cord slices varies with the glucose concentration and insulin affects this incorporation (Brachet, 1973). Both increase of glucose concentration from 5 mM to 20 mM and presence of insulin augmented incorporation. This contradicts the report of increased in vivo production of GAG with increased water retention reflected in elevation of wet weight/dry weight ratio and thickness of the umbilical cord in diabetic pregnancies (Patrick and Walker, 1969). In 21 normal full term human pregnancies, the glucosamine consisted of 1.03 mg/ wet weight or 9.42 mg/g dry weight with wet weight/dry weight ratio of 9.73. These values were elevated to 1.32 mg/g wet weight and 16.41 mg/g dry weight for glucosamine and 13.07 for wet/dry weight ratio in 16 diabetic pregnancies.

Focal inflammation of the connective tissue followed by ingrowth of granulation tissue and necrotic changes around the blood vessels affect GAG changes in the cord and the well-being of the fetus (Gille, 1977). Nothing is known about the conditions that cause the formation of the granulation tissue. Histochemical changes involving GAG have been reported in the umbilical cord in rheumatic heart disease (Beglarian and Ambardzharian, 1970).

9.3 Glycosaminoglycans in Allantoic fluid

Allantoic acid of normal embryonated eggs and those infected with influenza virus has been reported to contain sulfated glycosaminoglycans (Haukenes et al., 1965; Harboe and Haukenes, 1966). This sulfated material lacked both uronic acid and sialic acid and was antigenic. The composition and immunological reactions of sulfated mucopolysaccharide antigens from chicken chorioallantoic fluid with respect to the contribution of carbohydrate residues and proteins have been investigated (Lee et al., 1969).

A keratan sulfate-like GAG has been reported in chicken allantoic fluid (Meyer et al., 1967). This material resembled keratan sulfate in having several carbohydrate chains attached through O-glycosidic linkages to serine and threonine residues in a peptide backbone. It was isolated in a yield of 12.6 mg/l allantoic fluid by acetone precipitation of the supernatant obtained by centrifugation, papain-pronase digestion, fractionation by freeflow electrophoresis and purification on a Biogel P-2 column. Upon fractionation on DEAE Sephadex, a saccharide fraction eluted as a sharp peak in the 0.55-0.85 M NaCl fraction. Upon electrophoresis at pH 5 it showed a single Alcian blue-staining spot, but at pH 1.8 it showed an additional small slow moving spot. However, it differed from human cartilage keratan sulfate in the absence of galactosamine substitution at C-3 position as evidenced by appearance of no reduced chromogen upon alkaline-borohydride reduction. It contained galactose and N-acetylglucosamine; sialic acid and some methyl-pentose were present as non-reducing end groups. As compared to the 17-18 % sulfate content in cartilage keratan sulfate the allantoic fluid material contained only 6-7 % sulfate. The sulfate was found to be present as N-acetylglucosamine 6-sulfate from the results of partial hydrolysis and infrared spectra. It had a molecular weight less than 4,000 and was of epithelial origin unlike the keratan sulfate from cornea and cartilage which have a mesodermal origin.

9.4 Glycosaminoglycans in eggs

Besides glycoproteins, the egg white also contains a very small amount of acidic glycosaminoglycans, which can be detected by carbazole assay (Wood and Anastassiadis, 1965) and by histochemistry in the light and electron microscopes after staining of boiled egg-white with colloidal iron (Curran et al., 1966). Slices of boiled egg-white introduced into the peritoneum of experimental animals produced granuloma and both granulomatous tissue and the eggwhite slices showed the presence of acidic glycosaminoglycans (Curran et al., 1966).

A GAG related to heparan sulfate with a sulfated content between that of heparan sulfate and heparin has been reported in slices of boiled egg-white (Lovell et al., 1966). The GAG was isolated by papain digestion, deproteinization with trichloroacetic acid and ethanol precipitation of GAG. The ethanolic precipitate was fractionated by elution of GAG-cetylpyridiniumchloride complex from a cellulose column with magnesium chloride solutions. The GAG was present in 0.3 M magnesium chloride fraction and upon electrophoresis in

0.1 M phosphate buffer of pH 7.2 migrated as a single spot between hyaluronic acid and chondroitin sulfate. It was insensitive to hyaluronidase digestion and contained glucuronic acid, 0.9 sulfate groups per disaccharide unit and of the total hexosamine 75 % was glucosamine and 25 % galactosamine. The presence of heparan sulfate, chondroitin 4-sulfate, dermatan sulfate in chick oviduct stimulated with estrogen has been reported (Schiller, 1959).

Results on analysis of GAG in hen's egg white by another group (Johnson and Barker, 1971) differed from those of the aforementioned work. The material was isolated from 1.7 kg of egg white by papain digestion, deproteinization with trichloroacetic acid, dialysis and ethanolic precipitation of GAG. The total material obtained contained only 19.7 mg hexosamine (60.5 % of glucosamine and 39.5 % galactosamine), 2.6 mg glucuronic acid and 18.1 mg of neutral sugars. Upon anion-exchange chromatography on a column packed with Dowex 1-X2 (Cl^- form), only 10.4 % of the total uronic acid was present in 1.25 M NaCl fraction corresponding to heparan sulfate, while the uronic acid was found to be concentrated in 1.5 M and 2.0 M NaCl fractions.

A sulfur containing protein-mucopolysaccharide complex was formed by mollusc eggs during oogenesis which was present primarily in the egg jelly and the capsular membrane (Tapaswi, 1976). Embryos undergo abnormal development when they are cultured in sulfate-free artificial sea water indicating that mucopolysaccharides are indispensable for early development of embryos. Sea urchin embryos cultured in ^{35}S-sulfate-containing medium accumulate labeled GAG (Oguri and Yamagata, 1978). The presence of proteoglycan has been reported in sea urchin embryos by extraction with 0.05 M sodium pyrophosphate or 1.0 M guanidinium hydrochloride (Immers, 1961; Kinoshita, 1974). The mucopolysaccharide moiety of this proteoglycan was resistant to chondroitinase ABC (Kinoshita, 1974). A proteoglycan, in which half of the mucopolysaccharide moiety was dermatan sulfate and the rest was chondroitinase-resistant, was isolated from the embryos of a Japanese sea urchin by mild extraction procedures. These GAG did not exist in free form nor were they O-glycosidically linked (Oguri and Yamagata, 1978). Premitotic desquamation of heparin-like GAG in fertilized sea urchin eggs (Kinoshita, 1969, 1971) suggested that the sulfated GAG regulate growth by acting as a negative element of growth and a decrease in the relative amount of these macromolecules is important for cell differentiation and cellular transformation for the faster growth of neoplastic cells (Chiarugi, 1976). In contrast with the growth inhibitory effect of GAG present on the cell-coat, in vitro somite chondrogenesis is induced by exogeneous extracellular proteoglycans of embryonic chick and collagens leading to an increased synthesis of proteoglycans by the induced somites (Lash and Vasan, 1978).

9.5 Glycosaminoglycans of cardiac jelly

The space between the inner endocardial and outer epimyocardial layers of the tubular heart in the chick embryo is filled with elastic, semisolid cardiac jelly. The cardiac jelly helps the tubular heart to function as a pump and helps its moulding during development. Injection of ^{35}S-sulfate into a chick egg produced an accumulation of labeled GAG in the cardiac jelly, detected by radioautography (Johnston and Comar, 1957). The GAG could be detected in the cardiac jelly by histochemical technics and metachromasia was lost upon digestion with hyaluronidase (Ortiz, 1958). Other workers (Franco-Browder et al., 1963) have identified sulfated GAG in early chick embryos. Sulfated GAG consisting mainly of chondroitin sulfates and heparin-like compounds have been isolated from the cardiac jelly of 16-19 stage chick embryos (Gessner et al., 1965). The procedure involved removal of cardiac tubes under a dissecting microscope, defatting with acetone, NaOH treatment, dialysis, pronase digestion combined with deproteinization repeated twice, dialysis and concentration. The GAG isolated were also detected in intact embryonic hearts by histochemical staining with basic dyes and their presence was confirmed by ^{35}S-sulfate incorporation studies.

9.6 Bibliography

1. Adams, J. B. (1959): Biosynthesis of chondroitin sulfates. Nature 184, 274-275

2. Antonopoulos, C. D., Axelsson, I., Heinegard, D. and Gardell, S. (1974): Extraction and purification of proteoglycans from various types of connective tissue. Biochim. Biophys. Acta 338, 108-119

3. Balazs, E. A. (Ed.) (1970): Chemistry and Molecular Biology of the Intercellular Matrix. Academic Press, New York, Vol. I

4. Beglarian, A. G. and Ambardzharian, R. S. (1970): Histochemistry of mucopolysaccharides of placenta and umbilical cord in rheumatic heart diseases. Akush Ginekol. (Mosk.) 46, 75-76 (Russ)

5. Blumberg, B. S. and Ogston, A. G. (1958): Physícochemical studies on hyaluronic acids. In: G. E. W. Wolstenholm and M. O'Connor (Eds.). Ciba Foundation Sympos., Chemistry and Biology of Mucopolysaccharides. Little, Brown and Company, Boston, 22-41

6. Brachet, E. A. (1972): Glucose uptake and production of pyruvate and lactate by human umbilical cord in vitro. Biol. Neonate 21, 69-75

7. Brachet, E. A. (1973): Utilization of glucose-U-^{14}C by the human umbilical cord in vitro. FEBS Lett. 31, 277-279

8. Castellani, A. and Torlone, V. (1954): Photodynamic effect and chromotropism of acid mucopolysaccharides of the umbilical cord. Riv. Istochim. norm. Pat. 1, 107

9. Chiarugi, V. P. (1976): Cell-coat glycosaminoglycans in cellular transformation and differentiation. Exp. Cell. Biol. 44, 251-259

10. Cifonelli, J. A. and King, J. (1970): The isolation and properties of heparitin sulfate from human umbilical cord tissues. Biochim. Biophys. Acta 215, 273-279

11. Curran, R. C., Lovell, D. and Clark, A. E. (1966): Mucopolysaccharides in peritoneal granulomas in the rat. J. Path. Bact. 91, 429-439

12. Danishefsky, I and Abraham, G. (1971): Isolation and purification of chondroitin 6-sulfate protein from umbilical cord. Prep. Biochem. 1, 133-140

13. Danishefsky, I. and Bella, A., Jr. (1965): The sulfated mucopolysaccharides in human umbilical cord. Fed. Proceed. Abstracts 24, 355

14. Danishefsky, I. and Bella, A. (1966): The sulfated mucopolysaccharides from human umbilical cord. J. Biol. Chem. 241, 143-146

15. Fessler, J. H. (1960): A structural function of mucpolysaccharide in connective tissue. Biochem. J. 76, 124-132

16. Franco-Browder, S., De Rydt, J. and Dorfman, A. (1963): The identification of a sulfated mucopolysaccharide in chick embryos stages 11-23. Proc. Nat. Acad. Sci. U. S. 49, 643-647

17. Fransson, L. A. (1968): Structure of dermatan sulfate. 3. The hybrid structure of dermatan sulfate from umbilical cord. J. Biol. Chem. 243, 1504-1510

18. Gessner, I. H., Lorincz, A. E. and Bostrom, H. (1965): Acid mucopolysaccharide content of the cardiac jelly of the chick embryo. J. Exp. Zool. 160, 291-298

19. Gille, J. (1977): Granulation tissue in the umbilical cord. J. Reprod. Med. 18, 35-37

20. Hadidian, Z. and Pirie, N. W. (1948): The preparation and some properties of hyaluronic acid from human umbilical cord. Biochem. J. 42, 260-265

21. Hamilton, W. J., Boyd, J. D. and Mossman, H. W. (1962): Human Embryology, Williams and Wilkins, Baltimore, Chap. 16, 435-468

22. Harboe, A. and Haukenes, G. (1966): A uronic and sialic acid free chick allantoic mucopolysaccharide sulfate which combines with influenza virus HI-antibody to host material 3. A high content in chicken bile, and an inactivating enzyme in chicken faeces. Acta Path. Microbiol. Scand. 68, 98-102

23. Haukenes, G. A., Harbor, K. and Mortenson-Egmund, K. (1965): A uronic acid and sialic acid free chick allantoic mucopolysaccharide sulfate which combines with influenza virus hi-antibody to host material. 1. Purification of the substance. Acta Pathol. Microbiol. Scand. 64, 534-542

24. Immers, J. (1961): Comparative study of the localization of incorporated ^{14}C-labeled amino acids and $^{35}SO_4$ in the sea urchin ovary, egg and embryo. Exp. Cell Res. 24, 356-378

25. Inoue, S. and Iwasaki, M. (1976): Dermatan sulfate-chondroitin sulfate copolymers from umbilical cord. Isolation and characterization. J. Biochem. (Tokyo) 80, 513-524

26. Jeanloz, R. W. and Forchielli, E. (1950): Studies on hyaluronic acid and related substances. I. Preparation of hyaluronic acid and derivatives from human umbilical cord. J. Biol. Chem. 186, 495-511

27. Jeanloz, R. W. and Forchielli, E. (1951): Studies on hyaluronic acid and related substances. IV. periodate oxidation. J. Biol. Chem. 190, 537-546

28. Johnson, A. H. and Baker, J. R. (1971): Glycosaminoglycans of hen's egg white. Biochem. J. 122, 16 P

29. Johnston, P. M. and Comar, C. L. (1957): Autoradiographic studies of utilization of S-sulfate by the chick embryo. J. Biophys. Biochem. Cytol. 3, 231-238

30. Kasamanly, T. Yu (1970): Polysaccharides in the umbilical cord of a karakul ewe. Dokl. Akad. Nauk SSSR 191, 957-959

31. Kaye, M. A. G. and Stacey, M. (1950): Observations on the chemistry of hyaluronic acid. Communication Proc. 283rd Meeting of Biochemical Society, London, Biochem. J. 46, p. xiii

32. Kinoshita, S. (1969): Periodic release of heparin-like polysaccharides within cytoplasm during cleavage of sea urchin egg. Exp. Cell Res. 56, 39-43

33. Kinoshita, S. (1971): Heparin as a possible initiator of genomic RNA synthesis in early development of sea urchin embryos. Exp. Cell Res. 54, 403-411

34. Kinoshita, S. (1974): Some observations on a protein-polysaccharide complex found in sea urchin embryos. Exp. Cell Res. 85, 31-40

35. Lash, J. W. and Vasan, N. S. (1978): Somite Chondrogenesis in vitro. Stimulation by exogenous extracellular matrix components. Dev. Biol. 66, 151-171

36. Laurent, T. C. (1957): A comparative study of physicochemical properties of hyaluronic acid prepared according to different methods from different tissues. Arkiv for Kemi 11, 487-496

37. Lee, S. S. (1973): Studies on glycosaminoglycans in tissues I. On the method of extraction and preparation of acidic glycosaminoglycans from the human umbilical cord. Fukushima J. Med. Sci. 19, 25-32

38. Lee, L. T., Howe, C., Meyer, K. and Choi, H. U. (1969): Precipitin analysis of influenza virus host antigen and of sulfated mucopolysaccharides of chicken embryonic allantoic fluid. J. Immunol. 102, 1144-1155

39. Levene, P. A. and López-Suárez, J. (1916): The conjugated sulfuric acid of funis mucin (mucoitinsulfuric acid) II. J. Biol. Chem. 26, 373-378

40. Levene, P. A. and López-Suárez, J. (1918): Mucins and mucoids. J. Biol. Chem. 36, 105-126

41. Lovell, D., Schorah, C. J. and Curran, R. C. (1966): Isolation and identification of a mucopolysaccharide from egg-white. Nature (London) 212, 860-861

42. Mathews, M. B. (1958): Isomeric chondroitin sulfates. Nature 181, 421-422

43. Meyer, K., Bhavanandan, V. P., Yung, D., Lee, L. T. and Howe, C. (1967): The keratosulfate-like mucopolysaccharide of chick allantoic fluid. Proc. Nat. Acad. Sci. USA 58, 1655-1659

44. Meyer, K., Davidson, E., Linker, A. and Hoffman, P. (1965): The acid mucopolysaccharide of connective tissue. Biochim. Biophys. Acta 21, 506-518

45. Meyer, K. and Palmer, J. W. (1936): On glycoproteins: II. The polysaccharides of vitreous humor and of umbilical cord. J. Biol. Chem. 114, 689-703

46. Meyer, K. and Rapport, M. M. (1951): The mucopolysaccharides of the ground substance of connective tissue. Science 113, 596-599

47. Moore, R. D. and Schoenberg, M. D. (1957): Studies on connective tissues. I. The polysaccharides of human umbilical cord. Arch. Pathol. 64, 39-45

48. Oguri, K. and Yamagata, T. (1978): Appearance of a proteoglycan in developing sea urchin embryos. Biochim. Biophys. Acta 541, 385-393

49. Oritz, E. C. (1958): Estudio histoquimico de la gelatina cardiaca en el embrion de pollo. Arch. Inst. Cardiol. Mex. 28, 244-262

50. Patrick, R. S. and Walker, F. (1969): The glucosamine content of human umbilical cord jelly in normal and diabetic pregnancy. Diabetologia 5, 29-33

51. Plentl, A. A. (1961): Transfer of water across the perfused umbilical cord. Proc. Soc. Exp. Biol. 107, 622-624

52. Saliichuk, L. I. (1967): Luminescence microscopy of acid mucopolysaccharides in human umbilical cord. In: M. Ya. Izd. Subbotin (Ed.). Gistkhim-Norm. Patol. Morfol., Nauka Sib. Otd. 52-54 (Russ)

53. Schiller, S. (1959): Mucopolysaccharides of the estrogen-stimulated chick oviduct. Biochim. Biophys. Acta 32, 315-319

54. Shermatov, Kh. Kh. (1974): Preparation of hyaluronic acid from umbilical cords. Lab Delo (5), 311 (Russ)

55. Stefanovich, V. and Levental, M. (1972): Lipids and acid mucopolysaccharides in arteries and veins of human umbilical cord. Int. J. Biochem. 3, 383-386

56. Sundberg, R. D., Schaar, F. E., Powell, M. J. S. and Denboer, D. (1954): Tissue mast cells in human umbilical cord, and the anticoagulant activity of dried extracts of cords and placentae. Anat. Record 118, 35-56

57. Suzuki, M. (1938): Carbohydrates (XL): Preparation of mucoitin from umbilical cords. J. Biochem. (Tokyo) 28, 479-486

58. Szirmai, J. A. (1965): Glycosaminoglycans in umbilical cord. In: E. A. Balazs and R. W. Jeanloz (Eds.). The Amino Sugars Vol. IIA. Distribution and Biological Role, Academic Press, New York, Chap. 28, 219-227

59. Tapaswi, P. K. (1976): Investigations on the biosynthesis of sulfur-containing compounds in Limnae (mollusc) oocytes, capsules and jelly during oogenesis period. Acta Embryol. (3), 267-271

60. Varma, R., Varma, R. S., Allen, W. S. and Wardi, A. H. (1975): Human umbilical cord hyaluronate: Neutral sugar content and carbohydrate-protein linkage studies. Biochim. Biophys. Acta 399, 139-144

61. Varma, R., Vercellotti, J. R. and Varma, R. S. (1977): On arabinose as a component of brain hyaluronate. Confirmation by chromatographic, enzymatic and chemical ionization-mass spectrometric analyses. Biochim. Biophys. Acta 497, 608-614

62. Weissmann, B., Rapport, M. M., Linker, A. and Meyer, K. (1953): Isolation of the aldobionic acid of umbilical cord hyaluronic acid. J. Biol. Chem. 205, 205-211

63. Wharton, T. (1659): Adenographia: Sive Glandularum Totius Corporis descriptio, Amstelaedami Sumptibus Ravesteinii

64. Wood, D. F. and Anastassiadis, P. A. (1965): The mucopolysaccharides of avian oviduct, egg skin, and comb, and of porcine skin. Canad. J. Biochem. 43, 1839-1856

65. Wosicki, A. (1970): Quantitative and qualitative studies on hexosamines of human umbilical cord tissue. Acta Med. Pol. 11, 369-373

66. Wosicki, A. (1972): Hexosamines of human umbilical cord in first trimester of fetal development. Acta Med. Pol. 13, 359-361

67. Wosicki, A. (1972 a): Human chorionic hexosamine in the first trimester of embryonal development. Ginekol. Pol. 43, 669-672

68. Wosicki, A. (1972 b): Composition of hexosamines in human extraplacental membranes. Acta Med. Pol. 13, 411-415

69. Wosicki, A. (1975): Hexosamines of vessel walls of umbilical cord. J. Gynecol. Obestet. Biol. Reprod. 4, 209-215 (Fr)

Chapter 10

Glycosaminoglycans of Fluids and Fluid-Associated Tissues of Ear

10.1 General consideration

From the anatomical point of view, the organ of hearing is divisible into three compartments:

- the external ear,
- the middle ear and
- the inner ear.

Each of these three compartments is associated with a fluid under physiological or pathological states. The glycosaminoglycan composition of these fluids and of the delicate membranes in vicinity of these fluids will be described in the next sections. Figs. 10-1, 10-2 and 10-3 show anatomical relationships of the ear.

From the physiological and biophysical point of view, the hearing process may be divided into:

- the transmission of acoustic energy (sound) to the sensory end-organ, the organ of Corti,
- transduction, which involves the conversion of mechanical deformation of the sensory hair into electrochemical events,
- the electrical transmission of the impulse in the eighth nerve and the auditory responses. The physiology and pathology of the glycosaminoglycans of the fluids of the ear may affect only the first two processes. Since besides the hearing function, the ear has a vestibular function, the pathology of the fluids affects both the hearing and body equilibrium.

10.2 Secretion in external ear: Cerumen

The external ear consists of the cartilaginous pinna and the external auditory meatus leading to the ear drum (tympanic membrane). The external auditory meatus contains a waxlike secretion called the earwax or cerumen. Histochemical analysis indicates that the cerumen is the combined secretion of the sebaceous and ceruminous glands present in the external auditory meatus (Ham, 1974). Although cerumen is of a waxy nature, a small quantity of uronic acid and sulfate are also present (Masuda et al., 1978, 1979), which indicate the presence of glycosaminoglycan-related material in cerumen.

Masuda et al., (1978) reported the isolation and partial characterization of a glycopeptide from human cerumen obtained from 50 normal subjects of both sexes between 16-55 years. It contained 3.5 % sulfate and 0.1 % hexuronic acid in addition to hexoses (glucose + galactose) 29.3 %, glucosamine 1.7 %, galactosamine 12.6 %, fucose 1.5 %, sialic acid 3.5 % and peptide 48.1 %. Looking at this composition it appears this material is, perhaps, heterogenous and could possibly be separated into a uronic acid-containing fraction and other fractions lacking uronic acid.

This isolation procedure employed by Masuda et al. (1978) consisted of acetone-defatting of cerumen, papain digestion at 65 °C for 48 hours, trypsin digestion for 48 hours at 37 °C with simultaneous dialysis against 0.1 M phosphate buffer of pH 7.8. The supernatent obtained upon precipitation of the residual protein with 5 % trichloroacetic acid was dialyzed against water and the glycopeptide was recovered from the retentate inside the dialysis bag by precipitation with four volumes of ethanol containing acetic acid and potassium acetate. The yield of the glycopeptide was 9.1 mg/g of cerumen and 12.5 mg/g of the defatted material.

In a later publication (Masuda et al., 1979) the complete carbohydrate composition of the uronic acid-containing glycopeptide isolated from human cerumen has been reported.

The composition of cerumen has been reported to be changed in cystic fibrosis (Kopito et al., 1972). The concentrations of all electrolytes except sodium change in cerumen from patients with cystic fibrosis. However, the changes in the macromolecular polyelectrolytes (glycosaminoglycans) have not been investigated in ceruminous secretions of these patients.

About 1.2 % of the proteoglycan content in rabbit ear cartilage consists of low molecular weight chondroitin sulfate as protein-chondroitin sulfate complex (Serafini-Fracassini and Stimson, 1972). The glucosamine content of the ear cartilage relative to the total hexosamine were higher in senescence. However, the uronic acid and moisture contents of the ear cartilage decreased with aging (Muthiah et al., 1973).

10.3 Pathological middle ear fluid

The middle ear is separated from the external ear by the tympanic membrane and contains three bones collectively known as the bony ossicles arranged in a lever arrangement in a mucosal sac. The anatomical relationships of these bones and the airsac to the eustachian tube are shown in Fig. 10-1. Under normal healthy conditions the middle ear contains little fluid. However, in otitis media, the inflammation is usually accompanied by an accumulation of a mucoid fluid resulting from the obstruction of the eustachian tube. In a clinical study (Shurin et al., 1979) the persistence of middle ear effusion was found to be 42 % in children with otitis media in the age group of two months to 12 years. In children under two years the relative risk of the persistence of the middle ear fluid was 3.8 times higher and it was three times higher in white children than in black children.

We have not come across any report on the glycosaminoglycan composition of the mucoid middle ear fluid accumulated in pathological states. However, a relationship to glycosaminoglycan metabolism appears suggestive from the common occurrence of middle ear infections in Hurler syndrome (McKusick et al., 1978). In Hurler syndrome the deafness is seen and is in part due to the deformity of the ossicles.

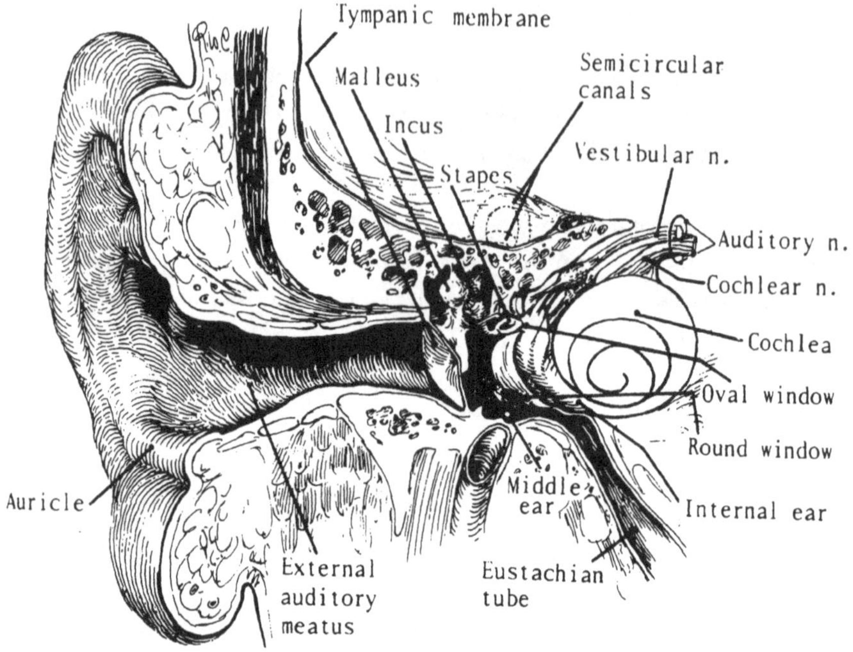

Fig. 10.1. Structure of ear. Reproduced by permission from Stedman's Medical Dictionary, 23rd Edition, Pg. 436, 1976. (Courtesy of Williams & Wilkins Co., Baltimore)

Conduction deafness is usually seen in some types of otitis media. Deafness has also been noted in patients with cystic fibrosis (Tayler et al., 1974). Although cystic fibrosis is a generalized abnormality of the exocrine glands (Mangos and Talamo, 1976) and the middle ear is a mucosal sac drained by a duct, functional and anatomical abnormalities of the middle ear surprisingly are not very frequent in cystic fibrosis.

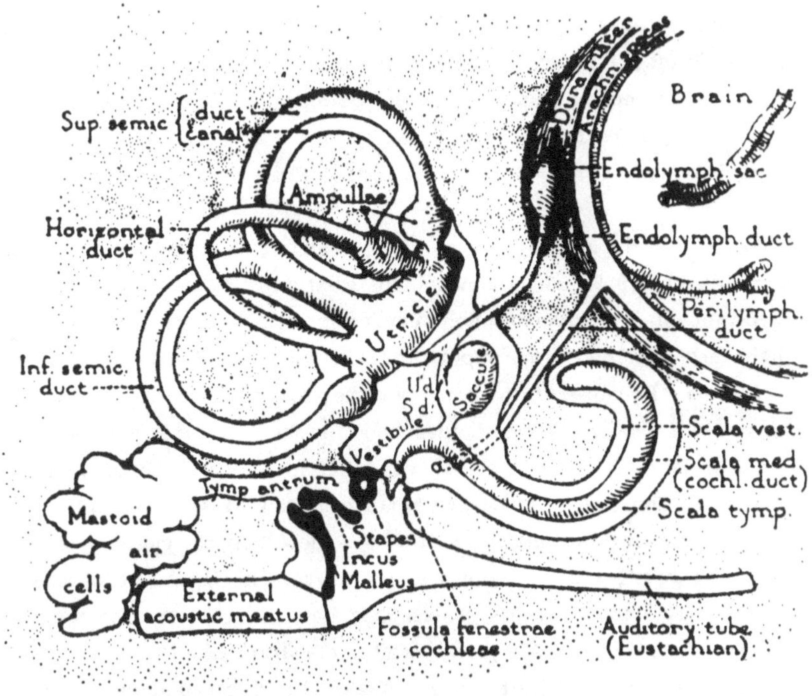

Fig. 10-2. General relationship of the various parts of the ear. Reproduced with permission from Bast, T. H. and Anson, B. J. (1949). The Temporal Bone and the Ear. Charles C. Thomas, Springfield, Illinois

10.4 Inner ear fluids: Perilymph and endolymph

The inner ear consists of a system of intercommunicating cavities or canals called the labyrinth, which consists of the membranous and osseous (bony) labyrinth. The mambranous labyrinth itself is a system of intercommunicating epithelial sacs and ducts, including the endolymphatic duct, utricle and the semicircular canals lodged within the bony labyrinth (Fig. 10-2).

The inner ear contains a mucinous system, which is associated with two fluids: endolymph and perilymph. The endolymph (or endolympha) is contained in the membranous labyrinth. The space separating the membranous labyrinth from the osseous labyrinth contains the other fluid called perilymph (or perilympha). Perilymph and endolymph together are called the labyrinthine fluids. The two labyrinthine fluids do not inter-communicate and bear an equal and opposite electrical potential of d. c. 80 mV. The production and electrochemical role (Sellick and Johnstone, 1975) and physiology (Bosher, 1976) of the labyrinthine fluid have been reviewed.

Both endolymph and perilymph have been reported to contain glycosaminoglycans (Poulsen, 1959, 1966; Jensen and Vilstrup, 1960). The majority of the information on the glycosaminoglycan composition of the labyrinthine fluids has been obtained from experiments in animals like shark, cod, guinea pig, etc. Because of the small volumes of these fluids, methods for collection on a microscale from cochlea and for microchemical analyses using only 1-2 µl fluid have been devised (Ulrich et al., 1966). Makimoto et al. (1967) have studied the glycosaminoglycan composition of perilymph. The perilymphatic and endolymphatic systems are mesenchymal cavities similar to the fluid-filled cavities of the joints and the eye. In fact, a parallelism between glaucoma and Meniere's disease has been noted. Patients having both glaucoma and Meniere's disease exhibit elevation of intraocular tension during Meniere's attacks (Godtfredson, 1949). In glycosaminoglycan composition, too, these labyrinthine fluids resemble the vitreous humor of the eye and the synovial fluid of the joints. Hyaluronic acid is the major hyaluronidase sensitive non-sulfated, metachromatic substance and mast cells are also present in perilymph and endolymph. However, Jensen and Vilstrup (1960) using histochemical methods could detect metachromasia only in perilymph, which was sensitive to hyaluronidase both from testicular and bacterial sources. Thus, according to these authors the cod perilymph contained only hyaluronic acid, which was not present in the endolymph. In shark, the endolymph contained hyaluronic acid at a concentration of 73 mg/dl (Vilstrup and Jensen, 1954). By using radioactive sulfate ($^{35}SO_4^{2-}$) and autoradiography, sulfated mucopolysaccharides have been demonstrated in the cupule, the otolith membrane and the tectorial membrane (Jensen and Vilstrup, 1960 a)

The endolymph of dog fish differs from other fishes like cod in having a very highly viscous, hyaluronidase-sensitive material. The former has a water-insoluble translucent nature, while the latter has a viscosity only slightly higher than that of water. The translucent endolymph of the dog fish liquified upon digestion with hyaluronidase. This formation of the translucent, water-insoluble gel has been attributed to the interaction of noncollagenous, serum globulin-like protein and glycosaminoglycans, both of which are present in high concentration in dog fish (Jensen and Vilstrup, 1960). In cod fish, the low viscosity is due to lower concentrations of both protein and glycosaminoglycans.

The mucopolysaccharide material from fish endolymph was prepared (Jensen and Vilstrup, 1954) by precipitation as a mucin clot by treatment with 1 % acetic acid for four days, dissolving the clot in water followed by precipitation of the carbohydrate-containing material from this solution with ethanol. The material so isolated contained 3.36 % nitrogen and exhibited decrease in viscosity upon digestion with hyaluronidase. From its osmotic pressure measurement, a molecular weight of 779.10^3 daltons was assigned to it.

Experiments in guinea pig ear have been more useful in investigating changes in various pathological states. Both perilymph and endolymph in guinea pig contain glycosaminoglycans (Poulsen, 1959; Jensen and Vilstrup, 1960; Poulsen, 1966). The viscosities of both perilymph and endolymph and glycosaminoglycans of these fluids are low. Because of the small volume of these fluids the majority of the investigations have been carried out by histochemical methods. The few direct analyses done on these labyrinthine fluids have measured hexosamine contents, rather than the characteristic hexuronic acid component of glycosaminoglycans. Hexosamines may also be contributed by glycoproteins, if present in these fluids. The values reported are therefore unreliable. Citron and Exley (1975) reported a 5.0-8.0 mg/dl hexosamine content in utricular endolymph and perilymph. However, Makimoto et al. (1978) found a higher total protein in perilymph as compared to endolymph. Poulsen (1966) from colorimetric hexosamine analysis of the mixture of endolymph and perilymph from 30 guinea pig labyrinths gave an average value of 6.2 mg/dl. Assuming a 39.4 % content of hexosamine for hyaluronic acid, this value corresponds to hyaluronic acid approximating 15 mg/dl. Histochemical studies, however, suggest lower glycosaminoglycan values.

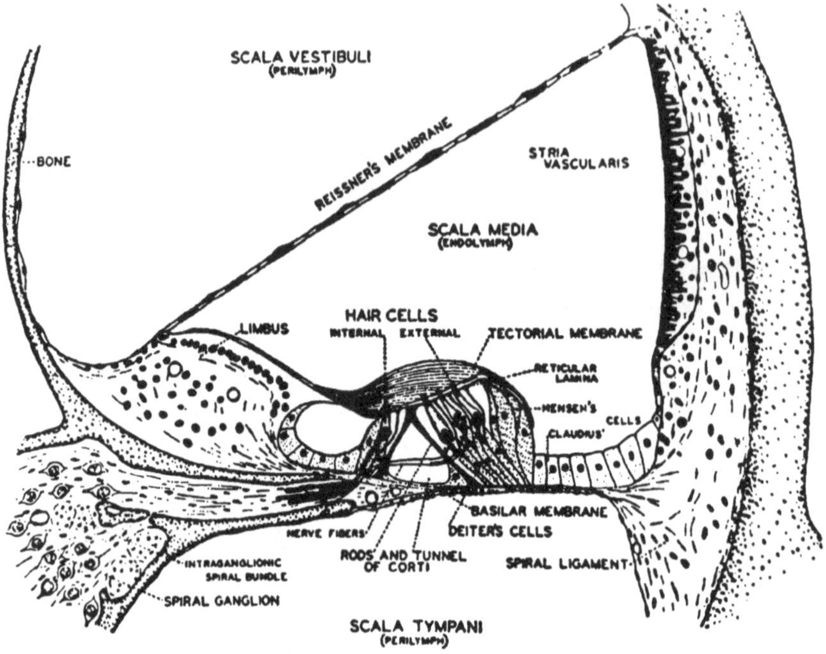

Fig. 10-3. Diagrammatic drawing of a cross section of the cochlear duct in the second turn of the guinea pig cochlea. Reproduced with permission from Davis, H. (1957): Physiol. Rev. 37, 1 (courtesy of American Physiological Society, Bethesda, MD)

By histochemical methods, the presence of hyaluronic acid and its selective hydrolysis with bacterial hyaluronidase have been confirmed. Almost all of the metachromasia was lost (Dohlman, 1959). This suggested that in all probability hyaluronic acid is the major, if not the only, glycosaminoglycan present in the guinea pig labyrinthine fluid. The presence of sulfated glycosaminoglycans in these fluids has been investigated by $^{35}SO_4$ uptake study. Forty-eight hours after the injection of the radioactive sulfate, the combined endolymph and perilymph were collected and wet-ashed. The sulfate was precipitated and measurement revealed only a trace radioactivity in the precipitate (Moltke, 1957). This also indicates that the sulfated glycosaminoglycans may be present in trace amounts. In experiments with pigeons and guinea pigs, radioactivity was found (24-hour after ^{35}S injection) in the coagulated endolymph and in the cupule. This also indicated the presence of sulfomucopolysaccharides in the inner ear fluids (Ormerod and Dohlman, 1958). The biosynthesis of the sulfated mucopolysaccharides (glycosaminoglycans), as revealed by ^{35}S incorporation, in the developing inner ear is reduced in manganese deficiency (Shrader et al., 1973).

Hughes and Chou (1963) reported an average of 6.5 mg/dl of hexosamine value in perilymph and endolymph. Juhn and Niederwieser (1968) using DEAE-cellulose thin-layer chromatography (TLC) could not detect any glycosaminoglycans in perilymph and endolymph. Ishiyama (1969) used the same TLC procedure and trypsin-pronase digestion of perilymph and endolymph withdrawn in vivo from guinea pigs by a freezing technique. Acidic glycosaminoglycans amounting to < 2.4 mg/dl were detected in perilymph and < 3.0 mg/dl in endolymph, based on detection limits of the procedure used. Removal of protein by protease digestion made the glycosaminoglycans more easily detectable with 0.5 of Alcian blue in glacial acetic acid/methanol (1:1, v/v).

Saito et al. (1976) reported the complete glycosaminoglycan composition of the lateral wall of the cochlea of guinea pigs. The inner ear contained dermatan sulfate, chondroitin 4-sulfate and hyaluronic acid in relative percentages of 75, 12 and 13 %, respectively.

Along with the glycosaminoglycans about 300 mast cells are also present in the endolymphatic space and the pars superior of the perilymphatic system. The pars inferior of the guinea pig contains little or no mast cells. The high molecular weight glycosaminoglycans like hyaluronic acid are synthesized locally by these mast cells. Histochemically, the mast cell counts at different sites correlate with metachromasia of the amorphous ground substance at these sites. In myxedema the increase in metachromatic substances in perilymph and endolymph accompanies an increase in mast cell counts, their size and granularity (Poulsen, 1959).

10.5 Source of inner ear fluids

The origin of perilymph and endolymph is not exactly known. The information on this aspect is based upon inorganic and organic composition of these fluids as compared with those of the cerebrospinal fluid (CSF) and blood in guinea pigs (Bosher, 1976; Prado, 1977; Scheibe et al., 1977). Results of the analytical comparisons revealed that perilymph from scala tympani and scala vestibule had 3-7 fold lower concentrations of individual components than those in C. S. F. Differences in protein, lactate, pyruvate and lactate dehydrogenase in perilymph were 5-10 times higher than in blood (Scheibe et al 1977). These findings suggest that the labyrinthine fluid does not originate from C. S. F. nor is it a plasma ultrafiltrate. In contrast to these results Giebel and Keppler (1977) using four different methods have reported no difference in total protein in perilymph from scala tympani and serum from guinea pigs. Although some serum proteins were absent in perilymph, no differences in single proteins were seen.

Hughes and Chou (1963) have reviewed the origin of the perilymph. The composition of the perilymph differs from that of the endolymph. The endolymph and perilymph contained 1.8 and 1.2 % protein, respectively (Vilstrup and Jensen, 1954). Makimoto et al. (1978) reported a higher concentration of glucose and protein in perilymph, which contradicts the earlier finding of Vilstrup and Jensen (1954). In view of these contradictory reports it is difficult to conclude if the compositions are different in perilymph and endolymph. If these were different, the two fluids may have a different origin. Gershbein et al. (1974) have reported definite differences in perilymph from scala tympani and scala vestibuli and these differences in metabolites were independent of acoustic treatment. These authors have suggested C. S. F. as the possible source of perilymph. As evidence to this they noted similarities in glucose contents and LDH isoenzyme patterns in both the fluids.

10.6 Glycosaminoglycans in delicate membranes of inner ear

By histochemical procedures, acidic glycosaminoglycans can be seen in the tectorial membrane, basilar membrane, spiral ligament, limbus spiralis and the Reissner's membrane (Schaetzle, 1971).

The tectorial membrane is normally at a zero potential between +80 mV in the endolymph on one side and -80 mV in the organ of Corti on the other side. The indication of the presence of glycosaminoglycans by histochemical methods prompted several investigators to study the amount and nature of the glycosaminoglycans in the tectorial membrane that could possibly polarize this membrane. Both histochemical and biochemical methods have been used. The histochemical techniques used for studies on inner ear mucopolysaccharides have been reviewed (Schaetzle, 1971; Mizukoshi and Saito, 1973, 1974; Thalmann, 1975, 1976).

Iurato (1960) performed chemical analysis of the tectorial membrane and the spiral ligament using the carbazole reaction of Dische. He failed to detect any hexuronic acid-containing material in the tectorial membrane and reported 0.3 % mucopolysaccharides in the spiral ligament. This was confirmed by Naftalin et al. (1964) who reported only a trace amount of hexosamine in the tectorial membrane. Some of the earlier literature on the glycosaminoglycans of the ear has been reviewed elsewhere (Balazs, 1965).

Saito (1967), in a biochemical and histochemical study of the cochlea of the guinea pigs, identified the hyaluronidase-sensitive acid mucopolysaccharide as hyaluronic acid. Juhn and Niederwieser (1968) have presented the results of their thin layer chromatography analysis of the glycosaminoglycans. They found only a small amount and also the glycosaminoglycans present in the stria vascularis with spiral ligament were similar to keratosulfate. Ishiyama (1969) using the same thin layer chromatographic procedure carried out the semiquantitative analysis of the guinea pig cochlea and reported a small amount of glycosaminoglycans after protease digestion of the cochlear material.

Lotz and Kuhl (1970, 1970 a), from their quantitative analysis of the mucopolysaccharides in the whole membranous labyrinth, reported the presence of these macromolecules in large amounts. However, their indirect analysis was based on hexosamine contents rather than the hexuronic acid contents. It is important to bear in mind that the hexosamine present in glycoproteins also contributes to such indirect quantitations of glycosaminoglycans. The results are, therefore, unreliable unless the characteristic component of mucopolysaccharides, hexuronic acid, is determined.

Using a modified turbidimetric method for microanalysis of insoluble complexes between glycosaminoglycans and cetyltrimethylammonium (CTAB) and enzymatic analysis, Saito and Daly (1970) analyzed the glycosaminoglycan composition of the membranous cochlea of the guinea pigs. The turbidimetric interference by nucleic acids was removed by precipitation of nucleic acids with trichloroacetic acid and by their digestion with deoxyribonuclease. The membranous cochlea was divided into four sections:

- stria vascularis with spiral ligament
- tectorial membrane
- basal membrane with the organ of Corti and limbus spiralis
- Reissner's membrane. The glycosaminoglycans (GAG) were isolated from these parts by dehydration with acetone, pronase digestion, deproteinization with 10 % trichloroacetic acid, ethanolic precipitation of GAG, removal of nucleic acids and turbidity measurement with CTAB-Li at 400 nm. By this procedure very little difference was seen in GAG distribution between these four parts. All contained hyaluronidase-sensitive hyaluronic acid and hyaluronidase resistant dermatan sulfate (chondroitin sulfate B). Both the tectorial and Reissner's membranes contained 0.1 % of GAG per dry weight. The stria vascularis contained 0.6 % GAG per dry weight, which consisted of 61 % hyaluronidase-digestible (hyaluronic acid) and 39 % hyaluronidase-resistant (dermatan sulfate). On the other hand, the basal membrane with the organ of Corti and limbus spiralis contained 0.5 % GAG per dry weight, which consisted of 74 % hyaluronic acid and 26 % dermatan sulfate. With such a low content of 0.1 % GAG in tectorial membrane it is difficult to visualize how this small amount of glycosaminoglycans could bring about polarization of the tectorial membrane.

Although the presence of the sulfated glycosaminoglycans in the cochlea has been indicated by several previous studies (Juhn and Niederwieser, 1968; Saito and Daly, 1970; Kuttner, 1977), the unequivocal proof of this was provided by a study of the sulfated acid mucopolysaccharides in the tectorial membrane from 20 guinea pigs (Tachibana et al., 1973). These authors used electron microscopic histochemical identification of glycosaminoglycans with ruthenium red staining, microspectrophotometric quantitation after Alcian blue staining and analysis of the sulfur element by electron probe analysis, digestion with testicular hyaluronidase and chondroitinase ABC. After digestion with the enzymes, the sulfur element was not detected, in spite of the possible presence of the sulfur-containing amino acids, present though in much smaller amounts as compared to the quantity of the sulfated glycosaminoglycans in the tectorial membrane. The enzymatic sensitivity to hyaluronidase and chondroitinase ABC, as revealed by the EMX technique, suggested the presence of chondroitin 4- and/or 6-sulfate diffusing on the tectorial membrane. The authors (Tachibana et al., 1973) suggested that these sulfated mucopolysaccharides would have a polarizing effect between the positive potential in the endolymph and the negative potential in the organ of Corti.

Mizukoshi and Saito (1973, 1974) have reviewed their work on the glycosaminoglycans on the cochlea and vestibule in guinea pigs. Morowitz and Porubsky (1971) have investigated the glycosaminoglycans in the otic capsule and the ossicles of the fetal and newly-born rats. In a study of the glycosaminoglycans of the inner ear, Saito et al. (1976) reported a content of about 0.4 % per dry weight in the lateral wall of the cochlea of the guinea pig. Electrophoretic microanalysis identified the presence of chondroitin sulfate B (dermatan sulfate), chondroitin sulfate A (chondroitin 4-sulfate) and hyaluronic acid in relative percentages of 75, 12 and 13 %, respectively.

The glycosaminoglycans associated with the sensory hair cells and the hair cell membrane have been investigated by histochemical studies (Bagger-Sjoback and Wersall, 1973; Saito et al., 1977; Flock et al., 1977; Kuttner et al, 1977).

Saito et al. (1977) investigated the characteristics of the hair cell membrane by freeze-fracture, electron microscopic histochemistry, radiochromatography, thin layer chromatography and single hair cell observations. The burst of the single outer hair cell in artificial endolymph and the presence of very tight junctions at the level of reticular membrane revealed by the freeze-fracture method clarified that the cochlear hair cells are sealed from the endolymph. Electron microscopic histochemistry revealed that in the presence of cationic substances the hair cell membrane undergoes changes (Saito et al., 1977). This suggested an affinity of aminoglycoside antibiotics and other related ototoxic agents for the hair cell membrane. The changes induced by these antibiotics in the hair cell membrane may be responsible for their ototoxicity.

Kuttner (1977) reported the presence of acid mucopolysaccharides (glycosaminoglycans) in the organ of Corti by electron microscopy following iron-binding reaction with ferrisol and ruthenium red staining of pellicle preparations or ultra-thin sections of the organ of Corti. On the external surface of the plasmalemma at the sensory and the supporting epithelia there is an acid mucopolysaccharide coat. This mucopolysaccharide coat is called the glycocalyx or the cell coat. Histochemical examination of this cell coat shows higher staining (due to the presence of an abundance of mucopolysaccharides) in the region of stereocilia and the cell-membrane (Kuttner 1977). The ground substance of the basilar membrane gives a ruthenium red precipitate like the one obtained with proteo-chondroitin sulfate.

The negative charges on the cell-coat (the glycocalyx) are due to the presence of mucopolysaccharides. These negative charges by virtue of their repulsion appear to be important for maintaining the normal distances between the stereocilia. Any damage to the cell coat would result in the loss of the repulsive forces between the cilia that keeps them apart at an appropriate distance. Such a damage could result from genetic and metabolic factors (Konigsmark and Gorlin, 1976), aminoglycoside antibiotics or acoustic traumas (Flock et al., 1977; Bagger-Sjoback and Wersall, 1973; Thalmann, 1976 a).

10.7 Effects of hormones and vitamins in inner ear

Thyrotropic hormone (thyrotropin) produced by the anterior pituitary gland (adenohypophysis) affects the mesenchymal tissues and this effect is the most conspicuous in the ground substance in the inner ear and in the retrobulbar connective tissue. Thyrotropin stimulates mucopolysaccharide production in the inner ear. It increases the mast cells to double the count and also increases their size and granularity. Along with these changes, there is an increase in the total content of hyaluronic acid in the labyrinth. However, since there is a concomitant influx of water, the hyaluronic acid increase is not apparent and the hexosamine concentration per unit volume of perilymph and endolymph does not change (Poulsen, 1959).

Thyroidectomy causes similar increase in size and number of mast cells and hyaluronate content in the inner ear. This effect is, probably, due to the fact that under normal circumstances thyroid hormone produced by the thyroid gland acts as an inhibitor of the thyrotropin release from the adenohypophysis, the anterior pituitary gland (Iversen and Asboe-Hansen, 1952). Similar changes are observed in hypoactivity of the thyroid gland producing myxedema in adult human subjects and endemic cretinism in young ones. The antithyroid drug, propylthiouracil, produces abnormal histology in the cells of the cochlea and its spiral ganglion (Bargman and Gardner, 1972).

Thyroid hormone and thyrotropin, by means of the aforementioned effects, seem to control the inner ear function. Abnormal inner ear functions like hearing defects are commonly seen in myxedema. Several other reports are available about the relationships of hypothyroid functions and Meniere's disease and inner ear defects in patients with thyrotoxicosis (Graves disease) treated by surgery, x-ray radiations or methyluracil (Soffer, 1956). Along with inner ear defects, vestibular dysfunction affects the body equilibrium in these patients.

Some other hormones may affect the tissues of the ear. Parathyroid hormone injection in guinea pigs can induce the osteolysis of the ossicles. Instillation of the aqueous solution of calcitonin three days prior to the parathyroid hormone injection inhibits the osteolytic effect of parathyroid hormone (Rudman and Parsons, 1979). Cortisone has been found beneficial in counteracting the effects of ototoxic agents on sensory epithelium (Savary, 1968; Brummet, 1976). Progesterone has been reported to stimulate regeneration of excised ear tissue in both male and female rabbits (Dyson and Joseph, 1971). Among the vitamins, vitamin A has been implicated in inner ear function (Chole, 1977). Vestibular dysfunction has been noted in human vitamin A deficiency. Histological examination of control and vitamin A deficient young guinea pigs revealed progressive metaplastic changes in the macular utriculi (Cakiere and Lachance, 1974). These structural alterations could account for the vestibular dysfunctions observed in human vitamin A deficiency. Vitamin A also has been reported to have a protective effect against the ototoxic effects of antibiotics and other drugs (Savary, 1968; Brummett, 1976). Histochemical examination of inner ear in six pigeons receiving only streptomycin and 25,000 units vitamin A compared to six pigeons receiving only streptomycin revealed more intense metachromatic staining in the former group. Observations on the vestibular apparatus indicated a protective effect of vitamin A (Zygmunt, 1967).

10.8 Experimental studies in myxedema (hypothyroidism)

The association between congenital and acquired hypothyroidism and hearing loss has been known for a long time, but the reports on the nature of this association have been conflicting. Thyroid hypofunction and hearing loss have been found to be coincidental in significant percentages of patients with full-blown myxedema, Méniére's disease and non-genetic endemic cretinism. An endocrinological survey of 250 patients with Méniére's disease showed a thyroid deficiency in 89.2 % of the cases (ref. 97). Several studies have been carried out in experimental hypothyroidism (induced in experimental animals by thyroidectomy or administration of radioactive iodine^{131} or propylthiouracil) for elucidating the changes responsible for the hearing loss. These are described below.

A recent electron microscopic and histochemical analysis in guinea pigs made hypothyroid with propylthiouracil or radioactive iodine showed morphological changes both in the middle and inner ear. The middle ear had crystallinity or cartilaginous consistency of the bone of the bullae and cochlea and fusion and distortion of incus and stapes. These could explain conductive hearing loss. On the other hand, the sensorineural loss in hypothyroidism appeared to result from degeneration of the cells of the stria vascularis and outer and inner hair cells damage and debris in cochlear duct (due to precipitation of the accumulated glycosaminoglycans, during histochemical staining) and irregularities in the tectorial membrane (ref. 98).

Poulsen (1959, 1966) and others (ref. 99) have also reported an accumulation of glycosaminoglycans in the cochlear duct of hypothyroid guinea pigs. Another group of investigators (ref. 33, 92) have noted an increased concentration of hyaluronic acid in the endolymph of thyroidectomized sharks. In the newly-born hypothyroid chicks (born to animals treated for two generations with thiourea or propylthiourea, an elevated auditory threshold, distortion of hair cells and edema in the fluid spaces around the hair cells and the supporting cells in the inner ear were present. Thyroid hormone therapy relieved the increased auditory threshold (100). Similar deformities in the hair cells and the tectorial membrane were also noted in the congenitally hypothyroid mice (101). The presence of edema and other deformities suggest an abnormality of the glycosaminoglycan metabolism in these hypothyroid animals.

In guinea pigs made hypothyroid by thyroidectomy followed by thyrotropin administration, histochemical staining of the labyrinth sections with Alcian blue after fixation with lead subacetate showed an increase in metachromatic substances in perilymph and endolymph and an increase to double the number of mast cells (Poulsen, 1959, 1966). The increased metachromatic staining was found to be due to an excess of hyaluronic acid on the basis of digestion of the sections with bacterial hyaluronidase and by chemical estimation of its content (1.5 mg/ml) in a mixture of perilymph and endolymph from 30 guinea pigs. The increased hyaluronic acid molecules bind salts and water producing increases in the volumes of the perilymph and endolymph. The increased pressure of the endolymph on the membranous labyrinth, which is surrounded by unyielding bony structures and contains the sensory epithelium, produces hearing and vestibular disturbances.

Some earlier studies (Dohlman, 1944; Ross, 1936) had attributed the abnormal vestibular function in myxedema to an increase in the viscosity of the endolymph, which could result from the increased hyaluronic acid contents observed by several other workers. The increased hyaluronic acid appears to be due to a deficiency of hyaluronidase activity, which is affected by thyroid hormone, since hyaluronic acid accumulated in all the tissues of hypothyroid pigs. Upon thyroid administration, hyaluronidase appeares for normal degradation of the accumulated hyaluronic acid (ref. 102).

Histochemical staining of the labyrinth section from normal human sujects and Me'nière's disease (with confirmed hypothyroidism) with toluidine blue revealed an accumulation of glycosaminoglycans (hyaluronic acid) in the patients. Also, electrophoresis on cellulose acetate strips (in pyridine/formic acid buffer of pH 2.8, at 150 volts/3.5 mA for 30 min.) followed by staining with acridine orange and contrastaining with toluidine blue showed that hyaluronic acid was present in excess in CSF, endolymph and blood of the patients (ref. 97). Evidence was presented that the rise in hyaluronic acid was due to an increase in antihyaluronidase and the presence of an excess of this hydrophilic hyaluronate macromolecule produced increased osmotic pressure (hyperosmolality), which reduced the outflow at the venous end of the blood loops in the stria vascularis and led to endolymphatic hypertension, which in turn resulted in clinical symptoms due to expansion and compression of the labyrinth wall.

Attempts to relieve the clinical symptoms in hypothyroid patients by depolymerization of hyaluronic acid, through direct infusion of hyaluronidase into the endolymphatic space or indirectly through the intravenous route, have failed (ref. 97). Symptoms returned soon after this treatment; sometime one hour after the treatment. This ineffectiveness of the etiological therapy may be due to the rapidly rising antihyaluronidase titers or rapid distribution of hyaluronidase in the extracellular compartment.

10.9 Pathological glycosaminoglycan changes in the inner ear

Disorders of the ear function are encountered in a large number of conditions including thyroid gland disorders, cystic fibrosis, and mucopolysaccharidoses.

Hypothyroidism. Defects in hearing and vestibular functions are seen in all those situations which involve hypothyroid activity. These include myxedema in adults, endemic cretinism in young children and in thyrotoxicosis patients treated with x-ray irradiation, thyroidectomy or antithyroid substances like methyluracil. The changes in mucopolysaccharides observed in human adult myxedema are similar to those seen in experimental myxedema in guinea pigs. (section 10.8). Hearing loss occurs in about 30 % of the human myxedema patients and thyroid hormone administration has been found useful in vestibular and acoustic disorders in patients with reduced basal metabolic rate.

Histochemical examination in endemic cretinism reveals increased formation of mesenchymal connective tissue structure in labyrinth. A congenital increase in mast cells and hyaluronic acid is accompanied by increase in the volume of the endolymphatic system which enlarges the petrous portion. Other changes include: atrophy of spiral ganglion and the organ of Corti and ectatic changes in semicircular canals, cochlear canal and saccus endolymphaticus (Poulsen, 1966).

Meniere's Syndrome, also called endolymphatic hydrops or labyrinthine hydrops is accompanied by deafness, tinnitus, recurrent vertigo attacks and dizziness. A relationship between Meniere's disease and hypothyroid function has been noted. Poulsen (1966) has attributed this to increased mast cells and hyaluronate in the endolymph which occurs in hypothyroidism. Increased hyaluronate holds more water and salts, producing increase in the volume of the endolymph, which in turn results in pressure changes on the membranes of the labyrinth. These lead to changes in hearing and vestibular function. There have been several suggestions that alterations in physicochemical constitution of the endolymph may be a major factor in the pathogenesis of Meniere's disease. Experimental loss of cochlear function can be brought about by changes in endolymph composition (Bosher, 1976).

Otosclerosis. It is the most common of the conductive type of hearing loss in adult life. It is hereditary and it is probably an autosomal dominant condition of variable expressivity. Its incidence in U.K. and USA is 1 %. The basic defect is formation of a spongy bone in the capsule of the labyrinth. A closely related condtition in children is *osteogenesis imperfecta,* which rarely leads to profound deafness. Arslan and Ricci (1960) have demonstrated histochemical changes in the otosclerotic foci. The active destructive phase of otosclerosis (osteospongiosis) has been considered to be intiated by the hydrolytic enzymes released from histiocytes and their lysosomes into the perilymph (Friedman, 1976). Calcitonin can arrest the bony pathology by inhibiting the parathyroid action and has been found useful for otosclerotic patients (Rudman and Parsons, 1979). Although the mode of inheritance of otosclerosis is similar to other connective tissue disorders like Marfan syndrome and rheumatoid arthritis, no change in hexosamine excretion (Schoenheyder et al., 1966) and no change in 24-hour urine glycosaminoglycans excretion (Soifer et al., 1970) have been observed. These observations suggest that otosclerosis is only a local mesenchymal (glycosaminoglycan) disorder and there is no generalized effect on glycosaminoglycan metabolism. Increase of proteins in perilymph and appearance of two additional bands in the alpha and beta globulin regions are seen (Rauch, 1971).

Acoustic traumas. Noise of sufficient sound levels is produced in many industrial processes and it can cause impairment of hearing. The biochemical and histochemical changes induced in the ear upon acoustic stimulation have been studied in guinea pigs (Schneider, 1971; Thalmann, 1976 a). The physical and metabolic stress exerted on the cochlea results in damage to cochlea (Lim and Mellnick, 1975). Changes in glycogen and mucopolysaccharides have been reported in experimental acoustic trauma (Ciges et al., 1975). Besides ionic shifts in perilymph, acoustic stimulation produces diminished function of the organ of Corti, with injury to the hair cells. The quantitative histopathological and biochemical changes induced in response of the auditory organ to excessive sound have been reviewed (Falk, 1977; Thalmann, 1976 a).

Light and scanning electron microscopy of the spiral organ of Corti after exposure of various animals to the effect of high intensity pure tones revealed derangement of the hairs of both outer and inner hair cells and fusion of stereocilia (Friedman, 1976). As mentioned above, these stereocilia are kept apart at suitable distances by the negative charges of mucopolysaccharides. The nature of the damage to these macromolecules has not been investigated.

While high intensity sound is an industrial and public hazard, it has been used in ultrasonic irradiation therapy of Meniere's disease (see above, this section). This therapy relieves the recurrent vertigo in Meniere's disease patients by selective vestibular labyrinthine destruction while preserving the hearing function (Igarashi, 1976). Transient post-traumatic facial palsy has been seen in 1 out of 234 such ultrasonic treatments.

Ototoxicity. A large number of drugs including aminoglycoside antibiotics, salicylates, quinine, nicotinic acid, aspirin, ethacrynic acid, etc. cause functional defects of the auditory system. The effects of these drugs and histochemical changes revealed in metachromasia due to mucopolysaccharides have been reviewed (Savary, 1968; Bronen and Feldman, 1978; Jolicoeur, 1972; von Ilberg, 1976; Brummett, 1976; Matz, 1976; Banovetz, 1976). Cortisone and vitamin A, both of which are known to affect mucopolysaccharide metabolism (Cakiere and Lachance, 1974; Chole, 1977; Savary, 1968), have a protective effect against these ototoxins.

Out of the aforementioned ototoxic agents an extensive study of effect on mucopolysaccharides in the inner ear has been carried only in case of kanamycin. Quantitative analysis of acid mucopolysaccharides in the kanamycin-intoxicated guinea pig cochlea (Saito and Daly, 1971) revealed decreased quantity of acid mucopolysaccharides in the stria vascularis and spiral ligament. This suggested that the main path of kanamycin into the endolymph is from the stria vascularis and the spiral ligament and that the change of the ground substance in the connective tissues may affect a static equilibrium state and dynamic transport process between endolymph and stria vascularis. Formation of kanamycin-acid mucopolysaccharide complexes in the hair cells also occurs, which may be more serious than presence of kanamycin in labyrinthine fluids.

10.10 Role of glycosaminoglycans in inner ear function

There appears to be a connection between the glycosaminoglycans in the inner ear and the inner ear functions of hearing and body equilibrium control. Deafness is one of the common clinical manifestations of the various types of the mucopolysaccharide storage diseases called the mucopolysaccharidoses. Deafness is especially common in Hurler syndrome, Hunter syndrome, Scheie syndrome and Morquio syndrome (McKusick et al., 1978). The loss of hearing may be progressive in these patients. Deafness in Hurler syndrome is partly due to abnormality of the inner ear and partly it is the result of the deformity of the bony ossicles in the middle ear.

The two inner ear fluids are essential for the hydrodynamic processes responsible for hair cell activation. An appropriate pressure of these labyrinthine fluids is essential for normal function of the inner ear. Measurements of pressure differences between cerebrospinal fluid and perilymph in cats lead to the conclusion that reduced pressure impedes hearing (Miriszlai et al., 1971). The pressure is directly related to the fluid volume, which in turn is related to the mucopolysaccharide (glycosaminoglycan) contents of these fluids. In conditions like hypothyroidism the increase in the hyaluronic acid in the inner ear is accompanied by an influx of water and ions in the labyrinthine fluids and impaired function of the ear. The mucopolysaccharides of perilymph and endolymph may have a role partly by their regulating effect on volume and ion concentrations of these fluids. The increased pressure due to the labyrinthine hydrops brings about displacement of the vestibular membrane towards the scala vestibuli throughout all the coils of the cochlea (Bosher, 1976). Also, the influx of water and ions in hydrops affects the electrochemical properties (Bosher, 1970; Naftalin, 1968) of the perilymph and endolymph.

The glycocalyx or cell coat on the external surface of the plasmalemma at the sensory and supporting epithelia of the organ of Corti is a coat of acid mucopolysaccharides. The negative charges on the mucopolysaccharides in the cell coat seem to be important for maintaining normal distances between the stereocilia (Kuttner, 1977). Damages to the cell coat due to hereditary and metabolic factors (Konigsmark and Gorlin, 1976), aminoglucoside antibiotics or acoustic traumas result in loss of repulsive forces between the cilia (sensory hairs), their fusion and impaired inner ear function (Bagger-Sjoback and Wersall, 1973; Flock et al., 1977).

Hanson (1971) has discussed the role of mucopolysaccharides in inner ear function. Some relevant discussion has been provided by others (Thalmann, 1975; Kuijpers, 1976, 1976 a). The role of hyaluronic acid in acoustic mechanoelectric transduction has been reviewed (Naftalin, 1968; Plattig, 1975; Casella and Taglietti, 1976).

The conversion of the mechanical shearing force in hair processes into electrochemical energy can occur either by depolarization of a membrane or by a change in configuration of some macromolecule. The molecular transduction has been explained on the basis of displacement potentials of the hyaluronate molecules (Jensen et al., 1954 a; Vilstrup and Jensen, 1961). When a 0.5% hyaluronic acid was moved by suction or blowing in a U-tube, an induced potential of 20-25 mV was generated. The small positive ions moving faster than the more negatively charged hydrogen-bonded polyanion lattice adhering to the glass capillary wall produced this flow or displacement potential. Similarly, Christiansen (1961) showed that when a hyaluronate solution was moved lengthwise in a capillary tube an electric signal of a polarity was generated. Upon changing the direction of the hyaluronate solution flow, the polarity was reversed. On this basis, it has been suggested that in response to an acoustic energy (sound) hyaluronate could undergo a conformational change or some form of resonance and produce an electric signal. Another suggestion (Vinnikov and Titova, 1964) related to this is the possibility that each acoustic signal may *tune* hair processes or bring about matching of the constituents of the macromolecule to a particular acoustic signal.

According to a theory of the function of the vestibular apparatus, hyaluronate molecules covering the sensory cells of the inner ear give a displacement potential. In this capacity they act as mechanoelectrical transducers and as molecular motors acting as resonators in the labyrinth (Christiansen,

1964). However, a later investigation (ref. 103) presented evidence that there was no electrochemical activity gradients or displacement potentials of hyaluronate solutions in glass capillaries. In view of this contradictory report, the inner ear function remains unexplained.

10.12 Bibliography

1. Arslan, M. and Ricci, V. (1960): Can otosclerosis be considered as a collagen disease? Results of histochemical research. Otorinolaring. Ital. 30, 81-86

2. Bagger-Sjöbäck, D. and Wersall, J. (1973): The sensory hairs and the tectorial membrane of the basilar papilla in the lizard Calotes versicolor. J. Neurocytol. 2, 329-350

3. Balazs, E. A. (1965): Amino sugar-containing macromolecules in the tissues of the eye and the ear. In: E. A. Balazs and R. W. Jeanloz (Eds.). The Amino sugars, Vol. II A:Distribution and Biological Role, Academic Press, Chap. 37, For ear, 458-460

4. Banovetz, J. D. (1976): Drugs in otolaryngology. In W. Modell (Ed.). Drugs choice 1976-1977 (Rev. Ed.), Mosby, St. Louis, 657-668

5. Bargman, G. J. and Gardner, L. L. (1972): Experimental production of otic lesions with antithyroid drugs. Adv. Exp. Med. Biol. 30, 305-323

6. Bosher, S. K. (1970): Possible importance of the labyrinthine fluids in the pathogenesis of sensorineural deafness. In: G. E. Wolstenholme and E. W. Gordon (Eds.): Sensorineural Hearing Loss, Ciba Found. Sympos. 1969 (Publ. 1970), 199-212

7. Bosher, S. K. (1976): Physiology of the labyrinthine fluids. In: R. Hinchcliffe and D. Harrison (Eds.). Scientific Foundations of Otolaryngology, Heinemann Medical Books, London, 290-303

8. Bronen, R. D. and Feldman, A. M. (1978): Pharmacology of hearing and ototoxicity. Annu. Rev. Pharmacol. Toxicol. 18, 233-252

9. Brummett, R. E. (1976): Ototoxicity: its detection and prevention. In: Mod. Pharmacol.-Toxicol. 6 (Horiz. Clin. Pharmacol.), 31-46

10. Cakiere, O. M. and Lachance, P. A. (1974): Effect of avitaminosis A on the inner ear of guinea pigs. Nutr. Rep. Int. 9, 383-392

11. Casella, C. and Taglietti, V. (1976): Aspects of mechano-electrical transduction in frog semicircular canals. Rend. Ist. Lomb. Accad. Sci. Lett. B. 110, 78-103

12. Chole, R. A. (1977): Experimental studies on the role of Vitamin A in the inner ear. Dissertation, Univ. Minnesota, Minneapolis, Minn. Diss. Abstr. Int. B. 38 (3), 1135

13. Christiansen, J. A. (1961): An attempt to explain the microphonic effect of the inner ear by means of displacement potentials. Acta Otolaryngol. Suppl. 163, 76-79

14. Christiansen, J. A. (1964): On hyaluronate molecules in the labyrinth as mechanoelectrical transducers, and as molecular motors acting as resonators. Acta Otolaryngol. 57, 33-49

15. Ciges, M., Quesada, P., Oloriz, J. et al. (1975): Glycogen and mucopolysaccharides in experimental acoustic trauma. An. Otorhinolaryngol. Ibero Am. 2, 29-46

16. Citron, L. and Exley, D. (1975): Recent work on the biochemistry of the labyrinthine fluids. Proc. Roy. Med. 50, 697-701

17. Dohlman, G. (1944): Investigations in function of semicircular canals. Acta Otolaryngol. Suppl. 51, 211

18. Dohlman, G. (1959): Modern Views on Vestibular physiology. J. Laryng. Otol. (London), 143-153

19. Dyson, M. and Joseph, J. (1971): Effects of female sexhormones on tissue regeneration in the rabbit's ear. J. Endocrinol. 51, 685-697

20. Falk, S. A. (1977): Pathological responses of the auditory organ to excessive sound. In: D. H. K. Lee (Ed.): Handbook of Physiology, Sect. 9: Reactions of Environmental Agents, Am. Physiol. Soc., Bethesda, 17-30

21. Flock, A., Flock, B., and Murray, E. (1977): Studies on the sensory hairs of receptor cells in the inner ear. Acta Otolaryng. (Stockholm) 83, 85-91

22. Friedman, I. (1976): Ultrastructure of ear in normal and diseased states. In: R. Hinchcliffe and D. Harrison (Eds.): Scientific Foundations of Otolaryngology, Heineman Medical Books, London, 202-211

23. Gershbein, L. L., Manshio, D. T. and Shurrager, P. S. (1974): Biochemical parameters of guinea pig perilymph sampled according to scala and following sound presentation. Enviro. Health Prospect. 8, 157-164

24. Giebel, W. and Keppeler, T. (1977): Quantitative determination of serum proteins in guinea pig perilymph obtained by different preparation methods. Colloq. Inst. Natl. Sante Rech. Med. 68 (Inner Ear Biol.), 301-309

25. Godtfredsen, E. (1949): Concurrence of glaucoma and Meniere's disease. Acta Otolaryng. 37, 533-538

26. Ham, A. W. (1974): Histology, J. B. Lippincot, Philadelphia, 973-987

27. Hanson, H. (1971): Biochemical aspects of inner ear research. Wiss. Z. Martin-Luther Univ., Halle-Wittenberg, Math. Naturwiss. Reihe 20, 3-24 (Ger)

28. Hughes, D. E. and Chou, J. T. Y. (1963): The origin of the perilymph of the ear. Life Sci. 2, 107-111

29. Igarashi, M. (1976): Vestibular system - Experimental Pathophysiology. In: R. Hinchcliffe and D. Harrison (Eds.). Scientific Foundations of Otolaryngology. Heinemann Medical Books, London, 361-371

30. Ishiyama, E. (1969): Semiquantitative analysis of acid mucopolysaccharides in the guinea pig's cochlea. Pract. Oto-rhino-laryng. 31, 274-282

31. Iurato, S. (1960): Submicroscopic structure of the membranous labyrinth. 1. The tectorial membrane. Z. Zellforsch. 52, 105-128

32. Iversen, K. and Asboe-Hansen, G. (1952): Studies on fat mobilizing factor of anterior pituitary gland: Suppressive action of thyroxin. Acta Endocrinol. 11, 111-118

33. Jensen, C. E. and Vilstrup, T. (1954): Hyaluronic acid III. Potassium hyaluronate from the endolymph of sharks. Acta Chem. Scand. 8, 292-294

34. Jensen, C. E., Kofoed, J. and Vilstrup, T. (1954 a): Flow potentials in hyaluronate solutions. Nature 174, 1101-1102

35. Jensen, C. E. and Vilstrup, T. (1960): Demonstration of a hyaluronidase-sensitive substance in the perilymph of cod. Acta Otolaryng. (Stockh.) 51, 585-592

36. Jensen, C. E. and Vilstrup, T. (1960 a): On the chemistry of human cupulae. Acta Otolaryng. (Stockh) 52, 383

37. Jolicoeur, G. (1972): Ototoxic changes due to drugs. Drug-Induced Disorders 4, 534-542

38. Juhn, S. K. and Niederwieser, A. (1968): Trials of thin-layer chromatographic demonstration of acid mucopolysaccharides in the inner ear. Arch. Klin. Exp. Ohr. Nas. Kehlkopfheilk. 190, 7-15

39. Konigsmark, B. W. and Gorlin, R. J. (1976): Genetic and Metabolic Deafness. Saunders, Philadelphia, 1976, 1-419

40. Kopito, L., Brand-Auraban, A., Vawter, G. F. and Schwachman, H. (1972): Cerumen and ceruminous glands in cystic fibrosis. Pediat. Res. 64, 427

41. Kuijpers, W. (1976): Biochemistry of the hearing organs. Chem. Weekly, Mag. (10), 529-532 (Neth)

42. Kuijpers, W. (1976 a): Biochemistry of the auditory organs. Ned. Tijdschr. Natuurkd. 42, 185-188 (Neth)

43. Kuttner, K. (1977): Ultrahistochemical studies to demonstrate mucopolysaccharides in the organ of Corti of the guinea pig. In: M. Portmann and J. M. Aran (Eds.). Les Colloques de L'Institut National de la Santé et de La Recherche Médicale, Inserm, Sept. 5-7, 1977, Vol. 68 (Inner Ear Biology), 201-208

44. Lim, D. J. and Mellnick, W. (1975): Mode of cochlear damage by excessive noise. An overview. AGARD Conf. Proc., AGARD-CP-171 C1, 6

45. Lotz, P. and Kuhl, K. D. (1970): Das quantitative Mucopolysaccharid-Eiweiss-Verhältnis in der Cochlea. Arch. Klin. Exp. Ohr. Nas. Kehlkopfheilk. 197, 122-128

46. Lotz, P. and Kuhl, K. D. (1970 a): Differenzierung der Mucopolysaccharide des Innenohres durch säulenchromatographische Hexosamintrennung. Arch. Klin. Exp. Ohren Nasen Kehlkopfheilkd. 197, 129-133

47. Makimoto, K., Takeda, T. Ibusuki, T. (1967): Mucopolysaccharide in perilymph. Ann. otol. 76, 885-895 (Jap)

48. Makimoto, K., Takeda, T. and Silverstein, H. (1978): Chemical composition in various compartments of inner ear fluid. Arch. Oto-Rhino-Laryngol. 220, 259-264

49. Mangos, J. A. and Talamo, R. C. (1976): Cystic Fibrosis: Projections into the Future. Stratton Intercontinental Medical Book Corp., New York, 115-124

50. Marovitz, W. F. and Porubsky, E. S. (1971): Mucopolysaccharides and calcium in the otic capsule and ossicular chain of fetal and neonatal rats. Laryngoscope 81, 273-296

51. Masuda, H., Shichijo, S., Goya, T. and Takeuchi, M. (1978): Isolation and characterization of a glycopeptide from cerumen. Kurume Med. J. 25, 203-205

52. Masuda, H., Shichijo, S. and Takeuchi, M. (1979): Carbohydrate composition of glycopeptides from the human cerumen. Biochem. Med. 22, 256-263

53. Matz, G. J. (1976): The ototoxic effects of ethacrynic acid in man. Laryngoscope 86, 1065-1086

54. McKusick, V. A., Neufeld, E. F. and Kelly, T. E. (1978): The mucopolysaccharide storage diseases. In: J. B. Stanbury, J. B. Wyngaarden and D. S. Fredrickson (Eds.). The Metabolic Basis of Inherited Disease, 4th Ed., McGraw-Hill, New York, Chap. 53, 1282-1307

55. Mirisziai, E., Ordogh, M. and Bodanszky, H. (1971): Fluid losses in the perilymph in mammals. Wiss. Z. Martin-Luther Univ., Halle-Wittenberg, Math. Naturwiss. Reihe 20, 140-144 (Ger)

56. Mizukoshi, O. and Saito, H. (1973): Glycosaminoglycans in otorhinolaryngology. Kagaku No Ryoiki Zokan 99, 119-134 (Japan)

57. Mizukoshi, O. and Saito, H. (1974): Mucopolysaccharides in the otolaryngological division. In: Y. Otaka (Ed.). Biochemistry and Pathology of Connective Tissues, Igaku Shoin, Tokyo, Chap. VIII, 111-126

58. Moltke (1957): Uptake of ^{35}S-sulfate by healing wounds. Effect of thyroxine and ascorbic acid deficiency. Acta Endocr. (Kobenhaven) 25, 179-186

59. Muthiah, P. L. Braeumer, K. and Kuehn, K. (1973): Glycosaminoglycan composition of proteoglycans of cartilage from young and old animals. Carbohyd. Res. 30, 211-214

60. Naftalin, L. (1968): Acoustic transmission and transduction in the peripheral hearing apparatus. Progr. Biophysics. Mol. Biol. 18, 1-27

61. Naftalin, L., Harrison, M. S. and Stephens, A. (1964): The character of the tectorial membrane. J. Laryng. 78, 1061-1078

62. Ormerod, F. C. and Dohlman, G. (1958): Secretion of the liquids of the internal ear. Cong. Franc. oto-rhino-laryngol. 56, Paris, Compt. rend., 241-244 (Fr)

63. Plattig, K. H. (1975): Electrophysiology and biochemistry of the inner ear. In: W. D. Keidel (Ed.): Physiol. Gehörs, Akust. Informationsverarb., Thieme, Stuttgart, 90-101 (Ger)

64. Poulsen, H. (1959): Influence of Myxedematous Connective Tissue Changes on the Function of the Labyrinth. Experimental Studies on the Influence of Thyrotropic Hormone on Mast Cells in the Inner Ear of Guinea Pigs. Thesis, University of Copenhagen, Denmark

65. Poulsen, H. (1966): Thyrotrophic hormone and thyroid hormone control of the inner ear with special reference to myxedema and Meniere's disease. In: G. Asboe-Hansen (Ed.). Hormones and Connective Tissue, Williams and Wilkins, Baltimore, 239-256

66. Prado, S. A. (1977): Osmotic relationships between blood, cerebrospinal fluid and inner ear fluids. PhD. Dissertation, Univ. Minnesota, Minn. Diss. Abstr. Int. B. 38 (10) 4671

67. Rauch, S. (1971): Pathological biochemistry of the inner ear fluids. Wiss. Z. Martin-Luther Univ., Halle-Wittenberg, Math.-Naturwissensch. Reihe 20, 25-30 (Ger)

68. Ross, D. A. (1936): Electrical studies on the frog's labyrinth. J. Physiol. (London) 86, 117

69. Rudman, C. G. and Parsons, J. A. (1979: Transdermal uptake of a pep hormone: inhibition by calcitonin eardrops of induced osteolysis in pig ossicles. Experientia 35, 521-522

70. Saito, H. (1967): Biochemical and histochemical studies on acid mucopolysaccharides in the cochlea of guinea pigs. J. Otorhinolaryng. Soc. Japan 70, 662-690 (Jap)

71. Saito, H. and Daly, F. (1970): Quantitative analysis of acid mucopolysaccharides in the normal guinea pig cochlea. Acta Otolaryngol. 69, 333-340

72. Saito, H. and Daly, J. F. (1971): Quantitative analysis of acid mucopolysaccharides in normal and Kanamycin intoxicated cochlea. Acta Oto-laryng. (Stockh) 71, 22-26

73. Saito, H., Nakamura, A. and Lida, T. (1976): A comparison of acidic glycosaminoglycans in the inner ear and other organs: electrophoretic microanalysis. Arch. Oto-Rhino-Laryngol. 214, 149-155

74. Saito, H., Mizukoshi, O., Toyoda, Y., Tachibana, M., Nishimura, H. (1977): Membrane biology of the inner ear hair cell. Colloq. Inst. Natl. Sante Rech. Med. 68 (Inner Ear Biol.), 179-192

75. Savary, P. (1968): Ototoxicity of drugs. Laval Med. 39, 604-605 (Fr)

76. Schaetzle, W. (1971): Histochemie des Innenohres. Verlag Urban und Schwarzenberg, München, 1-202

77. Scheibe, F., Haupt, H., Hache, U. and Bergman, K. (1977): Analytical comparison of some chemical components in perilymph, cerebrospinal fluid and blood of guinea pigs. Colloq. Inst. Natl. Sante Rech. Med. 68 (Inner Ear Biol.), 209-215 (Ger)

78. Schnieder, E. A. (1971): Concentration shifts in guinea pig perilymph after acoustic stimulation. Wiss. Z. Martin-Luther Univ., Halle-Wittenberg, Math. Naturwiss. Reihe 20, 148-153 (Ger)

79. Schoenheyder, F., Zimmermann-Nielsen, C. and Anderson, H. C. (1966): Urinary excretion of amino acids and hexosamine in otosclerotic patients. Arch. Otolaryng. 84, 495-496

80. Sellick, P. M. and Johnstone, B. M. (1975): Production and role of inner ear fluid. Progr. Neurobiol. 5, 337-362

81. Serafini-Fracassini, A. and Stimson, W. H. (1972): Synthesis of low molecular weight proteoglycans in rabbit ear cartilage. FEBS Lett. 26, 336-340

82. Shrader, R. E., Erway, L. C. and Hurley, L. C. (1973): Mucopolysaccharide synthesis in the developing inner ear of manganese deficient and pallid mutant mice. Teratology 8, 257-266

83. Shurin, P. A., Pelton, S. I., Donner, A and Klein, J. O. (1979): Persistence of middle ear effusion after acute otitis media in children. New England J. Med. 300, 1121-1123

84. Soffer, L. (1956): Diseases of the endocrine glands. Lea & Febiger, Philadelphia, Chap. 25, 829-868

85. Soifer, N., Waltner, J. G., Endahl, G. L. and Holdsworth, C. E. (1970): Biochemical studies on otosclerosis: Urinary acid mucopolysaccharides. Arch. Klin. Exp. Ohr-Nas-Kehlk. Heilk. 197, 1-6

86. Tachibana, M., Saito, H. and Machino, M. (1973): Sulfated acid mucopolysaccharides in the tectorial membrane. Acta Otolaryng. 76, 37-46

87. Taylor, B., Evans, J. N. and Hope, G. A. (1974): Upper respiratory tract in cystic fibrosis, ear-nose-throat survey of 50 children. Arch. Dis. Child. 49, 133-136

88. Thalmann, R. (1975): Biochemical studies of the auditory system. In: D. B. Tower (Ed.). Nervous System, Raven, New York 3, 31-44

89. Thalmann, R. (1976): Quantitative histochemistry and cytochemistry of the ear. In: C. A. Smith and J. A. Vernon (Eds.) Handbook of Auditory and Vestibular Research Methods, Thomas, Springfield, pp. 358-418

90. Thalmann, R. R. (1976 a): Quantitative biochemical techniques for studying normal and noise-damagèd ears. In: D. Henderson, R. P. Hamernick and D. S. Dosanjh (Eds.). Effects of Noise on Hearing (1975 Symp., published 1976), Raven, New York, 129-155

91. Ulrich, M. R., Mundi, J. R. and Margen, S. (1966): Biochemical properties of inner ear fluids: Methods of microchemistry and fluid withdrawal. U.S. Govt. Res. Develop. Rept. 41 (20), 34-35

92. Vilstrup, T. and Jensen, C. E. (1954): Chemical composition of the fluids of the labyrinth. I. Determination of hyaluronic acid. II. Determination of total protein in the endolymph and perilymph of the ear labyrinth. Ann. Otol. Rhinol. Laryngol. 53, 151-156; 156-159

93. Vilstrup, T. and Jensen, C. E. (1961): On the displacement potentials in acid mucopolysaccharides. Acta Otolaryng. (Stockh) Suppl. 163, 42-46

94. Vinnikov, J. A. and Titova, L. K. (1964): The Organ of Corti. Consultants Bureau, New York

95. Von Ilberg, C. (1976): Impairment of audition caused by aminoglycoside antibiotics. Diagnostik 9, 573-575 (Ger)

96. Zygmunt, U. (1967): Protective importance of Vitamin A on toxic action of streptomycin on the inner ear. Otolaryngol. Pol. 21, 715-719 (Pol.)

97. Godlowski, Z. (1972): Hyperomosis of endolymph as primary pathogenic mechanism in Ménière's disease and its clinical management. Acta Otolaryngologica (Stockholm) Suppl. 299, 1-36

98. Meyerhoff, W. C. (1979): Hypothyroidism and the ear: electrophysiological, morphological and chemical considerations. Laryngoscope 89, (Suppl. 19), 1-25

99. Schätzle, W. and Haubrich, J. (1967): Histochemical changes in the guinea pig cochlea in experimental hypothyroidism. Arch. Klin. Exp. Ohr. Nas. Kehlkopfheilk. 188, 224-231

100. Ritter, F. N. (1967): The effects of hypothyroidism upon the ear, nose and throat. Laryngoscope 77, 1427-1479

101. Deol, M. S. (1973): Congenital deafness and hypothyroidism. Lancet II, 105-106

102. Asboe-Hansen, G. (1966): Hormone and Connective Tissue, Munksgard, Copenhagen

103. Comper, W. D. (1977): Electric potentials generated by connective tissue polysaccharides in glass capillaries. Biochim, Biophys. Acta 497, 816-819

104. Tachibana, M., Saito, H., Yamamichi, I. and Morioka, H. (1978): A possible involvement of acidic glycosaminoglycans in Kanamycin ototoxicity. Acta Otolaryngol. 86, 15-21

Note:

Masuda et al., 1979 (Ref. 52) later appeared as follows:

Shichijo, S., Masuda, H. and Takeuchi, M. (1979): Carbohydrate composition of glycopeptide from the human cerumen. Biochem. Med. 22, 256-263

Chapter 11

Amniotic Fluid (Liquor amnii)

11.1 Introduction

The amniotic fluid fills up the amniotic cavity and surrounds the fetus developing inside the uterus (Fig. 11-1). In this position it protects the fetus from external trauma and provides an environment for the fetus to float in, assimilate nutrients by swallowing the fluid and to dispose of fetal secretions and excretions. Cells of fetal origin, especially the fetal skin and membranes and the amnion are continually shed into this fluid. Analyses of the glycosaminoglycans and the glycosaminoglycan-degrading enzymes in the soluble portion and studies of the cells in the particulate portion of the fluid have been very useful for prenatal diagnosis of the mucopolysaccharide storage and other related diseases. These parameters used for intrauterine detection vary with the gestational age and the size of the fetus and an appropriate age is selected so that the required amount of the amniotic fluid can be obtained without any risk to the fetus and that the GAG or their degradation enzymes and the cells are present in sufficient, measurable amounts. Since these analyses are based on the assumption that these parameters originate from the fetus and are measures of the state of fetal well-being and fetal maturity, it is essential to understand the formation, disposal and variation of the amniotic fluid volume and composition with age.

11.2 Origin, volume and fate of amniotic fluid

The amniotic cavity is present in the $7^1/2$ day old human embryo but becomes well established by the twelfth day, when amniotic fluid begins to fill up this cavity (Hamilton et al., 1962). Both the mother and the fetus contribute to the formation of the amniotic fluid. The relative contribution of the two varies with the gestational age. In early pregnancy the amnion is separate from the embryo and the contribution from the embryo is mini-

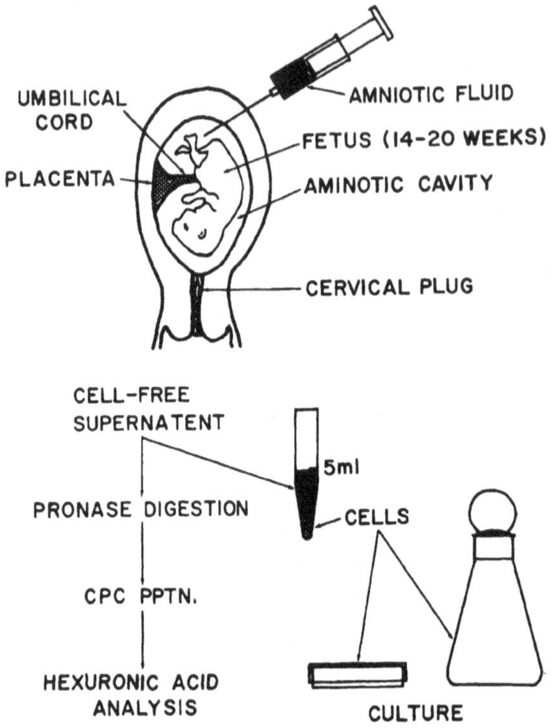

Fig. 11-1. Amniocentesis and analysis of the amniotic fluid

mum. The early pregnancy amniotic fluid is considered a transudate of the maternal serum across the placental membranes or as a transudate across the umbilical membrane (Ostergard, 1970); its composition resembles the maternal extracellular fluid. As pregnancy continues, gradual modification of its chemical composition occurs due to contribution from the micturition, defecation, deglutition and the respiratory system of the fetus.

The volume of the amniotic fluid varies with the fetal size and the gestational age. Up to 20 weeks the volume is directly related to the body surface of the fetus (Lind et al., 1972). Abramovich (1968), from a correlation between the logarithm of the volume and fetal size, has suggested that the fetus is the major contributor to the amniotic fluid. The volume of the amniotic fluid increases as the pregnancy progresses. At 8 weeks it is only 7.7-22 ml, 25 ml at 10 weeks and increases to the highest volume of 500-1355 ml at 34 weeks when it begins to decline to 800-1000 ml at full term pregnancy (Dito et al., 1975). It is important to bear this variation in mind, since the results of glycosaminoglycans analyses are expressed per 100 ml. During the second half of the pregnancy, the total amniotic fluid volume may be determined by dilution methods using a dye (like Evans blue, Congo red, phenol red) or a high molecular weight substance like ^{131}I-iodinated albumin, inulin, ^{51}Cr-tagged red blood cells. Deuterated water can also be used for this purpose (Votta et al., 1974).

The fluid is continuously formed and removed and therefore it is in a state of dynamic equilibrium. Experiments with deuterated and tritiated water suggest extremely rapid exchange of water molecules between the maternal circulation, the fetal circulation and amniotic fluid (Plentl, 1958). Although the mechanisms involved are not known, the fetal kidneys, intestine, skin and trachial and bronchial secretions and placenta are believed to participate in this exchange. The fetus continuously swallows the amniotic fluid and at full term swallows 500-1500 ml/hour (Fiereck, 1976). The evidence for this comes from the well-known fact that in anencephaly due to the absence and/or reduced size of cranial vault and cerebral hemispheres, there is a lack of control of the swallowing mechanism. This leads to impaired circulation of the fluid through the fetal gut producing the hydraminos condition, where the fetus is surrounded by an unusually large amount of amniotic fluid. Higher amniotic fluid volumes (hydraminos) are also seen in cases of severe hydrops fetalis, congenital obstruction of the oesophagus or upper gastrointestinal tract (e. g. duodenal atresia, pyloric stenosis) and in central nervous system defects of the fetus, including the absence of meninges and lack of antidiuretic hormone formation in the defective posterior pituitary gland of the fetus.

11.3 Relationship of amniotic fluid to urine

Fetal urine contributes significantly to the amniotic fluid and this varies with the age of the fetus. For this reason, increased acid mucopolysaccharide excretion in urine of mucopolysaccharidoses patients attracted the attention of the investigators for similar analyses in amniotic fluid surrounding an afflicted fetus (Section 11.7). The laboratory interrelations of amniotic fluid and urine have been reviewed (Free and Free, 1974, 1975; Guibaud et al., 1976; Pirani et al., 1976). Mitchell et al. (1977) have also reported biochemical differences between amniotic fluid and maternal urine.

The first signs of the functional fetal kidney appear in the 8th week of gestation. Although urine formation is noticeable in the 14th week, its contribution is not significant till the 20th week of gestation. Using ^{131}I-labeled dyes fetal urine has been shown to be a significant contributor in near-term rhesus monkeys (Dito et al., 1975). Also, decreased amniotic fluid volumes (oligohydraminos) are associated with bilateral renal agenesis, congenital urinary tract obstructions like absence of urethral orifice and in placental insufficiency (Hamilton et al., 1962).

The prenatal diagnostic tests based on glycosaminoglycan analysis of the amniotic fluid actually measure *the glycosaminoglycans of fetal urine in amniotic fluid*. Therefore, in such prenatal analysis for diagnosis, the gestational age and the aforementioned conditions affecting the fetal urine output and the volume of the amniotic fluid should be given due consideration.

11.4 Methods of analysis and fractionation

The analysis of the amniotic fluid glycosaminoglycans provides the basis for intrauterine detection of mucopolysaccharidoses and related disorders. Such a prenatal diagnosis helps to allay the fear of bearing a deformed or mentally defective child in pregnancies at such possible risks. Roughly, one birth in 25,000 is associated with a disorder of mucopolysaccharide metabolism. If the fetus is found to be affected, the parents are recommended a therapeutic abortion.

For antenatal diagnosis of the mucopolysaccharidoses both analysis of glycosaminoglycans in amniotic fluid and glycosaminoglycan synthesizing activities of the cells cultured from this fluid are carried out. This involves the following procedures:

1. Transabdominal amniocentesis for collection of amniotic fluid which is then centrifuged for separation into the soluble and cellular fractions.
2. Transport of the two fractions to another laboratory for analysis and/or storage till the time of analysis.
3. Isolation and determination of the total glycosaminoglycan content in soluble fraction. Direct assay of the enzyme iduronate sulfatase in the cell-free amniotic fluid has been reported to be decreased in Hunter syndrome (Liebaers et al., 1977). This provides another approach for prenatal diagnosis of the Hunter syndrome.
4. Fractionation of the total glycosaminoglycans into individual components helps to identify the type of the mucopolysaccharide storage disorder.
5. Culture of the cellular fraction of the amniotic fluid and assessment of the glycosaminoglycan synthesizing capabilities of these cultured cells of fetal origin enables to make a more definitive antenatal diagnosis.

Amniocentesis: The patient is asked to void and then asked to roll over several times for mixing the amniotic fluid. After locating the fetal parts by ultrasound procedure, a needle is introduced transabdominally (under sterile conditions), with a quick thrust through the skin, subcutaneous tissue, abdominal and uterine muscles into the amniotic cavity. With a syringe connected to this needle a 5-10 ml sample of amniotic fluid is collected. Finer details of the amniocentesis procedure are given elsewhere (Scrimgeour, 1973; Doran, 1976; Fairweather and Eskes, 1978). The risks and problems of amniocentesis will be presented later in section 11.8.

Transport and storage: The soluble cell-free and cellular fractions may need to be sent to another laboratory if facilities are not available locally for analysis or these may need storage for a deferred another time analysis. The procedure and precautions involved for transport and storage of the amniotic fluid have been reviewed (Niermeijr et al., 1973). The glycosaminoglycans are stable in amniotic fluid and their transport presents no problem (Dorfman, 1972).

Analysis of total glycosaminoglycans: For analysis, these macromolecules first need to be isolated from 3-5 ml of amniotic fluid by precipitation with a compound of a basic nature after proteolytic digestion or without protein removal. Proteolytic digestion may be carried out with 1 mg pronase at 60 °C for 6 hours followed by dialysis against water, precipitation of glycosaminoglycans in the supernatent with cetylpyridinium chloride and reprecipitation from 2 M lithium chloride with 4 volumes of ethanol (Matalon and Dorfman, 1970, 1972; Matalon, 1974; Duncan et al., 1973). An overnight digestion with protease at pH 8.9 may replace pronase digestion (Omura et al., 1973). After drying the final precipitate with ether, hexuronic acid, N-sulfohexosamine and electrophoretic analyses are carried out.

The glycosaminoglycans may be precipitated without prior protein removal with Alcian blue (Whiteman, 1973; Whiteman and Henderson, 1977), with N,N-diethylpseudoisocyanin (Suschke and Kunze, 1971; Suschke, 1973) or with quinacrine (Mitra and Blau, 1978). The glycosaminoglycans may be recovered from the dissociated complex or these may be quantitated by the change in the spectrum of N,N-diethylpseudoisocyanin due to binding to glycosaminoglycans. Alternatively, quinacrine released from the complex by addition of a surfactant may be measured at 450 nm.

Cetylpyridinium chloride and the basic dyes may not precipitate the low molecular weight glycosaminoglycans. The low molecular weight fraction may be analyzed in the supernatent from precipitation of high molecular weight fraction. Methods using basic dyes which do not measure hexuronic acid precipitated but instead measure the change in spectrum upon binding with glycosaminoglycan (Suschke, 1973; Mitra and Blau, 1978) may yield inaccurate results. Hyaluronic acid does not make a complex with quinacrine a other component glycosaminoglycans interact with different affinities (Mitra and Blau, 1978). In this respect the ion-exchange chromatography (Lee and Shafer, 1974) of the proteolytic digest could be the method of choice. The value of total glycosaminoglycans per 100 ml of amniotic fluid obtained by the different methods are presented in Table I.

Table I. Total Glycosaminoglycan Contents of Normal Amniotic Fluid

Authors	Method used	Gestational age (weeks)	Number examined	Mg Hexuronic acid/100 ml fluid	
				Range	Mean + S.D.
Danes et al. (1970)	CPC precipitation	12–36	36	0.5–11.0	2.00
Matalon et al. (1970)	Proteolysis, CPC precipitation	12–22	41	0.60–3.50	2.0±0.82
Matalon and Dorfman (1972a)	Papain digestion, CPC precipitation	12–22	45	0.4–6.0	2.0±0.80
Duncan et al. (1973)	Pronase digestion, CPC precipitation	9–14	39	0.08–3.20	0.66±1.70
		15–18	27	0.25–2.60	1.33±0.50
		19–40	14	0.52–1.39	1.00±0.35
Omura et al. (1973)	Protease digestion, CPC precipitation	18–22	7	0.40–0.80	0.60
		37–40	10	0.15–0.60	0.40
Lee and Schafer (1974)	Anion exchange, enzyme digestion	12–21	20	–	2.22±0.58
Mitra and Blau (1978)	Quinacrine precipitation	14–22	123	0.61–5.14[a]	2.52±1.11[a]
		22–40	128	0.46–3.5[a]	1.76±0.72[a]

[a] The values given are as chondroitin sulfate

An increase in total glycosaminoglycan content of the amniotic fluid may indicate a case of mucopolysaccharidosis, while normal levels may not mean that the fetus is not affected with the conditions. As a result of this, some mucopolysaccharidosis may be messed out if the diagnosis is based only upon the amniotic fluid glycosaminoglycans levels.

Fractionation of the total glycosaminoglycan mixture: A better chracterization of the glycosaminoglycans isolated from the amniotic fluid is needed before an intrauterine diagnosis can be meaningful. The simplest approach to this is electrophoretic characterization on cellulose acetate membranes in 0.1 M barium acetate pH 7.0 at 12.5 V per cm for two hours followed by differential staining in Alcian blue $MgCl_2$ solutions (Duncan et al., 1973). Single dimensional electrophoresis carried out in a Beckman Microzone Cell at 120 V for 20 min in barbiturate buffer of pH 8.7 followed by Alcian blue staining (Varma et al., 1974) results in separation of the mixture into hyaluronic acid, keratan sulfate plus heparan sulfate, and chondroitin 4- plus 6-sulfates. Electrophoresis on Sepraphore III strips (Gelman) in pyridine-formic acid buffer followed by staining in 0.5 % aqueous acridine has also been used (Matalon and Dorfman, 1972).

Unless dermatan sulfate and/or heparan sulfate have been demonstrated in excess in addition to increased total glycosaminoglycans in amniotic fluid corrected for changes at that gestational age (section 11.7), the fetus cannot be said to be affected with mucopolysaccharidosis.

Two dimensional electrophoresis on cellulose acetate and anion-exchange chromatography are better methods for quantifitation of individual glycosaminoglycan components. These provide separation of dermatan sulfate from chondroitin sulfate not usually achieved in single dimensional electrophoresis. Electrophoresis was first run in pyridine: glacial acetic acid: water, 10:1:89(v/v) at a pH of 6.0 and a potential gradient of 7.5 v/cm for 75 min. After drying the membrane electrophoresis was run in the second direction 90 ° to the first in 0.1 M barium acetate for four hours at a potentional of 7.5 V/cm (Whiteman, 1973; Whiteman and Henderson, 1977). After dissolving the cut small strips of the membrane containing the separated Alcian blue-GAG complexes in 1 ml of dimethyl sulfoxide containing 25 mM $MgCl_2$ and 6.24 ml/l glacial acetic acid, absorbance was measured at 678 nm.

Anion-exchange chromatography of the glycosaminoglycan mixture isolated from pronase digested amniotic fluid has been carried out on small columns packed with ECTEOLA-cellulose (Duncan et al., 1973) and Dowex 1-X2 (Cl^-– form, 200 mesh size) resin (Matalon and Dorfman, 1972; Lee and Shafer, 1974). Elution with a stepwise gradient of 0.25 - 4.0 M NaCl, pooling the contents of the tubes corresponding to each peak and hexuronic acid determinations give values for individual glycosaminoglycans. However, more definitive identification of each pooled fraction can be achieved by desalting these fractions by dialysis, precipitation of the glycosaminoglycans in the retentates with ethanol followed by determination of their portions susceptible or resistant to the action of specific enzymes, streptomyces hyaluronidase and AC-II and ABC chondroitinases. This approach has been used by Lee and Shafer (1974) for determination of individual glycosaminoglycans in pooled amniotic fluid from pregnancies of 12-21 weeks gestation.

Culture of amniotic fluid cells: The aforementioned methods refer to the glycosaminoglycan analysis in the cell-free amniotic fluid. The cellular fraction obtained by centrifugation of amniotic fluid for 10 min at 1000 rpm contains cells derived from the fetus, especially the skin. These can be cultured in vitro and like the skin fibroblasts their glycosaminoglycan synthesizing and enzymic activities have been used for very precise in utero diagnoses of the mucopolysaccharidoses.

Three types of cells are present in the amniotic fluid cellular fraction. The epitheloid type of cells are difficult to work with because of their slow growth. The second type, fibroblast-like cells grow well. The third type amniotic fluid cells have characteristics in between the epitheloid and the fibroblast cells. The cellular sediment obtained by centrifugation of 5-10 ml of amniotic fluid is used for culture. The cell-free supernatent is frozen at -20 °C and can be stored or transported in the frozen or freeze-dried state for analysis of glycosaminoglycans. The amniotic fluid cells are cultured in petri dishes or flasks at 37 °C in an atmosphere of 5 % CO_2 and 95 % air. The culture medium is more far complex than that used for skin fibroblasts. The details of the cell culturing procedures have been desribed in several reviews and monographs (Dingle and Webb, 1965; Priest, 1971; Acton and Lynn, 1977).

A number of approaches have been used in using cultivated cells from amniotic fluid for antenatal diagnosis of mucopolysaccharidoses. Danes and Bearn (1966) demonstrated metachromasia in cultured skin fibroblasts and this has also been shown in cultured amniotic fluid cells. However, the diagnostic value of this metachromasia has been questioned (Milunsky and Littlefield, 1969) since this metachromasia is associated with a number of hereditary disorders and is seen even in some normal individuals.

Cultured amniotic fluid cells like the skin fibroblasts from normal pregnancies accumulate $^{35}SO_4$-labeled mucopolysaccharides at a linear rate for 1-2 days and then attain a steady state (Fratantoni et al., 1968). Cells from Hurler and Hunter syndrome continue to accumulate ^{35}S-labeled mucopolysaccharides due to defective degradation processes (Fratantoni et al., 1968; Matalon and Dorfman, 1972; Matalon et al., 1972; Neufeld, 1972). However the finding of abnormal ^{35}S kinetics in amniotic fluid cells requires sex determination to distinguish between affected hemizygotes and normal heterozygotes in the case of X-linked Hunter syndrome.

The prenatal diagnosis of Hurler or Hunter syndrome may be made if cultured amniotic fluid cells fail to correct the abnormal ^{35}S uptake upon addition to known Hurler's cells. The cross correction in these mixing cell culture experiments is due to the addition of the *correction factor*, the missing degradative enzyme, which becomes available when cells from a disorder are mixed with cells from normal subjects or with cells from another disorder (Fratantoni et al., 1968 a).

Another approach for prenatal diagnosis using cultured amniotic fluid cells is determination of enzymes. The enzymatic defects in most of the mucopolysaccharidoses are now known (Chapter 17). All those enzymes are detectable in normal cultured skin fibroblasts and normal amniotic fluid cells. Analysis of the deficient enzyme in cultured amniotic fluid cells may be the most reliable method for antenatal diagnosis. Harper et al. (1974) have used this approach for Sanfilippo disease by demonstrating a deficiency of the enzyme heparin sulfamidase in cultured amniotic fluid cells from an abortus with Sanfilippo A disease. Related aspects of the enzymic diagnosis of the genetic mucopolysaccharide storage disorders have been reviewed (Kaback and Howell, 1973; Hall et al., 1978; Neufeld and Cantz, 1973).

11.5 Normal glycosaminoglycan composition

There is wide variation in the total glycosaminoglycan contents as well as in the quantitative distribution of the individual glycosaminoglycan components reported by different investigators. The values reported in various studies using different types of methods are given in Table I. It can be seen from this table that the values obtained for amniotic fluid from normal pregnancies of even the same duration, say 12-22 weeks are considerably different due to variation in methodologies used. It is therefore essential to establish a normal range for any new method to be used for antenatal diagnosis of mucopolysaccharidoses cases which are likely to have abnormally elevated values. Also, since the glycosaminoglycans vary at different gestational ages (Section 11.7) it is important that the patients results be compared only with the values from normal pregnancies of the same gestational age.

The glycosaminoglycan contents of the normal pregnancies between 12-22 weeks gestation have been quoted (Lee and Shafer, 1974) to be in the range of 3-32 micrograms uronic acid per ml. The major glycosaminoglycans in the amniotic fluid are hyaluronic acid and chondroitin sulfate (Matalon et al., 1970; Matalon et al., 1972; Duncan et al., 1973; Whiteman, 1973; Lee and Shafer, 1974). In addition to these two glycosaminoglycans, Duncan et al. (1973) found a substance with an electrophoretic mobility similar to heparan sulfate in barium acetate buffer, which was present in all 80 normal amniotic fluid samples tested by them. The presence of heparan sulfate amounting to about one-tenth the chondroitin sulfate has been confirmed, especially after the 20th week of pregnancy by two dimensional electrophoresis (Whiteman, 1973, 1975; Whiteman and Henderson, 1977). Matalon and Dorfman (1972) could not detect heparan sulfate in 45 amniotic fluids obtained from normal pregnancies of 12-22 weeks and reported the normal glycosaminoglycans content of 0.02 ± 0.008 mg/ml distributed between hyaluronic acid 78 %, chondroitin 6-sulfate 13 % and dermatan sulfate 9 %. Only in one study (Duncan et al., 1973) keratan sulfate has been identified in many amniotic fluid samples by differential staining with Alcian blue-$MgCl_2$.

Whereas Whiteman (1973), using two dimensional electrophoresis on cellulose acetate, was able to detect heparan sulfate but no dermatan sulfate, the results obtained by others (Matalon et al., 1970; Matalon and Dorfman, 1972) were opposite to this. These latter investigators reported the presence of significant amounts (6 %) of dermatan sulfate and an absence of heparan sulfate in pooled normal amniotic fluid. This controversy has been resolved by analysis of quantitative distribution of specific glycosaminoglycans in human amniotic fluid from pregnancies of 12-22 weeks gestation (Lee and Shafer, 1974). The total uronic acid recovered from the anion-exchange chromatographic columns and subjected to specific enzymatic analysis was found to consist of uronic acid-containing glycopeptides 7 %, hyaluronic acid 34 %, chondroitin 6-sulfate 20 %, non-sulfated chondroitin 14 %, chondroitin 4-sulfate 13 %, heparan sulfate 6 % and dermatan sulfate 5 %. Thus, amniotic fluid contains all the glycosaminoglycans present in the connective tissue except keratan sulfate.

11.6 Source of glycosaminoglycans in amniotic fluid

For prenatal diagnosis, amniocentesis is carried out after the 14th week of gestation. The glycosaminoglycans present in the amniotic fluid in early pregnancy are probably those derived from the primary fetal membranes, primarily the amnion and secondarily the placenta and the umbilical cord.
It might surprise some readers why placenta and umbilical cord are mentioned as fetal membranes, although these lack the *true* membranous character. The embryological reasoning for this is that these structures arise from the fertilized egg and yet do not form a part of the embryonic body and are shed at birth. Both placenta and umbilical cords synthesize and contain glycosaminoglycans.

Amniocentesis for antenatal diagnosis of mucopolysaccharidoses is commonly performed after the 14th week of pregnancy. The glycosaminoglycans present in the amniotic fluid at that time are derived from fetal urine and amniotic cells. The human amnion is biologically active and has the ability of bringing about synthesis of hexosamines from glucose and incorporating these into glycoproteins and glycosaminoglycans (Mazurkiewicz-Kilezewska, 1971). The cell lines derived from the chick amnion synthesize sulfated and unsulfated acid mucopolysaccharides in vitro. The predominant glycosaminoglycan component secreted into the culture medium was hyaluronic acid (Bischoff, 1971). This is in agreement with the presence of hyaluronic acid as the major component and its replacement with chondroitin sulfate later with the progression of pregnancy (Section 11.5). Three types of cells are present in the amniotic fluid, which are believed to originate from fetal tissues like skin. Out of these, epitheloid type, fibroblast-like and amniotic fluid cell, the last two types can grow in vitro and produce glycosaminoglycnas (Section 10.4). A part of the glycosaminoglycans of amniotic fluid may arise from the biosynthesis in these cells. The glycosaminoglycan present in the fetal urine probably arise from the blood and kidney desquamation in fetus as it is in adult humans. Keeping aside the hyaluronic acid, the patterns of glycosaminoglycans in amniotic fluid of mucopolysaccharidoses patients resembled those found in the urine of patients affected with the same syndrome. This indicated that dermatan sulfate and heparan sulfate originated from fetal urine (Whiteman, 1973).

11.7 Glycosaminoglycans variation with gestational age

The total glycosaminoglycan content and the relative distribution of its individual components vary with the fetal age like the volume of this fluid. The values of the total glycosaminoglycan contents of normal amniotic fluid reported by different authors have been summarized in Table I. The variation of glycosaminoglycans at various gestational ages can be visualized by referring to Figs. 11-2 and 11-3. Lack of consideration of these fetal age-related changes have resulted in false negatives in early gestational age and false positives in amniotic fluid samples from late pregnancies of over 20 weeks (Whiteman, 1973; Whiteman and Henderson, 1977; Lee and Shafer, 1974; Duncan et al., 1973). Several authors (Duncan et al., 1973; Omura et al., 1973; Lee and Shafer, 1974; Matalon, 1974; Mitra and Blau, 1978) have suggested that the gestational age is an important variable affecting the results of the chemical analysis of the amniotic fluid drawn from normal and pathological pregnancies (Holten, 1978).

There is a disagreement on the total hexuronic acid contents of the amniotic fluid at various gestational ages (Tables I and II). The total hexuronic acid has been reported to fall throughout the pregnancy (Danes et al., 1970). Contrary to this, other workers (Duncan et al., 1973; Mitra and Blau, 1978) have found the hexuronic acid to rise with the gestational age till after the 17th week when it begins to fall. These changes are shown in Figs. 11-2 and 11-3 .

Duncan et al. (1973) analyzed 80 samples of amniotic fluid from normal pregnancies of 9-40 weeks. The variation of GAG values with gestational age is presented in Fig. 11-2, while the mean values for their analyses for groups of 9-14, 15-18 and 19-40 weeks of gestation are given in Table I. Based on their results, the total GAG content of the amniotic fluid begins to increase at 11th week, becomes maximum at 17th week and thereafter it decreases.

Besides the total glycosaminoglycans measured as hexuronic acid, the spectrum of the individual glycosaminoglycan components also changes with the gestational age. In early pregnancy hyaluronic acid is the major component and at 13 weeks chondroitin sulfate is present in barely detectable amounts. After the 14th week of gestation, hyaluronic acid decreases with increasing age of the fetus, whereas chondroitin sulfate increases and becomes the major amniotic fluid glycosaminoglycan after 20 weeks (Duncan et al., 1973). Whiteman (1973) analyzed amniotic fluid from 20 normal pregnancies of 9-25 weeks and found a similar trend of the decrease of hyaluronic acid with gestation with a concomitant increase of chondroitin sulfate to be the principal glycosaminoglycan after 21 weeks. By two-dimensional electrophoresis the hyaluronic acid and chondroitin sulfate expressed as percentages of the total glycosaminoglycan at 21 weeks and 25 weeks of normal pregnancy contained 63%, 13% and 37%, 87%, respectively (Whiteman 1973, 1975). Dermatan sulfate and heparan sulfate could not be detected by this procedure in amniotic fluids from normal pregnancies.

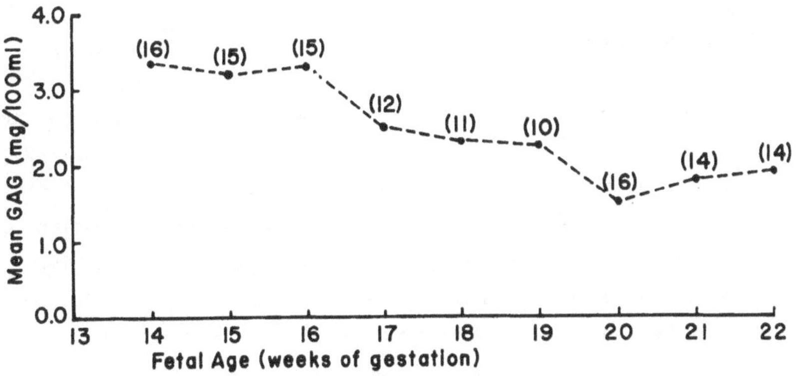

Fig. 11-2. Variation of amniotic fluid glycosaminoglycan concentration with fetal age. The numbers in parenthesis are the number of samples analyzed for a fetal age. (Drawn with permission from the data in Mitra and Blau, 1978)

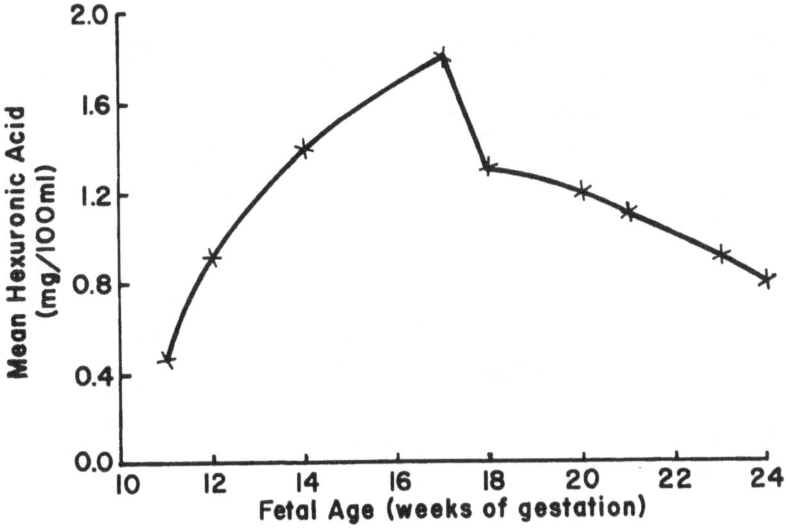

Fig. 11-3. Variation of hexuronic acid content of amniotic fluid with gestational age. (Drawn with permission from the data in Duncan et al., 1973)

11.8 Pathological changes in amniotic fluid glycosaminoglycans

The total glycosaminoglycan contents of the cell-free amniotic fluid are markedly elevated in the various mucopolysaccharidoses. The relative distribution of the individual components, hyaluronic acid, chondroitin sulfate and dermatan sulfate are also altered. These changes have been summarized in Table II. The most conspicuous changes observed in amniotic fluid glycosaminoglycans are in the ratios: dermatan sulfate/chondroitin sulfate and heparan sulfate/chondroitin sulfate, which are drastically elevated from > 0.1 to as high as 1.21 (Whiteman and Henderson, 1977).

Besides the mucopolysaccharidoses, the glycosaminoglycans of the liquor amnii exhibit changes in other genetic conditions. In five cases of anencephaly, the total glycosaminoglycans have been reported to increase from the control values of 2.52 ± 1.11 to 2.92-4.40 mg/100 ml with a mean value of 3.86 mg/100 ml (Mitra and Blau, 1978).

The elevation of glycosaminoglycans has been noted in a study of 14 cases of Rhesus incompatibility (Danes et al., 1970). In ten other cases (Mitra and Blau, 1978) the values decreased from 2.52 ± 1.11 mg/100 ml for the controls to 1.25-3.0 with a mean of 2.06. The latter authors also reported an increase of total glycosaminoglycans in three cases of Down's syndrome to 3.42-4.38 mg/100 ml.

Anton et al. (1973) have reported an increase of amniotic fluid total glycosaminoglycans from 3.49 ± 0.31 mg/100 ml in controls to 4.81 ± 0.52 mg/100 ml in pregnancies with placental insufficiency.

Abnormalities revealed by metachromasia and ^{35}S incorporation have been observed in patients with cystic fibrosis (Nadler et al., 1978).

Table II. Glycosaminoglycan Composition in Normal Pregnancies and Pregnancies at Risk for Mucopolysaccharidoses

Authors	Gestational age (weeks)	Number examined	Total GAG Mg hexuronic acid/ 100 ml (mean)	HA	CS(4&6)	DS	HS	Others
				%				
Matalon and Dorfman (1972a)	12–22	45 (Controls)	2.0	78	13	9	–	–
	14	1 (Hurler)	8.0	8	11	9	63	–
	22	1 (Hurler)	8.0	24	30	20	26	–
Omura et al. (1973)	18–22	7 (Controls)	0.60	86	14	–	–	–
	37–40	10 (Controls)	0.40	76	24	–	–	–
	20	1 (Hurler)	0.90	25	24	22	29	–
	20	1 (Hurler?)	0.80	90	10	–	–	–
Lee and Schafer (1974)	12–21	20 (Controls)	2.22	34	13 (4-S) 20 (6–S)	5	6	Ch 14, GP 7
	$18^1/_2$	1 (Hurler)	4.52	–	–	–	–	–
	16	1 (Hunter)	2.38	–	–	–	–	–
	$20^1/_2$	2 (Hunter)	3.65	–	–	–	–	–
Whiteman (1975)	21	? (Controls)	–	63	37	–	–	–
	21	1 (Hurler)	–	19	40	33	8	–
	25	? (Controls)	–	13	87	–	–	–
	25	1 (Sanfilippo)	–	5	43	–	52	–
Whiteman and Henderson (1977)	14	1 (Hurler)	–	39.5	33.7	18.8	8	–
	22	1 (Hurler)	–	15	42	30.4	12.6	–
	16	1 (Hurler)	–	25	39	25	11	–
	23	1 (Hurler)	–	16.6	32	30	21.4	–
	14	1 (Hurler)	–	33	35.7	23.7	7.7	–
	16	1 (Sanfilippo)	–	58.7	24.5	–	16.8	–
	25	1 (Sanfilippo)	–	5.4	42.9	–	51.7	–

Abbreviations GAG = glycosaminoglycans; HA = hyaluronic acid; CS (4&6) = chondroitin 4- and 6-sulfate; DS = dermatan sulfate; HS = heparan sulfate; Ch = chondroitin and GP = glycopeptide

11.9 Amniocentesis and intrauterine diagnosis of mucopolysaccharidoses: Problems and prospects

Transabdominal amniocentesis has become a commonly used procedure in the antenatal diagnosis of genetic and metabolic disorders. While this procedure provides reassurance to parents that they can selectively have unaffected offsprings, it is not free from risks and problems. Several authors have critically examined this clinical procedure in terms of risks and benefits (Nadler and Gerbie, 1970; Lind, 1971; Milunsky, 1975; Vrettos et al., 1975). The major risk involved is to the fetus rather than to the mother. Several cases of fetal loss due to in utero deaths and spontaneous abortions have been documented. The risk can be minimized if the procedure is carried out at a stage when the amniotic cavity is large enough to withdraw sufficient amount of the fluid and the position of the placenta and the fetus are localized by ultrasound technics to prevent any injury. Although at 14 weeks of gestation the fluid volume is about 155 ml and 10-15 ml of the fluid can be safely removed, yet at that stage the contribution of the fetal urine is not significant enough for intrauterine diagnosis of mucopolysaccharidoses. For direct analysis of glycosaminoglycans in cell-free amniotic fluid and for culture of amniotic fluid cells, 16-20 weeks of fetal age is appropriate.

Statistical analysis of the fetal loss between normal control pregnancies and pregnancies at risks with transabdominal amniocentesis has revealed a 1-2 % fetal loss in early pregnancies. In spite of this high risk of this fetal loss, amniocenteses have been performed in early stages (Niermeijr et al., 1976) so that the fetus can be aborted before the 20th week of gestation, if needed. Scrimgeour (1973) has reported only nine fetal losses in 12,960 amniocenteses performed after 20 weeks, when the volume of the amniotic cavitiy is large enough. A collaborative study in U.S.A., Canada and U.K. showed that with proper precautions the risk to the fetus (3.5 %) was not significantly different from that (3.2 %) in the control group (Milunsky, 1975).

The tests on amniotic fluid (Holten, 1978; Whitfield, 1976; Emery, 1973) for prenatal diagnosis of inherited metabolic disorders are based on the assumption that the changes in fetal metabolism are reflected in the amniotic fluid (Fairweather, 1973; Milunsky et al., 1970; Fuchs and Cederquist, 1973; Milunsky, 1973, 1975; Natelson et al., 1974). The use of amniotic fluid glycosaminoglycan analysis for prenatal diagnosis of mucopolysaccharidoses is based on the fact that fetal urine contributes to the amniotic fluid and carries with it the changes in the fetal tissue of glycosaminoglycans into this fluid. This is justified by the observations that (1) except for higher levels of hyaluronic acid in amniotic fluid the prenatal glycosaminoglycans patterns in amniotic fluid in a particular type of mucopolysaccharidosis resemble those found in the postnatal urine of the patients with that disease. (2) Increased dermatan sulfate and heparan sulfate levels in amniotic fluid have been found to be associated with increased contents of these glycosaminoglycans in aborted fetal tissues from Hurler and Hunter cases (Matalon and Dorfman, 1972; Crawfurd et al., 1973; Dean and Muir; 1975). Similar correlations have been found for glycosaminoglycans in amniotic fluids and tissues for pregnancies at risk for Sanfilippo syndrome (Whiteman, 1973; Whiteman and Henderson, 1977; Ferguson-Smith et al., 1973). (3) Tamm-Horsfall glycoprotein present in postnatal and adult urine has been identified in human amniotic fluid (Ross et al., 1975).

The earlier reports of significant elevation of amniotic fluid glycosaminoglycans (Danes et al., 1970; Matalon et al., 1970) appeared to be of a potential use for antenatal diagnosis of the mucopolysaccharidoses. However, failures to detect the presence of these disorders by glycosaminoglycan analyses of the amniotic fluid have been documented in several reports (Matalon, Dorfman and Nadler, 1972; Matalon, 1974). These problems may be the result of lack of normative data or lack of correlation of the value determined at a fetal age to the normal value at that gestational age. False negatives reported at or before the 16 weeks may be because of the fact that till that age the fetal urine contributes little to the amniotic fluid and therefore the abnormal glycosaminoglycan components from the fetal urine would represent only a small proportion of the total glycosaminoglycans. These small amounts of abnormal components are likely to go undetected when the total glycosaminoglycans or hyaluronidase-resistant glycosaminoglycan components are measured. In several instances abnormally high values of glycosaminoglycans have been obtained in the amniotic fluid from the pregnancies at 20 weeks which had normal values at 16 weeks of gestation (Lee and Shafer, 1974; Matalon, 1974; Matalon, Dorfman and Nadler, 1972). False positives have also been reported in normal pregnancies at 22 weeks gestation (Lee and Shafer, 1974; Matalon, 1974) probably due to normal age-related changes (Section 11.7) in glycosaminoglycan composition of amniotic fluid.

In spite of the aforementioned problems, analyses of the glycosaminoglycans may be of a great use in advanced pregnancies where there is no time for cell cultures or when cell cultures started earlier in pregnancy have failed to grow and a decision whether or not to terminate the pregnancy has to be made fast (Omura et al. 1973; Harper et al., 1974; Lee and Shafer, 1974; Whiteman and Henderson, 1977). The value of amniotic fluid glycosaminoglycan analysis could be enhanced if analysis is repeated upon finding normal glycosaminoglycan values at 16 weeks. Since chondroitin sulfate is a normal component of fetal urine, Whiteman (1973, 1975), Whiteman and Henderson (1977) have suggested the determination of individual GAG components by two dimensional electrophoresis. Dermatan sulfate and heparan sulfate are abnormally elevated in various mucopolysaccharidoses. The ratios of these abnormal components to the normal chondroitin sulfate (DS/CS and HS/CS) would be more meaningful parameters for discrimination between normal and affected pregnancies. The use of an external standard is unnecessary with this approach. All that needed was to compare the absorbance given by Alcian blue complexes of the separated fractions. The DS/CS and HS/CS ratios were found to be less than 0.1 in 30 control samples, but were elevated several times in Hurler and Sanfilippo syndromes (Whiteman and Henderson, 1977).

The use of amniotic fluid cell cultures has proved to be a useful tool for antenatal diagnosis of mucopolysaccharidoses. The most reliable test for this available today is the demonstration of the enzyme deficiency and its cross-correction in the presence of normal amniotic cells or amniotic cells from another disorder. The other approaches for the use of cultured cells like metachromasia in granules of cultured cells and ^{35}S-incorporation kinetics (see Section 11.4) are not very specific and are seen in some other disorders too (Milunsky and Littlefield, 1969; Neufeld, 1972).

The major limitation in the use of amniotic fluid cell cultures for in utero diagnosis is that for enzyme analysis a fairly large quantity of the cells are required. The cultures grow very slowly and starting from cells in 10-15 ml of amniotic fluid it takes 4-8 weeks to cultivate the required number of cells. This may be of utmost significance in those cases when decision for termination of pregnancy has to be made quickly. Micromethods have been developed which require only a small number of cells for assay of the enzymes. The amniotic fluid cells are grown directly on thin plastic foils and after 5-10 days these are freeze-dried and the number or dry mass is determined. After incubation with the substrate in a small volume of culture medium, the enzyme activity is measured by microspectrophotometry or microfluorometry (Hosli, 1974).

The following articles review the various aspects of the use of amniotic fluid cells for prenatal diagnosis:

Fuchs, F. and Cederquist, L. L. (1978): Use of amniotic fluid cells in prenatal diagnosis. In: D. V. I. Fairweather and T. K. A. B. Eskes (Eds.). Amniotic Fluid - Research and Clinical Application, 2nd ed. Excerpta Medica, Amsterdam, Chap. 14, pp. 277-295

Niermeijer, M. F., Fortuin, J. J. H., Koster, J. F., Jahodova, M. and Galjaard, H. (1974): Prenatal diagnosis of some lysosomal storage diseases. In: J. M. Tager, G. J. M. Hooghwinkel and W. Th. Daems (Eds.). Enzyme Therapy in Lysosomal Storage Diseases, North Holland Publ. Co., Amsterdam pp. 25-34

Di Natale, P. (1982): Diagnostic approaches for mucopolysaccharides, mucolipidoses and related disorders. In: R. S. Varma and R. Varma (Eds.). Glycosaminoglycans and Proteoglycans in Physiological and Pathological Processes of Body Systems, S. Karger, Basel, pp. 316-329

11.10 Bibliography

1. Abramovich, D. R. (1968): The volume of amniotic fluid in early pregnancy. J. Obstet. Gynecol. Br. Commonw. 75, 728-731

2. Acton, R. T. and Lynn, D. J. (Eds.) (1977): Cell Culture and its Applications. (Proceedings of the 1st International Cell Culture Congress, Birmingham, Alabama, 1975), Academic Press, New-York, 1977, 752 pp

3. Anton, W., Mende, H. E. and Schmidt, D. (1973): Relations between mast-cell function and the fetoplacental unit, particularly considering the histamine level and the metachromasia-producing substances in amniotic fluid. Deut. Gesundheitsw. 28, 1544-1548 (Ger)

4. Bischoff, R. (1971): Acid mucopolysaccharide synthesis by chick amnion cultures. Inhibition by 5-bromodeoxyuridine. Exp. Cell. Res. 66, 224-236

5. Crawfurd, M. d'A., Dean, M. F., Hunt, D. M., Johnson, D. R., MacDonald, R. R., Muir, H., Wright, E. A. P. and Wright, C. R. P. (1973): Early prenatal diagnosis of Hurler's syndrome with termination of pregnancy and confirmatory finding on the fetus. J. Med. Genet. 10, 144-153

6. Danes, B. S. and Bearn, A. G. (1966): Hurler's syndrome: a genetic study in cell culture. J. Exp. Med. 123, 1-6

7. Danes, B. S., Queenan, J. T., Gadow, E. and Cederquist, L. L. (1970): Antenatal diagnosis of mucopolysaccharidoses. Lancet 1, 946-947

8. Dean, M. F. and Muir, H. (1975): Glycosaminoglycans of foetal tissue in two cases of Hurler's syndrome. In: J. Holten and J. T. Ireland (Eds.). Inborn Errors of Skin, Hair and Connective tissue, University Park Press. Baltimore, pp. 61-65

9. Dingle, J. T. and Webb, M. (1965): Mucopolysaccharide metabolism in tissue culture, cells and tissues. In: E. N. Willmer (Ed.). Culture: Methods, Biology and Physiology, Academic Press, New York, Vol. I, pp. 353-396

10. Dito, W. R., Patrick, C. W. and Shelly, J. (1975): Clinical Pathological Correlations in Amniotic Fluid, American Society of Clinical Pathologists, Chicago

11. Doran, T. A. (1976): Amniotic fluid. In: J. W. Goodwin, J. O. Godden and G. W. Chance (Eds.). Perinatal Medicine. Williams and Wilkins, Baltimore, pp. 499-516

12. Dorfman, A. (Ed.) (1972): Antenatal diagnosis. University of Chicago Press, Chicago

13. Duncan, D. M., Logan, R. W., Ferguson-Smith, M. A. and Hall, F. (1973): Measurement of acid mucopolysaccharides (glycosaminoglycans) in amniotic fluid and urine. Clin. Chim. Acta 45, 73-83

14. Emery, A. E. H. (1973): Biochemical analysis of amniotic fluid. In: A. E. H. Emery (Ed.). Antenatal Diagnosis of Genetic Disease. Churchill-Livingstone, London, pp. 113-136

15. Fairweather, D. V. (Ed.) (1973): Amniotic Fluid. North Holland, Am. Elsevier, New York

16. Fairweather, D. V. I. and Eskes, Th. K. A. B. (Eds.) (1978): Amniotic fluid: Research and Clincal Application. Excerpta Medica, Amsterdam, 455 pp

17. Ferguson-Smith, M. A., Duncan, D. M., Logan, R. W., Hall, F., Wusteman, F. S. and Harper, P. S. (1973): Antenatal diagnosis of mucopolysaccharidoses. Lancet 2, 45-46

18. Fiereck, E. A. (1976): Amniotic fluid. In: N. Tietz (Ed.): Fundamentals of Clincal Chemistry, 2nd Edition. W. B. Saunders, Philadelphia, pp. 1163-1176

19. Fratantoni, J. C., Hall, C. W. and Neufeld, E. F. (1968): The defect in Hurler's and Hunter's syndromes: faulty degradation of mucopolysaccharides. Proc. Nat. Acad. Sci. U.S.A. 60, 699-706

20. Fratantoni, J. C., Hall, C. W. and Neufeld, E. F. (1968 a): Hurler and Hunter syndromes. Mutual correction of the defect in cultured fibroblasts. Science 162, 570-572

21. Free, A. H. and Free, H. M. (1974): Laboratory interrelations of amniotic fluid and urine. In: S. Natelson, A. Scommegno and M. B. Epstein (Eds.). Amniotic Fluid, Wiley and Sons, pp. 37-46

22. Free, A. H. and Free, H. M. (1975): Urinalysis in Clinical Laboratory Practice, CRC Press, Cleveland, Chap, 45, pp. 245-248

23. Fuchs, F. and Cederquist, L. L. (1973): Prenatal diagnosis based on amniotic fluid. In: D. V. I. Fairweather (Ed.). Amniotic Fluid, Excerpta Medica, Amsterdam, pp. 262-276

24. Guibaud, S., Bonnet, M. Dury, A. et al. (1976): Letter: Amniotic fluid or maternal urine? Lancet 1, 746

25. Hall, C. W., Liebaers, I., Di Natale, P. and Neufeld, E. F: (1978): Enzymic diagnosis of the genetic mucopolysaccharide storage disorders. Methods Enzymol. Vol. (50) Part C., 439-456

26. Hamilton, W. F., Boyd, J. D. and Mossmar, H. W. (1962): Human Embryology (Prenatal Development of Form and Function). 3rd Edition, Williams and Wilkins, Baltimore, 1962, Chap. 5, pp. 66-108

27. Harper, P. S., Laurence, K. M., Parkes, A., Wusteman, F. S., Kresse, H., Von Figura, K., Ferguson-Smith, M. A., Duncan, D. M., Logan, R. W., Hall, F. and Whiteman, P. (1974): Sanfilippo-A disease in the fetus. J. Med. Genet. 11, 123-132

28. Holten, J. B. (1978): Diagnostic tests on amniotic fluid. In: V. Marks and C. N. Hale (Eds.). Essays in Medical Biochemistry, Biochemical Society, London, Vol. 3, pp. 75-107

29. Hosli, P. (1974): Microtechniques for rapid prenatal diagnosis in early pregnancy. In. A. G. Motulsky and W. Lenz (Eds.). Birth Defects, Proc. 4th Int. Conf. Excerpta Medica, Amsterdam, pp. 234-239

30. Kaback, M. M. and Howell, R. R. (1973): Heterozygote detection and prenatal diagnosis of lysosomal storage diseases. In: H. G. Hers and F. Van Hoof (Eds.). Lysosomes and Lysosomal Storage Diseases. Academic Press, New York, 1973, Chap. 26, pp. 599-615

31. Lee, T. Y. and Shafer, I. A. (1974): Glycosaminoglycan composition of human amniotic fluid. Biochim. Biophys. Acta 354, 264-274

32. Liebaers, I., Di Natale, P. and Neufeld, E. F. (1977): Iduronate sulfatase in amniotic fluid: An aid in the prenatal diagnosis of the Hunter syndrome. J. Pediatr. (St. Louis) 90, 423-425

33. Lind, T. (1971): Diagnostic use of amniocentesis in late pregnancy. Proc. R. Soc. Med. 64, 1140-1141

34. Lind, T., Kendall, A. and Hytten, F. E. (1972): The role of the fetus in the formation of amniotic fluid. J. Obstet. Gynaecol. Br. Commonw. 79, 289-298

35. Matalon, R. (1974): Amniotic fluid and prenatal diagnosis of the mucopolysaccharidoses. In: S. Natelson, A. Scommegna and M. B. Epstein (Eds.). Amniotic Fluid: Physiology, Biochemistry and Clinical Chemistry. John Wiley and Sons, New York, pp. 343-352

36. Matalon, R. and Dorfman, A. (1970): Acid mucopolysaccharides in cultu tissue. Lancet 2, 838-841

37. Matalon, R. and Dorfman, A. (1972): The antenatal diagnosis of mucopolysaccharidoses. In: A. Dorfman (Ed.). Antenatal Diagnosis, University of Chicago Press, Chicago, pp. 213-216

38. Matalon, R., Dorfman, A. and Nadler, H. L. (1972): A chemical method for the antenatal diagnosis of mucopolysaccharidoses. Lancet 1, 798-799

39. Matalon, R., Dorfman, A., Nadler, H. L. and Jacobson, C. B. (1970): A chemical method for the antenatal diagnosis of mucopolysaccharidoses. Lancet 1, 83-84

40. Mazurkiewicz-Kilczewska, D. (1971): Hexosamines and their biosynthesis in human amnion in vitro. Amer. J. Obstet. Gynecol. 110, 355-357

41. Milunsky, A. (1973): The Prenatal Diagnosis of Hereditary Disorders. Charles C. Thomas, Springfield, Chap. IV, pp. 62-124

42. Milunsky, A. (1975): Risk of amniocentesis for prenatal diagnosis. N. Engl. J. Med. 293, 932-933

43. Milunsky, A. and Littlefield, J. W. (1969): Diagnostic limitations of metachromasia. New Engl. J. Med. 281, 1128-1129

44. Milunsky, A., Littlefield, J., Kaufer, J., Kolodony, E., Shih, V. and Atkins, L. (1970): Prenatal genetic diagnosis (in three parts) N. Engl. J. Med. 283, 1370-1381, 1441-1447 and 1498-1504

45. Mitchell, G., Field, B. and Kerr, C. (1977): Biochemical differences between amniotic fluid and maternal urine. Br. Med. J. 1, 811-812

46. Mitra, S. K. and Blau, K. (1978): An improved determination of total glycosaminoglycans in body fluids by formation of complexes with quinarcrine: changes in amniotic fluid total glycosaminoglycans during normal pregnancies and pregnancies at risk for mucopolysaccharidoses. Chlin. Chim. Acta 89, 127-134

47. Nadler, H. L. and Gerbie, A. B. (1970): Role of amniocentesis in the intrauterine detection of genetic disorders. N. Engl. J. Med. 282, 596-599

48. Nadler, H. L., Rao, G. J. S. and Taussig, L. (1978): Cystic fibrosis. In: J. B. Stanbury, J. B. Wyngaarden and D. S. Fredrickson (Eds.). The Metabolic Basis of Inherited Disease. McGrawhill, New York, 4th Ed., pp. 1683-1710

49. Natelson, S., Scommegna, A. and Epstein, M. B. (1974): Amniotic Fluid: Physiology, Biochemistry and Clinical Chemistry. Wiley, New York

50. Neufeld, E. F. (1972): Mucopolysaccharidoses. In: A. Dorfman (Ed.). Antenatal Diagnosis. University of Chicago Press, Chicago, pp. 217-228

51. Neufeld. E. F. and Cantz, M. (1973): The mucopolysaccharidoses studied in the cell culture. In: H. G. Hers and F. Van Hoof (Eds.). Lysosomes and Lysosomal Storage Diseases. Academic Press, New York, Chap. 9, pp. 261-275

52. Niermeijer, M. F., Halley, D. and Sachs, E. (1973): Transport and storage of amniotic fluid samples for prenatal diagnosis of metabolic diseases. Humangenetik 20, 175-178

53. Niermeijer, M. F., Sachs, E. S., Jahodova, M., Tichelaar-Klepper, C., Kleijer, W. J. and Galjaard, H. (1976): Prenatal detection of genetic disorders. J. Med. Genet. 13, 182-194

54. Omura, K., Higami, S., Issiki, G., Nishizawa, K. and Tada, K. (1973): Prenatal diagnosis of the Hurler syndrome. Mucopolysaccharide patterns in amniotic fluid. Tohoku J. Exp. Med. 111, 87-91

55. Ostergard, D. R. (1970): The physiology and clinical importance of amniotic fluid. A review. Obstet. Gynecol. Survey 25, 297-319

56. Pirani, B. B., Doran, T. A. and Benzie, R. J. (1976): Letter: Amniotic fluid or maternal urine? Lancet 1, 303

57. Plentl, A. A. (1958): The origin of amniotic fluid. In: C. A. Villee (Ed.). Gestation: Trans. 4th Conf. Macy Found., New York, pp. 71-114

58. Priest, J. H. (1971): Human cell culture in diagnosis of disease. Charles C. Thomas, Springfield, 1971, Chap. 5, pp. 125-136

59. Ross, N., Massuchi, N., Pecarovich, R. et al. (1975): Identification of Tamm-Horsfall urinary glycoprotein in human amniotic fluid. Am. J. Obstet. Gynecol. 122, 790-791

60. Scrimgeour, J. B. (1973): Amniocentesis: technique and complications. In: A. E. H. Emery (Ed.). Antenatal Diagnosis of Genetic Disease, Churchill-Livingstone, London, pp. 11-39

61. Suschke, J. (1973): Acid mucopolysaccharide in urine, amniotic fluid and serum. Fortschr. Med. 91, 485-487 (Ger)

62. Suschke, J. and Kunze, D. (1971): Quantitative determination of acid mucopolysaccharides in human amniotic fluid. Münch. Med. Wochenschr. 113, 1495-1496

63. Varma, R. S., Varma, R., Allen, W. S. and Wardi, A. H. (1974): Urinary excretion of acid mucopolysaccharides in schizophrenia. Biochem. Med. 11, 358-369

64. Votta, R. A., Gadow, E. C., Alvarez, O. H. et al. (1974): Amniotic fluid volume determinations in clinical obstetrics. Obstet. Gynecol. 44, 811-813

65. Vrettos, A. S., Koliopoulos, C. and Panayotou, P. P. (1975): Amniocentesis: indications and complications. Int. Surg. 60, 101-103

66. Whiteman, P (1973): Prenatal diagnosis of mucopolysaccharidoses. Lancet 1, 1249

67. Whiteman, P. (1975): Acid glycosaminoglycan excretion in the mucopolysaccharidoses. Determination of glycosaminoglycans in urine and amniotic fluid using new microanalytical techniques. In: J. Holten and J. T. Ireland (Eds.). Inborn Errors of Skin, Hair and Connective Tissue. University Park Press, Baltimore, pp. 251-265

68. Whiteman, P. and Henderson, H. (1977): A method for the determination of amniotic fluid glycosaminoglycans and its application to the prenatal diagnosis of Hurler and Sanfilippo diseases. Clin. Chim. Acta 79, 99-105

69. Whitfield, C. R. (1976): Amniotic fluid analysis. In: R. W. Beard and P. W. Nathanielsz (Eds.). Fetal Physiology and Medicine. Saunders, London, pp. 329-344

Chapter 12

12.1 Introduction

The synovial fluid fills the cavity of a diarthrodial joint. The diarthrodial (synovial) joint cavity (Fig. 12-1) is a specialized connective tissue compartment and it arises from a cleft in the embryonic mesenchyme beginning at 3-6 weeks after fertilization and completing by 7-8 weeks. Because of its mesenchymal origin, the synovial cavity is affected in a number of connective tissue (or mesenchymal) diseases. Analyses of the glycosaminoglycans present in the synovial fluid and the ground substance of the connective tissue of the synovial membrane are of diagnostic and prognostic value in those mesenchymal diseases.

The synovial fluid is the product of secretion from the highly vascular synovial membrane. It provides nutrition to the avascular articular cartilages in the joint cavity. Besides providing lubrication of the joint at low load interaction involving the synovial membrane, it acts as a shock absorber.

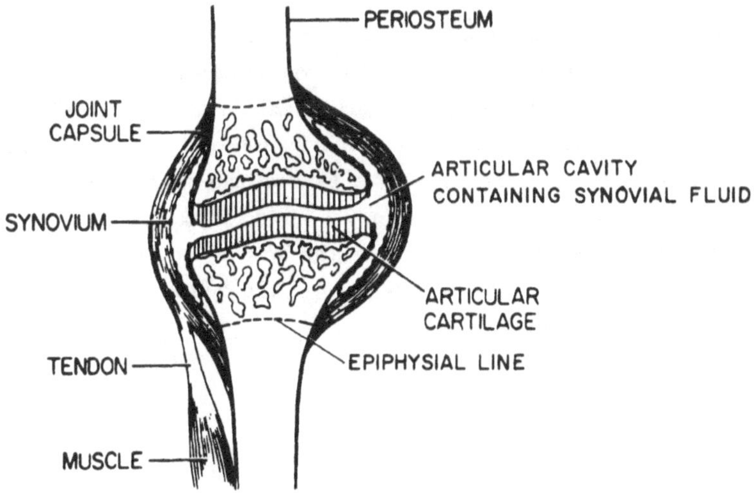

Fig. 12-1. A diarthrodial joint showing the articular cavity with synovial fluid

The structure (Wright et al., 1973) and function (Barnett et al., 1961; Hamerman, 1970; Sokoloff, 1978), the pathological (Mueller et al., 1971; Radin, 1973) aspects of the synovial joints have been the subjects of excellent reviews and monographs. The synovial cavities are large tissue spaces enclosed by a modified connective tissue present as the joint capsule (Fig. 12-1). During the embryonic development the capsule forms from the dense external tissue, which is continuous with the periosteum. The cells on the inner surface of the capsule flatten in the growing embryo into a false epithelium to form the synovial membrane. The synovial membrane lines the joints (and also the synovial tendon sheaths and bursae), has a smooth, glistening pink appearance and is fringed to provide a large surface area of 43 square inches for the knee joint. It is highly vascular and carries three important functions: as a barrier between blood and synovial fluid, phagocytosis and formation of the synovial fluid and biosynthesis of glycosaminoglycans. The membrane consists of a matrix of collagen and GAG in which the synovial cells of two types (called A or M cell and the B cell) are embedded. The synovial A cells contain numerous lysosomes, a small endoplasmic reticulum and vacuoles and these cells are concerned with phagocytosis. The synovial B cells have an abundant endoplasmic reticulum with only a few vacuoles and these appear to synthesize the GAG of the synovial fluid. The physiology of the synovial cell and its contribution to the disease processes have been reviewed (Castor, 1975).

The synovial fluid by acting as a biological lubricant allows the free movement of a diarthrodial joint. The lubricating mechanisms and molecular interactions involved in this have been reviewed (Radin and Paul, 1972; Swann et al., 1974; Swanson, 1979). The lubrication of the synovial joints is impaired in inflammatory conditions such as rheumatoid arthritis due to changes in the macromolecules of the synovial fluid. In case of the degenerative joint disorders such as osteoarthrosis the loss of lubrication is primarily due to changes in structure and integrity of the opposing articular cartilages of the diarthrodial joint.

12.2 Volume and general composition

Normal synovial fluid is viscous, clear, pale yellow and varies in amount from 0.2-3.5 ml with an average of 2 ml in a normal human adult knee joint. Because of the small volume, the majority of the studies on human synovial fluid have been carried out on pooled samples. The majority of the information on synovial fluid is derived from samples present in sufficient amounts (30-40 ml) in horse and (20 ml) in ox. The volume of the synovial fluid is greatly increased in patients with rheumatoid arthritis due to increased permeability of the synovial membrane brought about by inflammation. The specific gravity of the fluid is 1.008-1.015 and is higher in fluids from traumatic arthritis and osteoarthritis. Its pH ranges between 7.38 and 7.77 (Cohen, 1967; Cohen et al., 1975).

The synovial fluid is considered as a plasma dialysate and resembles plasma in regard to the low molecular weight organic molecules, inorganic electrolytes and nonelectrolytes. The fluctuations in plasma concentrations of these small molecules and electrolytes are correspondingly reflected in the synovial fluid. However, the synovial fluid differs from plasma in the composition of the macromolecules. Its protein content is only 2 g/100 ml as compared to 7 g/100 ml in plasma. Normal synovial fluid contains no fibrinogen or plasmin and so it does not clot but the fluid from patients with synovial infections or rheumatoid arthritis clots due to the presence of fibrinogen. The 2 % protein content is contributed by albumin, globulins, enzymes and the protein associated with glycoproteins and glycosaminoglycans. In inflammation of the synovial membrane, these macromolecules are increased and the total protein content is increased to over 8 g/100 ml. Lymphocytes and monocytes averaging 65/mm^3 are present in normal synovial fluid but in inflammation polymorphs also appear, thereby producing a great increase in the leukocyte count of the fluid. These and other physiological aspects of the synovial fluid have been reviewed (Robinson, 1978; McCarty, 1979 a; Sinkin, 1979).

12.3 Glycosaminoglycan composition

The high viscosity of the synovial fluid and its action as a biological lubricant are due to the presence of glycoproteins and glycosaminoglycans. The two types of the macromolecules are involved in two entirely different lubrication systems of the joint. Specific lubricating glycoproteins present in the synovial fluid are concerned with the high load interaction between the apposed articular cartilages while at low shear and low loads, the glycosaminoglycans lubricate the synovial joint (Swann and Radin, 1972; Swann et al., 1974).

The glycoproteins (Sunblad, 1965) and glycosaminoglycans (Caygill, 1968; Muir, 1971; Iwata, 1972; Swann, 1978) in normal synovial fluid and in disease states have been reviewed.

By fractionation of the bovine synovial fluid using ultrafiltration on Millipore filters, density gradient and column chromatography, a glycoprotein was isolated which acted as a lubricant in an in vitro joint system (Swann and Radin, 1972). This glycoprotein contained threonine, glutamic acid and proline as the major amino acids and the principal carbohydrate components consisted of galactose, glactosamine and sialic acid. In a later publication (Swann et al, 1977) this group of investigators reported partial characterization of the major lubricating glycoprotein, which they named *LGP-I*. This glycoprotein was present in bovine synovial fluid at a concentration of 20 mg/100 ml, had a molecular weight of 227,500 and contained galactosamine galactose and N-acetylneuraminic acid at 202, 162 and 114 residues, respectively per mole of LGP-I.

The synovial fluid, like other fluids of the fluid-filled mesenchymal cavities, contains hyaluronic acid as the major glycosaminoglycan (GAG), the sulfated GAG being less than 5 µg/ml (Sundblad, 1953). Since the isolation of hyaluronic acid from synovial fluid for the first time (Meyer et al., 1939), the work has been concentrated only on this macromolecule, although chondroitin sulfate has been detected in synovial fluid (Barker et al., 1966), especially from patients with arthritis and osteoarthritis. The chondroitin sulfate was isolated from three patients with rheumatoid arthritis and osteoarthritis by first freeing the fluid from protein by passing through a DEAE-Sephadex column followed by gel filtration on Sephadex G-200. Further investigation showed that the chondroitin sulfate was present in the synovial fluid as a proteoglycan (Seppala et al., 1972).

The dry weight of the mucin clot formed on acidification of synovial fluid collected from cases of trauma of the knee was 150-1,000 mg/100 ml; the ox synovial fluid contained 22-250 mg/100 ml (Fessler et al., 1954). Based upon the hexosamine content of the synovial mucin clots, normal human synovial fluid contained 360 mg/100 ml hyaluronic acid, and bovine synovial fluid contained 450 mg/100 ml hyaluronic acid (Hamerman and Schuster, 1958). Specific coupled bacterial hyaluronidase-colorimetric procedure revealed that only 66 % of the uronic acid content of normal synovial fluid was due to hyaluronic acid (Jourdian et al., 1979), whereas hyaluronic acid represented all the hexuronic acid in a sample from a patient with systemic lupus erythematosus. This implies that the remaining 34 % or less hexuronic acid content of normal synovial fluid may be present as GAG other than hyaluronic acid. This appears to be the case, since Kuwano (1969) reported that the hyaluronic acid/chondroitin sulfate ratio in normal human synovial fluid was 2.0. The average values of this ratio were changed to 2.5 in osteoarthritis and 3.0 in rheumatoid arthritis and in bovine synovial fluid it was 1.8. The decrease in chondroitin sulfate with the consequent elevation of hyaluronate/chondroitin sulfate ratio is supported by the reported decrease in infrared adsorption of sulfate groups of these compounds in the synovial fluids from these patients (Oshima et al., 1962). In rheumatoid arthritis, the concentrations of hyaluronic acid and chondroitin sulfate had a tendency to decrease with aging, but had no relation to pH or volume of the synovial fluid. In osteoarthritis, the concentration of chondroitin sulfate increased with aging but hyaluronic acid did not vary with age in these patients (Kuwano, 1969). The chondroitin sulfate in normal synovial fluid was present as a proteoglycan and was reduced by 50 % in the rheumatic arthritis patients (Wood et al., 1971; Seppala et al., 1972).

The hyaluronic acid content of the synovial fluid samples from different animal species is variable, being 3.89, 1.93, 0.45 and 0.56 mg/ml in rabbit, calf, ox and horse, respectively. Since these contents are related inversely to the synovial fluid volumes in these animals, the differences in hyaluronic acid contents may be related to different degrees of dilutions in these joint fluids (Sundblad, 1965).

Because of the problems in collection and the small amount (only 2 ml) of fluid in human biggest (knee) joint, only a few analyses have been carried out in normal subjects. The hyaluronic acid contents of the post mortem samples have been found to be variable due to varying degree of peripheral edema. In normal human synovial fluid samples, the hyaluronic acid contents reported by different investigators vary due to methodological differences in isolation. The reported values are given in Table 12-I. The average hyaluronate in normal human adult synovial fluid is 300 mg/100 ml.

Hyaluronic acid in the synovial fluid was found to have a particle weight of 10^6 and was associated with about 30 % protein (Ogston and Stanier, 1950). Undegraded hyaluronate in protein-free form could not be isolated using different methods (Ogston and Stanier, 1949), although the protein association with hyaluronic acid was through a process akin to salting out adsorption (Curtain, 1955). The ox synovial fluid hyaluronate prepared by four methods was found to have 27 % protein and attempts to remove additional proteins resulted in changes of its physicochemical properties. The molecule was a random coil and highly hydrated (Ogston and Stanier, 1953). Its axial ratio was low but it had a very large specific volume of the order of 10^3 ml/g (Ogston and Stanier, 1951). Preston et al. (1965) have presented detailed physicochemical measurements on hyaluronate from ox synovial fluid, which contained 21 % protein. The molecular configuration obtained from light scattering and viscosity measurements suggested a random coil structure with some degree of branching and cross-linking. The branching and cross-linking gave this hyaluronate the rigidity with respect to the expansion of molecular domain that would not be possessed by a random coil.

The size of hyaluronate from human synovial fluid varies in different samples. The average molecular weight in normal human subjects was $5 \cdot 10^6$ (Balazs et al., 1967) and it was reduced to $2 \cdot 10^6$ in synovial fluids from rheumatoid arthritis patients. Because of the association with protein it has been suggested that this hyaluronate may exist as several short chains, each of molecular weight $1 \cdot 10^6$ attached to a protein chain. Sandson and Hamerman (1962) showed that the association with protein was only a mild, loose one and probably occurred during the isolation procedures. Following column chromatography on hydroxyapatite only 2 % protein was found to be associated with hyaluronate separated by electrophoresis.

Human synovial fluid following dialysis and hydrolysis was found to contain nondialyzable neutral sugars (Sandson et al., 1960). Using a combination of zone electrophoresis in barbiturate buffer of pH 8.5 and an ascending paper chromatography, fucose, glucose, mannose, and galactose, were identified in the hydrolyzates. Determination of the nondialyzable hexose content of the synovial fluid samples gave a value of 0.5 mg/g for normal subjects. In osteoarthritis the hexose content was normal or lower, while in rheumatoid arthritis it was elevated to 0.7 mg/g (Sandson et al., 1960).

12.4 Source of Glycosaminoglycans

The composition of synovial fluid resembles that of a plasma dialysate with glycosaminoglycans added to it. Since the fluctuations in electrolytes and nonelectrolytes of blood-plasma and synovial fluid occur concomitantly, the fluid appears to originate as a plasma dialysate from the highly vascular synovium. The synovial cells (type A and B) having endoplasmic reticulum and Golgi apparatus synthesize hyaluronic acid and the sulfated GAG and secrete these into the fluid of the joint. This was supported by several pieces of indirect evidence from in vitro biosynthesis in cell cultures or tissue slices from glucose-^{14}C (Yielding et al., 1957; Hedberg and Moritz, 1958; Roy and Ghadially, 1967). Other investigators have also shown by use of histochemical and autoradiographic studies that the free synovial cells and histiocytes available in the synovia of early embryos and adult animals are viable cells and synthesize unsulfated GAG (Berezkin and Zhitnikov, 1976). The synovial sarcoma cells incorporate radioactive sulfate both in vivo and in tissue cultures into sulfated GAG related to heparin, which explains the presence of macromolecular GAG resistant to degradation by hyaluronidase and chondroitinases (Bounassisi and Ozzello, 1973). The biosynthesis of GAG in the inflammed synovial membrane such as seen in rhematoid arthritis is increased (Hirai, 1973). The in vitro biosynthesis of GAG from ^{14}C-labeled glucose using slices of porcine synovial membranes was inhibited by anti-inflammatory steroids. Acetylsalicylic acid had no inhibitory effect, while prednisone was more potent than dexamethasone (Mikio, 1975).

The presence of the sulfated glycosaminoglycans such as chondroitin sulfate in synovial fluid has been attributed to their infiltration into the synovial fluid from the ground substance of the cartilage (Barker et al., 1966). Thus the sulfated GAG of the synovial fluid, unlike the hyaluronate-protein, are not synthesized by the Type B lining cells of the synovial membrane. Suggestions have been made that a part of the GAG of the synovial fluid could be derived from fibroblasts found beneath the synovial cell layer.

The developing synovial membranes were found to start synthesizing the GAG from the 16th day of intrauterine life. By histochemical methods combined with specific hydrolytic enzymes, the GAG were found to consist of hyaluronic acid and chondroitin 4- and 6-sulfates (Usenko, 1967, 1971).

12.5 Analytical and fractionation methods

The procedures involved in analysis of the synovial fluid consist of two steps: Aspiration of synovial fluid (arthrocentesis) and physicochemical analysis. Arthrocentesis is carried under sterile conditions from the knee joint by shaving a part of the overlying skin, iodine painting, several alcohol washes followed by subcutaneous injection of a local anesthetic, procaine, xylocaine, etc. The anatomical considerations are described in several text books (McCarty, 1979). The fluid is aspirated with a needle of size 14 to 20 and transferred quickly into a tube without an anticoagulant. The biochemical studies carried out on the synovial fluid in relation to glycosaminoglycans include: Analysis of the GAG content and the variation in their molecular size by viscosity measurements, gel filtration or by other physical measurements. Since a limited amount of the synovial fluid is available, microadaptations of the procedures need to be used.

For viscosity measurement as little as 0.3 ml of synovial fluid was sufficient in cone-in-cone viscometer (Dintenfass, 1966) and Hess viscometer can also be used. For removal of the most associated protein, ultrafiltration through graded pore size filters, DEAE-cellulose column chromatography and gel filtration have been used. Zone electrophoresis on agarose gel (How and Long, 1969) and other supports (Hamerman and Sandson, 1960) have been used for isolation of hyaluronic acid from human synovial fluid.

A number of different methods have been employed for analysis of hyaluronate content of the synovial fluid. Turbidimetric method based on measurement of the turbidity formed at an acidic pH (Mathews, 1966) may not be accurate since the degree of polymerization affects the turbidity. Methods based on the weight of washed and dried mucin clot or determination of the hexosamine contents of these mucin clots are inaccurate. Precipitation of the hyaluronate with cetylpyridinium chloride and analysis of the precipitate have been used (Seppala, 1964). A better method used by some workers (Sundblad, 1953; Hamerman and Schuster, 1958) consisted of indirect measurement of hyaluronate from the hexosamine content before and after hyaluronidase digestion. These combined chemical-enzymic methods were based on the assumption that hyaluronate was the only GAG present in the synovial fluid and also a nonspecific testicular hyaluronidase (EC 3.2.1.35) was used for its hydrolysis.

The aforementioned methods of analysis are not specific. More sensitive and specific analysis of the hyaluronate content of the synovial fluid has been achieved by using the microbial hyaluronidase (EC 4.2.2.1). The hydrolytic products formed are the oligosaccharides containing terminal Δ 4,5-unsaturated glycopyrauronic acid units, which are measured by their characteristic absorption at 232 nm (Greiling, 1974) or by colorimetric quantitation at

585 nm of the red chromogen formed by condensation of the N-acetylglucosamine end groups with p-dimethylaminobenzaldehyde in strong acid (Greiling, 1974). Since the biological fluids contain proteins or nucleic acid which may interfere with the UV absorption at 232 nm, the colorimetric measurement at 585 nm may be more suitable for synovial fluid analysis. Another method (Jourdian et al., 1979) is based on the periodic acid oxidation of the Δ 4,5-unsaturated glycopyrauronic acid residues (present at the non-reducing termini of the oligosaccharides formed upon bacterial hyaluronate digestion) to formyl pyruvic acid and measurement at 549 nm of the chromophore formed by reaction with thiobarbituric acid. All these three specific methods have been used directly on synovial fluid and the values obtained are given in Table 12-I. Only 25 μl of the synovial fluid is required for the last described coupled-enzyme colorimetric procedure (Jourdian et al., 1979). The absorbance at 549 nm with 5 TRU* hyaluronate lyase and without this enzyme is read and the difference is proportional to hyaluronate content of the synovial fluid.

12.6 Alterations in physiological and pathological states

Variations in hyaluronic acid contents and viscoelastic properties of synovial fluid samples have been found in a number of physiological and pathological states.

12.6.1 Physical activity

The hyaluronic acid content may vary in the joint fluid from the same subject sampled at different times and depending upon the degree of function performed by the joint, such as movement. Increased amounts are found during heightened physical activity (Ekholm and Norback, 1951). This is inconsistent with the role of hyaluronic acid as a mechanical lubricant.

*TRU = Turbidity reducing units. One TRU is the amount of enzyme which will reduce the turbidity produced by 0.2 mg hyaluronic acid to that produced by 0.1 mg.

12.6.2 Edema

The hyaluronic acid concentration of the synovial fluid has been found to be very low in patients with edema and this appears to be attributable to the dilution of synovial fluid occurring in these patients. The absolute amount of hyaluronic acid was found to increase in myxedema (Ropes et al., 1947).

12.6.3 Aging

Several studies have reported a decreased viscosity of samples from subjects of increasing ages. Some studies have reported lower concentrations of hyaluronic acid by measurements involving formation of mucin clots in an acidic medium (Hammerman and Schuster, 1958). Since both mucin clot formation and viscosity measurements are decreased with the depolymerization of hyaluronic acid, it is not clear if the decrease observed in aging is due to an increased dilution, actual decrease in hyaluronic acid concentrations or depolymerization of hyaluronic acid present in normal amounts.

The hyaluronate concentration of synovial fluid in the age group 18-29 years was 2.17 ± 0.3 mg/ml as compared to the value of 2.45 ± 0.04 mg/ml in the age group 27-39 years. The major change occurred between the ages of 22 and 27 years. The values were even slightly lower (2.37 ± 0.17 mg/ml) for 52-78 years (Balazs, 1977). Schiavetti et al. (1970) discussed the alterations in GAG metabolism as a function of age and arthrosis.

The bovine synovial fluid from carpal and hock joints contained increasing volumes of the fluid and hyaluronic acid contents from one month to two years of postnatal development and later remained unchanged (Seppala and Balazs, 1969).

12.6.4 Tuberculoarthritis and purulent arthritis

The carbazole reaction for hexuronic acid carried out directly on the diluted synovial fluid revealed that the hyaluronic acid content of the fluid was decreased in purulent and tuberculous types of arthritis (Ihogi et al., 1962).

12.6.5 Traumatic arthritis and effusions

The hyaluronate content of synovial fluid is reduced in traumatic arthritis (see Table 12-I for values and references). In recent traumatic arthritis, the hyaluronic acid concentration varies inversely with the degree of effusion, the hyaluronate is only slightly elevated after high effusion in osteoarthritis.

12.6.6 Synovial carcinoma

Cancerous tumors sometimes occur in the synovial tissues. Synovial sarcome siothelioma (also called synovial sarcoma or synovioma), hemangioma of the synovial membrane in knee joint and osteochondromatosis are known. Alterations in glycosaminoglycan composition and function of the synovial tissues are seen in these conditions. Meyer (1947) reported isolation of a large excess of hyaluronic acid from two cases of synovioma.

12.6.7 Synovitis

The hyaluronate contents of the synovial fluid in synovitis are, probably, the lowest, especially in the severe cases (Ropes, 1953). In granular synovitis, recurring mild hydroarthrosis is seen in children. The joint fluid becomes less and the synovial membrane becomes thick. In exudative synovitis, inflammation of the soft tissues and bone produces abscess which ruptures externally.

Table 12-I. Glycosaminoglycan concentrations (mg/100ml) of Human Synovial Fluid in Health and Disease

References	Normal	Rheumatoid arthritis	Osteoarthritis	Lupus erythematosus	Traumatic arthritis	Postmortem
Blix (1948)	100–200	–	–	–	–	–
Sundblad (1953)	297 (2)[a]	61.40	66 (42)	–	57 (40)	–
Fessler et al. (1954)	–	7.0 (7)	7.0 (1)	–	5.9 (14)	4.7 (1)
Bollet et al. (1956)	–	96 (23)	190 (3)	–	67 (6)	182 (8)
Hamerman and Schuster (1958)	360	–	–	–	–	–
Decker et al. (1959)	321±38 (8)	115±37 (28)	188±53 (9)	180 (1)	170±99 (7)	–
Stafford et al. (1964)	400 (8)	70 (21)	61 (40)	–	115 (15)	245 (13)
Balazs et al. (1967)	145–312 (268) (Ages 21–45 yrs)	<190 (85) (Ages 17–72 yrs)	–	–	–	–
Seppälä et al. (1972)	HA = 243 (9) ChS = 2.2% of total	HA = 18–235 (25) = 89.5±11.8 S.E. Ch-4S = 2.1–8.7 (25) = 4.1±0.6 S.E.	–	–	–	–
Greiling (1974)	255 (hyaluronate only) Enzymatic-U.V. at 232 nm	–	–	–	–	–
Greiling (1974)	174.7±8.84 (hyaluronate only) Enzymatic-colorimetry at 585 nm	–	–	–	–	–
Balazs (1977)	217±30 (Ages 18–29 yrs) 245±40 (Ages 27–39 yrs) 237±17 (Ages 52–78 yrs)	–	–	–	–	–
Jourdian et al. (1979)	202 (2) (hyaluronate only)	–	–	392 (1)	–	–

[a] The numbers in brackets denote the number of samples analyzed. HA = hyaluronic acid. ChS = chondroitin sulfate.

12.6.8 Lupus erythematosus

Jourdian et al. (1979) used a specific bacterial hyaluronidase-colorimetric procedure and also the borate-carbazole reaction on the same sample. They noted that in two normal synovial samples only 66 % of the total hexuronic acid was present as hyaluronic acid while in one Lupus erythematosus patient all hexuronic acid was due to hyaluronic acid.

12.6.9 Rheumatoid arthritis

Reports regarding changes in GAG composition of the synovial fluid have been conflicting. Some reports claimed an increase in the concentration of synovial fluid in patients with rheumatoid arthritis (Ragan and Meyer, 1949) and degenerative joint disease (Sundblad, 1953), which was present in an incompletely polymerized form (Blix, 1948) and probably resulted from accelerated synthesis with incomplete polymerization of hyaluronic acid (Sundblad, 1965). On the contrary, other reports (Hamerman and Schuster, 1958 a; Kofoed et al., 1973) have presented evidence for a decrease in hyaluronic acid. The chondroitin sulfate present in the synovial fluid of these patients as a proteoglycan (Wood et al., 1971) and the hyaluronic acid were both decreased to one-half the values (Seppala et al., 1972). The volume of synovial fluid is increased several folds in rheumatoid arthritis, the intrinsic viscosity and hyaluronate contents per unit volume are reduced, but the total hyaluronate of the joint cavity is increased. The low viscosity and the resulting loss of lubrication of joints is due to the fact that the hyaluronic acid present in the synovial fluid of rheumatoid patients is of lower molecular weights (Burke and Sonnenscheim, 1952; Barnett, 1958; Blumberg, 1958; Johansen and Sylvest, 1961; How and Long, 1969). This was intitially shown by comparison with viscosity of hyaluronidase degraded hyaluronic acid and later confirmed by fractionation of the cetylpyridinium complexes of this hyaluronic acid (Nanto et al., 1962) and by fractionation of the isolated hyaluronic acid by agarose gel chromatography (How and Long, 1969).

Suggestions have been made that the hyaluronic acid-hyaluronidase system may be important in rheumatoid arthritis (Rossing, 1952). However, the hyaluronidase activitiy was not abnormal in the synovial fluid (Ropes, 1957), and it was suggested that the presence of low molecular weight hyaluronic acid in the synovial fluid of the rheumatoid arthritis patients might be due to an enhanced synthesis of incompletely polymerized hyaluronate molecules by the β-type mesenchymal cells lining the synovium (Robinson, 1978). On the other hand, the synovial fluid from rheumatoid arthritis patients has been reported to contain abnormally high activities of the lysosomal enzymes, including collagenase (Lazarus et al., 1968), acid proteinase and acid hyaluronateglycan hydrolase (Fil'chagin et al., 1977), β-N-acetylglucosaminidase, β-glucuronidase and β-galactosidase (Bartholomew, 1972), hydrolases for chondroitin 4- and 6-sulfates and keratan sulfate (Greiling, 1972). The activities of some of these enzymes varied in positive correlation with the intensity of inflammation (Fil'chagin et al., 1977). The proteoglycan hydrolizing enzymes apparently were formed by the polymorphonuclear leukocytes (Greiling, 1972). An exoglycosidase, β-D-acetamidodeoxyglucosidase (EC 3.2.1.30) has also been reported to be present in abnormally high amounts in the synovial membranes of the patients (Hendry and Carr, 1963) and could degrade the hyaluronic acid. The reduction of the chondroitin sulfate proteolglycan to one-half the normal level and its presence in a degraded form in the synovial fluid of the patients with rheumatoid arthritis has been found to be due to the degradation of the chondromucoprotein by a neutral protease present in the synovial fluid of these patients (Wood et al., 1971). The activities of this protease and the chondroitin sulfate hydrolase (Greiling, 1972) may be responsible for increasing the ratio of hyaluronic acid/chondroitin sulfate from 2.0 in normal subjects to 3.0 (Kuwano, 1969) in the synovial fluid from the rheumatoid arthritis patients (also see Section 12.3).

Since ascorbic acid and other reducing agents depolymerize hyaluronic acid in vitro by an oxidation-reduction reaction (Sundblad and Balazs, 1966) it was suggested that such a reaction may be responsible for the presence of hyaluronic acid in a degraded form in the synovial fluid of patients with inflammatory joint disease. However, Caygill and Ali (1969) showed that this does not happen in vivo due to the presence of some inhibitors in the synovial fluid acting as scavangers for the free radicals. It must be mentioned here that a free radical type of depolymerization appears to have some significance in the pathogenesis process or in aggravating it, since topical application of a free-radical scavanger, dimethyl sulfoxide, has been found to inhibit the advance of the arthritis process.

It appears now that the changes in hyaluronic acid in the synovial fluid are associated with inflammation of the synovial tissue, which results in increased permeability of the synovial membrane and accelerated synthesis of incompletely polymerized hyaluronic acid by the inflammed membrane. On this basis a number of antiinflammatory drugs have been used with varying clinical improvements in patients. Intraarticular treatment with hydrocortisone acts either by the antiinflammatory action by directly or indirectly affecting hyaluronic acid synthesis in the synovial membrane for bringing about increased viscosity and in restoration of the changes in hyaluronic acid in rheumatoid arthritis patients (Jessar et al., 1953; Sundblad et al., 1954). Administration of the anti-inflammatory agent brings about a partial relief in reducing pain, swelling and tenderness. Examination of the hyaluronic acid isolated from synovial fluids by chromatography on agarose gel showed that the relief brought about by aspirin, indomethacin and prednisone was due to a decrease in the low molecular content of hyaluronic acid in the synovial fluids of the arthritis patients (How and Long, 1969). However, repeated articular injections of hydrocortisone have been reported to cause degeneration of the cartilage due to the loss of proteoglycans (hydrocortisone arthropathy).

Sulfated glycosaminoglycans (chondroitin sulfate, heparin and Arteparon (an oversulfated heparinoid) stimulate hyaluronic acid production in vitro. Injection of Arteparon into the synovial fluid of rheumatoid arthritis patients has been reported to inhibit hyaluronidase, β-glucuronidase and other enzymes, which increases hyaluronic acid content of synovial fluid by inhibiting breakdown and stimulating hyaluronic acid synthesis (Verbruggen and Veys, 1977; Momberg et al., 1976).

12.7 Alterations in Viscosity

The synovial fluid is a non-Newtonian fluid and shows the property of thixotropy, whereby its viscosity decreases as the shear rate (movement in joint) increases. The synovial fluid obtained from knee joints of normal persons and of persons with traumatic arthritis was thixotropic, while the fluid from rheumatoid arthritis cases was Newtonian and the viscosity was independent of time and shear rate or movement of joint (Dintenfass, 1966).

Several reports (Johnston, 1955, 1955 a) have shown a decrease in the viscosity of synovial fluid to be the cause of the loss of lubrication properties of synovial fluid in rheumatoid arthritis patients. It is agreed that this loss of lubrication is responsible for enhanced friction between opposing cartilages and their degeneration.

The viscosity of the normal synovial fluid is high because of the presence of the hyaluronic acid in this fluid. Since the viscosity is not altered upon proteolysis with papain or trypsin, the protein content does not contribute to viscosity. However, hyaluronidase digestion reduces the viscosity of the synovial fluid to that of water. Proteolysis reduces the boundary lubrication but hyaluronidase digestion does not reduce the lubrication ability as suggested by measurement of coefficient of friction. The decrease in boundary lubrication due to papain or trypsin digestion is probably due to partial breakdown of the lubricating glycoproteins such as LGP-I isolated by Swann et al. (1977, see section 12.3).

Analysis of viscosity and the nature of synovial mucin (or its content) is useful in differential diagnosis (Hirthe and Muhlbach, 1968) of traumatic and rheumatioid arthritis. The mucin (hyaluronate) content measured by acidification and washing and weighing of the dry mucin clot did not change in traumatic arthritis but was low in inflammation. Addition of acetic acid to normal synovial fluid results in formation of tight, ropy clump, which does not break upon agitation (Ropes' test). The precipitate formed in mild inflammation is soft and breaks on·shaking, while in severe inflammation the precipitate formed is cloudy. These differences in precipitation behavior are due to differences in GAG composition and molecular sizes of these macromolecules, being of a smaller size and of higher concentrations in the inflammatory states. These alterations in GAG also affect the viscosity which can be measured with a Hess viscometer. The synovial fluid from patients with inflammation is non-viscous and flows freely from the syringe (the syringe drip test).

The various laboratory diagnostic procedures for synovial fluid analysis in diseases of the joints have been reviewed (Ropes, 1953, 1957; Cohen, 1967; Chapchal, 1967; Greiling, 1974 a; Cohen et al., 1975).

12.8 Synovial membrane in rheumatoid arthritis

The glycosaminoglycans of the synovial membrane have been studied by electron microscopic-histochemical technics, when electron-opaque aggregates were observed in the intracellular matrix upon intravital perfusion with Alcian blue (Ohkura, 1966). The accumulation of hyaluronic acid in the synovial cells suggested that the lining cells were responsible for production of GAG (Maldyk and Kalczak, 1970) in rheumatoid arthritis. The intact synovial membranes and also cell cultures of these membranes showed characteristic differences, probably, due to the persistence of a virus in the rheumatoid cells (Hamerman and Smith, 1970). Rheumatoid synovial fibroblasts synthesized the same amount of hyaluronic acid but less sulfated GAG than did the normal synovial fibroblasts. In both cases, hydrocortisone added to culture had no effect on molecular weight but decreased hyaluronate synthesis by 50 % (Saarni et al., 1977). The use of hyaluronidase and papain in combination with ruthenium red staining indicated the presence of chondroitin sulfate and hyaluronate protein (Myers et al., 1969). The pattern of GAG distribution in patients was different, hyaluronate decreased while dermatan sulfate increased 1.5-2-fold as compared to normal subjects (Kittlick and Mueller, 1974).

12.9 Bibliography

1. Balazs, E. A. (1977): Intercellular matrix of connective tissue. In: C. E. Finch and L. Hayflick (Eds.): Handbook of the Biology of Aging. Van Nostrand Reinhold, New York, Chap. 9, pp. 222-240

2. Balazs, E. A., Watson, D., Duff, I. F. and Roseman, S. (1967): Hyaluronic acid in synovial fluid. Arthritis Rheum. 10, 357-376

3. Barker, S. A., Hawkins, C. F. and Hewins, M. (1966): Mucopolysaccharides in synovial fluid. Detection of chondroitin sulfate. Ann. Rheum. Dis. 25, 209-213

4. Barnett, C. H. (1958): Measurement and interpretation of synovial fluid viscosities. Ann. Rheum. Dis. 17, 229-233

5. Barnett, C. H., Davies, D. V. and MacConaill, M. A. (1961): Synovial Joints; their Structure and Mechanism. Thomas, Springfield

6. Bartholomew, B. A. (1972): Synovial fluid glycosidase activity. Scand. J. Rheumatology 1, 69-74

7. Berezkin, A. G. and Zhitnikov, A. Ya. (1976): Functional activity of cells of the synovial fluid. Dopov. Akad. Nauk Ukr. RSR, Ser. B. Geol. Khim. Biol. Nauki, 742-745

8. Blix, G. (1948): Some chemical observations on joint physiology and pathology. Acta Orthop. Scand. 18, 19-20

9. Blumberg, B. S. (1958): Synovial fluid in arthritis. Rheumatism 14, 37-41

10. Bollet, A. J., Gospodarek, M. and Rozhin, J. (1956): Intrinsic viscosity of synovial-fluid hyaluronic acid. J. Lab. Clin. Med. 48, 721-728

11. Bounassisi, V. and Ozzello, L. (1973): Sulfated mucopolysaccharide production by synovial sarcoma cells in vivo and in tissue culture. Cancer Res. 33, 874-881

12. Burke, W. and Sonnenscheim, A. (1952): Occurrence and distribution of mucopolysaccharide in the synovialis in pathological states of the knee joints. Arch. Pathol. Anat. Physiol. 322, 442-451

13. Castor, C. W. (1975): The physiology of the synovial cell and its contribution to disease processes. In: P. J. L. Holt (Ed.). Current Topics in Connective Tissue Disease. Churchill Livingstone, London, Chap. 1

14. Caygill, J. C. (1968): Glycosaminoglycans in joint Disorders. In: F. Dickens, P. J. Randle and W. J. Whelan (Eds.). Carbohydrate Metabolism and its Disorders. Academic Press, New York, Vol. II, Chap. 10, 240-263

15. Caygill, J. C. and Ali, Z. (1969): Degradation of synovial fluid and hyaluronic acid by serum and testicular hyaluronidase and by L-ascorbic acid and the effect of inhibitors. Clin. Chim. Acta 26, 395-400

16. Chapchal, G. (Ed.) (1967): Synovectomy and Arthroplasty in Rheumatoid Arthritis. 2nd Int. Symp. Jan. 27-29, 1967, Basel, Switzerland, Stuttgart, Thieme

17. Cohen, A. S. (1967): Synovial Fluid. In: A. S. Cohen (Ed.), Laboratory Diagnostic Procedures in Rheumatic Diseases. Little, Brown and Company, Boston, Chap. 1, 2-50

18. Cohen, A. S., Brandt, K. D. and Krey, P. R. (1975): Synovial Fluid. In: A. S. Cohen (Ed.). Laboratory Diagnostic Procedures in the Rheumatic Diseases. Little, Brown and Company, Boston, 2nd Ed., Chap 1, 1-62

19. Curtain, C. C. (1955): The nature of the protein in the hyaluronic complex of bovine synovial fluid. Biochem. J. 61, 688-696

20. Decker, B., McGuckin, W. F., McKenzie, B. F. and Slocumb, C. H. (1959): Concentration of hyaluronic acid in synovial fluid. Clin. Chem. 5, 465-469

21. Dintenfass, L. (1966): Rheology of complex fluids and some observations on joint lubrication. Fed. Proc. 25, 1054-1060

22. Ekholm, R. and Norback, B. (1951): On relationship between articular changes and function. Acta Orthopaed. Scand. 21, 81-98

23. Fessler, J. H., Ogston, A. G. and Stanier, J. E. (1954): Some properties of human and other synovial fluids. Biochem. J. 58, 656-660

24. Fil'chagin, N. M., Agababova, E. R., Nazarenko, V. M. and Chepoi, V. M. (1977): Lysosomal enzyme activities in the blood serum and synovial fluid in the articular cartilage. Ter. Arkh. 49, 120-123

25. Greiling, H. (1972): Biochemistry of rheumatoid arthritis. Verh. Deut. Ges. Rheumatol. 2, 26-37 (Ger)

26. Greiling, H. (1974): Hyaluronic acid. In: H. U. Bergmeyer (Ed.). Methods of Enzymatic Analysis, Vol. 3, 2nd English Ed., Academic Press, New York, pp 1158-1164

27. Greiling, H. (1974 a): Clinical biochemistry in rheumatoid arthritis. In: K. Seidel (Ed.). Immunologie und Biochemie der Rheumatoiden Arthritis. Jena, Friedrich-Schiller-Univ., pp. 143-159

28. Hamerman, D. (1970): Synovial joints: Aspects of structure and function. In: E. A. Balazs (Ed.). Chemistry and Molecular Biology of the Intercellular Matrix, Vol. 3, Academic Press, New York, pp. 1259-1277

29. Hamerman, D. and Sandson, J. (1960): The isolation of hyaluronate from human synovial fluid by zone electrophoresis. Nature 188, 1194-1195

30. Hamerman, D. and Schuster, H. (1958): Hyaluronate in normal human synovial fluid. J. Clin. Invest. 37, 57-64

31. Hamerman, D. and Schuster, H. (1958 a): Synovial fluid hyaluronate in rheumatoid arthritis. Arthritis Rheum. 1, 523-531

32. Hamerman, D. and Smith, C. (1970): Normal and rheumatoid synovial membrane cells in tissue culture, In: E. A. Balazs (Ed.). Chemistry and Molecular Biology of the Intercellular Matrix, Vol. 3, Academic Press, New York, pp. 1503-1506

33. Hedberg, H. and Moritz, U. (1958): Biosynthesis of hyaluronic acid in tissue cultures of human synovial membrane. Proc. Soc. Exptl. Biol. Med. 98, 80-84

34. Hendry, N. G. C. and Carr, A. J. (1963): A glycosidase abnormality in synovial membrane in joint disease. Nature 199, 392

35. Hirai, K. (1973): Acid mucopolysaccharide synthesis in the inflamed synovial membranes (with special reference to chronic rheumatoid arthritis). Ryumachi 12, 355-365 (Jap)

36. Hirthe, D. and Muhlbach, R. (1968): Differential diagnosis of congenital systemic diseases of the skeleton based on biochemical changes in the body fluids. Beitr. Orthop. Trauma 15, 653-658

37. How, M. J. and Long, W. J. (1969): Separation and fractionation of hyaluronic acid from human synovial fluid on agarose gels. Clin. Chim. Acta 23, 251-256

38. Ihogi, M., Matsunaga, M. and Aoki, T. (1962): Chemical estimation of hyaluronic acid in synovial fluid and chondroitin sulfate in articular cartilage. I. Hyaluronic acid in synovial fluid and II. Chondroitin sulfuric acid in articular cartilage. In. F. Egami and Y. Oshima (Eds.). Biochemistry and Medicine of Mucopolysaccharides, Maruzen, Tokyo, pp. 183-186 and 186-189

39. Iwata, H. (1972): Chronic rheumatoid arthritis and mucopolysaccharides. Ryumachi 12, 10-12 (Jap)

40. Jessar, R. A., Ganzell, M. A. and Ragan, C. (1953): The action of hydrocortisone in synovial inflammation. J. Clin. Invest. 32, 480-482

41. Johansen, P. E. and Sylvest, O. (1961): Synovial fluid changes in degenerative joint disease, rheumatoid arthritis, and traumatic arthritis. Acta Rheumatol. Scand. 7, 240-248

42. Johnston, J. P. (1955): A comparison of the properties of hyaluronic acid from normal and pathological human synovial fluids. Biochem. J. 59, 626-633

43. Johnston, J. P. (1955 a): The viscosity of normal and pathological human synovial fluids. Biochem. J. 59, 633-637

44. Jourdian, G. W., Wolfman, M., Sarber, R. and Distler, J. (1979): A specific, sensitive method for the determination of hyaluronate. Anal. Biochem. 96, 474-480

45. Kittlick, P. D. and Mueller, P. (1974): Nonspecific variations in the distribution patterns of acid mucopolysaccharides in the synovial membrane in rheumatoid arthritis. Immunol. Biochem. Rheumatoiden Arthritis, Symp. Int. Beteiligung Ges. Rheumatol. DDR 4th 1973, pp. 234-242 (Ger)

46. Kofoed, J. A., Tocci, A. A. and Barcelo, A. C. (1973): The acidic glycosaminoglycans of the synovial fluid in rheumatoid arthritis. Experientia 29, 680-681

47. Kuwano, S. (1969): Study of acid mucopolysaccharides in the pathological synovial fluid. Igaku Kenkyu 39, 199-221

48. Lazarus, G. S., Decker, J. M. and Oliver, C. H. (1968): Collagenolytic activity of synovium in rheumatoid arthritis. N. Eng. J. Med. 279, 914-919

49. Maldyk, E. and Kalczak, M. (1970): Content and distribution of acid mucopolysaccharides in the articular synovium in rheumatoid arthritis. Pol. Tyg. Lek. 25, 520-522

50. Mathews, M. B. (1966): Animal Mucopolysaccharides. Methods in Enzymol. 8, 654-662

51. McCarty, D. J. (Ed.) (1979): Arthritis and Allied Conditions. A Textbook of Rheumatology, 9th Ed., Lea and Febiger, Philadelphia

52. McCarty, D. J. (1979 a): Synovial fluid. In: McCarty (1979): Chap. 5, pp. 51-70

53. Meyer, K. (1947): The biological significance of hyaluronic acid and hyaluronidase. Physiol. Revs. 27, 335-359

54. Meyer, K., Smyth, E. M. and Dawson, M. H. (1939): The isolation of a mucopolysaccharide from synovial fluid. J. Biol. Chem. 128, 319-327

55. Mikio, I. (1975): Effects of several anti-inflammatory steroids on the biosynthesis of acid mucopolysaccharides in pig synovial membranes in vitro. Showa Igakkai Zasshi 35, 325-331 (Japan)

56. Momburg, M., Stuhlsatz, H. W., Voegeli, H., Voitisek, O., Eylau, O. and Greiling, H. (1976): Clinical-chemical changes in the synovial fluid after intraarticular injection of glycosaminoglycan polysulfates. Verh. Dtsch. Ges. Rheumatol. 4, 383-390

57. Muir, H. (1971): Structure and metabolism of proteoglycans and glycosaminoglycans in relation to rheumatoid arthritis. In: W. Mueller (Ed.). Rheumatoid Arthritis, Proc. Int. Symp., Academic Press, pp. 45-67

58. Mueller, W., Harwerth, H.-G. and Fehr, K. (Eds.): Rheumatoid Arthritis, Academic Press, New York, 1971

59. Myers, D. B., Highton, T. C. and Rayns, D. G. (1969): Acid mucopolysaccharides closely associated with collagen fibrils in normal human synovium. J. Ultrastruct. Res. 28, 203-213

60. Nanto, V., Seppala, P. and Kulonen, E. (1962): Effect of hydrocortisone administration on the hyaluronic acid fractions of synovial fluid in rheumatoid arthritis. Clin. Chim. Acta 7, 794-799

61. Ogston, A. G. and Stanier, J. E. (1949): The state of combination of hyaluronic acid in synovial fluid. First Int. Congr. Biochem. Cambridge, Eng. Abstracts, pp. 144-145

62. Ogston, A. G. and Stanier, J. E. (1950): On the state of hyaluronic acid in synovial fluid. Biochem. J. 46, 364-376

63. Ogston, A. G. and Stanier, J. E. (1951): Dimensions of the particle of hyaluronic acid complex in synovial fluid. Biochem. J. 49, 585-590

64. Ogston, A. G. and Stanier, J. E. (1953): Composition and properties of hyaluronic acid complex of ox synovial fluid, Discussions Faraday Society No. 13, 275-280

65. Ohkura, T. (1966): Electron microscopic-histochemical study of acidic mucopolysaccharides in the dog synovial membrane. J. Electron Microsc. (Tokyo) 15, 167-178

66. Oshima, Y., Mizushima, Y., Sasaki, S., Haruke, F., Toyama, M., Yokoyama, Y., Furuichi, M., Yoshizaki, T., Yokohari, R., Umezono, K., Otsuki, A., Noda, M. and Miyamoto, T. (1962): Clinical and experimental studies on mucopolysaccharides. Mucopolysaccharide and rheumatic diseases. In: F. Egami and Y. Oshima (Eds.). Biochemistry and Medicine of Mucopolysaccharides, Maruzen, Tokyo, pp. 198-212

67. Preston, B. N., Davies, M. and Ogston, A. G. (1965): The comparison and physicochemical properties of hyaluronic acids prepared from ox synovial fluid and from a case of mesothelioma. Biochem. J. 96, 449-474

68. Radin, E. L. (1973): The physiology and degeneration of joints. Seminars Arthritis Rheum. 2, 245-257

69. Radin, E. L. and Paul, I. L. (1972): A consolidated concept of joint lubrication. Bone Joint Surg. 54A, 607-613

70. Ragan, C. and Meyer, K. (1949): The hyaluronic acid of synovial fluid in rheumatoid arthritis. J. Clin. Invest. 28, 56-59

71. Robinson, D. R. (1978): Connective tissue and bone diseases. In: J. H. Shaw, E. A. Sweeney, C. C. Cappuccino and S. M. Miller (Eds.). Textbook of Oral Biology. W. B. Saunders, Philadelphia, Chap. 28, pp. 917-927

72. Ropes, M. W. (1953): Synovial Fluid Changes In Joint Diseases. Harvard University Press, Cambridge, 151 pp

73. Ropes, M. W. (1957): Examination of synovial fluid. Bull. Rheum. Dis.6 21-22

74. Ropes, M. W., Roberston, W. v. B., Rosmeisl, E. C., Peabody, R. B. and Bauer, W. (1947): Synovial fluid mucin. Acta Med. Scand. 128, (Suppl. 196), 700-744

75. Rossing, P. (1952): The importance of the hyaluronic acid-hyaluronidase system in rheumatoid joint diseases. Z. Rheumaforsch. 11, 176-180

76. Roy, S. and Ghadially, F. N. (1967): Synthesis of hyaluronic acid by synovial cells. J. Pathol. Bacteriol. 93, 555-557

77. Saarni, H., Tammi, M. and Vuorio, E. (1977): Effects of cortisol on glycosaminoglycans synthesized by normal and rheumatoid synovial fibroblasts in vitro. Scand. J. Rheumatol. 6, 222-224

78. Sandson, J. and Hamerman, D. (1962): Isolation of hyaluronateprotein from human synovial fluid. J. Clin. Invest. 41, 1817-1830

79. Sandson, J., Hamerman, D. and Schuster, H. (1960): Nondialyzable hexose of human synovial fluid. J. Clin. Invest. 39, 782-789

80. Schiavetti, L., Porzio, F. and Tuzi, T. (1970): Mucopolysaccharide metabolism as a function of age and arthrosis. J. Belge. Rheumatol Med. Phys. 25, 72-81 (Fre)

81. Seppala, P. (1964): Synovial fluid in rheumatoid arthritis. Physicochemical and chemical properties of hyaluronic acid and proteins with reference to the effect of corticosteroids. Scand. J. Clin. Lab. Invest. 16 (Suppl. 79), 1-107

82. Seppala, P. and Balazs, E. A. (1969): Hyaluronic acid in synovial fluid. III. Effect of maturation and aging on the chemical properties of bovine synovial fluid. J. Gerontol. 24, 309-314

83. Seppala, P. O., Karkkainen, J., Lehtonen, A. and Makisara, P. (1972): Chondroitin sulfate in the normal and rheumatoid synovial fluid. Clin. Chim. Acta 36, 549-553

84. Sinkin, P. A. (1979): Synovial Physiology. In: D. J. McCarty (Ed.). Arthritis and Allied Conditions, 9th Ed., Lea and Febiger, Philadelphia, pp. 167-178

85. Sokoloff, L. (Ed.) (1978): The joints and synovial fluid. Vol. I., Academic Press, New York, 491 pp.

86. Stafford, C. T., Niedermeier, W., Holley, H. L. and Pigman, W. (1967): Hyaluronic acid in synovial fluid. I. Molecular parameters of hyaluronic acid in normal and arthritic human fluids. Arthritis and Rheumatism 10, 357-376

87. Sundblad, L. (1953): Studies on hyaluronic acid in synovial fluids. Acta Soc. Med. Upsalien. 58, 113-228

88. Sundblad, L. (1965): Glycosaminoglycans and glycoproteins in synovial fluid. In: E. A. Balazs and R. W. Jeanloz (Eds.). The Amino Sugars. Vol. IIA. Academic Press, New York, Chap. 29, pp. 229-250

89. Sundblad, L. and Balazs, E. A. (1966): Chemical and physical changes of glycosaminoglycans and glycoproteins caused by oxidation-reduction systems and radiation. In: E. A. Balazs and R. W. Jeanloz (Eds.). The Amino Sugars, Vol. IIB. Academic Press, New York, Chap. 47, pp. 229-250

90. Sundblad, L., Egelius, N. and Jonsson, E. (1954): Action of hydrocortisone on hyaluronic acid of joint fluids in rheumatoid arthritis. Scand. J. Clin. Lab. Invest. 6, 295-302

91. Swann, D. A. (1978): Macromolecules of synovial fluid. In: L. Sokoloff (Ed.). Joints and Synovial Fluid. Academic Press, New York, pp. 407-435

92. Swann, D. A. and Radin, E. L. (1972): The molecular basis of articular lubrication. I. Purification and properties of a lubricating fraction from bovine synovial fluid. J. Biol. Chem. 247, 8069-8073

93. Swann, D. A., Radin, E. L., Nazimee, M., Weisser, P. A., Curran, N. and Lewinnek, G. (1974): Role of hyaluronic acid in joint lubrication. Ann. Rheum. Dis. 33, 318-326

94. Swann, D. A., Sotman, S., Dixon, M. and Brooks, C. (1977): The isolation and partial characterization of the major glycoprotein (LGP-I) from the articular lubrication fraction from bovine fluid. Biochem. J. 161, 473-485

95. Swanson, S. A. V. (1979): Friction, Wear and Lubrication. In: M. A. R. Freeman (Ed.). Adult Articular Cartilage. Pittman Publ. Co., Kent, 2nd Ed., Chap. 6, pp 415-460

96. Usenko, V. P. (1967): Polysaccharide metabolism in the cells of the developing synovial membrane. Tsitol. Genet., Akad. Nauk Ukr. SSR 1, 56-60

97. Usenko, V. P. (1971): Peculiarities of polysaccharide metabolism in the cells of the synovial membrane being formed. In: A. D. Soboleva (Ed.). Svrem. Biokhim. Morfol. Probl. Soedin. Tkani, Mater. Soveshch. Soedin. Tkani, 3rd 1968. Nauka Sib. Otd.; Novosibirsk, USSR, pp. 171-174

98. Verbruggen, G. and Veys, E. M. (1977): Influence of sulfated glycosaminoglycans upon proteoglycan metabolism of the synovial lining cells. Acta Rheumatol. Belg. 1, 75-92

99. Wood, G. C., Pryce-Jones, R. H., White, D. D. and Nuki, G. (1971): Chondromucoprotein-degrading neutral protease in rheumatoid synovial fluid. Ann. Rheum. Dis. 30, 73-77

100. Wright, V., Dowson, D. and Kerr, J. (1973): The structure of joints. Int. Rev. Connective Tissue Res. 6, 105-125

101. Yielding, K. L., Tomkins, G. M. and Bunin, J. J. (1957): Synthesis of hyaluronic acid by human synovial tissue slices. Science 125, 1300

Chapter 13

Mucopolysaccharides of Physiological and Pathological Fluid Accumulations (Effusions)

13.1 Introduction

Accumulations of fluids occur in physiological and pathological states in lined or unlined cavities in the body. Such accumulations include the physiological edematous exudate from the sexual skin of some animals; puncture fluids such as cyst fluids, blister fluids, edema fluids and infiltrates from the leprous skin. Serous fluids accumulate as pathological effusions in serous cavities. These pathological effusions include the pleural effusions, ascitic or peritoneal fluids and pericardial fluid.

13.2 Glycosaminoglycans in extracellular fluids of tumors

Several types of advanced carcinomas of different tissues are associated with tumor fluids which are rich in glycosaminoglycans. During tumor growth, the tumor-host interzone shows an increased degradation of the ground substance in the peritumoral connective tissue. The destruction of the peritumoral connective tissue by the invading tumor cells has been related to the action of hyaluronidase in the tumor extracts, the presence of which was suspected by the spreading of Indian ink and diptheria toxin (Chain and Duthie, 1939; Pirie, 1942; Fiszer-Szafarz, 1968). The release of the degradation products into the blood resulted in the elevation of GAG in blood plasma. In the blood of the tumor bearing animals, these degradation products of hyaluronic acid act as competitive inhibitors of hyaluronidase, thereby preventing further degradation of these products (Fiszer-Szafarz and Guillino, 1970). Since the exudates and transudates are formed from plasma, the GAG changes in plasma are also reflected in these pathological fluids. Simultaneous measurement of the glycosaminoglycans in plasma and ascitic fluid of rat during experimental tumor growth gave a significant positive correlation between GAG concentration in the two fluids but a negative correlation was found between the glycosaminoglycan content of the ascitic fluid and tumor size (Skyvova et al., 1973).

In the connective tissue adjacent to the epithelial tumors of the skin, the increased levels of acidic glycosaminoglycans and the mucolytic action of the neoplastic cells were the main factors in the invasiveness of the tumors. The accumulation of the acidic glycosaminoglycans in the peritumoral tissue may be a defense reaction of the adjacent connective tissue caused by a variability in the neoplastic tissue structures and proteolytic enzymes (Wozniak, 1965). In an extensive review on the mechanism of the invasiveness of cancer (Coman, 1947), the invasiveness of the cancer was reported to depend on decreased adhesiveness of cancer cells, amoeboid movement of cancer cells and the liberation of hyaluronidase. The enzyme so liberated hydrolyzed the hyaluronic acid of the intercellular cement substance of the connecteive tissue to open the way for penetration of malignant cells.

Although the peritumoral tissue contains, besides hyaluronic acid, other sulfated glycosaminoglycans, hyaluronic acid is characteristically affected and elevated in physiological (urine, blood) and pathological fluids (exudates, transudates). Since the first demonstration of a large quantity of hyaluronic acid in the viscous pleural fluid of a patient with a malignant tumor of the pleura and peritoneum (Meyer and Chaffee, 1940), several other reports have appeared. Hyaluronic acid has been found in increased amounts in tumor extracellular fluids of two fowl tumors, Rous No. 1 and Fujinami myxosarcoma (Pirie, 1942), in tumor fluid of Walker carcinoma of rat (Fiszer-Szafarz and Guillino, 1970 a) and other tumor extracellular fluids (Fiszer-Szafarz and Szafarz, 1974), in transplantation tumors (Buddecke and Sittner, 1966), in meningiomas (Aruna et al., 1973), nervous tissue (Smith and Butler, 1973), in dermal carcinoma (Majewska and Majewski, 1966; Asboe-Hansen and Levi, 1959). Acidic glycosaminoglycans were histochemically demonstrated in greater amounts in malignant tumors than in benign tumors (Hieronymi, 1954). In soft and hard fibromas the neutral mucopolysaccharides predominated, whereas the fibrosarcomas contained acidic glycosaminoglycans (Zhuravleva, 1974).

All the normal tissues contained two or more acidic glycosaminoglycans in contrast to tumors of mesodermal origin which produced only hyaluronate (Davidson et al., 1956). The electrophoretic patterns of all the spontaneous neoplasias differed in patterns of the GAG from the tissue of their origins (Chiarugi et al., 1978). In metastasizing carcinomas the glycosaminoglycan composition of a cancerous tissue may be different from that of the tissue in the healthy state, since most tumors retain the capacity of forming polysaccharide material of their tissue of origin (Hakomori, 1959; Nigam and Cantero, 1972). By histochemical analysis on 60 tumors, the connectivum tumors were found to be richer in glycosaminoglycans than the epithelial tumors and these macromolecules were concentrated in the peripheral part of the metastasis or tumor tissues and the parenchymal cells adjacent to the tumor (Giarelli, 1955).

Although hyaluronic acid is the only glycosaminoglycan in some tumors like myxoma (Houngbadji et al., 1974) or the predominant glycosaminoglycan in the extracellular fluid of the majority of tumors, sulfated glycosaminoglycans are also present in some types of tumors. The Rous fowl sarcoma contains a sulfated glycosaminoglycan in addition to hyaluronic acid (Asboe-Hansen and Levi, 1959) and Chiba chicken sarcoma contained 11.4 % H_2SO_4 in addition to 31.2 % hexosamine and 37 % glucuronic acid (Fujita, 1954). The carcinoma of the cervix contained a small amount of sulfated macromolecules in addition to an appreciable amount of hyaluronic acid (Bradbury et al., 1970). A human carcinoma of the pancreas contained glycosaminoglycans consisting of 90 % hyaluronic acid and 10 % sulfated glycosaminoglycans (Cudkowicz, 1956). Sulfated glycosaminoglycans are characteristically produced in malignant vascular tumors of the skin and this is used as a diagnostic test (Girard et al., 1970). In vinyl chloride-induced hepatic angiosarcoma the acidic glycosaminoglycans were drastically elevated in the angiosarcoma tissue and this increase in various fractions was in the order of heparin > hyaluronic acid > chondroitin sulfate (Kupchella and Tamburro, 1978). The presence of increased amounts of hyaluronic acid and the sulfated glycosaminoglycans in various tumors and tumor fluids have been attributed to the presence of the hyaluronidase acitivity in an inactive form in the lysosomes and the extracellular fluid of tumor cells (Fiszer-Szafarz and Szafarz, 1974). This, probably accounts for the apparent non-detection of hyaluronidase by histochemical methods. The dialyzable factor present in normal lysosomal extracts can activate this inactive enzyme.

The studies on mucoproteins and mucopolysaccharides (Hakomori, 1959), on glycolipids and glycoproteins (Nigam and Cantero, 1973) and on mucopolysaccharides (Nigam and Cantero, 1972) in relation to cancer have been reviewed extensively.

13.3 Blister fluids

The blister fluids obtained from certain skin diseases and from blisters artificially induced on skin with CO_2 contained 11.4-13.4 mg/100 ml carbohydrates (measured as galactose plus mannose) in the mucoprotein fraction. The total protein-bound carbohydrates in the blister fluid were 67-165 mg/100 ml as compared to that in blood serum of 120-184 mg/100 ml (Caccalianza, 1954). Another report (Zannini and Scanu, 1955) on the determination of mucopolysaccharides in 14 normal and pathological cases of various types of chronic edema of the limbs confirmed lower values in blister fluid than in serum. In cantharidin blisters, produced by application of cantharidin plaster on the skin of patients with pleural exudates, the blister fluid contained a protein mixture which was very similar in its hexose content to the protein present in the pleural exudate (Lustig and Nassau, 1941). We have not come across any report on the presence of acidic glycosaminogylcans in blister fluids. The available information has been presented in this section to serve as a starting point for an investigator desiring of exploring this avenue.

13.4 Cyst fluids

Pathological collections of fluids occur in some tissues in cavities which are usually lined. These fluid-containing pathological cavities are called cysts. Several different types of blood-group substances, other glycoproteins and glycosaminoglycans have been reported in fluids from cysts of different origins.

The fluids from some cysts have been analyzed for glycosaminoglycans. This fluid is very viscous and loses viscosity upon digestion with hyaluronidase (Skaug and Hofstad, 1972). Some cysts formed in the oral cavity (see Section 5.4) are lined by a cyst wall made up of granulation tissue. The mucosal cysts of the maxillary sinus are detected as well-outlined dome-shaped radiopaque areas in panoramic radiographs (Halstead, 1973) due to mucus retention. In the maxillary jaw cysts this collected fluid contains the glycosaminoglycans originating from osteolysis of the maxillary bone and contains the hyaluronidase-digestible and hyaluronidase-resistant electrophoretic bands corresponding to hyaluronic acid and an acid glycoprotein, respectively (Skaug and Hofstad, 1972). In fluids collected from non-keratizing jaw cysts glycosaminoglycans are present as complexes with glycoproteins (Skaug, 1975).

A number of studies have been reported on the fluids from cysts of human ovaries. From the ovarian cyst mucoids, pseudomucin and paramucin, chitosamine was isolated in 1900 by Zängerle and in 1902 by Steudel, respectively (Gottschalk, 1966). Odin (1959) reviewed the chemistry of the cyst fluids and presented analyses of 79 cyst fluids belonging to different types: Pseudomucinous cystadenoma, serous cystadenoma, cystadenocarcinoma; follicular, lutein, endometrial, and dermatoid cysts; cystic fibroma and myoma; par-ovarian and tubo-ovarian cysts; appendeal mucocele; and pseudomyxoma peritonii. No hyaluronic acid was detected on the basis of viscosity-reducing action of added testicular hayluronidase, yet a small amount of uronic acid-containing glycosaminoglycan was present in the precipitate obtained with ethanol.

The pathological fluid from pseudomucinous ovarian cysts is far richer a source of blood group substances than any normal body fluid and has provided yields in grams (Watkins, 1966). A glycoprotein of molecular weight 260,000 having blood group A activity was isolated from human ovarian cyst fluid (Aminoff et al., 1950). It contained 11 amino acids, fucose 18 %, galactose 17 % and N-acetylhexosamine 46 %. The pseudomucinous ovarian cyst fluid was found to contain two types of glycoproteins: Neutral glycoproteins (fucomucins) and acidic glycoproteins or sialomucins (Odin, 1959). The fucomucins were water-miscible, contained the blood-group activity and the carbohydrate components were fucose, hexose (mainly galactose and a small amount of mannose), hexosamine (glucosamine/galactosamine = 0.8-> 10) and sialic acid (N-acetylneuraminic acid) in proportions of 1:1:1:0.1. The sialomucins were water-insoluble gels and contained the corresponding carbohydrate components in proportions of 0.2:0.6:1:0.6 (Odin, 1959).

In contrast with Odin (1959) who detected glucuronic acid but no hyaluronic acid, Jensen (1954) isolated a hyaluronic acid of molecular weight $4.53 - 4.61 \cdot 10^5$ in yields of 42.9 mg/65 ml and 37.9 mg/98 ml from two pseudomucinous ovarian cysts, Cystadenoma pseudomucinosum ovarii I and Cystis pseudomucinosum ovarii II, respectively. No hyaluronic acid was detected in another pseudomucinous ovarian cyst and three non-pseudomucinous ovarian cysts.

Three glycopeptides and a glycosaminoglycan were isolated from a human pancreatic cyst by DEAE-cellulose chromatography of the protease-digested material. All these four fractions contained fucose, mannose, glucose, galactose, glucosamine, galactosamine, sialic acid and amino acids. The glycosaminoglycan, which eluted in the 0.45 M NaCl fraction, contained hexuronic acid in addition to the above components (Masuda et al., 1977). A proteoglycan of hyaluronic acid was isolated from a cyst of cystic mucoid degeneration (Endo et al., 1973). Hyaluronic acid and dermatan sulfate were detected in the fluid from an ocular cyst (Blumenkrantz, 1959). These macromolecules were identified by their susceptibility and resistance to hyaluronidase digestion, respectively, and by their infrared spectra. From turbidimetric measurement, the hyaluronidase-digestible and hyaluronidase-resistant materials accounted for 77.5 % and 22.5 % of the total precipitable material, respectively.

13.5 Glycosaminoglycans of edema fluids

Under certain pathological states, abnormally large amounts of fluids are accumulated in the intercellular tissue spaces of the body producing edema. Edema can result from factors that directly or indirectly disturb the fluid dynamics of the body (Guyton et al., 1975). Although the edema fluids are ultrafiltrates of plasma, yet some of the edematous fluids contain a large amount of hyaluronic acid. Such is the case with myxedema (Sisson, 1968; Watson and Pearce, 1947), hydrops in inner ear (Poulsen, 1966, also see Section 10.9), hydrops in other tissues seen in hypothyroidism (Asboe-Hansen, 1966), and pulmonary edema.

In pulmonary edema, serous fluid effuses into the air vesicles and interstitial tissue spaces in the lungs due to an abnormality in the interstitial fluid equilibrium (Guyton et al., 1975; Dickinson, 1976). Protein-glycosaminoglycan complexes (proteoglycans) which make only 1 % of the dry weight of the lung are present in the connective tissue filling the minute spaces between the epithelial linings of the adjacent alveoli (Hance and Crystal, 1975). These proteoglycans act as a matrix for accumulation of the fluid in the lung in pulmonary edema (Prockop, 1979). The fluid from the various types of hydrops and myxedema is very rich in hyaluronic acid due to an increased synthesis of these anionic macromolecules in hypothyroidism which then bind a large excess of water (Poulsen, 1966; Asboe-Hansen, 1966).

13.6 Edema fluid from sexual skin

The skin around the external genital of some genera of primates (macaque, baboon, chimpanzee, etc.) undergoes changes of tumescence (swelling) and involution in response to changes in increased levels of estrogens during the tumescent phase on sexual maturation and decreased blood estrogen levels during involution. During the tumescent sexual phase, the sex skin becomes red, smooth, shiny and so edematous that on squeezing or cutting a highly viscous fluid drips. The exudate of edematous sexual skins of Rhesus monkeys, pig-tail monkeys and baboons contained hyaluronic acid, which lost viscosity upon digestion with streptococal and testicular hyaluronidase, contained equimolar amounts of hexuronic acid and glucosamine, but no sulfate (Rienits, 1960). In the three species the swelling was not related to the degree of polymerization of hyaluronic acid but was due to the increased content of hyaluronic acid during the sexual phase. A correlation between increased hyaluronic acid and water contents during the swelling of the sexual skin was noted. The chemical analyses for glycosaminoglycans in this tumescent fluid have been reviewed (Szirmai, 1966). Its carbohydrate content is 3.7 % with glucosamine being the only hexosamine at a concentration of 2-3 % and glucuronic acid was detected by paper chromatography. Recently the cyclic rejuvenation of the sexual skin has been used as a model for aging studies (Freiherr, 1979).

13.7 Leprous infiltrates

Leprosy is a communicable disease caused by Mycobacterium leprae, which produces various granulomatous lesions in the skin, the mucous membranes and the peripheral nervous system. Hyaluronidase activity is absent in the skin of patients with lepromatous leprosy and this activity is inversely proportional to the amount of Mycobacterium leprae present in the leprous skin (Mom and Basombio, 1944). This results in a persistent accumulation of hyaluronic acid in lepra cells of lepromas. In contrast, the tuberculoid epithelial cells contain a minimum hyaluronic acid which decreases as these age, thereby suggesting involvement of hyaluronic acid in host/parasite interaction in leprosy (Skinsnes and Matsuo, 1974). Hyaluronic acid accumulates in the limiting membranes of lepra cells, while β-glucuronidase is present in Mycobacterium leprae but is absent in lepra cells. These facts indicate that hyaluronic acid may originate in the histiocytes and the absence of the degradative enzyme β-glucuronidase in lepra cells may enable the bacilli to utilize GAG as a nutrient (Matsuo and Skinsnes, 1974).

13.8 Mucopolysaccharides of serous effusions

13.8.1 Serous cavities and their fluids

The heart, lungs and the visceral organs are enclosed by and contained in the pericardial, pleural and peritoneal cavities, respectively. These cavities are serous cavities lined by serous membranes, which are thin layers of connective tissue covered by a single layer of mesothelium. The cells of the mesothelium are oval or round cells, 10-20 μm in diameter and in normal health secrete serous fluids in quantities just enough to act as a lubricating agent for movements of the organs contained in those cavities. However, these mesothelial cells are very sensitive and in response to the slightest irritation, chemical agents, trauma, radiation or malignancy are transformed into large, active forms of cells with multinuclei, which produce increased amounts of serous fluids. The serous fluids accumulated in pathological states in the serous cavities are the pleural, pericardial and peritoneal (ascitic) fluids. These fluids can be called transudates or exudates based on the properties and the etiological origin of the fluid.

13.8.2 Nature and source of malignant and benign transudates and exudates

Transudates are ultrafiltrates of the plasma. These fluids accumulate in the serous cavities as a result of increased hydrostatic pressure in capillaries (as in congestive heart failure) or decreased osmotic pressure (as in nephrotic syndrome). A combination of the two factors causes the accumulation of the ascitic fluid in the peritoneal cavity and involves an increased portal vein pressure caused by fibrotic scarring (hepatic cirrhosis) and a low plasma osmotic pressure accompanying hypoalbuminemia. The pores in the capillary membranes do not allow filtration of molecules of molecular weights higher than 200,000 daltons. For this reason the transudates do not contain fibrinogen (340,000 M_r) and hence do not clot on standing. Their protein contents are less than 2 g/dl and their specific gravities are below 1.015. The various causes of the transudate accumulation have been summarized

(Takahashi, 1971) as cardiac or renal insufficiency, hypoproteinemia and hepatic cirrhosis. In chronic renal disease the transudation of pleural fluid occurs not because of the circulatory failure but due to the changes in composition which facilitate filtration through the capillary walls.

Exudates are the fluids that accumulate when the serous membranes are injured due to benign or malignant inflammatory processes. The benign inflammation could result from direct trauma to the serous cavities or from the spreading of infeciton from the organ inside the serous cavity or infection from an organ outside the serous cavity (Oizima et al. 1969) carried into the cavity through the circulating blood or lymph. In each case the serous membrane becomes more permeable to molecules of higher molecular weights giving a protein content of more than 3 g/dl and specific gravity more than 1.018. Since fibrinogen is present, these fluids clot on standing. The composition of these exudates varies with the nature and virulence of the infecting agent, duration of the infection (acute or chronic), response of the host, etc. Examples of exudates from acute inflammation are myocarditis with pericardial effusion due to a viral infection; purulent pleural fluid in pleurisy (empyema) from pneumococcal infection in pneumonia; chronic pleural effusion in chronic pulmonary diseases, bronchopneumonia, pulmonary tuberculosis; pericardial effusion in rheumatic heart disease; pleural effusion in rheumatoid arthritis; ascites in chronic peritonitis etc. The inflammatory exudate persists till the etiological agent has been eliminated, following which the exudates may be absorbed or these find a natural route of escape such as through perforation into a bronchus.

Malignant exudates are the fluids accumulating in the serous cavities as a result of the primary tumors of the serous cavities (called mesotheliomas) or metastasis of a malignant tumor in any other organ to serous cavities, usually through lymphatic or blood flow. Mesotheliomas can occur in pericardial, pleural and peritoneal cavities. These mesotheliomas are present as nodules or as thick plaque-like tumor masses which interfere with the movements of the lungs or the heart. Carcinomas (adenocarcinomas) of the breast and the lungs metastasize to the pleural cavities. The gastrointestinal carcinomas (carcinomas of esophagus, stomach, large intestines) metastasize to pleural and peritoneal cavities. The carcinomas of the liver, gall bladder and pancreas sometimes metastasize to the peritoneal cavity producing ascites with clumps of tumor cells; hapatoma cells from ascites in patients with liver cirrhosis is an example. Carcinomas of the genitourinary system (e. g. ascites in renal tubule carcinoma) cervix, uterus, prostate and testis also, rarely, metastasize to the serous cavities. Pleural and peritoneal exudates are frequently encountered in malignant blood dyscrasias as a result of metastasis of lymphomas, lymphosarcomas, thymoma, multiple myelomas, etc.

More than one serous cavity may be involved in a disease state and the unrelated mesotheliomas may bear very close resemblances. For example, synovial tumors resemble pleural mesotheliomas. The simultaneous involvement of the various serous cavities and their morphological resemblances are related to their similar embryogenic origin. All the serous cavities except the peritoneal cavity are slit-like spaces and arise in an analogous manner by splitting of the embryonic mesenchyme. The pericardial cavity is formed first from the embryonic coelom and the pleural and peritoneal cavities are formed thereafter from the pleuroperitoneal fold and the pleuropericardial fold.

13.8.3 Glycosaminoglycans of pleural effusions

The space between the two serous membranes which line the thorax and envelope the lungs is the pleural cavity which in normal health contains only a very small quantity of the pleural fluid to act as a lubricant during the movement of the lungs. Abnormally large quantities of this fluid accumulate in a number of disease states including exudative pleurisy, pleural empyema, malignant tumor (mesothelioma), congestive heart failure, thoracic operation, hepatitis with lung infarct, primary amyloidosis, and polyserositis. The changes in the perchloric acid soluble, phosphotungstate precipitable (mucoprotein) fraction of pleural fluid in these disease conditions have been summarized by Stacey and Barker (1962). Elevations of acidic glycosaminoglycans have also been reported in some of these disease states.

Meyer and Chaffee (1940) isolated hyaluronic acid at a concentration of 0.18 % from the viscous pleural and peritoneal fluids of a patient with a malignant tumor of pleura. The tumor cells were suggested to be the source of the hyaluronic acid in the pleural mesothelioma. Since then several other authors have provided evidence for the presence of hyaluronic acid by decrease of the viscosity of the pleural and peritoneal effusions upon digestion with hyaluronidase enzyme from various sources (Massimiliano, 1942). In five pleuritic patients the lowering of viscosity indicated a hyaluronic acid content of only 0.0125 % (Cutinelli, 1951). On the other hand, a case of a malignant tumor of pleura with a hyaluronic acid content of 0.7 % in pleural effusion was reported (Blix, 1951). The Rivalta reaction (addition of a drop of exudate to 0.05 % acetic acid) precipitated a mucopolysaccharide which contained equimolar amounts of hexosamine and glucuronic acid (Berkes et al., 1954). Pleural effusions due to adenocarcinoma and liposarcoma also contained hyaluronic acid (Meyer et al., 1956). Chemical analyses of serous effusions from 14 patients suggested the presence of hyaluronic acid (Bollet et al., 1957). A component was noted in the perchloric acid soluble, phosphotungstate-precipitable material from pleural fluid which possessed staining and agar electrophoretic properties similar to those of glycosaminoglycans from urine (Fischel, 1960).

Different malignant pleural mesotheliomas secrete hyaluronic acid, which is not secreted by alveolar cell carcinomas (Wagner, 1962). On that basis, histochemical characterization of hyaluronic acid in the tissue of the pleural mesothelioma was suggested as a means of differentiating pleural mesothelioma from peripheral adenocarcinoma of the lung (Wagner et al., 1962). Effusions from pulmonary tuberculosis and malignancies did not contain hyaluronic acid, whereas in six histochemically proved cases of diffuse pleural mesothelioma its content ranged between 0.012 % and 0.79 % depending on the severity of the disease (Harrington et al., 1963). Higher levels of hyaluronic acid in pleural fluids from patients with repeated pleural effusion were diagnostic of mesothelioma (Havez et al., 1967; 1971).

Hyaluronic acid could be detected and quantitated at a concentration of more than 1.5 mg/dl in the cysteine-treated fluid by Alcian blue staining and hyaluronidase sensitivity and in lesser concentrations by anion-exchange chromatography on ECTEOLA-cellulose and carbazol/orcinol ratio on the eluate (Havez et al., 1971). Out of 81 patients with repeated pleural effusions, 20 were hyaluronic acid positive and 13 of these positives corresponded to mesothelioma which was further corroborated on histological and radiological grounds. Hyaluronic acid was not detected in a fibrous mesothelioma and secondary tumors of the pleura (Havez et al., 1971). Rasmussen and Faber (1967) using turbidity measurement on mixing hyaluronic acid and serum reported that hyaluronic acid at a concentration of less than 0.2 mg/ml was found in all types of pleural effusions, whereas a concentration above 0.8 mg/ml was found only in cases of mesothelioma. Thompson et al. (1969) reported a simple colorimetric method for acid mucopolysaccharide measurements at concentrations above 120 µg/ml. By this method, they observed little acidic mucopolysaccharides in 7 cases of effusion without mesothelioma and significant elevations in three patients with untreated mesothelioma.

Other methods of analysis of acidic glycosaminoglycans from pleural fluids were described (Arai et al., 1973; Ritchie et al., 1977). Histochemical methods on surgical specimens taken from a case of pleural mesothelioma (Arai et al., 1975) and localized pleural mesothelioma (Arai et al., 1976) involved confirmation by means of routine histochemical methods and digestion with hyaluronidase from Streptomyces hyalurolyticus, which is strictly specific for hyaluronic acid. Biochemical analyses of the biopsy specimens from pleural mesotheliomas were reported (Havez et al., 1971; Motomiya et al., 1975; Arai et al., 1976). Tumor fragments of pleural mesothelioma contained dermatan sulfate in addition to hyaluronic acid (Havez et al., 1971). The tissue of a benign, localized, pleural mesothelioma was digested with pronase and fractionated on Dowex-1 (Cl^-form). Analyses of the various eluted fractions by electrophoresis and digestion with hyaluronidase from Streptomyces hyalurolyticus and chondroitinase-AC showed the presence of 82.7 % hyaluronic acid, 3.3 % chondroitin 4-(6-) sulfate and 8.8 % dermatan sulfate (Motomiya et al., 1975).

Degand et al., (1972) have also emphasized the importance of detecting hyaluronic acid in establishing the diagnosis of pleural mesothelioma. A systematic analysis of glycosaminoglycans was carried on aspirated effusions freed from cell components with a view to find a clue to the etiology of pleurisy. Pleurisy (inflammation of the two pleural serous membranes) accompanies pneumonia and tuberculosis with initial formation of a shaggy rough exudate layer on the smooth pleural surfaces and later on accumulation of serous fluid or effusion, both of which interfere with movement of lungs during respiration producing a sharp stabbing pain. Besides this, if serous effusion is large, it presses the air out of lungs causing partial or complete collapse of the lungs. To understand the biochemical mechanism involved, the aspirated pleural effusions studied included: 2 pleural effusions due to malignancy, 2 tuberculous pleural effusions, 10 pleural effusions of unknown diagnosis, 1 pericardial effusion due to malignancy and 3 ascites fluid due to malignancy (Arai, 1972). The glycosaminoglycans concentrations in these effusions ranged between 2.76 and 70.7 μg/ml and contained mainly hyaluronic acid with small amounts of chondroitin sulfate. Chondroitin sulfate could have been derived from plasma. The differences in glycosaminoglycans or the polydisperse eultion profiles from Dowex-1 (Cl^-form) for the various effusion fluids did not proivide any clue to the etiology of pleurisy (Arai, 1972). In another study (Arai et al., 1974) the glycosaminoglycans in tuberculous and malignant pleural effusions were found to be similar qualitatively and quantitatively. Both contained a total of glycosaminoglycans in the range of 2.76-70.7 μg/ml and the various glycosaminoglycan components present were: Free and protein-bound forms of hyaluronic acid, free and protein-bound forms of chondroitin sulfate and dermatan sulfate. However, when pleural fluids from 10 cases of pleurisy were analyzed by anion-exchange chromatography on DEAE-cellulose, glycosaminoglycans ranged from 4.0 to 60.3 μg/ml (Arai et al., 1975 a). The glycosaminoglycan spectrum obtained by cellulose acetate electrophoresis used in conjunction with bacterial hyaluronidase was similar to that obtained in the aforementioned earlier study (Arai, 1972) and, probably, reflected the metabolism of the pleural cavity.

Several reports have confirmed that demonstration of a large quantity of hyaluronic acid by histochemical tests on pleural tissue or by chemical analysis on pleural fluid is one of the most accepted and reliable criteria for diagnosis of mesothelioma with effusion (Thompson et al., 1969) or for differentiating pleural mesothelioma from lung cancer (Wagner et al., 1962). However, Castor and Naylor (1967) upon analysis of serous effusions from 100 patients found strikingly high concentrations of hyaluronic acid present not only in effusions from patients with mesothelioma but also in effusions from patients with other neoplasms. In 15 out of 23 fluids analyzed in detail, 20-90 % of the total acidic glycosaminoglycans were present as a type of chondroitin sulfate. Similarly, large quantity of hyaluronic acid was also found in a pleural effusion due to alveolar cell carcinoma (Izumi and Kurokane, 1970). Thus the presence of a large quantity of hyaluronic acid in the pleural effusion was not specific for mesothelioma.

The treatment of mesothelioma has been, however, directed at reducing the hyaluronic acid content in the pleural effusion. The two approaches used for this purpose have been the use of antimitotic agents and the use of hyaluronidase. By chemotherapy with antimitotics higher levels of hyaluronic acid in mesothelioma patients were reduced (Havez et al., 1971). Thompson et al. (1969) also failed to detect any acidic glycosaminoglycans by their colorimetric method (detection limit 120 µg/ml) in one patient with mesothelioma who had undergone chemotherapy. Testicular hyaluronidase (E.C. 3.2.1.35) injected intraperitoneally in a case of mesothelioma resulted in decrease of viscosity, which helped easy removal of pleural fluid by paracentesis, hyaluronic acid production was considerably reduced (Meyer, 1947). Hyaluronidase treatment in the case of experimental aseptic pleurisy was found to affect the adhesive process in the pleural cavity. The adhesiveness was lesser in experimental subjects than the controls (Hitrina, 1968) along with decreased hyaluronic acid. Hyaluronate concentrations, probably, increased in a later terminal stage.

13.8.4 Glycosaminoglycans of ascitic/peritoneal fluid

Ascitic fluid is the serous fluid in the peritoneal cavity. Ascitic fluid accumulation occurs in patients with hepatic cirrhosis, congestive heart failure, tuberculous peritonitis, nephrotic nephritis, malignant peritoneal tumor, disseminating lupus erythematosus and polyserositis. Stacey and Barker (1962) have reviewed the changes in the perchloric acid-soluble, phosphotungstate-precipitable (mucoprotein) fractions in these disease states. The various glycoprotein studies in ascitic fluid in relation to inflammation and cancer haven been reviewed (Werner and Odin, 1965).

Glycosaminoglycans have been known to be present in ascitic fluid and peritoneal fluid. Methods for analysis of glycosaminoglycans in ascitic fluid have been desribed (Kanaya et al., 1974; Ritchie et al., 1977). In patients with mesothelioma, the peritoneal fluid had the same appearance as the chest fluid and hyaluronic acid at a concentration of 0.18 % was isolated from viscous peritoneal fluid (Meyer and Chaffee, 1940). Other workers confirmed the presence of hyaluronic acid in peritoneal exudate by the decrease in viscosity upon digestion with hyaluronidases from different sources (Massimiliano, 1942; Blix, 1951; Baggi, 1952; Caselli and Caputo, 1952). Hyaluronidase did not produce any decrease of viscosity in transudates and the exudative character appeared in transudates when the concentration of hyaluronic acid exceeded 0.25 % (Caputo, 1948). This was confirmed by the acquisition of exudative characteristics by transudates on addition of hyaluronic acid (Caselli, 1952). Peritoneal exudate could be experimentally produced by intraperitoneal injection of 0.2 ml of croton oil in olive oil and this exudation could not be produced if 10 % hyaluronidase was concomitantly used (Baggi and Favilli, 1953). The accumulation of hyaluronic acid was also noted in experimental silicosis (Collet and Daniel-Mussard, 1956) and asbestos-induced silicosis (Preger et al., 1978; Stanton and Wrench, 1972; Gross and Harley, 1973).

About 72-96 % of the asbestos-induced tumors secreted hyaluronic acid (Pyler 1976). In pseudomucinous cysts from the peritoneal material sialomucins (sialic acid-rich glycoproteins) are present as water-insoluble gels which can be solubilized by treatment with alkali, urea or trypsin (Odin, 1959).

A sialomucopolysaccharide was isolated from a peritoneal pseudomyxoma peritonei by pronase digestion, fractional precipitation with ethanol and Sephadex G-50 and G-100 gel filtration (Tomoda and Kitamura, 1966). This sialomucopolysaccharide was a Le^a-specific substance and it contained 3.7 % fucose, 13.7 % galactose, 13.4 % glucosamine, 18.2 % galactosamine and 31.8 % N-acetylneuraminic acid.

Experimental production of aseptic inflammation in guinea pigs resulted in elaboration of hyaluronic acid-rich ascitic fluid (Baggi, 1952). Bollet et al. (1957) on analysis of 14 serous effusions from different patients showed the presence of both hyaluronic acid and sulfated glycosaminoglycans. The ascitic fluid of a patient with nephritis contained uronic acid-containing material (Nagai et al., 1959). Higher levels of hexosamine, glycoproteins and acidic glycosaminoglycans were present in the ascitic fluid from patients with tumor than in hydrothorax fluid of patients with heart failure (Jozsa and Pinter, 1964). During the active development of tumor hyaluronic acid content was maintained at 40 mg/100 ml but increased abruptly at later stages (Hano and Nishino, 1965). Hyaluronidase activity in serum and ascitic fluid increased during the growth of Zajdela ascitic hepatoma in rat. The average activities in Yoshida ascitic fluid and Zajdella hepatoma were 74 % and 54 %, respectively, of their serum activity (Kolarova and Brada, 1969). During the experimental tumor growth, acidic glycosaminoglycans in both blood and ascitic fluid were significantly elevated as compared to those in controls (Skyvova et al., 1973). However, hexosamine content measurements in another study (Agostini, 1976) gave contradictory results during tumor development. The hexosamine content of serum progressively decreased, while that of the ascitic fluid was unrelated to the number of neoplastic cells. Upon X-ray irradiation, the mucopolysaccharide in rat ascitic fluid decreased from the values of 6.10 mg/100 ml in 20 nonirradiated control rats at 8-20 days of tumor development to 2.20-3.40 mg in 30 rats irradiated with 500 R (Illeni and Gelsomino, 1971).

13.8.5 Glycosaminoglycans in Pericardial fluid*

Of all the effusion fluids this is the least studied one. A systematic analysis was carried out on cell-free aspirated pericardial effusion along with other pleural and ascitic samples for investigating a clue to the etiology of pleurisy in the case of a 46-year-old male with metastatic lung cancer and pleural and pericardial effusion. The pericardial fluid contained GAG equivalent to glucuronic acid value of 3.48 µg/ml and had a specific gravity of 1.036 and a protein content of 4.6 g/dl (Arai, 1972). Hyaluronic acid was identified as the major GAG on the basis of its elution pattern from a Dowex-1 (Cl^-form) column.

Pericardial effusions appear most frequently in association with infection (in particular, tuberculosis and viral disease), ischemia (transmural infarction), tumor, immunological disorders (systemic lupus erythematosus, etc.), uremia and myxedema. Technical difficulty of pericardiocentesis as compared with thoracentsis may partly be responsible for the fact that there have been so few biochemical studies of glycosaminoglycans in pericardial effusions.

In a case of primary adenocarcinoma of the thyroid gland plus metastases to the lung and pleuropericardial effusion, the level of GAG in terms of hyaluronic acid (3.48 µg/ml) was much lower in the pericardial fluid as compared to that (19.8 µg/ml) in the pleural fluid of the same patient (Arai, 1972). Another study by Castor and Naylor (1967) indicated that hyaluronic acid was the major GAG in the serous effusions, including the pericardial fluids. However, since the data are not classified according to the type and source of effusions (pericardial, pleural or ascitic), no definite conclusion can be drawn from that study concerning the characteristics of GAG in the pericardial effusions. The level of GAG in a case of effusion associated with systemic lupus erythematosus was the highest out of 14 cases of effusions studied by Bollet et al. (1957). Also, the presence of hyaluronic acid and a sulfated GAG (chondroitin sulfate and/or heparin) was suggested in their study.

---

* By M. Motomiya, H. Arai and K. Konno, Research Institute for Chest Diseases and Cancer, Tohoku University, Sendai 980, Japan

13.8.6 Bone Cyst Fluid

Anghileri and Hossfeld (ref. 107) analyzed cystic fluid from a 49-year old female patient with marked generalized calcification, which occurred due to hyperparathyroidism secondary to renal insufficiency. The GAG concentration found was 7.3 mg/g of this fluid and consisted of 94.1 % hyaluronic acid and 4.8 % chondroitin sulfate. These GAGs were characterized on the basis of their electrophoretic migration and the presence of glucosamine and galactosamine, respectively, in their hydrolysates.

13.9 Bibliography

1. Agostini, C. (1976): Hexosamine content found in serum and ascitic fluid of the rat during the development of Yoshida hepatoma cells. IRCS Med. Sci. Libr. Compend. 4, 400

2. Aminoff, D., Morgan, W. T. J. and Watkins, W. M. (1950) Studies on immunochemistry 8. The isolation and properties of the human blood-group A substance. Biochem. J. 46, 426-438

3. Arai, H. (1972): A study on acidic mucopolysaccharides in pleural and pericardial effusions, and in ascites. Sci. Rep. Res. Inst. Tohoku Univ. (Med). 19, 26-37

4. Arai, H., Endo, M., Sasai, Y., Yokosawa, A., Sato, H., Motomiya, M. and Konno, K. (1975): Histochemical demonstration of hyaluronic acid in a case of pleural mesothelioma. Am. Rev. Resp. Dis. 111, 699-702

5. Arai, H., Endo, M., Yokosawa, A., Sato, H., Motomiya, M. and Konno, K. (1975 a): On acid glycosaminoglycans (mucopolysaccharides) in pleural effusion. Am. Rev. Resp. Dis. 111, 37-42

6. Arai, H., Ogata, K., Haga, K. et al. (1976): Acid glycosaminoglycans in localized pleural mesothelioma - histochemical and biochemical study. Jpn. J. Thorac. Dis. 14, 196-205 (Japan)

7. Arai, H., Yokosawa, A., Sato, H. et al. (1973): A study on methodology of fractionation of acid glycosaminoglycans (mucopolysaccharides) from pleural effusion. Sci. Rep. Res. Inst. Tohoku Univ. (Med) 20, 9-15

8. Arai, H., Yokosawa, A., Sato, H., Motomiya, M. and Konno, K. (1974): Acid glycosaminoglycans in pleural effusion. Kosankinbyo Kenkyu Zasshi 26, 149-155 (Japan)

9. Aruna, R. M., Balasubramanian, K. A., Mathai, K. V. et al. (1973): Isolation and characterization of glycolipids and glycosaminoglycans from meningiomas. Indian J. Med. Res. 61, 1688-1693

10. Asboe-Hansen, G. (1966): Hormone control of connective tissue. Fed. Proc. 25, 1136-1140

11. Asboe-Hansen, G. and Levi, H. (1959): S^{35} autoradiographic studies of sulfomucopolysaccharides in the connective tissue of induced skin tumors in mice and Rous fowl sarcoma I. Lab. Invest. 8, 251-259

12. Baggi, G. (1952): The chemical composition of inflammatory fluids. The presence of mucopolysaccharides in experimentally produced peritoneal exudates. Boll. Soc. Ital. Biol. Sper. 28, 1853-1856

13. Baggi, G. F. and Favilli, G. (1953): Factors controlling chemical composition and cytological characteristics of exudates experimentally induced by phlogistic agents. Rev. Pathol. Gen. et Comparee 53, 1331-1337

14. Berkes, I., Devic-Mikac, D. and Karas, V. (1954): Paper electrophoretic studies on the internal effusions, classified by the Rivalta reaction. Hoppe-Seyler's Z. Physiol. Chem. 294, 142-148

15. Blix, G. (1951): Hyaluronic acid in the pleural and peritoneal fluids from a case of mesothelioma. Acta Soc. Med. Upsalien 56, 47-50

16. Blumenkrantz, N. (1959): Hyaluronic acid and chondroitin sulfate B in the fluid of an ocular cyst. Laboratorio 27, 319-325 (Span)

17. Bollet, A. J., Serayadarian, M. W. and Simpson, W. F. (1957): Acid mucopolysaccharide in pleural, pericardial and ascitic fluids. J. Lab. Clin. Med. 50, 795

18. Bradbury, S., Wiernik, G., Williams, E. A. et al., (1970): Preliminary observations on the histochemistry of the cell surface of the carcinoma of the cervix. Br. J. Cancer 24, 741-745

19. Buddecke, E. and Sittner, B. (1966): Glycosaminoglycane and glycosaminoglycane hydrolases in transplantation tumors. Z. Klin. Chem. 4, 157-163

20. Caccalianza, P. (1954): Mucoproteins and polysaccharides in blister fluid. Giorn. Ital. Dermatol. E. Sifiol. 95, 48-54

21. Caputo, A. (1948): Chemical composition of pathological fluids. Sperimentale 99, 571-576

22. Caselli, P. (1952): Acquisition of exudative characteristics by transudates on addition of hyaluronic acid. Riv. Ist. Sieroterap. Ital. 27, 196-202

23. Caselli, P. and Caputo, A. (1952): Chemical composition of pathological fluids. I. Neoplastic fluids. Riv. Ist. Sieroterap. Ital. 27, 256-278

24. Castor, C. W. and Naylor, B. (1967): Acid mucopolysaccharide composition of serous effusions: Study of 100 patients with neoplastic and nonneoplastic conditions. Cancer 20, 462-466

25. Chain, E. and Duthie, A. (1939): A mucolytic enzyme in testis extract. Nature, 144, 977-978

26. Chiarugi, V. P., Vannucchi, S., Cella, C., Fibbi, G., Del Rosso, M. and Cappeletti, R. (1978): Intercellular glycosaminoglycans in normal and neoplastic tissues. Cancer Res. 38, 4717-4721

27. Collet, A. and Daniel-Moussard, H. (1956): Experimental peritoneal silicosis in rats. Bull. Microscop. Appl. 6, 1-8

28. Coman, D. R. (1947): Mechanism of the invasiveness of cancer. Science 105, 347-388

29. Cudkowicz, G. (1956): The polysaccharides of human carcinoma. Brit. J. Cancer 10, 758-62

30. Cutinelli, C. (1951): Presence of hyaluronic acid in exudates. Riv. Ist. Sieroterap. Ital. 26, 17-26

31. Davidson, A., Linker, A. and Hoffman, P. (1956): Acid mucopolysaccharides of connective tissue. Biochim. Biophys. Acta 21, 506-518

32. Degand, P., Boersma, A., Muh, J. P., Taquet, A. and Havez, R. (1972): Apport de la recherche et du dosage de l'acide hyaluronique dans le diagnostic des mesotheliomes pleurax. Rev. Tuberc Pneumol. 36, 119

33. Dickinson, C. J. (1976): Clinical disorders of lung liquid. In: Lung Liquids. Ciba Sympos. on Lung Liquids. Ciba Foundation, London 22-24 April, 1975, Elsevier, Amsterdam

34. Endo, M., Tamura, S., Minakuchi, S., Ouchi, H. and Yosizawa, Z. (1973): Isolation and identification of proteohyaluronic acid from a cyst of cystic mucoid degeneration. Clin. Chim. Acta 47, 417-424

35. Fischel, C. W. (1960): Mucopolysaccharides of serum and pleural fluid. Nature 186, 804-805

36. Fiszer-Szafarz, B. (1968): Demonstration of a new hyaluronidase inhibitor in serum of cancer patients. Proc. Soc. Exp. Biol. Med. 129, 300-302

37. Fiszer-Szafarz, B. and Guillino, P. M. (1970): Hyaluronidase activity of normal and neoplastic interstitial fluid. Proc. Soc. Exp. Biol. Med. 133, 805-807

38. Fiszer-Szafarz, B. and Guillino, P. M. (1970 a): Hyaluronic acid content of the interstitial fluid of Walker carcinoma. Proc. Soc. Exp. Biol. Med. 133, 597-600

39. Fiszer-Szafarz and Szafarz, D. (1974): Hyaluronidase acitivity in lysosomes and extra-cellular Fluid of tumor cells. In: R. Fricke and F. Harman (Eds.). Connective Tissue: Biochemistry and Pathophysiology, Springer-Verlag, New York, pp. 261

40. Freiherr, G. T. (1979): Two models for aging studied. Res. Resources Reporter (NIH, Bethesda) 3 (8) 1-5

41. Fujita, K. (1954): Properties of a polysaccharide in chicken sarcoma. Nagoya J. Med. Sci. 17, 98-101

42. Giarelli, L. (1955): Behavior of acid mucopolysaccharides in tumors. Riv. Anat. Patol. E. Oncol. 10, 519-540

43. Girard, D., Johnston, W. C. and Grahm, J. H. (1970): Cutaneous angiosarcoma. Cancer 25, 868-883

44. Gottschalk, A. (Ed.) (1966): Glycoproteins, Elsevier, Amsterdam, p. 6

45. Gross, P. and Harley, R. A. (1973): Asbestos-induced intrathoracic tissue reactions. Arch. Pathol. 96, 245-250

46. Guyton, A. C., Taylor, A. E. and Granger, H. J. (1975): Physiology: Dynamics and Control of the Body Fluids. Saunders, Philadelphia

47. Hakomori, S. (1959): Mucoproteins and mucopolysaccharides in relation to cancer. Tampaku Shitsu Kakusan Koso 4, 14-28

48. Halstead, C. L. (1973): Mucosal cysts of the maxillary sinus: report of 75 cases. J. Am. Dent. Assoc. 87, 1435-1441

49. Hance, A. J. and Crystal, A. G. (1975): The connective tissue of the lung. Am. Rev. Resp. Dis. 112, 657-711

50. Hano, K. and Nishino, Y. (1965): Effects of some drugs on hexosamine synthesis in Ehrlich ascites carcinoma cells and on hyaluronic acid content in the ascitic fluid. Gann 56, 367-375

51. Harrington, J. S., Wagner, J. C. and Smith, M. (1963): The detection of hyaluronic acid in pleural fluids of cases with diffuse pleural mesotheliomas. Br. J. Expt. Pathol. 44, 81-83

52. Havez, R., Degand, P., Boersma, A. and Richet, C. (1971): Caracterisation et dosage de l'acide hyaluronique dans le liquide pleural de mesotheliome. Clin. Chim. Acta 33, 443-454

53. Havez, R., Tacquet, A., Biserte, G., Foissac-Gegoux, Ph., Degand, P., Lelievre, G., Gosselin, B. and Duthoit, A. (1967): Le diagnostic biochimique du mesotheliome pleural. Lille Med. 12, 460-466

54. Hieronymi, G. (1954): Occurrence and distribution of acid mucopolysaccharides in tumors. Frankfurt Z. Pathol. 65, 409-434

55. Hitrina, G. B. (1968): Action of hyaluronidase on adhesive process in the pleural cavity under experimental conditions. Bull. Exp. Biol. Med. (Moscow) 65, 115

56. Houngbadji, L., Laumonier, R., Dalion, J. and Vairel, E. G. (1974): Histophysiological study of the cells of a human myxoma of the soft tissues. Comparison of the data on the incorporation of N-acetyl-D-glucosamine-1-^{14}C, the immediate precursor of hyaluronic acid, and composition of the myxoma extract. C. R. Seances, Soc. Biol. Ses. Fil. 168, 83-86 (Fr)

57. Illeni, M. T. and Gelsomino, M. R. (1971): Effects of x-rays on the mucopolysaccharide content of serum and ascites fluid in rats with Yoshida ascites hepatoma. Boll. Soc. Ital. Biol. Sper. 47, 249-250 (Ital)

58. Izumi, H. and Kurokane, K. (1970): Hyaluronate-rich pleural effusion: A biochemical approach in a case of alveolar cell carcinoma. Haigan 10, 161-163 (Japan)

59. Jensen, C. E. (1954): Isolation of hyaluronic acid from pseudomucinous ovarian cysts. Acta Pharmacol. Toxicol. 10, 83-88

60. Jozsa, L. and Pinter, A. (1964): Separation of the origin of ascites and hydrothorax fluids on the basis of their mucopolysaccharide content. Brit. J. Cancer 18, 484-489

61. Kolarova, M. and Brada, Z. (1969): Host-tumor relation XXV. Hyaluronidase activity in blood serum and ascitic fluid of rats during growth of transplanted tumors. Neoplasma 16, 377-387

62. Kanaya, M., Nakamura, M., Kojima, J. et al. (1974): Simple determination of acid mucopolysaccharides in the ascitic fluid. Jpn. J. Clin. Pathol. 22, 160 (Jpn)

63. Kupchella, C. E. and Tamburro, C. H. (1978): Urinary and tissue glycosaminoglycan patterns in hepatic angiosarcoma. In: H. E. Nieburgs (Ed.). Detection and Prevention of Cancer, Vol. I, Marcel Dekker, New York, pp. 915-926

64. Lustig, B. and Nassau, E. (1941): Carbohydrate content of proteins in blood serum, cantharidin blisters and pleural effusions. Am. Rev. Tuberc. 6, 817-824

65. Majewska, H. and Majewski, C. (1966): Acid mucopolysaccharides of the ground substance of the connective tissue in cancer of the skin. Pol. Med. J. 5, 1288-1294

66. Massimiliano, C. (1942): Chemical composition of the serous transudates and exudates. Med. Sper. Arch. Ital. 10, 305-320

67. Masuda, H., Schichijo, S. and Takeuchi, M. (1977): Isolation and characterization of the glycopeptide from a pancreatic cyst. Int. J. Biochem. 8, 627-631

68. Matsuo, E. and Skinsnes, O. K. (1974): Acid mucopolysaccharide metabolism in leprosy. 2. Subcellular localization of hyaluronic acid and beta-glucuronidase in leprous infiltrates suggestive of a host-Mycobacterium leprae metabolic relationship. Int. J. Lepr. 42, 399-411

69. Meyer, K. (1947): The biological significance of hyaluronic acid and hyaluronidase. Physiol. Rev. 27, 335-339

70. Meyer, K. and Chaffee, E. (1940): Hyaluronic acid in the pleural fluid associated with a malignant tumor involving the pleura and peritonium. J. Biol. Chem. 133, 83-91

71. Meyer, K., Davidson, E. A., Linker, A. and Hoffman, P. (1956): The acid mucopolysaccharides of connective tissue. Biochim. Biophys. Acta 21, 506-518

72. Mom, A. M. and Basombrio, G. (1944): The diffusion factor in leprous skin. Intern. J. Leprosy 12, 49-59

73. Motomiya, M., Endo, M., Arai, H., Yokosawa, A., Sato, H. and Konno, K. (1975): Biochemical characterization of hyaluronic acid from a case of benign, localized, pleural mesothelioma. Am. Rev. Res. Dis. 111, 775-780

74. Nagai, Y., Takahashi, S., Saito, Y. and Turumi, K. (1959): Fukushima J. Med. Sci. 6, 257 Quoted by Werner and Odin (1965)

75. Nigam, V. N. and Cantero, A. (1972): Polysaccharides in cancer. Adv. Cancer Res. 16, 1-96

76. Nigam, V. N. and Cantero, A. (1973): Polysaccharides in cancer: glycoproteins and glycolipids. Adv. Cancer Res. 17, 1-80

77. Odin, L. (1959): The chemistry of ovarian cyst contents. Acta Soc. Med. Upsalien. 64, 25-91

78. Oizima, S., Hara, K. and Nakano, S. (1969): Pleural effusion due to infections. Rinsho-to-Kenku. 46, 12

79. Pirie, A. (1942): A hyaluronidase and a polysaccharide from tumors. Brit. J. Exp. Biol. 23, 277-284

80. Poulsen, H. (1966): Thyrotrophic hormone and thyroid hormone control of the inner ear with special reference to myxedema and Meniere's disease. In: G. Asboe-Hansen (Ed.). Hormones and Connective Tissue, Williams and Wilkins, Baltimore, pp. 239-256

81. Preger, L., Arai, D. T. (Eds.) (1978): Asbestos Related Disease. Grune and Stratton, New York

82. Prockop, D. J. (1979): Collagen, elastin and proteolglycans: matrix for fluid accumulation in the lung. In: A. P. Fishman and E. M. Renkin (Eds.). Pulmonary Edema, Am. Physiol. Soc., Bethesda, pp. 125-135

83. Pyler, L. N. (1976): Histochemical study of rat pleural mesotheliomas induced by asbestos. Vopr. Onkol. 22, 53-57 (Russ)

84. Rasmussen, K. N. and Faber, V. (1967): Hyaluronic acid in 247 pleural fluids. Scand. J. Resp. Dis. 48, 366-371

85. Rienits, K. G. (1960): Acid mucopolysaccharides of the sexual skin of apes and monkeys. Biochem. J. 74, 27-38

86. Ritchie, J. C., Smith, S. F. and Castor, C. W. (1977): Measurement of urinary and serous-fluid glycosaminoglycans and urinary hydroxyproline. Am. J. Clin. Pathol. 67, 585-590

87. Sisson, J. C. (1968): Hyaluronic acid in localized myxedema. J. Clin. Endocr. 28, 433-436

88. Skaug, N. (1975): Glycoproteins in fluid from nonkeratinizing jaw cysts. Scand. J. Dent. Res. 83, 159-170

89. Skaug, N. and Hofstad, T. (1972): Demonstration of glycosaminoglycans in fluid from jaw cysts. Acta Pathol. Microbiol. Scand. Section A. 80, 285-286

90. Skinsnes, O. K. and Matsuo, E. (1974): Acid mucopolysaccharide metabolism in leprosy. 1. Storage of hyaluronic acid and its possible significance in the pathogenesis of leprosy. Int. J. Lepr. 42, 392-398

91. Skyvova, M., Kocent, A. and Vermousek, I. (1973): Host-tumor relation. XXXI. Acid glycosaminoglycans in the plasma and in the ascitic fluid of rats during experimental tumor growth. Neoplasma 20, 181-188

92. Smith, B. and Butler, M. (1973): Acid mucopolysaccharides in tumors of the myelin sheath cells, the oligodendroglioma and the neurilemmoma. Acta Neuropathol. (Berl) 23, 181-185

93. Stacey, M. and Barker, S. A. (1962): Carbohydrates of Living Tissues. D. Van Nostrand Co., London, pp. 175-176

94. Stanton, M. F. and Wrench, C. (1972): Mechanism of mesothelioma induction with asbestos and fibrous glass. J. Natl. Cancer Inst. 48, 797-821

95. Szirmai, J. A. (1966): Effect of steroid hormones on the glycosaminoglycans of target connective tissues. In: E. A. Balazs and R. W. Jeanloz (Eds.). The Amino Sugars, Vol. II B, Academic Press, New York, Chap. 41, pp. 129-154

96. Takahashi, M. (1971): Color Atlas of Cancer Cytology. Igaku Shoin, Tokyo, pp. 258-292

97. Thompson, M. E., Bromberg, P. A. and Amenta, J. S. (1969): Acid mucopolysaccharide determination: A useful adjunct for the diagnosis of mesothelioma with effusion. Am. J. Clin. Pathol. 52, 335-339

98. Tomoda, M. and Kitamura, M. (1966): Mucous substance in pseudomyxoma peritonei. II. Isolation and properties of a sialomucopolysaccharide. Jap. J. Exp. Med. 36, 335-339

99. Wagner, J. C. (1962): Experimental production of mesothelial tumors of the pleura by implantation of ducts in experimental animals. Nature 196, 180-181

100. Wagner, J. C., Munday, D. E. and Harrington, J. S. (1962): Histochemical demonstration of hyaluronic acid in pleural mesotheliomas. J. Pathol. Bact. 84, 73-78

101. Watkins, W. M. (1966): Blood group substances. In: A. Gottschalk (Ed.). Glycoproteins, Elsevier, Amsterdam, pp. 462-515

102. Watson, E. M. and Pearce, R. H. (1947): The mucopolysaccharide content of the skin in localised (pretibial) myxedema. Am. J. Clin. Pathol. 17, 507-512

103. Werner, I. and Odin, L. (1965): Amino sugar-containing compounds in pathological effusions. In: E. A. Balazs and R. W. Jeanloz (Eds.) The Amino Sugars, Vol. II A, Academic Press, New York, Chap. 26, pp. 205-207

104. Wozniak, L. (1965): Histochemical investigation of the acid mucopolysaccharide in peritumoral connective tissue. Patol. Pol. Supl. 1, 364-369 (Pol)

105. Zannini, G. and Scanu, A. (1955): The behavior of proteins and non-glucosamine polysaccharides in the serum and the blister fluid of patients with different types of chronic edema of the limbs. Minerva Cardioangiol. 3, 108-109

106. Zhuravleva, M. V. (1974): Histochemical investigation of some connective tissue tumors. Arkh. Patol. 36, 19-24 (Russ)

107. Anghileri, L. J. and Hossfeld, D. K. (1978): Evidence of phospholipids and mucopolysaccharides in the cystic fluid from a patient with generalized tissue calcification. Klin. Wschr. 56, 1013-1017

Chapter 14

Fluids of the Reproductive System

14.1 Introduction

The male reproductive system contains only one fluid, the semen, which itself is a mixture of secretions from the gonads and accessory sex glands. The female reproductive system elaborates secretions in the ovary (ovarian follicular and cyst fluids) during periodic discharge of ripe ova for fertilization with concomitant cyclic structural changes in the connective tissue of the endometrium of the uterus, in preparation for reception of the ova, if fertilized. Upon fertilization of the ovum and its implantation in the uterus, the fetal membranes form a sac containing the amniotic fluid (Chapter 11) in which the fetus is suspended for development. The birth canal through which the mature fetus leaves the mother, consists of the cervix uteri and vagina. These produce secretions which lubricate the passage of the mature fetus at parturition, while in the nongravid state these secretions act as barriers to sperm penetration during the anestrous phase. The skin around the external genitals of some genera of primates baboon, chimpanzee, etc. (sex skin) undergoes cyclic edematous swellings (tumescence) and its involution in response to sex hormone levels during the estrous and anestrous phases (see sections 3.3 and 13.6 in Chaps. 3 and 13).

The glandular epithelium of the female genital tract, uterus and cervix uteri, and the gonads and their accessory sex glands produce secretions depending upon their functional states, regulated by the endogenous secretion of the hormones. The physiology of the connective tissue of the reproductive tract (Harkness, 1964), the hormonal control of the acid mucopolysaccharides in the female genital tract (Zachariae and Thorsoe, 1966; Schumacher, 1977) and the role of GAG of the female genital system in the mechanism of ovulation (Zachariae, 1960) have been reviewed. The types, properties of GAG and their changes in response to gestation, hormone and metals have been discussed (Yoshizaki, 1973).

The glycosaminoglycans (GAG) and glycoproteins in the male and female genital tract and the accessory sexual organs are steroid sensitive and act as the target connective tissues for the actions of the gonadal steroids. These sexual and accessory organs are maintained by the action of the gonadal steroid hormones on the intercellular ground substance of these organs. In females the estrogens induce an increased synthesis and an accumulation of GAG and rise in water content of these tissues while progesterone and testosterone have the opposite effects. In males the formation and accumulation of GAG in the ground substance of the target connective tissues in the accessory sexual organs is regulated by testosterone (Szirmai, 1966). The development of the secondary sexual characteristics in the males and the females is mediated by the regulatory actions of the gonadal hormones on the connective tissue.

Several germ cell tumors of the ovary and testis (Kurman et al., 1979) and carcinoma of the cervix are known and involve metabolic alterations in the connective tissue. Alterations in the secretions of the genital tract of males and females with aging have been reviewed (Talbert, 1977). In contrast with changes observed in aging of other tissues, these alterations of the genital tract are strictly hormone-dependent and follow corresponding alterations in the secretion of hormones from the gonads, anterior pituitary and hypothalmus at different ages.

The discussion presented in this chapter has been restricted to the male genital tract secretion originating from the gonads and the accessory sex organs and the various female secretions consisting of the ovarian follicular and cyst fluids, uterine secretion and the secretions of the birth canal (cervical and vaginal secretions). A brief summary of the glycosaminoglycans/glycoproteins present in the tissues that elaborate these secretions has been included.

14.2 Semen or Seminal fluid

About 2.5-3.5 ml of this male genital secretion is ejaculated in spurts at the time of orgasm. This fluid is a mixture of secretions of the epididymus, seminal vesicles, prostate glands, Cowper's glands and bulbourethral glands. The total fluid consists of the spermatozoa and the seminal plasma. In a normal human adult approximately one hundred million sperms per ml of the ejaculate are present and these are reduced to less than twenty million per ml in the sterile males. Since the fluid is ejaculated in spurts, three different fractions can be collected in separate containers. The first fraction of the ejaculate contains the most of the sperms and is a mixture of the epididymal and prostatic secretions. The second fraction is a mixture of prostatic and seminal vesicles' secretion, while the third fraction originates mainly in the seminal vesicles (Mawson and Fischer, 1953).

The biochemistry of semen (Mann, 1964, 1975), carbohydrate composition (Mann and Rottenberg, 1966) the functional anatomy of spermatozoon (Afzelius, 1975) and the role of the blood-testis barrier (Setchel and Waites, 1975) have been reviewed. The biochemical and functional aspects of the testis and the testicular fluid have been discussed in a monograph (Johnson and Gomes, 1977). The secretory function of the testis is altered in tumors of the testis, the various types of which have been recently classified (Mostofi, 1979).

The spermatozoa contain the enzyme hyaluronidase and this is believed to be involved in human fertility. Viscosimetric procedures for assay of hyaluronidase in human serum from infertile men (Wilson and Aronson, 1950) and for bull semen in veterinary practice (Dalgaard-Mikkelsen et al., 1947) have been desribed. During fertilization the hyaluronidase present in the head of the sperm hydrolyzes the hyaluronic acid layer on the zona pellucida of the ovum (da Silva Sasso, 1959, 1963). A proteinase *Acrosin* required for fertilization is present in the freshly ejaculated mammalian spermatozoa as an inactive zymogen, proacrosin. A GAG factor present in the uterine flushings activates proacrosin to acrosin (Wincek et al., 1979). Electron microscopy revealed that upon fusion of the sperm head with the egg membrane chemical changes take place, since the sperm head plasma membrane before fusion does not bind colloidal iron hydroxide but does so upon fusion with the egg mem-

In vitro studies have shown that the kallikrein-kinin-system is important in the penetration of sperm through cervical mucus. A close correlation between the qualitative and quantitative increase in sperm motility and an increase of cervical mucus penetration was observed in kallikrein-stimulated sperm (Wallner et al., 1975). Also, the acrosomal region of human sperm head contains lactoferrin-related fluorescent material whose concentration decreases upon penetration through the cervical mucus (Broer et al., 1977).

14.2.1 Seminal plasma and seminal vesicles

Fructose is the principal free sugar in the mammalian semen. It is secreted by the seminal vesicles and plays an important role in the survival and motility of the spermatozoa (Foquet, 1971). Three zinc-binding glycoproteins have been separated from human seminal plasma by agar gel electrophoresis followed by staining with dithiozone (Herrmann et al., 1975). Variations in the glycoproteins present in the human seminal plasma from normal subjects and patients with male sterility have been investigated (Kajio, 1967).

The seminal vesicles contain acidic glycosaminoglycans (GAG) in the epithelial cells. Administration of testosterone (0.5 mg/day) for 7 days produced no changes in the seminal vesicles of the white rat but after 30 days a slight increase in concentration of the GAG was noted by microphotometric estimation (Rzeszowska, 1967). The levels of GAG in the glandular epithelium of the seminal vesicles increased with age and with the intensity of the sexual function. The spermatogenic elements of the gonadal tubules contained low levels of GAG (Rzeszowska, 1966).

14.2.2 Semen in cystic fibrosis

Alterations in the composition of semen have been noted in 98 % of the males with cystic fibrosis. The seminal vesicles excrete less fructose and testes produce immature and abnormal sperms. These changes produce decreased amounts of semen, aspermia and infertility. Obstruction and secondary obliteration or atrophy of the body and the tail of the epididymus, vas deferens and seminal vesicles occur (Nadler et al., 1978).

14.2.3 Semen changes in aging

There is a decline in the sperm formation with age from 90 % in 20-39 year old men to 50 % at 70 and to 10 % after 80 years (Sasano and Ichijo, 1969). The capacity of the seminal vesicles to produce seminal fluid increases from 1.75/ml in the upper teens to 5 ml at 21-60 years and thereafter decreases to 2.25 ml in aged men (Nilsson, 1962). Hypertrophy of the seminal vesicles occurred in 60 % of the mice over the age of 20 months, which was due to fluid accumulation caused by occlusion of the ejaculatory duct (Finch and Girgis, 1974).

14.2.4 Prostate and prostatic fluid

Human prostate glands from 5 subjects contained 48.3 ± 4.04 mg hexuronic acid/g dry tissue and hyaluronic acid and another fraction similar to hyaluronic acid was identified in this material (Sato and Gyorkey, 1972).

Sulfated GAG have been detected in the bulbourethral gland of camel (Ali, 1976). The formation, secretion and mucopolysaccharide composition of the prostatic fluid are regulated by nervous and hormonal factors. Prostate is the target organ for the action of testosterone whose continuous stimulus is essential for regulating the mucopolysaccharides in its connective tissue in the normal state. The mechanism of gonadal hormone action in the maintenance of prostate and secretion of prostatic fluid in dogs have been reviewed (Smith, 1975). The physiological and pathological aspects of the prostate and prostatic fluid have been reviewed (Senge et al., 1975). Until puberty very little secretory activity is seen in human prostate gland, but from puberty till 50 years of age the tubulo-alveolar glands become developed and produce the secretion which makes an important contribution to semen. Besides the prostate, at puberty, the seminal vesicles and bulbourethral glands also enlarge and start secreting and these changes are related to a spurt in the formation of testosterone. Along with these changes the external genitalia also increase and their further maintenance depends upon a continued stimulation from testosterone secreted in the normal adult male at a rate of 4-9 mg/day.

Kajio (1967) has investigated the mucopolysaccharides of the prostatic fluid and the effect of hormones on these macromolecules of this fluid.

4.2.5 Prostate and prostatic fluid in aging

The fibrous stroma contains large amounts of glycosaminoglycans. Histochemical investigations in prostate glands from men aged 4-80 years revealed that the GAG decrease with age in the stroma and blood vessels (Lugovaya, 1969). This is contradicted by another report (Leutert and Klaus, 1970) which claimed an increase in GAG with advancing age of 17-90 years. Beginning after 40 years of age atrophy of the glandular function and secretions of prostatic fluid is reflected in the gradual change of columnar epithelial cells into cuboidal cells and the replacement of muscles in the stroma with dense collagen (Moore, 1952). These degenerative changes lead to a complete atrophy of the acini and a loss of secretory activity by the epithelium of the gland. With an advancing age there is a decrease in the blood supply to the prostate gland and the gland shows hypertrophy. This age-related prostatic hypertrophy appears to be connected to the increased binding capacity of this tissue for 5 α-dihydrosterone and higher contents of the dihydrosterone in aging.

4.2.6 Alterations in prostatic carcinoma

Histochemical study of the mucus in 50 cases of the carcinoma of prostate showed that one-third of the tumors produced non-sulfated mucopolysaccharides, one-third produced only the sulfated form, while only two tumors produced both sulfated and nonsulfated mucopolysaccharides. Normal prostate contained primarily neutral mucopolysaccharides while the cancerous prostate was rich in acidic mucopolysaccharide (Hukill and Vidone, 1967). The prostatic tumors may be the result of a shift in the androgen/estrogen ratio (Gasparyan and Portnoy, 1970). The increased amounts of androgens stimulate the epithelium, whereas the decreased estrogens contents inhibit the fibromuscular tissue.

14.3 Female reproductive system

The glycosaminoglycans and glycoproteins of the female reproductive system undergo cyclic changes during the reproductive cycle. The hexuronic acid content of the various reproductive tissues was found in the descending order oviduct > cervix > uterine body > parametrium > vagina > ovary in the proliferative phase of the menstrual cycle. No such differences were present in the secretory phase (Ohki, 1976). These changes can be mimicked by administration of estrogens. The mammalian genital tract produces mucus, which is also under hormonal control (Gibbons, 1978). Zachariae and Thorsoe (1966) have reviewed the hormonal control of acidic glycosaminoglycans in the female genital tract. The acidic glycosaminoglycans present in the female genital secretions have important biological functions. These are involved in the ovary for maturation of follicles, ovulation and ovarian cyst formation and in the secretion of Fallopian tube epithelium. These affect the fate of ovum. Abnormal secretion during endometrial growth affects the implantation of the ovum in the uterus and development of the placenta. The mucopolysaccharides in the secretion of the cervix affect the penetration of the sperm into the cervical mucus.

14.3.1 Secretion of the oviduct

The secretory cells of the rabbit oviduct secreted mucoitinsulfuric acid (da Silva Sasso, 1960). In the bovine oviduct, the secretory products were identified as mucoprotein and GAG and these were produced under the influence of estrogen. Progesterone was important in the release of these products and in inhibiting their replenishment (McDaniel et al., 1968).

14.3.2. Ovarian follicular fluid

The ovaries consist of a pair of spheroidal bodies in the abdomen. Each ovary consists of connective tissue with spherical cavities (follicles) filled with a fluid (follicular fluid). The follicular fluid surrounds an ovum having a thick coat (zona pellucida) around it. The vitelline membrane of the bird's egg is similar to the mammalian zona pellucida. Before ovulation the ovarian follicles increase in size due to accumulation of follicular fluid. The mammalian follicle is 75-150 μ in diameter and the ripened ovum is expelled out due to rupture of the follicle, when the follicular fluid also oozes out. These changes in follicular fluid are associated with alterations in the metachromatic, testicular hyaluronidase-susceptible and also the PAS-staining material present in this fluid.

The concentrations and distribution of the acidic and neutral mucopolysaccharides in follicular structures are related to different stages of ovulation Histochemical studies revealed that hyaluronic acid, sulfated GAG other than isomeric chondroitin sulfates were concentrated in the follicular fluid and intercellular matrix of the granulosa layer and sialic acid and hyaluronic acid to a minor extent were present in the zona pellucida (thick outer cell coat) of oocytes (Tadano and Yamada, 1978): In another investigation (da Silva Sasso, 1963) most of the acidic mucopolysaccharides were found in the inner layer of the zona pellucida, while its outer layer contained mainly neutral mucopolysaccharides. On the other hand, Takeva and Donchev (1971) reported only hyaluronic acid in the zona pellucida of the younger follicles and this was considered to be a precursor of follicular fluid. The follicular fluid is formed by the granulosa cells of ovary along with the synthesis of the GAG by these cells stimulated by estrogen (Zachariae and Thorsoe, 1966). Rat ovarian slices synthesized in vitro heparan sulfate, chondroitin 4- and 6-sulfates and dermatan sulfate (Gebauer et al., 1978). Thyroidectomy resulted in an increased GAG content of the rabbit ovary due to stimulation of GAG synthesis by estrogens synthesized in response to enhanced TSH release in thyroidectomy.

The ovarian follicular fluid contains blood group substances and neutral and acidic mucopolysaccharides. The presence of glycosaminoglycans (GAG) in the follicular fluid from bovine and human ovaries was observed by metachromasia with toluidine blue, which was decreased during the preovulatory phase due to the depolymerizing activity of a substance present in the follicul fluid (Zachariae and Jensen, 1958). Takamine et al. (1951) reported that the depolymerizing hyaluronidase was located in the intercellular fluid while the GAG were present in the follicular fluid. The GAG in this fluid were present at a concentration of 0.3 % in small follicles and 0.2 % in large bovine Graafian follicles. The GAG content and molecular size of the GAG decreased during maturation due to the hydrolysis by hyaluronidase (Jensen and Zachariae, 1958). Hyaluronic acid and chondroitin 4- or 6-sulfates were identified in the bovine follicular fluid on the basis of electrophoretic mobility, chemical composition and hyaluronidase digestion. In the small follicles both these macromolecules were present in equal amounts while in large follicles, chondroitin sulfate predominated.

14.3.3 Ovarian cyst fluids

Defects in connective tissue function are involved in the failure of ovulation and the formation of cystic ovaries. In the cystic ovaries, fluid accumulates in pathological cavities (cysts). The ovarian cyst fluids are often fluid accumulations as a result of carcinoma. These fluids contain fucomucins, sialomucins and acidic mucopolysaccharides. The ovarian cyst mucin was found to be rich in L-fucose, D-galactose, N-acetyl-D-glucosamine and N-acetyl-D-galactosamine and consisted mostly of blood group substances (Morgan, 1963; Fukushima, 1966). This fluid has provided a source of gram quantities of blood group substances for structural studies (Hashimoto, 1967, 1967 a).

Jensen (1954) has reported isolation of a hyaluronic acid of molecular weight of $4.54\text{-}4.61 \cdot 10^5$ in yields of 42.9 mg/65 ml and 37.9 mg/98 ml of the fluid from two pseudomucinous ovarian cysts. The acidic GAG in effusion fluids of ovarian tumors have been studied (Grudzien et al., 1975). A proteohyaluronic acid has been isolated and identified from a cyst of cystic mucoid degeneration (Endo et al., 1973). A more detailed discussion on the mucopolysaccharide composition of various types of ovarian cyst fluids is given in Chapter 13 (section 13.4).

14.3.4 Uterus and uterine secretions

The uterus consists of an outer muscular wall (myometrium) and the inner connective tissue lined with epithelial cells (endometrium). The presence of GAG in the endometrium of normal uterus was initially shown by histochemical tests including enzymatic digestion. Besides acidic GAG, glycogen and neutral polysaccharides were also present (Hoynck van Papendrecht, 1965). Fluorescence histochemistry revealed that the GAG were absent in the branched cells of the non-gravid uterus (Gudev, 1973). The human uterus (endometrium and cervix) contained 0.2-0.3 % of the dry, defatted weight and consisted of dermatan sulfate 42-53 %, heparan sulfate 10-13 % and other GAG 35-42 % (Loewi and Consden, 1962). A slight rise in the chondroitin sulfate fraction occurred in the postmenopausal uteri. Along with GAG mast cells were also present. The hyaluronic acid fraction paralleled the increase or decrease of the number of mast cells during the menstrual cycle, suggesting that these cells are involved in the production of hyaluronic acid (Likar et al., 1964). These changes in biosynthesis of GAG are mediated by variations in estrogen levels in various phases of the menstrual cycle, since in vitro formation was increased in guinea pigs' uteri when they were injected with 1 mg estradiol benzoate prior to the experiment (Ferrando and Zipper, 1967). The stimulation of biosynthesis was suppressed with progesterone.

The histochemical changes seen in the uterus during various phases of menstrual cycle are reflected in the secretions collected as smears from the cervix of the uterus and posterolateral formix vaginae. Only a small quantity of mucopolysaccharides and high glycogen contents were present on the surface and intermediate epithelial cells in 26 healthy women in the proliferation phase. In the secretion phase of the menstrual cycle in 33 healthy women mucopolysaccharides increased considerably while glycogen decreased (Botnar, 1968).

The fluid in the uterine lumen during the estrogenic phase is a transudate and it contains acidic mucopolysaccharides. Estrogens stimulated formation while progesterone blocked the synthesis of these macromolecules (Zachariae and Thorsoe, 1966). Studies with intravenously injected $Na_2{}^{35}SO_4$ in female rabbits showed the greatest sulfate incorporation in the uterine tube secretion 12 hours after mating or at the moment of fertilization (Gothie, 1961). Since the sperm meets the ovum for fertilization in the Fallopian tube, disturbances in the anatomy or physiology affecting tubal secretion could lead to sterility. The dried defatted powder obtained from the human Fallopian tube mucosa contained hexosamine, hexuronic acid, fucose and galactose (Werner, 1953). This epithelium elaborated a secretion containing acidic mucopolysaccharides and the production and release of these mucopolysaccharides was regulated by estrogen and progesterone (Zachariae and Thorsoe, 1966).

The endometrial secretion was found to have a hexosamine content of 2.5 mg/ml, while it contained hexuronic acid at 0.15-0.25 mg/ml suggesting the presence of an appreciable amount of GAG. The secretion was stimulated by estrogen and inhibited by androgen, while progesterone produced alterations in chemical composition of this secretion (Homberger et al., 1963).

The quantity and degree of polymerization of GAG and mast cell content and their degree of granulation are inherently related to estrogen and progesterone. Estrogen stimulates the production of GAG in mast cells; progesterone acts as an antiestrogen (Lebherz, 1969). Ovarian hormones (estrogen) increase the uterine sulfomucopolysaccharides (Bo et al., 1965). Estrogen benzoate given to rabbits, stimulated the metabolism of uterine acidic glycoprotein, hyaluronic acid, sulfated glycoprotein, low sulfated chondroitin, heparan sulfate, chondroitin 4-sulfate and dermatan sulfate. Progesterone diminished these increases but had no effect on the estrogenic stimulation of a neutral glycoprotein and oversulfated chondroitin sulfate in the uterus (Endo and Yosizawa, 1973). The specific radioactivity of hexosamine in the subfractions indicated that the metabolic rate of the uterine complex polysaccharides under estrogenic stimulation was in the order of: hyaluronic acid > sulfated glycopeptide > heparan sulfate > chondroitin 4-sulfate > chondroitin 6-sulfate > dermatan sulfate (Endo and Yosizawa, 1975, 1976). Histochemical study of the estrogen-stimulated uterus showed that the estradiol decreased the levels of nonsulfated GAG in the endometrial stroma but increased the sulfated GAG in the uterine stroma (Wurgaft and Tchernitchin, 1976).

In female rabbits, administration of daily doses of 20-30 μg of 17-hydroxyprogesterone acetate decreased the hexuronic acid content in every fraction from the anion-exchange chromatographic column but progesterone had no effect (Kofoed et al., 1977). Norethisterone acetate administration in 26 human females showed early secretory effects which soon progressed to drug-induced changes (Shahani, 1964). In proliferation induced with folliculin in rats, the GAG content of the uterus increased but was less marked than those with progesterone (Riminiceanu et al., 1966). Positive histochemical tests for GAG were seen in the secretions from Fallopian tubes, endometrium, cervix and follicular fluid upon administration of gonadal hormones to rabbits (Mitsugu, 1966).

14.3.5 Functional uterine bleeding

Besides the gonadal hormones, thyroidectomy in rabbits also led to an increase in the GAG of the ovaries (Thorsoe, 1962). In functional uterine hemorrhage, a lower content of GAG was observed in the stroma of the proliferative endometrium (Koreneva, 1962; Ghosh and Sengupta, 1967).

14.3.6 Pregnancy and postpartum stage

Histochemical alterations in the amount of GAG of uterus of rats during pregnancy, spontaneous and delayed labor and initial and late post-partum period correlated directly to the amount of estrogen and progesterone in these stages (Paraventi et al., 1960). Histochemical postcalving changes observed in the endometrium of cows were stabilized in 50 days after calving (Emel'yanova, 1973).

14.3.7 Effect of intrauterine devices

The presence of the intrauterine device (IUD) increased the hexuronic acid content of endometrium (Cooper and Hawk, 1968). The IUD accelerated the effect of estrogen on stimulation of GAG increase but suppressed the progesterone effect (Abe et al., 1972). The antifertility effect of IUD might be owing to changes induced in the metabolic pattern of mucosubstances in the endometrium.

14.3.8 Hyperthecosis and Stein-Leventhal syndromes

Histochemical examination of the ovaries obtained from women with hyperplasia associated with excessive luteinization of the cells of the theca interna showed GAG accumulation in the perifibrillary cement substances and the media of the arteries of the cortex and of the medulla, probably resulted from the suppression of the tissue hyaluronidase activity (Marsalek and Obrucnik, 1958). A similar increase of GAG occurred in the media of the arteries and in the tunica albuginea of the ovaries in Stein-Leventhal syndrome, characterized by bilateral polycystic ovaries, amenorrhea and anovulation (Marsalek and Obrucnik, 1958 a).

14.3.9 Uterine carcinomas

In folliculinic hyperplasia (Melica and Papadia, 1964) and initial adenocarcinomatous transformation (Maggioni and Maimone, 1969) of the uterus, histochemical investigations using Alcian blue showed a substantial increase of GAG with a simultaneous decrease of glycogen in the epithelium component as it passed gradually from simple glandular hyperplasia to carcinomatous appearance. In progestogen-treated endometrial cancers, a marked intracytoplasmic accumulation of GAG and glycogen was inversely proportional to the mitotic index (Hustin et al., 1977).

14.3.10 Cervix uteri and cervical secretion

Most of the cervix is muscular but at the vaginal end called the *portio*, instead of a single layer of cells it contains many layers of squamous type of cells and resembles vagina in secretory functions. The muscle portion, the myometrium, of the human cervix, has been reported to contain GAG equivalent to 1.5 % of dried defatted tissue distributed as 55 % hyaluronic acid and 45 % chondroitin 4- and 6-sulfates. The mast cells present decreased in aging, pregnancy and were more in the secretory phase than in the proliferative phase (Iversen, 1960). The glycoprotein decreased and GAG markedly increased in the post-partum cervix and from electrophoretic analysis contained hyaluronic acid, chondroitin 4-sulfate, chondroitin 6-sulfate and dermatan sulfate (Danforth et al., 1974).

Fractionation of cervical mucopolysaccharides from nonpregnant controls revealed that 13.6, 16.4, 61.6 and 8.4 % were recovered as chondroitin 4-sulfate, chondroitin 6-sulfate, dermatan sulfate and hyaluronic acid, respectively (Nakaya, 1973). Histochemical examination revealed that the anterior, posterior and left lateral walls contained similar amounts while the lower portion contained less GAG than the middle or upper portions of the cervix uteri (Kondo, 1972). In 29 pregnant women the GAG content of the anterior wall increased in the first trimester of pregnancy but the highest amounts were seen in second trimester in subjects with cervical incompetence (Kondo, 1972 a) or at term. At the time of ovulation or immediately before, there was an increase in cervical mucus and shortly afterwards these changes were reversed (Moghissi, 1966).

The synthesis of sulfate-containing mucopolysaccharides by guinea pig cervical cells was found to be under the influence of estrogen, while the secretion was influenced by estrogen-progesterone equilibrium (Gothie et al., 1966). The GAG content of the human cervix uteri increased with estrogen treatment, whereas progesterone or chorionic gonadotrophin produced no changes. The increase in GAG during pregnancy, at term and with estrogen treatment were due to a marked increase in hyaluronic acid and a slight increase in chondroitin 4- and 6-sulfates (Nakaya, 1973).

The cervical secretion consists of the cervical mucus, which is the most studied of all the genital secretions because of the ease of collection during the middle of the menstrual cycle. However, in human females this midcycle mucus is very little in quantity and for this reason the studies reported in literature have been on pooled human cervical mucus from different subjects. The majority of the chemical structural investigations have been carried out on bovine mid-cycle cervical mucus, which is usually 300-500 ml per collection.

The cervical mucus present as a viscous secretion in the vagina is produced by the columner epithelial cells of the cervix and is expelled toward the vagina by the beating action of the cilia present on the ciliated columner epithelial cells. It acts as a lubricant and as a barrier against sperms and microorganisms. The physicochemical properties (Platt, 1966) and functions of this fluid have been reviewed (Elstein et al., 1973; Blandau and Moghissi, 1973; Doehr, 1977; Elstein, 1978).

The secretion of estrogen stimulates a flow of cervical mucus, while progesterone has an opposite effect. Similar hormonal changes occurring during the menstruation and pregnancy are responsible for the cyclic variations in the physicochemical characteristics of the cervical mucus (Zachariae and Thorsoe, 1966). Estrogens stimulate the cervical glands to produce a thin, watery and alkaline secretion, while the progesterone reduces the amount of salt and water in the secretion (MacDonald, 1969) making it viscous and impermeable to sperms. Thus during the ovulatory or midcycle stage estrogen levels are at peak and this produces a secretion receptive to the sperm. Depolymerization of the mucopolysaccharides and glycoproteins of the cervical mucus occurs at the time of ovulation. Detection and determination of free glucosamine in the cervical mucus has been found as a positive test for ovulation in 96 % of the women (Kurzrok, 1958). The high viscosity leading to a sperm-impermeable medium throughout the estrous cycle may be the cause of sterility in some women (Tateno, 1977).

The cervical mucus is very viscous due to its high carbohydrate content of 70-80 %. The salic acid in glycoproteins and sulfate groups of the sulfated mucopolysaccharides are partly responsible for this high viscosity. However, cross-linkages of the glycoproteins units through disulfide bridges are more significant in making this fluid viscous (Meyer, 1977). The concentrations of the glycoproteins and differences in the extent of disulfide bridges in the glycoproteins in the cervico-vaginal mucus may be responsible for differences in fertility among different breeds of animals (Agrawal et al., 1977). The oral contraceptives (synthetic progestogens) act by producing highly viscous mucus, which is impermeable to sperm. Foreign bodies like intrauterine devices act similarly, while copper wire implantation in the uterus releases copper ions for disrupting the disulfide bridges between the glycoprotein chains, thereby reducing the viscosity of the cervical mucus. Disulfide reducing agents like dithiothreitol, mercaptoethanol, also liquefy the viscous cervical mucus in vitro (Elstein et al., 1973).

The cervical mucus consists of glycoproteins, sulfated glycoproteins and possibly a small amount of GAG. The 70-80 % carbohydrates are present as oligosaccharide chains attached through alkali-labile O-seryl/O-threonyl glycosidic linkages between N-acetyl-D-galactosamine on the oligosaccharide and serine/threonine residues on the peptide core. Differences in the total carbohydrate/total amino acid ratios in the fragments obtained by cleavage with cyanogen bromide suggested that some portions of the peptide lacked oligosaccharide chains (Elstein, 1978). The monosaccharide units of the oligosaccharide chains are the same as in epithelial secretions and consist of hexosamine, galactose, fucose and sialic acid in a ratio of 1.0:0.94:0.27:0.29 (Meyer, 1977). The sialic acid content of the cervical mucus varies with estrogen/progesterone levels so that the preovulatory sialic acid/fucose ratio of 0.05 changes to 1.1 and 1.3 in the postovulatory and pregnancy stages, respectively. These increases are due to increased sialic acid incorporation brought about by induction of biosynthesis of the enzyme sialyltransferase.

Studies with intravenously injected $^{35}SO_4$ showed that the radioactivity appeared in a large amount in the cervical muciparous cells of rabbits five days after the injection of estradiol. Progesterone addition resulted in a massive secretion of mucus rich in $^{35}SO_4$ (Gothie et al., 1966). Zachariae and Thorsoe (1966) have reviewed the contradictory reports about the presence and absence of uronic acid in cervical mucus and about the radioactive sulfate incorporation studies and have stated that uronic acid-free sulfated acidic mucopolysaccharides such as keratan sulfate were present in the cervical mucus. Bhushana Rao et al. (1973) noted a component with anodal mobility upon cellulose acetate electrophoresis, which stained with Alcian blue or periodic acid Schiff reagent. The mucus present in the vagina but synthesized in the cervix has been reported to contain chondroitin sulfate (Takamine et al., 1951).

14.3.11 Cervical mucus in carcinoma

Histochemical analysis revealed that in neoplastic cells of epithelial origin the mucopolysaccharide content decreased. Immature carcinoma had a little mucopolysaccharide while invasive forms of the carcinoma or late mature type carcinomas produced mucopolysaccharides in excess (Llusia and Nogales, 1960). A close relationship existed between the presence of mast cells and the occurrence of mucopolysaccharides in 60 human uteri (Gupta and Schueller, 1967). The squamous cell carcinomas of the cervix contained appreciable levels of hyaluronic acid in the cell coat and intercellular matrix (Bradbury et al., 1970). Increased values of mucopolysaccharides in undifferentiated tumors of the cervix were seen (Kasab'yan and Sgibneva, 1972).

14.3.12 In cystic fibrosis

Women with cystic fibrosis have decreased fertility due to a decreased water content of the vaginal mucus leading to high viscosity of this mucus which inhibits penetration of the sperm (Rosenow and Lee, 1968; Kopito et al., 1973). Histochemical examination revealed that the cervical mucus glands were dilated although the ovaries and uterus were normal (Oppenheimer and Esterly, 1970).

14.3.13 Vagina and vaginal secretion

Like other reproductive tissues the vagina and vaginal fluid also undergo cyclic changes during menstruation and pregnancy in response to the altered levels of estrogen and progesterone in these states. The vaginal secretion consists primarily of the desquamated cells of the vaginal epithelium with minor contribution from the actual holocrine secretion from the vaginal epithelium. The sialic acid content of the vaginal secretion did not vary in response to estrous cycle, castration or administration of sex steroids (Maignan, 1967), while hexosamine showed cyclic changes (Morgan, 1963 a). The sex steroids induced metaplasia of the vaginal mucosa with alterations in acid mucopolysaccharides (Bartels, 1968). Chondroitin sulfate was detected in the vaginal mucus originating from the cervix (Takamine et al., 1951).

The bovine vaginal mucosa and vaginal secretion contained sialomucins and sulfomucins, including dermatan sulfate and neutral mucopolysaccharides. The acidic polysaccharides predominated in the active phase of the estrous cycle, whereas the reverse was true in the non-active phase (Golvinov and Biryukov, 1973). The changes during estrous cycle appear in the uterus prior to appearing in the vagina. Detection of radioactivity in epithelial cells and secretions of the rabbits' vagina indicated the presence of sulfated mucopolysaccharides (Zachariae, 1958) which were resistant to hyaluronidase. Estrogens caused a reduction of Alcian blue positive mucopolysaccharides in the vagina of prepuberal rat (Verne et al., 1957). A similar decrease was noted in infection with Trichomonas vaginalis (Zinca et al., 1959).

During the estrous cycle in animals the GAG content of vaginal epithelium increased (Sandroglio, 1967, 1967 a), the electrical resistance of the vaginal mucus decreased with the lowest values corresponding with serum LH peaks (Schams et al., 1977). With the serum luteinizing hormone peak at ovulation the water content of the vaginal mucus increased tremendously. The increased water content detected by insertion of a hydrophilic cylinder in human vagina has been used for detection of ovulation (Polishuk and Ron, 1976).

14.4 Bibliography

1. Abe, T., Endo, M. and Yosizawa, Z. (1972): Uterine mucosubstances accompanying the infertility effect of an intrauterine device. Clin. Chim. Acta 42, 29-35

2. Afzelius, B. (Ed.) (1975): The Functional Anatomy of the Spermatozoon. Wenner-Gren Center International Symposium Series Vol 23, Pergamon Press, Oxford, 1975, 393 pp

3. Agrawal, S. C., Agrawal, R. G. and Datta, I. C. (1977): Glycoprotein concentration in cervico-vaginal mucus in cows and buffaloes at estrus. Indian J. Dairy Sci. 30, 346-348

4. Ali, H. A. (1976): Some histochemical studies on the prostate, urethral and bulbourethral glands of the one-humped camel (Camel dromedarius). Histochem. J. 8, 565-578

5. Bartels, C. (1968): Histochemical and enzyme topo-chemical studies of hormone-induced metaplasia of rat vaginal epithelium. Acta Histochem. 31, 82-97

6. Bhushana Rao, K. S. P., Van Roost, E., Masson, P. L., Heremans, J. F. and Andre, F. (1973): Studies of bovine cervical mucin. I. The amino acid composition and N-terminal amino acid of bovine oestrus cervical mucin. Biochim. Biophys. Acta 317, 286-302

7. Blandau, R. J. and Moghissi, K. (1973): The Biology of the Cervix. University of Chicago Press, Chicago, p. 287

8. Bo, W. J., Smith, S., Reiter, R. and Pizzarello, D. J. (1965): The effect of ovarian hormones of the sulfomucopolysaccharides of the uterus. J. Histochem. Cytochem. 13, 461-469

9. Botnar, E. R. (1968): Cytochemical features of the epithelium of the cervix of the uterus in various phases of the menstrual cycle. Lab. Delo 6, 352-354

10. Bradbury, S., Wiernik, G., Williams, E. A. and Cowdell, R. H. (1970): Preliminary observations on the histochemistry of the cell surface of carcinoma of the cervix. Brit. J. Cancer 24, 741-745

11. Broer, K. H., Dauber, U., Herrmann, W. P. and Hirschhauser, C. (1977): The presence of lactoferrin (SCA) on the human sperm head during in vitro penetration through cervical mucus. IRCS Med. Sci. Libr. Compend. 5, 362

12. Cooper, B. S. and Hawk, H. W. (1968): Variation in mucopolysaccharide content of the endometrium of the ewe caused by ovarian hormones and intrauterine devices. J. Anim. Sci. 27, 1373-1377

13. da Silva Sasso, W. (1959): Existence of hyaluronic acid at the zona pellucida of the rabbit's ovum. Acta Anat. 36, 352-357

14. da Silva Sasso, W. (1960): Estimation of mucopolysaccharides in mucosa of the oviduct of rabbit. Anals. Fac. Farm. Odontol. Univ. San Paulo 17, 55-60

15. da Silva Sasso, W. (1963): Histochemical detection of mucopolysaccharides in the zona pellucida of the ova of some mammals. Ann. Histochim.8, 395-400

16. Dalgaard-Mikkelsen, S., Kvorning, S. A. and Rasbech, N. O. (1947): On hyaluronidase concentration in bull semen. Skandinavisk Veterinartidskrift 42, 661-676

17. Danforth, D. N., Veis, A., Breen, M., Weinstein, H. G., Buckingham, J. C. and Manalo, P. (1974): Effect of pregnancy and labor on the human cervix. Changes in collagen, glycoprotein and glycosaminoglycans. Am. J. Obstet. Gynecol. 120, 641-651

18. Doehr, S. A. (1977): Cervical mucus glycoproteins. In: M. I. Horowitz and W. Pigman (Eds.). Glycoconjugates. Academic Press, New York, Vol. 1, pp. 239-257

19. Elstein, M. (1978): Functions and physical properties of mucus in the female genital tract. Br. Med. Bull. 34, 83-88

20. Elstein, M., Moghissi, K. S. and Borth, R. (Eds.) (1973): Cervical mucus in human reproduction, WHO Colloquium, Scriptor, Copenhagen

21. Emel'yanova, M. F. (1973): Postcalving changes in the levels of glycogen and mucopolysaccharides in the endometrium of cows. Zhivotonovodstvo (12), 84-85

22. Endo, M., Tamura, S., Minakuchi, S., Ouchi, H. and Yosizawa, Z. (1973): Isolation and identification of proteohyaluronic acid from a cyst of cystic mucoid degeneration. Clin. Chim. Acta 45, 73-83

23. Endo, M. and Yosizawa, Ż. (1973): Hormonal effect on glycoproteins and glycosaminoglycans in rabbit uteri. Arch. Biochem. Biophys. 156, 397-403

24. Endo, M. and Yosizawa, Z. (1975): Glycosaminoglycans and acidic glycoproteins in rabbit uterus under estrogenic conditions. Biochim. Biophys. Acta 404, 274-280

25. Endo, M. and Yosizawa, Z. (1976): Metabolic rate of acidic complex saccharides in rabbit uterus under estrogenic condition. J. Biochem. (Tokyo) 79, 1-4

26. Ferrando, G. and Zipper, J. (1967): Biosynthesis of uterine endocervical mucoprotein in the guinea pig. Biochim. Biophys. Acta 136, 404-

27. Finch, C. E. and Girgis, F. G. (1974): Enlarged seminal vesicles of senescent C57BL/6J mice. J. Geront. 29, 134-138

28. Fouquet, J. P. (1971): Secretion of free glucose and related carbohydrates in the male accessory organs of rodents. Comp. Biochem. Physiol. 40A, 305-317

29. Fukushima, Y. (1966): Biochemical studies of blood group mucopolysaccharides 18. Properties of group active mucopolysaccharides isolated from pseudomucinous fluid of ovarian cysts in group AB human. J. Jap. Biochem. Soc. 38, 291-299

30. Gasparyan, A. M. and Portnoy, A. S. (1970): Reaction of the prostate to experimental endocrine disorders. Int. Urol. Nephrol. 2/3, 277-285

31. Gebauer, H., Lindner, H. R. and Amsterdam, A. (1978): Synthesis of heparin-like glycosaminoglycans in rat ovarian slices. Biol. Reprod. 18, 350-358

32. Ghosh, B. K. and Sengupta, K. P. (1967): Histochemistry of the endometrium in functional uterine bleeding. I. Glycogen and acid mucopolysaccharides. Bull. Inst. Post Grad. Med. Educ. Res. (Calcutta) 9, 81-88

33. Gibbons, R. A. (1978): Mucus of the mammalian genital tract. Br. Med. Bull. 34, 34-38

34. Golvinov, D. D. and Biryukov, V. G. (1973): Polysaccharide compounds of cow vaginal secretory compounds. Veterinariya (Moscow), 84-85

35. Gothie, S. (1961): Metabolism of ^{35}S in the genital tract under the action of sex hormones. Pathol. et Biol., Semaine hop. 9, 655-661

36. Gothie, S., Moricard, R. and Jannel, J. (1966): Semi-quantitative kinetic study of mucopolysaccharide sulfate metabolism in guinea pig cervical cells. Ann. Histochim. 11, 29-40

37. Grudzien, M., Waga-Rzucidlo, M. and Tynecki, J. (1975): Serum acid mucopolysaccharides (MPS) in tissues and the effusion of ovarian tumors in women. Ann. Univ. Mariae Curie Sklodowska (Med.) 30, 209-215 (Pol)

38. Gudev, K. (1973): Pigments of the nongravid uterus and the fetal placenta of sheep. II. Fluorescence-histochemical study of acid mucopolysaccharide in the pigment. Zentralbl. Veterinärmed., Reihe A 20, 305-313

39. Gupta, R. K. and Schueller, E. S. (1967): Acid mucopolysaccharide and mast cell variations in endometrium and some uterine tumors. Obstet. Gynecol. 30, 510-517

40. Harkness, R. D. (1964): The physiology of the connective tissues of the reproductive tract. Int. Rev. Connective Tissue Res. 2, 155-211

41. Hashimoto, K. (1967): Blood group mucopolysaccharides. XIX. Action of Pronase on blood group substances from human ovarian cyst fluids. I. Seikagaku 39, 79-89

42. Hashimoto, K. (1967 a): Blood-group mucopolysaccharides. XX. Action of Pronase on blood-group substances from human ovarian cyst fluids. Seikagaku 39, 152-160

43. Hermann, W. P., Weyland, I. and Hartmann, I. (1975). Demonstration of zinc glycoproteins in human seminal plasma. Andrologia 7, 329-339 (Ger)

44. Homberger, F., Bernfield, P., Treiger, A., Harpel, P. and Grossman, M. S. (1963): Endometrial secretions. Ann. N. Y. Acad. Sci. 106, 683-691

45. Hoynck van Papendrecht, H. P. C. M. (1965): Histochemistry of the endometrium. Ned. Tijschr. Verlosk. Gynaecol. 65, 5-20 (Dut)

46. Hukill, P. B. and Vidone, R. A. (1967): Histochemistry of mucus and other polysaccharides in tumors. II. Carcinoma of the prostate. Lab. Invest. 16, 395-406

47. Hustin, J., Binard, J. M. and Lambotte, R. (1977): Effect of synthetic progestogens on endometrial cancer. Cytochemical criteria. Ann. Anat. Pathol. 22, 323-336

48. Iversen, O. H. (1960): Isolation of acid mucopolysaccharides in the myometrium of the human cervix uteri. Acta Path. Microbiol. Scand. 50, 25-28

49. Jensen, C. E. (1954): Hyaluronic acid. IV. Isolation of hyaluronic acid from pseudomucinous ovarian cysts. Acta Pharmacol. (Kbh)10, 83-88

50. Jensen, E. and Zachariae, F. (1958): Isolation and analysis of acid mucopolysaccharides in bovine follicular fluid. Acta Endocrinol. 27, 356-368

51. Johnson, A. D. and Gomes, W. R. (Eds.). (1977): Advances in Physiology, Biochemistry and Function, Vol. 4: The Testis. Academic Press, New York, N. Y., 637 pp

52. Kajio, K. (1967): Studies on glycoprotein in the fluid of the male accessory sexual organs. 3. Human seminal plasma, especially in patients with male sterility. Acta Urol. Jap. 13, 200-206 (Jap)

53. Kasab'yan, S. S. and Sgibneva, O. V. (1972): Histochemical changes in cancer of the cervix uteri. Vop. Onkol. 18, 29-36

54. Kofoed, J. A., Houssay, A. B., Tocci, A. A., Curbelo, H. M. and Barcelo, C. (1977): Effects of acetoxyprogesterone upon glycosaminoglycans in the uterus and salivary glands. Horm. Metab. Res. 9, 85-89

55. Kondo, T. (1972): Histochemical studies on the human cervix uteri. 1. Acid mucopolysaccharides of the human cervix uteri in non-pregnancy. Nagoya Med. J. 17, 111-112

56. Kondo, T. (1972 a): Histochemical studies on the human cervix uteri. 2. Acid mucopolysaccharides of the human cervix uteri during pregnancy. Nagoya Med. J. 17, 111-112

57. Kopito, L. E., Kosasky, H. J. and Shwachman, I. (1973): Water and electrolytes in cervical mucus from patients with cystic fibrosis. Fertil. Steril. 24, 515-516

58. Koreneva, G. P. (1962): Histochemical studies of the endometrium in functional uterine bleeding. Tr. Saratpvsk. Med. Inst., 184-188

59. Kurman, R. J., Scardino, P. T., McIntire, K. R., Waldmann, T. A., Javadpour, N. and Norris, H. R. (1979): Malignant germ cell tumors of the ovary and testis. An immunohistologic study of 69 cases. Ann. Clin. Lab. Sci. 9, 462-466

60. Kurzrok, R. (1958): The presence of glucosamine in cervical mucus as a test for the date of ovulation. Intern. J. Fertility 3, 80-85

61. Lebherz, T. B. (1969): Changes in endometrial ground substance. In: E. T. Tyler (Ed.). Progress in Conception Control. Lippincott, Philadelphia, pp. 30-33

62. Leutert, G. and Klaus, J. (1970): Age dependent histological and histochemical findings in the human prostate. Acta Histochem. 37, 136-147

63. Likar, I. N., Likar, L. J. and Robinson, R. W. (1964): Mast cells and hyaluronic acid in the bovine endometrium. Nature 203, 730-733

64. Llusia, J. B. and Nogales, F. (1960): The polysaccharide content of carcinoma in situ and invasive carcinoma of the portio (vaginalis uteri). Ginecol. Y. Obstet. Mex. 15, 169-174

65. Loewi, G. and Consden, R. (1962): Acid mucopolysaccharides of the human uterus. Nature 195, 148-150

66. Lugovaya, A. Z. (1969): Morphological and histochemical changes in the prostate gland in men in different age groups. Onkologiya (Kiev) 25, 73-76

67. MacDonald, R. R. (1969): Cyclic changes in cervical mucus. J. Obstet. Gynaecol. Brit. Commonw. 76, 1090-1099

68. Maggioni, G. and Maimone, G. (1969): Histochemistry of simple endometrial glandular hyperplasia in initial adenocarcinomatous transformation and with evident adenocarcinomatous aspects. Biol. Lat. 22, 127-134

69. Maignan, C. (1967): Activity of sex steroids on vaginal sialic acid levels in rats. C. R. Soc. Biol. 161, 1818-1822

70. Mann, T. (1964): The Biochemistry of Semen and of the Male Reproductive Tract. Methuen, London

71. Mann, T. (1975): Biochemistry of Semen. Handbook of Physiology, Sect. 7. Endocrinology. (R. O. Greep and E. B. Astwood, Eds.). Am. Physiological Society, Washington, D. C., pp. 461-471

72. Mann, T. and Rottenberg, D. A. (1966): The carbohydrate of human semen. J. Endocr. 34, 257-264

73. Marsalek, J. and Obrucnik, M. (1958): Histochemical changes in ovaries in the hyperthecosis syndrome. Acta Univ. Palackianae Olomucensis 253-259

74. Marsalek, J. and Obrucnik, M. (1958 a): The topochemistry of carbohydrates in the human ovary in the Stein-Leventhal syndrome. Acta Histochem. 5, 325-336

75. Mawson, C. A. and Fischer, M. I. (1953): Zinc and carbonic anhydrase in human semen. Biochem. J. 55, 696-700

76. McDaniel, J. W., Scalzi, H. and Black, D. L. (1968): Influence of ovarian hormones on histology and histochemistry of the bovine oviduct. J. Dairy Sci. 51, 754-761

77. Melica, F. and Papadia, S. (1964): Histochemical analysis of mucopolysaccharides in pathologic endometrium. I. Simple folliculinic hyperplasia. Monit. Ostet.-Ginecol. 35, 50-56

78. Meyer, F. A. (1977): Comparison of structural glycoproteins from mucus of different sources. Biochem. Biophys. Acta 493, 272-282

79. Mitsugu, Y. (1966): Distribution and metabolism of acid mucopolysaccharides in the female rabbit genital organs by hormonal influence. Nippon Funin Gakkai Zasshi 11, 49-61 (Japan)

80. Moghissi, K. S. (1966): Cyclic changes of cervical mucus proteins. Fert. Steril. Proc. World Congr., 5th, Stockholm, pp. 723-725

81. Moore, R. A. (1952): Male secondary sexual organs. In: A. I. Lansing (Ed.). Cowdry's Problems of Aging. 3rd Ed. Williams and Wilkins, Baltimore, pp. 686-707

82. Morgan, C. F. (1963): A study of oestrogenic action on the collagen, hexosamine and nitrogen content of skin, uterus and vagina. Endocrinology 73, 11-19

83. Morgan, C. F. (1963 a): Temporal variations in the collagen, noncollagen protein and hexosamine of the rat uterus and vagina. Proc. Soc. Exptl. Biol. Med. 112, 690-694

84. Mostofi, F. K. (1979): Classification of tumors of testis. Ann. Clin. Lab. Sci. 9, 455-461

85. Nadler, H. L., Rao, G. J. S. and Taussig, L. M. (1978): Cystic fibrosis, In: J. B. Stanbury, J. B. Wyngaarden and D. S. Fredrickson (Eds.). The Metabolic Basis of Inherited Disease. 4th Ed. McGraw Hill, New York, Chap. 73, pp. 1683-1710

86. Nakaya, T. (1973): Acid mucopolysaccharides in the human cervix uteri. Nagoya Med. J. 18, 295-319

87. Nilsson, S. (1962): The human seminal vesicle. A morphogenetic and gross anatomic study with special regard to changes due to age and to prostatic carcinoma. I, II, III. Acta Chir. Scand. Suppl. 296, 5-96

88. Ohki, M. (1976): Composition and distribution of acidic glycosaminoglycans in human female reproductive organs. Nippon Funin Gakkai Zasshi 21, 86-93

89. Oppenheimer, E. H. and Esterly, J. R. (1970): Observations on cystic fibrosis of the pancreas. VI. The uterine cervix. J. Pediat. 77, 991-995

90. Paraventi, H., da Silva Sasso, W. and Mastroianni, E. (1960): Histochemical study of the uterus of white rats during pregnancy and in the postpartum stage. Rev. Paulista Med. 57, 73-90

91. Platt, H. A. (1966): The application of biophysical chemistry to the study of human cervical mucus. Ann. N. Y. Acad. Sci. 130, 925-934

92. Polishuk, W. Z. and Ron, M. (1976): Detection of ovulation by water content of vaginal mucus. Isr. J. Med. Sci. 12, 1207-1208

93. Riminiceanu, C., Dragan, M., Tudor, L., Litvac, E., Schneider, F. and Dema, E. (1966): Histochemical and histoenzymic aspects of experimentally induced proliferations of the uterine mucosa. Morfol. Norm. Pathol. (Bucharest) 11, 499-508

94. Rosenow, E. C. III and Lee, R. A. (1968): Cystic fibrosis and pregnancy. J. Am. Med. Assoc. 203, 227-229

95. Rzeszowska, G. (1966): Histochemical analysis of acid mucopolysaccharides in the epithelium of seminal vesicles and gonads during postembryonic development. Folia Morphol. 25, 463-469

96. Rzeszowska, G. (1967): Investigations of acid mucopolysaccharides in the epithelial cells of the seminal vesicles of white rats under experimental conditions. Endokr. Pol. 18, 29-35 (Pol)

97. Sasano, N. and Ichijo, S. (1969): Vascular patterns of the human testis with special reference to its senile changes. Tohoku J. Exptl. Med. 99, 265-272

98. Sato, C. S. and Gyorkey, F. (1972): Glycosaminoglycans of normal human prostate. Proc. Soc. Exp. Biol. Med. 140, 70-83

99. Scandroglio, R. (1967): On changes of the mucopolysaccharide content of the cells of the epithelium of the vagina in relation to the estrus cycle. (Studies in the rat). Atti Accad. Fisiocr. Siena (Medicofis) 16, 657-670 (Ita)

100. Scandroglio, R. (1967 a): Histochemical obeservations on rat vaginal epithelium. Rass. Med. Sarda 70, 471-476

101. Schams, D., Schallenberger, E., Hoffmann, B. and Karg, H. (1977): The estrous cycle of the cow. hormonal parameters and time relationships concerning estrus, ovulation, and electrical resistance of the vaginal mucus. Acta Endocrinol. (Copenhagen) 86, 180-192

102. Schumacher, G. F. Ḃ. (1977): Cervical secretions - a product or target organ for estrogens and gestagens. Clin. Orthop. Relat. Res. 129, 258-278

103. Senge, T., Neumann, F. and Richer, K. D. (Eds.) (1975): Physiology and Pathophysiology of the Prostate. Georg Thieme Verlag, Stuttgart, (Ger), 108 p.

104. Setchell, B. P. and Waites, G. M. H. (1975): The blood testis barrier. In: R. O. Greep and E. B. Astwood (Eds.). Handbook of Physiology, Sect. 7: Endocrinology, American Physiological Society, Washington, D. C. 5, 143-172

105. Shahani, S. M. (1964): Effect of norethisterone acetate (Anovlar) on human endometrium, and cervical mucus. Bull. Nat. Inst. Sci. India (27), 84-87

106. Smith, E. R. (1975): Canine prostate and its secretion. Adv. Sex. Horm. Res. 1 (Mol. Mech. Gonadal Horm. Action), 167-204

107. Szirmai, J. A. (1966): Effect of steroid hormones on the glycosaminoglycans of target connective tissues. In: E. A. Balazs and R. W. Jeanloz (Eds.). The Amino Sugars. Vol. IIB. Academic Press, New York, Chap. 41, pp. 129-154

108. Tadano, Y. and Yamada, K. (1978): The histochemistry of complex carbohydrates in the ovarian follicles of adult mice. Histochemistry 57, 203-215

109. Takamine, H., Ono, S. and Notsuki, I. (1951): Histochemical studies on acid polysaccharides. III. Distribution in female genital organs. Igaku to Seibutsugaku 20, 144-146

110. Takeva, Z. and Donchev, N. (1971): Histochemical investigations concerning polysaccharide material in the human embryonic ovary. Histochem. Cytochem. 9, 377-380

111. Talbert, G. B. (1977): Aging of the repoductive system. In: C. E. Finch and L. Hayflick (Eds.). Handbook of the Biology of Aging. Van Nostrand, New York, Chap. 13, pp. 318-356

112. Tateno, M. (1977): Cervical mucus in female sterility. Sanfujinka Chairyo 34, 568-573

113. Thorsoe, H. (1962): Development of polycystic ovaries following throidectomy (in rabbits). Role of acid mucopolysaccharides. Acta Endocrinol. 40, 161-174

114. Verne, J., Gabe, M. and Schramm, B. (1957): Histochemical changes caused by estrogens in the vagina of the prepuberal rat. Ann. Histochim. 2, 187-196

115. Wallner, O., Schill, W. B., Groesser, A. and Fritz, H. (1975): Participation of the kallikrein-kinin-system in sperm penetration through cervical mucus - in vitro studies. Kinnogenases, Symp. Physiol. Prop. Pharmacol. Ration., Reprodu. 2nd 1974, 63-70

116. Werner, I. (1953): Studies on glycoproteins from mucous epithelium and epithelial secretion. Acta Soc. Med. Upsalien. 58, 1-55

117. Wilson, L. and Aronson, W. (1950): Assay of hyaluronidase in human semen. Fertility and Sterility 1, 254-258

118. Wincek, T. J., Parrish, R. F. and Polakoski, K. L. (1979): Fertilization: a uterine glycosaminoglycan stimulates the conversion of sperm proacrosin to acrosin. Science 203, 553-554

119. Wurgaft, R. and Tchernitchin, A. (1976): Histochemical study of the rat uterus after estrogen stimulation. IRCS Med. Sci. Libr. Compend. 4, 85

120. Yanagimachi, R., Nicolson, G. L., Noda, Y. D. and Fujimoto, M. (1973): Electron-microscopic observations of the distribution of acidic anionic residues on hamster spermatozoa and eggs before and during fertilization. J. Ultrastruct. Res. 43, 344-353

121. Yoshizaki, H. (1973): Glycosaminoglycan in gynecology. Kagaku No Ryoiki, Zokan 99, 269-280

122. Zachariae, F. (1958): Autoradiographic and histochemical studies of sulfomucopolysaccharides in the rabbit uterus, oviducts, and vagina. Acta Endocrinol. 29, 118-134

123. Zachariae, F. (1960): Acid mucopolysaccharides in the female genital system and their role in the mechanism of ovulation. Acta Endocrinol. 33, 471-474

124. Zachariae, F. and Jensen, C. E. (1958): Mechanism of ovulation. Histochemical and physiochemical investigations on genuine follicular fluids. Acta Endocrinol. 27, 343-355

125. Zachariae, F. and Thorsoe, H. (1966): Hormonal control of acid mucopolysaccharides in the female genital tract. In: G. Asboe-Hansen (Ed.). Hormones and Connective Tissue. Munksgard, Copenhagen, pp. 257-281

126. Zinca, V., Sreimbeanu, I. and Perju, Al. (1959): Cytochemical, histochemical, and histopathological changes of the vaginal mucosa in infestations with Trichomonas vaginalis. Obstet. si ginecol. (Bucharest) 6, 347-354

Chapter 15

Cerebrospinal Fluid

15.1 Introduction

The cerebrospinal fluid (CSF), as the name implies, is the fluid associated with the brain and spinal cord and is contained within and surrounds these neural tissues. Embryologically, these tissues develop from a hollow, fluid-filled neural tube and retain this character throughout the life. The CSF acts as a cushion for the brain within the solid vault and acts as a *drain* for the central nervous system parenchyma.

Four macromolecules, proteins, glycoproteins, gangliosides and glycosaminoglycans have been investigated in cerebrospinal fluid as a measure of their metabolism in brain. The majority of these macromolecules are known to originate from blood, but because of the proximity of this fluid to the central nervous system increased quantities of normal and abnormal components appear in the fluid in conditions involving inflammatory, degenerative and neoplastic changes in the nervous tissue. These conditions directly or indirectly alter the blood-brain barrier or the blood-CSF barrier (Bakay, 1972). Protein elevation in CSF observed in inflammatory, atrophic and degenerative diseases of brain and in demyelinating diseases has been attributed to a change in permeability of the blood-brain barrier. The rate of the passage of ^{131}I-labeled albumin into serum following injection into the cisternal fluid was found to change in such conditions.

The discussion in this chapter will be restricted to the components of glycosaminoglycans (hexosamine, hexuronic acid and sulfate), the intact macromolecules and the enzymes involved with their metabolism and present in the CSF. Since hexosamine is a component common to both glycoproteins and glycosaminoglycans, some reference will also be made to glycoproteins of CSF in normal and pathological states. In order to understand these changes brief information on the physiology will first be presented.

15.2 Location, origin and fate of CSF

The physiological and pathophysiological aspects of CSF have been reviewed in several excellent articles and monographs (Davson, 1969; Prockop, 1973; Eliasson et al., 1974; Cserr et al., 1975; Katzman, 1975; Guyton et al., 1975; Bannister, 1978). The chemical pathophysiology of CSF has been reviewed by Lowenthal (1972).

In adult human subjects approximately 150 ml of CSF is present in two major fluid compartments of the central nervous system, the ventricular compartment and the subarachnoid space and in a minor compartment consisting of the cisterns outside the brain. These compartments communicate with each other through one or more of the foramina. The ventricular compartment consists of the two lateral ventricles, a medial third ventricle and the fourth ventricle and only one-sixth (25 ml) of the total CSF in an adult is present in this compartment. The remaining fluid is present in the subarachnoid space and the extraventricular cisterns.

The CSF is formed by secretion at a rate of 520 μl per minute in man. About two-thirds is secreted by the choroidal plexuses, which are highly vascular cauliflower-like structures formed from the walls of the ventricles lined by specialized ependymal epithelium cells. The remaining one-third is derived from the extra-choroidal sources. The CSF so formed is also continously drained out by absorption at arachnoid villi, which have a kind of one-way valvular system permitting flow only from the subarachnoid space into their lumens. The rate of fluid formation is not dependent on the intraventricular CSF pressure but its absorption depends on this pressure. The CSF pressure is regulated by a combination of the rate of fluid formation and the resistance to fluid absorption in the subarachnoid villi. An increase in either of these factors raises the CSF pressure and vice versa. An increase in CSF pressure resulting from increased CSF volume may cause the head to enlarge producing hydrocephalous. Congenital overdevelopment of the choroid plexuses in some newly-born babies do produce excessive CSF pressure and hydrocephalous. In meningitis, increased CSF pressure results from the blockage of the pores for the arachnoidal villi with large quantities of tissue debris and leukocytes. Similarly increased pressure could result from compression of the arachnoid space by a brain tumor.

In the condition called hydrocephalous, excess fluid accumulates in the cranial vault. This condition can be of two types: communicating and non-communicating hydrocephalous. In communicating type, the CSF can flow freely from the ventricles into the subarachnoid space and is due to production of more fluid than can be drained out into venous system through the arachnoid villi. This is caused by congenital overdevelopment of choroid plexuses in the newborn infants and results in swelling of the head and damage to the brain. In noncommunicating hydrocephalous, CSF outflow from one or more of the ventricles is blocked. Usually the block occurs in the aqueduct of the Sylvius with the result that the fluid formed by the choroid plexuses in the two lateral ventricles and the third ventricle cannot escape and grows in volume.

The composition of the CSF may vary with the region from where this fluid is withdrawn. For mucopolysaccharide and glycoprotein analyses the fluid is usually obtained by lumbar puncture of the subarachnoid space.

15.3 Glycosaminoglycans in normal CSF

Only a small amount of glycosaminoglycans (acid mucopolysaccharides) are present in the CSF from normal subjects. Using a spot test and cellulose acetate electrophoresis of 160 µl of CSF concentrated to 25 µl and also spectrophotometric analysis of the precipitate obtained with protamine sulfate, Friman (1967) failed to detect any mucopolysaccharides in three girls of 5,6 and 8 years suffering from Hurler's syndrome.This investigator reported the absence of acid mucopolysaccharides and stated if these are present, would be less than 3 µg/ml. The failure to detect any GAG in CSF was due to the use of less sensitive methods (Constantopoulos and Dekaban, 1970). It was surprising that whereas mucopolysaccharides accumulate in many tissues including the brain and meninges and proteins of high molecular weights can pass through the blood-CSF barrier, why could not the small molecular weight mucopolysaccharides like dermatan sulfate (M_r 15,000) and heparan sulfate (M_r 2,700-5,500)? Margolis (1969) suggested that the reported absence of these macromolecules may be due to their highly charged nature which might make them impermeable to the blood-CSF barrier.

Using more highly sensitive methods with a lower detection limit of less than 3 µg/100 ml (as compared to that used by Friman 1967, lower detection limit 300 µg/100 ml) Constantopoulos and Dekaban (1970) reported the presence of a small quantity of mucopolysaccharides in CSF from normal subjects. In five unaffected controls the CSF contained about 10 µg of hexuronic acid per 100 ml which increased 8-20 times in four patients with Hunter-Hurler's syndrome. In one patient the acid mucopolysaccharides isolated from the large sample of CSF obtained at autopsy upon paper electrophoresis showed equal proportions of heparan and dermatan sulfates. The molecular weight distribution obtained by gel filtration through Sephadex G-200 revealed molecular weights of mucopolysaccharides below 3,000; only a small fraction of dermatan sulfate had a relatively high molecular weight.

15.4 Source of CSF glycosaminoglycans

The source of glycosaminoglycans in CSF is not known exactly. These may be derived largely from the serum. It appears that the conditions that affect the synthesis and degradation of glycosaminoglycans (GAG) in brain cells also affect the changes in composition of these macromolecules in CSF. The indirect evidence for brain as a source of GAG in CSF is provided by elevation of GAG levels in CSF of some patients with mucopolysaccharidoses when their serum GAG levels are normal (Dekaban and Constantopoulos, 1973 a). If it is established that some GAG in normal CSF are derived from the central nervous system, this contribution may become more prominent in diseases that involve inflammatory, degenerative and neoplastic processes in the nervous tissue. It is interesting to note in this context that when the CSF proteins are elevated in these conditions of the central nervous system, proteins of brain origin appear in the CSF (Lowenthal, 1972). The CSF contains cells which stain metachromatically for mucopolysaccharides (Consoli et al., 1955) and these may also be the minor contributor to the GAG in CSF. The proteins in CSF are derived from three sources: plasma, neural tissue and local synthesis by cells in CSF (Nakajima, 1959). The same may possibly be true for the CSF glycosaminoglycans.

15.5 Relationship of CSF glycosaminoglycans to age

This relationship has not been investigated since there have been analysis on only a few CSF samples from normal subjects and patients. It is possible that there may be age-related changes in CSF like those observed in other biological fluids. In this context other carbohydrate components like sialic acid in CSF have been reported to be age-dependent (Papadopoulos et al., 1960; Rossi, 1960). The levels of sialic acid (0.54 mg/dl) in children are lower than adult levels of 0.70 mg/dl and these levels are the highest between 35-45 years of age (Papadopoulos et al., 1960).

15.6 Glycosaminoglycans of CSF in pathological states

The variations in the CSF composition of hexosamine, sulfate, glycosaminoglycans and the enzymes involved in degradation of these macromolecules have been studied in several disease states.

The protein bound hexosamine in the CSF represents both the glycoproteins and glycosaminoglycans (GAG); the major portion of it being from the glycoproteins. The protein-bound hexosamine in CSF is 6.7-12.2 μmoles per 100 ml and consists of about 15 % galactosamine and 85 % glucosamine. In 14 cases of non-demyelinating diseases the hexosamine content of the ethanol-precipitated fraction corresponded to 4.2 ± 1.8 mg per liter (Stacey and Barker, 1962). The hexosamine values are also elevated in febrile states and infection of the meninges (Bauer, 1957). Variations in protein-bound hexosamines in CSF have been reported in various psychiatric states including mood changes (Bogoch et al., 1960), and an elevation to twice the normal value in depressed patients (Kuhn, 1959). An increase of the galactosamine in CSF occurred in alcoholic psychosis and chronic schizophrenics (Burtin, 1959) and cerebral sclerosis and some schizophrenics (Kuhn, 1959). Some of these changes may be the result of the reported changes of the cell permeability of the brain in mental states (Eichorn and Taschev, 1961). An earlier report contradicted the aforementioned work and hexosamine was reported to vary only slightly, if at all, in psychiatric disorders (Yoshida, 1957) but the hexosamine levels do not parallel the protein levels (Corridori and Pacini, 1965) in psychiatric disorders. Hexosamine may stay normal or increase in multiple sclerosis (Dyrbye and Fog, 1962). The variations in CSF hexosamine in various pathological states have been reviewed (Canal and Frattola, 1963). Progressive increase in CSF mucoprotein levels occurred during the entire 3-5 months of electroshock therapy (Stary and McNichol, 1959).

The sulfate concentration is lower in CSF than in plasma and the choroid plexus transports sulfate out of the CSF for uptake and incorporation into glycosaminoglycans in brain (Robinson et al., 1968, 1968 a). The entry of sulfate from CSF to brain tissue is also affected by neural activity (Lorenzo, 1969) and the incorporation of ^{35}S sulfate into GAG is affected by vitamin A (Barker et al., 1964). In vitamin A deficiency, the synthesis of sulfated GAG decreases and that of nonsulfated GAG increases. An increase in CSF pressure in calves has been reported as a consequence of vitamin A deficiency (Cousins et al., 1969) and its cause has been traced to an increase in nonsulfated GAG in the intracellular connective tissue of cerebral dura mater, which affects water binding (Mukerjee and Bachhawat, 1965) and resistance to CSF absorption thereby producing hydrocephalus. This is further supported by the observation that hypervitaminosis A leads to a decline in GAG content of the dura mater and a decreased CSF pressure (Gorgacz et al., 1971) and that chronic vitamin A treatment of rats produces a drop in CSF pressure (Maddux et al., 1974).

The glycosaminoglycans of CSF have been investigated in mucopolysaccharidoses, since these macromolecules accumulate in brain and meninges and other tissues in patients with these disorders. Only a few such reports have appeared. Friman (1967) reported lack of glycosaminoglycans in the CSF from three girls with Hurler syndrome. Using more sensitive methods, including ECTEOLA-cellulose ion exchange chromatography, paper electrophoresis and gel chromatography on Sephadex G-200, the GAG were found to be elevated in four patients with Hunter-Hurler syndrome (Constantopoulos and Dekaban, 1970). Whereas the CSF form 5 unaffected controls contained only 8.3-10 µg of hexuronic acid per 100 ml CSF, these values were elevated to 75-76 µg/100 ml in two patients with Hurler's syndrome, who were heparan sulfate excretors and were in advanced state of the disease. In two autosomal recessive mixed excretors in advanced state, the values of hexuronic acid were even higher, 179 and 400 µg/100 ml CSF. On the basis of the carbazole/orcinol ratio, paper electrophoresis and elution behavior on ECTEOLA-cellulose and Sephadex G-200, the GAG in CSF from a decreased mixed excretor contained equal quantities of heparan sulfate and dermatan sulfate. The urinary excretion pattern in this patient was different from that in CSF and contained 15 % heparan sulfate and 85 % dermatan sulfate.

Although the source of glycosaminoglycans in CSF is not definitely known, the relation of elevated GAG in CSF of 14 mucopolysaccharidoses patients to brain involvement and mental retardation has been investigated (Dekaban and Constantopoulos, 1973, 1973 a). The GAG were markedly elevated in three Hurler (Type I) patients, in two patients with Hunter (Type II) syndrome and six with Sanfilippo (Type III) syndrome. The hexuronic acid content of GAG isolated by ECTEOLA-cellulose chromatographic method of Di Ferrante (1967) was in the range of 5-15 µg/100 ml in 11 control subjects, which increased 8-25 times (68-179 µg/100 ml) in Hurler, Hunter and Sanfilippo patients. However, the GAG levels in CSF of two patients with Scheie (Type V) syndrome were (7.0 and 9.0 µg/100 ml) within the normal range and in one patient with Scheie syndrome the values increased only three fold (41 µg/100 ml) above the normal values. Only these three patients with Scheie syndrome had unimpaired intellect and had no neurological symptom even though true mucopolysacchariduria and involvement of organs was present in these patients.

In Hurler, Hunter and Sanfilippo patients, elevation of GAG in CSF had a positive correlation with mental impairment. In patients who had a follow up study since early infancy and since the onset of symptoms, the GAG in their CSF increased with progression of time (Dekaban and Constantopoulos, 1973).

The GAG have not been investigated in the CSF of Marteaux-Lamy type mucopolysaccharidosis patients (where also like Scheie syndrome the intellect remains unimpaired) and in other types of mucopolysaccharidoses; one problem is that patients with some types of these diseases are scarce. Additional work on such patients will help to establish the relation of elevated GAG to mental retardation already suggested by the work of Dekaban and Constantopoulos (1973, 1973 a).

Analysis of the GAG isolated from the CSF of the mucopolysaccharidoses patients by chromatographic elution profile and by digestion with chondroitinase ABC (which degrades chondroitin 4- and 6- sulfates and dermatan sulfate but does not affect heparan sulfate) showed that heparan sulfate was elevated in Hunter, Hurler and Sanfilippo syndromes but not in Scheie's syndrome (Dekaban and Constantopoulos, 1973). In an earlier investigation (George and Bachhawat, 1971) heparan sulfate was also found as the predominant GAG in the CSF of a Sanfilippo patient. Heparan sulfate has been suggested to be the culprit for producing mental retardation through its deleterious effects on the nervous system in Hurler, Hunter and Sanfilippo patients, through an unknown mechanism. Normal intellect is preserved in Scheie type mucopolysaccharidosis, where heparan sulfate was not the major GAG in the CSF (Dekaban and Constantopoulos, 1973, 1973 a).

Communicating hydrocephlous (see Sec. 15.2) is a complication in Hurler, Hunter and Sanfilippo syndromes, which as described above, are accompanied by changes in GAG in CSF. Fowler et al. (1975) reported from pneumographic and radioisotope (RISA) cisternographic evidence that the communicating hydrocephalous in these patients has its pathogenesis in the storage of mucopolysaccharide material in the piaarachnoid causing an impairment in CSF absorption. The communicating hydrocephalous occurs in several other genetic metabolic defects in which mucopolysaccharide metabolism is also deranged, including achondroplasia (James et al., 1972; Fowler et al., 1975), Marfan's syndrome (McKusick, 1966) and osteogenesis imperfecta (Freda and Vosburgh, 1961). Although GAG in CSF have not been investigated in these conditions, yet there seems to be a possibility of alteration of these macromolecules in CSF, as also in acute radiation sickness in which GAG in the meninges are affected (Izmailova, 1968).

The thickening of the meninges in the ventricular region of the brain (ventriculomegaly) due to storage of the mucopolysaccharide material in Hurler's syndrome produces communicating hydrocephalous and changes in GAG in the CSF. A similar thickening of the dura mater or arachnoid in the cervical spinal cord or in the thoraco-lumbar region produces spinal cord compression (Kennedy et al., 1973), gait disorder and spinal pachymeningitis (Paulson et al., 1974). It is possible that the changes observed in GAG in the CSF of the patients with mucopolysaccharidoses represent these changes in the ventricular and extraventricular meninges. Manno et al. (1965) have reported the total polysaccharide changes in CSF in diseases of meninges, cerebrum, brainstem and cerebellum.

15.7 Lysosomal GAG hydrolyzing enzymes in CSF

Lysosomal enzymes related to GAG degradation present in CSF include: β-glucuronidase, β-hexosaminidase, aryl sulfatase and β-galactosidase. An increase in the CSF β-glucuronidase occurs in conditions involving destruction of the nervous tissue, leukocyte increase in CSF and abnormalities of the blood-brain barrier (Prockop, 1973). An increase in activity of this enzyme has also been reported in demyelinating diseases and tumors of the nervous tissue (Allen and Reagen, 1964), meningitis (D'Agostino et al., 1964), Uremia (Mueting, 1965). glioblastoma multiforme (Anlyan and Starr, 1952), diabetic neuropathy (Hildebrand and Levin, 1973), eclampsia and preclampsia (Toschi, 1964), cerebral anoxia in rabbits (Laborit and London, 1970) but not in epilepsy (Plum, 1967). A precipitous rise of β-glucuronidase and protein occurred independent of the response of the nervous system injury in experimental allergic encephalitis in rabbits (Paterson and Steele, 1970). The postmortem chemistry of CSF has been reviewed (Coe, 1977). The β-glucuronidase was significantly elevated in chronic neoplastic meningitis and the variably changed activity after death may be similar to the one when the brain is irreversibly damaged (Paulson and Stickney, 1971). The mechanism involved in the increase of this enzyme may differ in various disease conditions; it may be related to degree of brain damage and neuropathy in distemper-induced enecephalitis in dogs (Long et al., 1973).

The enzyme β-hexosaminidase is present in CSF and may vary in disease conditions (Viallard et al., 1972). The β-galactosidase and aryl sulfatase are also present in CSF and are not changed in cerebrovascular disease and presenile dementia (Yates et al., 1976).

15.8 Effects of hyaluronidase injection into CSF

Intracisternal administration of hyaluronidase (EC 3.2.1.35) disintegrates the choroid plexus producing a ten-fold increase in protein and appearance of cells of choroidal origin in CSF (Bairati, 1953). Intraventricular injection of hyaluronidase in cats (Custod and Young, 1968) produces neurological abnormalities and a rise in CSF pressure (Owens and Clark, 1956). Although the GAG in CSF were not investigated in these studies, the significance of GAG in brain function appears evident and in some studies there is an evidence of an increase in the permeability of the blood-brain barrier produced by hyaluronidase administration, which might change the GAG in CSF. Intrathecal injection of hyaluronidase has been used therapeutically for producing behavioral improvement in patients with thought block and excited states (Cargenello and Tischler, 1952) and in patients with senile arteriosclerotic behavioral disorders (Cameron et al., 1957). Similarly, intraventricular/subarachnoid injections of hyaluronidase in 15 cases of cranial archnoiditis/communicating hydrocephalous complicating tubercle meningitis produced improvements of sensorium and neurological deficit. It was without any side effects and was superior to shunt surgery (Gouri-Devi and Satish, 1980). Intrathecal hyaluronidase was likewise helpful in therapy of spinal arachnoiditis complicating tuberculous meningitis (Gouri-Devi, 1980).

Gouri-Devi, M. (1980): Use of intrathecal hyaluronidase in spinal arachnoiditis complicating tuberculous meningitis. Indian J. Med. Res. 71, 581-593

Gouri-Devi, M. and Satish, P. (1980): Hyaluronidase as an adjuvant in the treatment of cranial arachnoiditis (hydrocephalous and optochisamatic arachnoiditis) complicating tubercle meningitis. Acta Neurol. Scand. 62, 368-381

15.9 Bibliography

1. Allen, N. and Reagen, E. (1964): Beta-glucuronidase activities in cerebrospinal fluid. Arch. Neurol. 11, 144-154

2. Anlyan, A. J. and Starr, A. (1952): β-Glucuronidase activity of spinal and ventricular fluids in humans. Cancer 5, 578-580

3. Bairati, A. (1953): Spreading factor and mucopolysaccharides in the central nervous system of vertebrates. Experientia 9, 461-462

4. Bakay, L. (1972): Alterations of the brain barrier system in pathological states. In: A. Lajtha (Ed.). Handbook of Neurochemistry, Plenum Press, New York, Vol. 7, Chap. 17, pp. 417-427

5. Bannister, R. Sir (1978): Brain's Clinical Neurology. Oxford Univ. Press, New York, 5th Ed., pp. 239-245

6. Barker, S. A., Cruickshank, C. N. D. and Webb, T. (1964): The effect of vitamin A, hydrocortisone and citral upon sulfate metabolism in skin. Exp. Cell Res. 35, 255-261

7. Bauer, H. (1957): The question of the identity of cerebrospinal fluid proteins with those of the blood serum I. The fore fraction, γ-globulin, and lipoproteins. II. The protein-bound carbohydrates of the cerebrospinal fluid. Deut. Z. Nervenheilk. 175, 354-357 and 488-510

8. Bogoch, S., Dussik, K. T., Fender, C. and Couran, P. (1960): Longitudinal clinical and neurochemical studies on schizophrenic and manic-depressive psychoses. Am. J. Psychiat. 117, 409-420

9. Burtin, P. (1959): The proteins of cerebrospinal fluid. Clin. Chim. Acta 4, 72-78

10. Cameron, D. E., Levy, L. and Hunzinger, W. (1957): Intrathecal administration of hyaluronidase. Effects upon the behavior of patients suffering from senile and arteriosclerotic behavior disorders. Am. J. Psychiat. 113, 893-900

11. Canal, M. and Frattola, I. (1963): Constuenti Organici del Liquor Cefalor a Chidiano. Fondazione, Ganassini, Milano

12. Cargnello, D. and Tischler, A. (1952): Sistema nervosa (Milano) 4,2. Quoted by Cameron et al. (1957)

13. Coe, J. I. (1977): Postmortem chemistry of blood, cerebrospinal fluid and vitreous humor. Legal Medical Annual, 1976 (Publ. 1977), 55-92

14. Consoli, G., Pariante, F. and Schettini, F. (1955): The particular behavior of leukocytes of the spinal fluid in subjects with meningitis evaluated in connection with cytochemical studies on the localization of lipides. Boll. Soc. Ital. Biol. Sper. 31, 607-609

15. Constantopoulos, G. and Dekaban, A. S. (1970): Acid mucopolsaccharides in the cerebrospinal fluid of patients with Hunter-Hurler's syndrome. J. Neurochem. 17, 117-120

16. Corridori, F. and Pacini (1965): Determination of sialic acid and hexosamine contents of the CSF in different neurological diseases. Riv. Neurol. 11, 655-668

17. Cousins, R. J., Eaton, H. D., Rousseau, J. E. Jr. and Hall, R. C., Jr. (1969): Biochemical constituents of the dura mater in vitamin A deficiency. J. Nutr. 97, 409-418

18. Cserr, H., Fenstermacher, J. and Fencl, V. (Eds.) (1975): The Fluid Environment of the Brain. Academic Press, New York

19. Custod, J. T. and Young, I. J. (1968): Cat brain mucopolysaccharides and their in vivo hyaluronidase digestion. J. Neurochem. 15, 809-813

20. D'Agostino, S., Maddaluno, R. and Diversi, D. (1964): Glucuronic acid level in spinal fluid and in blood of healthy subjects and of patients with meningeal diseases. Aggiorn. Pediat. 15, 565-572

21. Davson, H. (1969): The cerebrospinal fluid. In: A. Lajtha (Ed.). Handbook of Neurochemistry, Vol. 2, Plenum Press, New York, Chap. 3, pp. 23-48

22. Davson, H. (1970): Physiology of cerebrospinal fluid. Williams and Wilkins Co., Baltimore

23. Dekaban, A. S. and Constantopoulos, G. (1973): Mucopolysaccharidoses: relation of elevated CSF mucopolysaccharides to brain involvement. Trans. Am. Neurol. Assoc. 98, 14-17

24. Dekaban, A. S. and Constantopoulos, G. (1973 a): Mucopolysaccharidoses. Relation of elevated cerebral spinal fluid to mental retardation. Arch. Neurol. 28, 385-388

25. DiFerrante, N. M. (1967): The measurement of urinary mucopolysaccharides. Anal. Biochem. 21, 98-106

26. Dyrbye, M. O. and Fog, T. (1962): The hexosamine concentration in the spinal fluid in patients with disseminating sclerosis. Arch. Neurol. Scand. 38, 1-2

27. Eichorn, O. and Taschev, T. (1961): Cell permeability of the brain in mental disease. In: J. Folch-Pi (Ed.). Chemical Pathology of the Nervous System, Pergamon Press, New York, pp. 638-640

28. Eliasson, S. C., Prensky, A. L. and Hardin, W. B. (Eds.) (1974): Neurological Pathophysiology, Oxford Univ. Press, New York

29. Fowler, G. W., Sukoff, M., Hamilton, A. and Williams, J. P. (1975): Communicating hydrocephalous in children with genetic inborn errors of metabolism. Childs Brain 1, 251-254

30. Freda, V. J. and Vosburgh, G. J. (1961): Osteogenesis imperfecta congenita: a presentation of 16 cases and review of the literature. Obstet. Gynecol. 18, 535-547

31. Friman, C. (1967): Hurler's syndrome. Lack of acid mucopolysaccharides in the cerebrospinal fluid. Clin. Chim. Acta 15, 375-378

32. George, E. and Bachhawat, B. K. (1971): Biochemical studies in Sanfilippo syndrome. Neurol. India 19, 97-101

33. Gorgacz, E. J., Rousseau, J. E. Jr., Frier, H. I., Hall, R. C. and Eaton, H. D. (1971): Composition of the dura mater in chronic bovine hypervitaminosis. Amer. J. Nutr. 101, 1541-1546

34. Guyton, A. C., Taylor, A. E. and Granger, H. J. (1975): Circulatory Physiology II. Dynamics and Control of the Body Fluids. W. B. Saunders, Philadelphia, Chap. 15, pp. 206-224

35. Hildebrand, J. and Levin, S. (1973): Enzymatic activities in cerebrospinal fluid in patients with neurological diseases. Acta Neurol. Belg. 73, 229-240

36. Izmailova, G. M. (1968): State of acid mucopolysaccharides in the meninges in acute radiation sickness. Med. Radiol. (Moskva) 13, 86-87

37. James, A. E. Jr., Dorst, J. P., Mathews, E. S. and McKusick, V. A. (1972): Hydrocephalous in achondroplasia studies by cisternography. Pediatrics 49, 46-49

38. Katzman, R. (1975): Cerebrospinal fluid physiology: Role of secreted and mediated transport systems. In: R. O. Brady (Ed.). The Basic Neuroscience. Vol. 1 of the Nervous System (Editor-in-chief D. B. Tower), Raven Press, New York, pp. 291-297

39. Kennedy, P., Swash, M. and Dean, M. F. (1973): Cervical cord compression in mucopolysaccharidosis. Develop. Med. Child Neurol. 15, 194-199

40. Kuhn, R. (1959): Sympos. I. Carbohydrate Chemistry of Substances of Biological Interest, Proceedings of Fourth International Congress of Biochemistry, Vienna, September 1-6, 1958, Pergamon Press, London, Vol. 1, p. 67

41. Laborit, H. and London, A. M. (1970): Variations de l'activité de la beta-glycuronidase du sang artériel du sang veineux jugulaire et du liquide céphalo-rachidien après une anoxie cérérbrale par clampage chez le lapin. Agressologie 11, 475-481

42. Long, J. F., Jacoby, R. O., Olson, M. and Koestner, A. (1973): Beta-glucuronidase activity, and levels of protein and protein fractions in serum and cerebrospinal fluid of dogs with distemper associated demyelinating encephalopathy. Acta Neuropathol. 25, 179-187

43. Lorenzo, A. V. (1969): Mechanisms of drug penetration in the brain. In: P. Black (Ed.). Drugs Brain, John Hopkins Press, Baltimore, pp. 41-59

44. Lowenthal, A. (1972): Chemical pathophysiology of the cerebrospinal fluid. In: A. Lajtha (Ed.). Handbook of Neurochemistry, Plenum Press, New York, Vol. 7, pp. 429-464

45. Maddux, G. W., Foltz, F. M. and Nelson, S. R. (1974): Effect of vitamin A intoxication on intracranial pressure and brain water in rats. J. Nutr. 104, 478-482

46. Manno, N. J., McGuckin, W. F. and Goldstein, N. P. (1965): Cerebrospinal fluid total polysaccharide in diseases of the nervous system. Neurology 15, 49-55

47. Margolis, R. U. (1969): Mucopolysaccharides. In: A. Lajtha (Ed.). Handbook of Neurochemistry, Plenum Press, New York, Vol. I, Chap. 12, pp. 245-260

48. McKusick, V. A. (1966): Heritable Disorders of Connective Tissue, Mosby, St. Louis, p. 354

49. Mueting, D. (1965): Pathogenesis of uremia. Comparative determinations of glucuronic acid, indican, free and bound phenols in the serum, cerebrospinal fluid, and urine of renal diseases with and without uremia. Clin. Chim. Acta 12, 551-554

50. Mukerjee, B. and Bachhawat, B. K. (1965): Role of vitamin A in sulfate metabolism. I. Studies on sulfate incorporation into mucopolysaccharides in vitamin A deficient rats. Indian. J. Biochem. 2, 90-93

51. Nakajima, K. (1959): Paper electrophoretic studies on cerebrospinal fluid protein. I. Normal cerebrospinal fluid protein fractions, and influence of serum protein on cerebrospinal fluid protein pattern in various pathophysiologic conditions. Naika Hokan 6, 221-230

52. Owens, C. and Clark, S. L. (1956): Some effects of injections of hyaluronidase in the subarachnoid space of experimental animals. J. Neurosurg. 8, 311-317

53. Papadopoulos, N. M., William, M. and Hess, W. C. (1960): Neuraminic acid in spinal fluid: identification and determination. Georgetown Med. Bull. 14, 15-18

54. Paterson, P. Y. and Steele, F. M. (1970): Cerebrospinal fluid beta-glucuronidase activity in rabbits with experimental allergic encephalomyelitis. Clin. Exp. Immunol. 7, 759-768

55. Paulson, G. W., Meagher, J. N. and Burkhart, J. (1974): Spinal pachymeningitis secondary to mucopolysaccharidosis. Case report. J. Neurosurg. 41, 618-621

56. Paulson, G. W. and Stickney, D. (1971): Cerebrospinal fluid after death. Confin. Neurol. 33, 149-162

57. Plum, C. M. (1967): Beta-glucuronidase activity in serum, cerebrospinal fluid and urine in normal subjects and in neurological and mental patients. Enzymol. Biol. Clin. 8, 97-112

58. Prockop, L. D. (1973): Disorders of cerebrospinal fluid and brain extracellular fluid. In: G. E. Gaull (Ed.). Biology of Brain Dysfunction, Vol. 1, Plenum Press, New York, Chap. 6, pp. 229-263

59. Robinson, R. J. and Cutler, R. W. P. (1968): Development of transport mechanisms for sulphate and iodide in immature choroid plexus. J. Neurochem. 15, 455-458

60. Robinson, R. J., Cutler, R. W. P., Lorenzo, A. L. and Barlow, C. F. J. (1968 a): Transport of sulphate, thiosulphate and iodine by choroid plexus in vitro. J. Neurochem. 15, 1169-1179

61. Rossi, F. (1960): The determination of sialic acid in the cerebrospinal fluid with Hess' reaction. Riv. Patol. Clin. 15, 146-149 (It)

62. Stacey, M. and Barker, S. A. (1962): Carbohydrates of Living Tissues. Van Nostrand, London, p. 174

63. Stary, Z. and McNichol, R. (1959): Studies on mucoproteins in cerebrospinal fluid. Fed. Proc. Abst. 18, 1306

64. Toschi, P. (1964): β-glucuronidase activity of serum, amniotic fluid, and cerebrospinal fluid in preeclampsia and eclampsia. Attuatlita Ostet. Ginecol. 10, 22-31

65. Viallard, J. L., Motta, C. and Dastugue, G. (1972): Activité de la N-acetyl-beta-glucosaminidase (E C 3.2.1.30) dans le liquide céphalo-rachidien humain. Pathol-Biol (Paris) 20, 655-659 (Fre)

66. Yates, C. M., Urquhart, A., Wilson, H. and Davidson, D. (1976): Lysosomal enzymes in cerebral atrophy. Clin. Chim. Acta 71, 215-219

67. Yoshida, T. (1957): On hexosamine contained in the cerebrospinal fluid of psychotics and neuropathics. Folia Psychiat. Neurol. Jap. 11, 259-300

Chapter 16

Glycosaminoglycans of Blood

16.1 Introduction

Blood contains only a small amount of glycosaminoglycans. These glycosaminoglycans of blood are present both in plasma and the formed elements (blood cells). In plasma, the GAG are present in free form and protein-bound forms. In addition to these, a large quantity of low molecular weight degraded species of GAG are also present. The values reported by different workers have varied, who by using different methods have measured the free, bound or both GAG fractions together in plasma/serum. Also, since the presence of proteins interferes with the determination of uronic acid (Mosher, 1962; Harris and Fraser, 1969), the values obtained vary with the extent of protein removal achieved in a procedure prior to uronic acid determination. Some of the methods used have involved the determination of hexuronic acid directly on the plasma protein precipitates (Stary and Yuvanadis, 1953), on GAG isolated from trichloroacetic acid precipitate of plasma (Taniguchi et al., 1974), on the material isolated from euglobulin fraction of plasma (Badin et al., 1955; Bassiouni, 1955; Bollet et al., 1957; Schiller, 1958; Endo et al., 1976) or on GAG isolated by proteolysis of serum/plasma (Niebes and Schifflers, 1975; Varma et al., 1979).

16.2 Source of Plasma Glycosaminoglycans

The glycosaminoglycans of plasma originate from the connective tissue (McClusky and Thomas, 1958). In tissues, the GAG are present as protein-carbohydrate complexes (proteoglycans). The carbohydrate and protein components of these composite macromolecules are formed and degraded concomitantly (Gross et al., 1960). On the basis of the total amount of GAG present in the body and the average half life of the GAG components of the proteoglycans, it has been estimated that approximately 250 mg of GAG are removed from the tissue proteoglycans by the action of proteases and carbohydrases (Leaback, 1970). A major portion of the GAG released from tissue proteoglycans, especially from the vascular tissue, enter the blood circulation for reutilization in the body and only a small fraction of it (8-10 mg) is filtered out by the kidney in urine (Di Ferrante and Rich, 1956; King and Boyce, 1963).

A part of the plasma GAG originates from the formed elements of blood (Kerby and Langley, 1959; Olsson, 1968). Besides the connective tissue cells the blood cells also synthesize GAG. The presence and synthesis of GAG in blood platelets and leukocytes have been reported by several authors (see section 16.6).

The contribution of GAG from the tissues or the formed elements may increase in different physiological and pathological states. Thus, the fully sulfated chondroitin 4-sulfate proteoglycan present in the thrombocytes (Kerby and Langley, 1959), is released into the blood during the early phase of coagulation (Olsson and Gardell, 1967). It is, therefore, possible that the blood plasma/serum GAG spectrum and total GAG could change in coagulation disorders, thrombosis, etc. Also, since thrombocytosis is known to be associated with rheumatoid arthritis (Selroos, 1972), the elevation of GAG reported in rheumatoid arthritis may partly be the consequence of increased contribution of chondroitin 4-sulfate from increased number of thrombocytes present in rheumatoid arthritis. Similarly, alterations in connective tissue metabolism resulting from variations in activities of proteinases or glycosaminoglycan hydrolases (as in lysosomal disorders) may affect the rates of breakdown of the tissue proteoglycans and the release of glycosaminoglycans into the bloodstream. Such may be the situation in various lysosomal disorders which are accompanied by elevated blood GAG levels and upon papain injection in experimental animals (Calatroni et al., 1969; Ali, 1969).

16.3 Relationship between glycosaminoglycans in blood and urine

The blood GAG are believed to be derived from degradation of the proteoglycans in the ground substance of connective tissue and from blood cells. The majority of the urinary GAG are derived from these blood GAG by glomerular filtration in the kidney and to a smaller extent from epithelial desquamation in kidney. A part of the urinary heparan sulfate fraction may originate in renal tissue (Wasteson and Wessler, 1971). In general, the GAG levels and individual GAG components in blood are reflected by corresponding changes in urine. However, in certain situations especially those which affect the addition of GAG to urine from the kidney tissue itself, there may not be any relationship between blood and urinary GAG (Erickson et al., 1975). This lack of relationship was observed in cases of mucopolysaccharidoses and mucolipidoses in which cases the urinary excretion of GAG expressed as clearance ratios to creatinine was found to vary significantly without corresponding changes in the serum GAG levels. Also, the clearance ratios of the lysosomal enzymes β-galactosidase, β-glucuronidase and N-acetyl-β-glucosaminidase did not increase with elevations of serum GAG levels, either. From these observations coupled with the fact that the molecular weights of the GAG excreted in urine were higher than the glomerular membrane allowed to pass, the suggestion was made that these urinary macromolecules and the lysosomal enzymes excreted in the aforementioned genetic disorders did not originate from plasma but from epithelial cell turnover in the kidney (Erickson et al., 1975).

Morse and Nussbaum (1967) observed an elevation of mucopolysaccharide (hyaluronic acid) in the blood but not in the urine of a patient with neuroblastoma. Similarly, whereas serum GAG were elevated in various leukemias and solid tumors, elevation of urinary GAG levels was present in only one-half of the cases of solid tumors (Bacchus, 1965). In addition to these, while there was a decreased excretion of GAG in urine in schizophrenia (Varma et al., 1974), in serum the GAG were elevated (Varma et al., 1979). Whereas the patients with mucopolysaccharidoses types I, II, IIIA and V excreted 10-25 times more GAG in their urine than the normal controls, the GAG levels in the sera of these patients were only 2.5 times or so higher than in the controls (Dekaban and Constantopoulos, 1977). Low-sulfated chondroitin sulfate was the main acidic glycosaminoglycan in blood of patients with Hurler's syndrome, though these patients excreted predominantly dermatan sulfate and heparan sulfate in urine (Hata et al., 1978). All these examples indicate that in some situations the serum and urine GAG levels and types may be independent of each other and that some factors other than glomerular filtration may be involved. On the other hand, the presence of low-sulfated chondroitin sulfate in plasma at all ages in the face of its excretion as the main urinary GAG component in young patients with uremia and nephritic syndrome and its increased excretion in the aged (Hata et al., 1978) do suggest that the renal dysfunction is involved in aging and the mentioned renal disorders.

The patterns of GAG in normal serum and normal urine differ in several ways: 1. The GAG content of serum is lower but it contains heparin which is absent in urine; 2. serum contains only three GAG components, chondroitin sulfate, heparan sulfate and heparin, while urine contains other GAG components too. However, hyaluronic acid in both serum and urine is present in diseased states; 3. the GAG of serum are derived from proteoglycans in the connective tissue, since papain injection increases the serum GAG levels (Calatroni et al., 1969) and also higher levels are seen in connective tissue disorders such as rheumatic diseases (see section 16.7.6). The urinary GAG are derived from glomerular filtration of serum GAG and also from cellular desquamation in the kidney (Erickson et al., 1975).

A low sulfated chondroitin sulfate has been isolated both from human blood and urine (Hata et al., 1978). Endo et al. (1979) have presented a comparison of GAG in normal human plasma and urine. Plasma contained a proteochondroitin sulfate A and hyaluronic acid which were absent in urine. The GAG in blood were associated with much larger peptides and the glycoprotein/glycosaminoglycan ratios were higher in plasma and chondroitin sulfate in plasma was of a higher degree of sulfation. On the other hand the two biological fluids resembled in having partially degraded chondroitin sulfate as the major component and the subfractions eluted from the anion exchange column with 1.25-2.0 M NaCl were identical.

16.4 Methods of analysis and fractionation

The major portion of the glycosaminoglycans are associated with proteins. The fraction of GAG not bound to proteins is the *free* GAG fraction which being highly sulfated can be recovered from undiluted plasma by adsorption on an anion exchange resin, ECTEOLA-Cellulose (Calatroni et al., 1969; Singh et al., 1977), diethylaminoethyl cellulose (Whatman, DE-52) (Friman, 1972; Friman et al., 1977, 1977 a). Since the first demonstration of protein bound glucuronic acid by naphthoresorcinol reaction on the trichloroacetic acid precipitate of plasma protein (Stary and Yuvanidis, 1953), the majority of the investigations on plasma and serum have been carried out on plasma protein fractions. It must, however, be realized that the protein-bound fraction represents only the 80 % of the total plasma GAG; 20 % being in the free GAG fraction (Calatroni et al., 1969).

A number of studies on plasma GAG have been carried out on the euglobulin fraction obtained by dialyzing serum against a citrate-phosphate buffer (Badin et al., 1955; Bollet et al., 1957; Schiller, 1958; Schiller and Dewey, 1956). The euglobulins are insoluble in water but soluble in dilute salt solutions and therefore these precipitate from their salt solutions upon dialysis against water. Badin and Schubert (1955) have reported the conditions for euglobulin precipitation, which involve dialysis of 30-40 ml plasma diluted with three volumes of aqueous 1 % sodium citrate in a cellophane bag against two liters of a 6.67 mM citrate-phosphate buffer of pH 5.8 for three days at 4 °C. These precipitates were freed from protein by proteolysis with trypsin, papain, etc., dialysis and precipitation of the residual protein with trichloroacetic acid. The GAG so liberated in the supernatant were precipitated either with protamine sulfate (Badin et al., 1955; Bollet et al., 1957) or by addition of ethanol containing sodium acetate (Schiller and Dewey, 1956; Schiller, 1958; Friman and Brunish, 1966). The GAG contents were determined either from the turbidity with protamine sulfate at 400 nm (Bacchus, 1965) or by carbazole reaction on the precipitate (Badin et al., 1955; Bollet et al., 1957).

In the earlier work the GAG components in the precipitate were identified by paper electrophoresis in 0.025 M phosphate buffer pH 5 at 8-9 V/cm for three hours (Schiller, 1958) or by cellulose acetate electrophoresis of dialysate of the alkaline solution of the alcoholic precipitate (Friman and Brunish, 1966). Upon staining with Alcian blue chondroitin 4-sulfate was found to make 40-100 % of GAG with a small amount of another component resembling hyaluronic acid. Friman and Brunish (1966) noticed two additional components, one situated between hyaluronic acid and chondroitin sulfate and another component slower than hyaluronic acid. Schiller (1958) further characterized the chondroitin 4-sulfate by chemical analysis and infra-red spectrum, optical rotation and testicular hyaluronidase digestion.

Kerby (1958) used Cohn fractions III and IV-1 and by paper chromatography of GAG isolated from the Cohn fractions by trypsin digestion and cetyltrimethylammonium bromide precipitation, found two fractions, one resembling chondroitin 4-sulfate and the other heparin, but no hyaluronic acid was detected. Endo et al. (1976) isolated a proteochondroitin sulfate A by extraction of human plasma euglobulin fraction with 0.01 M phosphate buffer pH 7.3 followed by DEAE-cellulose and Sephadex G-200 chromatography. Taniguchi et al. (1974) isolated GAG from protein fraction precipitated with trichoroacetic acid and from rabbit leukocytes by proteolytic digestion, nucleic acid removal and CPC precipitation. The GAG components were characterized by digestion with the specific enzymes, Streptomyces hyaluronidase and chondroitinases AC-II and ABC (see section 2.12).

Ali (1969) reported a method for direct precipitation of GAG from 1 ml plasma using cetyltrimethylammonium bromide followed by carbazole-borate reaction. However, from our experience this procedure is of a use only in metabolic studies like papain injections where large amounts of GAG may be present in blood plasma in the free form unbound to proteins.

The methods based on pronase digestion followed by cetylpyridinium chloride precipitation (Taniguchi et al., 1974 a; Emura and Fukuda, 1973; Niebes and Schifflers, 1975; Murata et al., 1972; Murata, 1974; Varma et al., 1979) measure the total GAG in plasma or serum. Fractional determination of free and protein-bond GAG of plasma has been carried out by adsorption of the free GAG fraction on an anionic-resin like ECTEOLA-cellulose followed by papain digestion of the eluate and adsorption of GAG from this digest on ECTEOLA-cellulose column (Calatroni et al., 1969; Singh et al., 1977). Friman et al. (1977) used DE-52 cellulose for direct adsorption of the free GAG fraction of plasma followed by adsorption of the protein-bound GAG in the papain digest of the eluate on an AG 1-X2 (Cl^-) resin. Chmiel et al. (1975) presented a modification of the method of Calatroni et al. (1969) for estimation of free and conjugated GAG of plasma after separation on ECTEOLA-cellulose columns with recoveries of 97.5 %. However, the values for free GAG obtained by this method were more than those for the conjugated fractions (see Table 16.1).

Methods for determination of heparin presnt in micro-amounts (10 µg/100 ml) in blood involve spectrophotometry or a biological property such as anti-coagulant activity or a combination of the two coupled with electrophoresis (Howell, 1925; Jaques and Monkhouse, 1949; Gibson et al., 1951; Bassiouni, 1954; Freeman et al., 1954; Engelberg et al., 1955).

Other methods for quantitative determination of hyaluronic acid (Greiling, 1974) and chondroitin sulfates A and C in plasma (Andreev, 1965) and a general method for microdetermination of GAG in biological fluids by formation of complexes with quinacrine (Mitra and Blau, 1978) were reported. A method similar to quinacrine method was based on precipitation with N, N-diethylpseudoisocyanin iodide (or chloride) and absorption at 570 nm (Suschke, 1973). Methods for isolation of a low sulfated chondroitin sulfate in acetone-defatted serum by pronase digestion and cetylpyridinium chloride precipitation were reported (Hata and Nagai, 1978; Hata et al., 1978). Methods for analysis of chondroitin sulfate isomers in human plasma were reported (Murata et al., 1972; Murata, 1974).

Methods used for isolation of GAG from platelets and leukocytes are similar to those for plasma or serum except that these cells need to be disrupted either by homogenization or by ultrasonication (Taniguchi et al., 1974; Olsson and Gardell, 1967; Olsson, 1969). The GAG can be isolated and characterized in the subcellular fractions obtained from disrupted cells by differential centrifugation or density gradient centrifugation (Olsson, 1969; Zeya and Spitznagel, 1971). Alternatively, the GAG in granules and sub-cellular fractions may be characterized in the intact cells by the methods desribed below.

Histochemical methods for the study of morphological and metachromatic characteristics due to GAG (Smith and Collins, 1977; Cosson et al., 1977; Szabo et al., 1973) and for localization of GAG by electron-microscope autoradiography (MacPherson, 1972) in the intact leucocytes have been described. The analysis of GAG components in the cytoplasmic granules of the normal and leukemia granulocytes (Catini et al., 1974) and ultrastructural localization of GAG in the lysosomal granules and other subcellular components of the intact pig blood platelets (Morganstern and Weber, 1974) have been carried out by cytochemical methods. Histochemical ultrastructural localization of GAG in senile platelets by use of hyaluronidase and lysozyme (Schiffler, 1957) and in vacuoles of chicken platelets by a combination of histochemical and electron probe X-ray microanalysis (Daimon et al., 1977) have been reported. Histochemical methods for investigation of GAG inclusions in cells (Mittwoch, 1961; Szabo et al., 1973; Horn and Spicer, 1964) have been used. Fluorescent methods following staining with acridine orange have been used for qualitative and quantitative determination of GAG in cells by spectrofluorophotometry and spectrofluoropolarimetry (West and Lorincz, 1973).

Anzai (1972) presented a comprehensive review of methods for isolation, separation and determination of serum GAG by liquid-liquid extraction, gel electrophoresis, column and thin-layer chromatography. Micromodifications of methods for GAG in serum have been reported (Enomura and Kurata, 1971; Emura and Koizumi, 1975). The carbazole reaction was performed after addition of sulfamate and borate buffer for more accurate determination of serum GAG isolated by pronase digestion, trichloroacetic acid deproteinization, ethanol, cetylpyridinium chloride and celite precipitations and finally NaCl fractionation.

Specific methods for analysis of glycosaminoglycan-hydrolyzing enzymes in serum have been described (Bownes and Tan, 1968; Goldstein, 1961; Wong, 1970; Salkie, 1971).

16.5 Glycosaminoglycans of plasma and serum

There is a disagreement on both the qualitative as well as the quantitative distribution of glycosaminoglycans in plasma. The values of glycosaminoglycans in plasma and serum reported by different investigators as mg of hexuronic acid per 100 ml are given in Table 16.1. It is evident from this table that the disagreement in reported values stems not only from the methodological differences but also partly from the use of different starting materials. Serum contains more GAG than plasma probably due to the release of chondroitin 4-sulfate from blood platelets and involvement of coagulation factor IX (which is a rich source of chondroitin 4-sulfate, Juvani et al., 1975) during the coagulation process.

In human plasma chondroitin 4-sulfate is the major glycosaminoglycan component in addition to heparan sulfate, keratan sulfate and heparin as the minor GAG components. Some workers have reported the presence of oversulfated chondroitin sulfate (Murata et al., 1972; Murata and Horiuchi, 1977) and hyaluronic acid (Murata and Horiuchi, 1977), while others (Bollet et al., 1957; Taniguchi et al., 1974; Greiling, 1974) failed to detect these. Normal human plasma also contains a small amount of heparin averaging 10 µg/100 ml (Jaques et al., 1949; Engelberg et al., 1955).

The presence of chondroitin 4-sulfate as the predominant glycosaminoglycan in human plasma has been reported by many investigators (Bollet et al., 1957; Schiller, 1958; Kerby, 1958; Friman and Brunish, 1966; Calatroni et al., 1969; Taniguchi et al., 1974). About 80 % of this chondroitin 4-sulfate is undersulfated (Calatroni et al., 1969; Juvani et al., 1975; Murata and Horiuchi, 1977) and represents almost 80 % of the macromolecular GAG present in normal human plasma in amounts equivalent to 300 µg hexuronic acid/100 ml (Juvani et al., 1975) or 24 nM hexosamine (Hata et al., 1978). The undersulfated chondroitin sulfate has been isolated and characterized from human plasma (Juvani et al., 1975; Hata et al., 1978), rat plasma (Hata and Nagai, 1978) and from rabbit plasma (Taniguchi et al., 1974 a). Heterogeneity of sulfation (Murata et al., 1972) and molecular species (Murata and Horiuchi, 1977) have been reported for this macromolecule by gel filtration and analysis with chondroitinases at the subunit level. Tani-

Table 16.1. Plasma/Serum Glycosaminoglycan Levels (Mean ± S.D.) in Milligrams per 100 ml

Source	Method	Normal Values (mg/100 ml)		No. of Subjects	Reference
Serum	Papain digestion, ECTEOLA-cellulose	0.310–0.592 (av. 0.525)		11	Dekaban and Constantopolus (1973)
Serum	Pronase, TCA, ethanol, CPC pptn.	(av. 0.560)		–	Emura and Koizumi (1975)
Plasma	ECTEOLA-cellulose, Pronase AG 1-X2	free bound total	0.060±0.018 0.238±0.116 0.298±0.128	16	Calatroni et al. (1969)
Plasma	Modified ECTEOLA-cellulose, method of Calatroni (1969)	free conjugated	1.23±0.11 0.36±0.07	–	Chmiel et al. (1975)
Plasma	DE-52 resin for free GAG, papain digestion and AG 1-X2 for bound GAG	free bound total	0.060±0.043 0.410±0.093 0.571±0.102	50	Friman et al. (1977a)
Plasma	ECTEOLA-cellulose, Pronase	free bound total	0.060±0.019 0.218±0.099 0.278	28 (20–29 yrs)	Singh et al. (1977)
Plasma	DE-52 resin for free GAG, TCA pptn. papain digestion and AG 1-X2 for bound GAG	free bound total	0.145±0.040 0.402±0.098 0.548±0.098	30	Friman et al. (1977)
Plasma	ECTEOLA-cellulose Dowex 1-X2 (Cl)	free bound low mol. wt. total	0.074±0.022 0.423±0.061 2.050±0.400 2.552±0.423	20	Neri et al. (1971)

Table 16.1. Plasma/Serum Glycosaminoglycan Levels (Mean ± S.D.) in Milligrams per 100 ml

Source	Method	Normal Values (mg/100 ml)	No. of Subjects	Reference
Serum	Pronase, CPC-alcohol pptn.	0.602±0.066	30	Niebes and Schifflers (1975)
Serum	Pronase, CPC-alcohol pptn.	0.520±0.015	50	Varma et al. (1979)
Plasma	Euglobulin pptn.	0.206±0.054	7	Badin et al. (1955)
Serum	Euglobulin pptn.	0.144±0.021	13	Kerby (1958)
Serum	CPC pptn.	0.228±0.112	18	Harris and Fraser (1969)
Serum	Euglobulin, protamine pptn.	0.277±0.042	34	Bollet et al. (1957)
Serum	Euglobulin pptn.	0.227±0.047	10	Winand (1968)
Serum	DE-52, AG 1-X2 resins	0.235±0.097	60	Friman (1972)
Serum	Pronase, TCA, alcohol-CPC pptn.	0.548±0.175	20	Emura and Mukuda (1973)
Serum	Quinacrine pttn.	1.17±0.57	24	Mitra and Blau (1978)
Serum	Euglobulin, protamine sulfate turbidity at 400 nm	0.150±0.010	30	Bacchus (1965)
Serum	N,N'-diethylpseudoisocyanin iodide (or chloride) pptn. 570 nm	–	–	Suschke (1973)

guchi et al. (1974) from such enzymatic analysis found that the undersulfated chondroitin 4-sulfate from plasma was composed of equimolar amounts of unsaturated nonsulfated disaccharides and unsaturated 4-sulfated disaccharides at the disaccharide subunit levels. These workers found no oversulfated unsaturated disaccharide indicating the absence of any oversulfated chondroitin sulfate in plasma. However, they detected a small amount of chondroitin 6-sulfate.

In human plasma the GAG are present in free and protein-bound forms (Calatroni et al., 1969; Friman et al., 1977). These are named so, since the free GAG fraction which forms about 20 % of the plasma GAG is adsorbable on an anion-exchange column directly from an undiluted plasma and the protein-bound fraction which comprises the remaining 80 % of the plasma GAG is adsorbable on that column only after removal of the protein by proteolysis with papain, pronase, etc. Chondroitin 4-sulfate is the major component of both the free and protein-bound plasma GAG, being fully sulfated in the free GAG and undersulfated in the protein-bound GAG. For this reason these fractions have also been mentioned in literature as high sulfated and low sulfated plasma GAG. The concept of these two separate fractions originated by Calatroni et al. (1969) is of a very great significance since the relative proportions of these two fractions may vary in physiological and pathological states without even affecting the total plasma GAG contents. Such a situation was encountered in rheumatic arthritis where the elevation of free fraction was associated with a decrease in the bound fraction with no change in total GAG of plasma (Friman et al., 1977). The relative proportions of these two fractions vary with age (Calatroni et al., 1969; Singh et al., 1977; Hata et al., 1978) and in different disease states like mucopolysaccharidoses (Calatroni et al., 1969), renal insufficiency (Friman et al., 1977 a), uremia and nephrotic syndrome (Hata et al., 1978), etc.

Besides the macromolecular GAG, plasma also contains low molecular weight glucuronic acid-containing species which are fragmented GAG products released from tissue proteoglycans. These constitute 40-60 % of glucuronic acid material and these are lost if plasma is dialyzed (Gazith and Schimmelpfennig, 1964). Neri et al. (1971) reported a method by which the effluent and 0.9 % NaCl washings from ECTEOLA-cellulose chromatography of the papain-digested bound GAG fraction were used to retain the low molecular weight species on a stronger anion-exchanger, Dowex 1-X2 (Cl^- form). Measurements of the low molecular weight degraded species of GAG may be useful in those conditions in which the degradation of GAG is impaired. Calatroni (1974) analyzed the GAG breakdown products of normal and pathological plasma and urine based on this concept. The plasma and urine of the mucopolysaccharidoses patients often had large amounts of fully sulfated, or even oversulfated polysaccharides, which suggested the role of sulfatases in the normal degradation of GAG. Neri et al. (1971) determined the free and bound GAG along with the low molecular weight degraded species in normal human plasma. The low molecular weight fraction was over four times (205.50 ± 40.13) the total macromolecular fraction (free + bound GAG) at 49.79 ± 7.39 µg hexuronic acid per 10 ml of plasma.

16.6 Glycosaminoglycans of formed elements

The platelets (thrombocytes), white blood cells (leukocytes) and the red blood cells (erythrocytes) are called the formed elements of blood. These cells of blood, notably the thrombocytes and leukocytes contain (Kerby, 1955; Bassiouni, 1955) and synthesize glycosaminoglycans. These are in part a source of the GAG of blood (Odell and Anderson, 1957; Kerby and Langley, 1959). The level of GAG in human serum exceeds that of the oxalated plasma sample due to the release of chondroitin 4-sulfate proteoglycan and other hexuronic acid-containing polysaccharides from thrombocytes during the early phase of clotting and from leukocytes entrapped during the clotting of blood (Kerby and Langley, 1959; Olsson and Gardell, 1967).

Leukocytes (leucocytes)

The leukocytes were initially shown to contain nondialyzable GAG (Bassiouni, 1955; Kerby, 1955). The material isolated from leukocytes resembled chondroitin sulfate by electrophoresis and in male and female human subjects corresponded to 1180 ± 51 µg and 910 ± 47.5 µg hexuronic acid per 10^{10} leukocytes, respectively (Kerby, 1955). The leukocyte GAG have been isolated and characterized by sophisticated methods. Human leukocytes contained 100 µg glucuronic acid per 100 mg of the dry weight of the defatted material (Taniguchi et al., 1974) and in rabbit leukocytes (Taniguchi et al., 1974 a) the GAG content was about 150 µg per 100 mg dry weight.

The analyses of the GAG in peripheral human leukocytes have been carried out by specific enzymatic analysis with Streptomyces hyaluronidase and bacterial chondroitinases. The chondroitin 4-sulfate was the predominant GAG in human leukocytes (Murata, 1974; Olsson and Gardell, 1967) and by electrophoresis in 0.1 M HCl it showed heterogeneity of the degree of sulfation (Taniguchi et al., 1974). Hyaluronic acid was reported earlier as a probable minor component of normal human leukocytes (Olsson and Gardell, 1967) and was also demonstrated in two cases of myeloid leukemia and one case of lymphoid leukemia (Clausen and Andersen, 1963). No chondroitin 6-sulfate, dermatan sulfate or hyaluronic acid was detected by electrophoresis and by enzymatic analysis at the disaccharide subunit level; the chondroitinase-resistant material was probably due to a minute amount of heparan sulfate (Taniguchi et al., 1974).

The GAG from leukocytes of experimental animals have been characterized. The rabbit leukocytes in peripheral blood (Fedorko and Morse, 1965) and polymorphonuclear leukocytes' granules from rabbit bone marrow (Zeya and Spitznagal, 1971) synthesize and contain GAG. The cytoplasmic granules stainable with Azure A above pH 4.0, Alcian blue and iron diamine were due to the presence of GAG. In the absence of leukocytosis these were present only in small numbers in the circulating heterophils but were greatly increased during endotoxin-induced leukocytosis (Horn and Spicer, 1964). By histochemical and autoradiographic studies, the presence of azurophilic granules due to sulfated GAG were found to vary in quantities in different types of rabbit leukocytes (Horn and Spicer, 1964 a). The sulfated GAG contents varied as: basophils > eosinophils > neutrophils.

The blood basophil is the equivalent of tissue mast cells and may take over the function performed otherwise by mast cells. Subcutaneous injection of insulin produced a significant fall in the number of circulating basophils. Variations in basophil count during menstruation and decreases during sexual stimulation and mating have been noted. The post-menopausal women over 60 years have lower basophil counts than men of the corresponding age. These fluctuations in basophil counts (for review see Boseila, 1966) indicate that these are influenced by hormonal changes and these may produce corresponding fluctuations of GAG of blood under these hormonal changes.

When the ^{35}S-labeled GAG synthesized in vivo and in vitro were characterized by the action of the specific chondroitinases, 55 % of the isolated material was chondroitin 4-/6-sulfate and degradable by chondroitinase-AC and an additional 30-35 % could be degraded by chondroitinase-ABC, indicating an equivalent amount of dermatan sulfate. In addition to these, 15 % of the GAG consisted of heparan sulfate on the basis of degradation with heparitinase (Orenstein et al., 1978).

The GAG in bovine leukocytes studied by enzymatic and electrophoretic analyses were found to consist of 75-80 % chondroitin 4-sulfate, 10-15 % hyaluronic acid and about 10 % heparan sulfate (Murata and Okuyama, 1974). The 150 μg hexuronic acid of bone marrow cells (leukocytes) in rabbit analyzed by specific hyaluronidase and chondroitinase was found to be distributed among chondroitin 6-sulfate as the major component with minor contributions of chondroitin 4-sulfate, dermatan sulfate and hyaluronic acid (Taniguchi et al., 1974 a). Fedorko and Morse (1965) analyzed the GAG from whole rabbit polymorphonuclear leukocytes and from cytoplasmic granules on the basis of solubility of cetylpyridinium chloride complexes in solutions of differing salt concentrations. Hyaluronic acid and chondroitin sulfate were present. However, the polymorphonuclear leukocyte granules contained 2.6 times the GAG content of the whole cells and also, the hyaluronic acid was concentrated in the granules in particular. Zeya and Spitznagel (1971 a) using a linear sucrose gradient centrifugation of the homogenates of peritoneal polymorphonuclear leukocytes of rabbits obtained three groups of granules. The GAG, cathepsins and β-glucuronidase were found to be concentrated in the heaviest and largest granules, 6 and III fractions at 41-48 % sucrose.

Whereas the chondroitin 4-sulfate has been reported to be ubiquitously present in human and animal leukocytes, reports about other GAG components have been confusing in spite of using the sophisticated and specific enzymatic analyses. Whereas, some reports claim the presence of the glucosamine-containing hyaluronic acid as a minor component, Kharchenko and Seits (1968) reported the presence (251 μg/g dry mass) of a galactosamine-containing GAG (chondroitin?) instead in leukocytes from 20 normal human blood donors. However, their analysis was based on cetyltrimethylammonium bromide fractionation of the aqueous phenol extracts of the leukocytes and chondroitin sulfate equivalent to 661 μg uronic acid per gram of the dry mass of these cells was present. No heparin or heparan sulfate was detected.

The biosynthesis of GAG in human leukocytes has been studied extensively. In both normal and leukemic human leukocytes, the major GAG synthesized is the chondroitin 4-sulfate proteoglycan (Olsson, 1969). In the subcellular fractions prepared by differential centrifugation and density gradient centrifugation, the granules contained the bulk of the chondroitin sulfate proteoglycan while only a small portion of it was present in the microsomal fraction. However, experiments with ^{35}S-sulfate and ^{14}C-glucosamine indicated that the biosynthesis and sulfation of the chondroitin sulfate took place in the microsomal fraction, while the storage of the completed GAG was mainly in the granules (Olsson, 1969 a). By in vitro radioactive pulse labeling it was shown that the chondroitin sulfate newly synthesized on the microsomal structures was gradually transferred to the granule fraction (Olsson, 1969 b). Whereas the biosynthesis of the proteoglycan macromolecule was initiated by formation of the core protein at the ribosome, the chain prolongation took place on the membranes where the glycosyltransferases for the specific and sequential addition of monosaccharide residues were present bound to the membranes (Olsson, 1972). By using endogenous and exogenous acceptors, the enzymes catalyzing incorporation of the sugar residues were found to be concentrated in a light subfraction of the smooth endoplasmic reticulum (Olsson and Gardell, 1971).

Platelets (Thrombocytes)

Platelets (thrombocytes) have been reported to contain sulfated GAG. Odell and Anderson (1957) found that most of the radioactive sulfate in the rat blood platelets was recovered in the alkali extract. Based on staining with Alcian blue and elctrophoresis and chromatography, chondroitin sulfate was identified in this material. This was confirmed (Anderson and Odell, 1958) by examination of the hydrolysis products of the isolated GAG. These authors also reported another mucopolysaccharide in rat blood platelets which was uronic acid-free. Histochemical methods and electron probe X-ray microanalysis have confirmed the existence of GAG in the vacuoles of chicken thrombocytes (Daimon et al., 1977). Senile platelets show metachromasia which is markedly reduced by the action of hyaluronidase (Schiffler, 1957).

The distribution of the GAG at the subcellular level has been studied by histochemical technics (Morganstern and Weber, 1974). The GAG in human and pig blood platelets were localized in lysosomal granules, concentrated more in the nucleoid than in the matrix, in the glycocalyx, and on the surface membrane. The synthesis and localization of sulfated mucopolysaccharides in rat blood platelets have been studied by autoradiography and electron microscopy (MacPherson, 1972). The subcellular localization of GAG probably involves a different biological role at the different subcellular sites. The GAG (heparinoids and other GAG) have been implicated in the in vitro process of platelet aggregation (Scharrer and Breddin, 1972) and it probably involves the membrane mucopolysaccharides of the platelets (O'Brien, 1972). The GAG have also been involved in storage of biogenic amines like 5-hydroxytryptamine and histamine in the storage organelles from the platelets by ultrasonic disruption of platelet suspensions (Bergendorff and Uvnas, 1972). The mechanism proposed for this storage involves an ionic linkage of the amine formed by a basic protein and a sulfated GAG, probably chondroitin sulfate (Aborg and Uvnas, 1971). Methods for the separation of platelets from blood (Dillard et al., 1951) and those for sulfomucopolysaccharide-containing granules involved in storage of serotonin in rat and rabbit blood platelets have been described (Anderson et al., 1976).

Thrombocytes contain a fully sulfated chondroitin 4-sulfate proteoglycan which is released as the platelet factor 4 during the early phase of the coagulation of blood (Olsson and Gardell, 1967). Kerby and Langley (1959) had also suggested the platelets as the source of GAG in normal serum and of the increased serum levels of GAG in patients with rheumatoid arthritis. Their suggestion may be true since thrombocytosis is known to be associated with rheumatoid arthritis (Selroos, 1972) and the increased platelet count in these patients may be the contributing factor for the observed elevation in serum GAG.

Radioactive sulfate uptake studies showed two mechanisms for the uptake of the sulfate for biosynthesis of the GAG by blood platelets of rats (Odell and McDonald, 1964). Blood platelets were of larger size in vitamin A-deficient rats and with hypervitaminosis A these were smaller. In both cases, the sulfate incorporation into GAG was increased but the hexosamine contents of the platelets were decreased by 27 % in the vitamin-deficient rats and were increased by 27 % in the vitamin-treated rats as compared to the controls (McDonald, 1966).

The GAG composition of the platelets has been analyzed in rats by radioactive studies (McDonald, 1966) and in human platelets by specific enzymatic analyses with bacterial chondroitinases (Murata, 1974 a). The rat platelets contained chondroitin 4-sulfate, 58 %; chondroitin 6-sulfate, 17 %; dermatan sulfate, 6 % and heparin, 13.5 %. An excess of vitamin A increased the sulfated GAG in rats (McDonald, 1966). An enzyme degrading heparan sulfate was demonstrated in platelets (Wasteson et al., 1976).

16.7 Physiological and pathological variations in blood glycosaminoglycans

The blood GAG levels vary in a diverse number of unrelated physiological and pathological states. These variations are nonspecific and the serum GAG changes are related in most cases to seromucoid changes and in diverse disease processes behave like *acute phase reactions*.

16.7.1 Age and sex related variations

The glycosaminoglycans measured as hexuronic acid in healthy subjects were not found to vary either with age or sex. The values in 17 younger subjects (20-30 years) were 0.596 ± 0.067 as compared to the values of 0.626 ± 0.07 mg hexuronic acid/100 ml in 17 older (35-54 years) subjects. Similarly the values of 0.608 ± 0.079 in 14 males were insignificantly different from the values of 0.612 ± 0.071 in 18 females (Niebes and Schifflers, 1975). Earlier investigations (Bollet et al., 1957) also found no apparent difference between males and females or between individuals of different ages. Significant quantitative differences ($P < 0.05$) were noted between plasma GAG of normal adults and children, but not between normal males and females, either adults or children (Calatroni et al., 1969).

When high and low sulfated GAG were measured in 79 normal individuals of various ages, no significant differences were observed between males and females of the same age group. However, within the same sex high/low sulfated GAG ratios in plasma of older individuals (50-59 years) were higher (0.41-0.45) as compared to the values (0.27-0.37) in the younger individuals (20-30 years). This increase in the ratio with age was due to an increase in the plasma GAG fraction which might interact with the circulating lipoproteins and might be responsible for the increased predisposition to atherosclerosis with advancing age (Singh et al., 1977). The free and protein-bound GAG fractions of plasma were affected differently in rheumatoid arthritis in males and females (Friman et al., 1977 a). Whereas free GAG levels were significantly elevated, the bound GAG decreased in the female but not in male patients.

16.7.2 Pregnancy

Nonspecific elevations of glycosaminoglycans occur in normal pregnancy and those involving complications. The serum hexosamines begin to rise in the seventh month, and by the ninth month reach 150 % of their initial levels (Stary, 1959). The increased levels are sustained and even increased after delivery and are normalized some weeks later. Higher elevations are observed in isoantigen incompatibility (Dryzhova, 1968), during labor (Grudzien, 1971), and in accidental and artificial abortions (Grudzien, 1973). Significant differences in serum GAG were observed in women during instrumental delivery and on the third and sixth days of puerperium as compared with spontaneous deliveries (Grudzien, 1973). The serum GAG were lower in spontaneous deliveries apparently because of aging of the placenta and depletion of mucoproteins at the end of full term pregnancy. Galambos and Dale (1976) found that the connective tissue components measured in plasma were similar during the third trimester of pregnancy and six weeks postpartum. However, plasma glycosaminoglucuronides and hydroxyproline were significantly increased indicating the origin of these components from increased rate of hydrolysis of extraskeletal connective tissue. These increases of GAG and hydroxyproline were suggested to involve involution of the uterus after the childbirth (Galambos and Dale, 1976).

16.7.3 Effect of stress

Although the amount of blood serum GAG seem to have constant values, certain fluctuations occur with changes in volume of the circulating blood like that observed in dehydration or water deprivation. Several types of stress situations may affect serum GAG values and components through ACTH and other hormones released in these states.

A number of different types of nonspecific body injuries like burns, bleeding, infectious diseases, ultrasound treatment, mechanical lesions, introduction of talcum, turpentine or other foreign substances, loading with saline or epinephrine solutions, etc., increase total mucopolysaccharides in blood (Stary, 1959; Winzler, 1965). Sustained changes in body temperature produced by exposure to external heat or cold or by injections of pyrogenic substances produced elevations of hexosamine in dog (Shetlar et al., 1953). Several other forms of acute clinical stress (Boas and Peterman, 1953; Boas et al., 1957), surgery (Bjornesjo et al., 1959), radiotherapy (Karcher, 1966) increase hexosamine levels in blood and urine, in addition to changes in the electrophoretic components of acid mucopolysaccharides (Karcher, 1966). Plasma levels of hyaluronic acid, chondroitin sulfate and heparin were higher in animals conditioned at 26 °C than in cold-torpid animals (Kupchella et al., 1977). Food deprivation, fasting and starvation lowered plasma hexosamine and urinary hexosamine (Boas and Peterman, 1953).

16.7.4 Marfan syndrome

Marfan syndrome is a disease of connective tissue (McKusick, 1972) in which alterations of the ground substance may be expected. A comparative investigation carried out in ten normal control subjects and five Marfan syndrome patients revealed a significant ($P < 0.01$) elevation of serum GAG (Bacchus, 1960). The normal control values expressed as mean ± S.E. of 147 ± 12.4 μg glucuronic acid per deciliter (100 ml) were increased to 374 ± 21.2 μg glucuronic acid per deciliter in the Marfan syndrome patients. In general, a parallel increase in seromucoids almost invariably accompanies increased serum GAG in diseased states. However, serum seromucoids instead of being increased showed a characteristic decrease in Marfan syndrome patients (Bacchus, 1960).

16.7.5 Infection and inflammation

Elevation of serum hexosamine occurs in infection and in both rheumatic and non-rheumatic inflammatory conditions (Winzler, 1965). In inflammatory reactions, serum GAG have been reported to be elevated in a nonspecific manner occurring in both rheumatoid and nonrheumatoid conditions to an equal degree (Kerby, 1958 a). These elevations were accompanied by increases in the euglobulin fraction of plasma from which glycosaminoglycans were isolated. Elevations of serum GAG levels were encountered in patients with viral hepatitis (Bednarski et al., 1970), in cirrhosis and alcoholic hepatitis patients (Galambos, 1975), in patients with chronic liver diseases (Loginov et al., 1972) and experimental peritonitis (Szulc et al., 1970). Bacterial infections affect metabolic processes causing elevations of GAG and glycoprotein components in serum. Characteristic changes in serum GAG occur at the early stages of infection and have been monitored by gas-chromatography as useful indications of in vivo metabolic processes affected by bacterial infection (Mitruka, 1971).

16.7.6 Rheumatic and related disorders

Serum protein-bound hexosamines are elevated in rheumatoid arthritis, rheumatic fever and related diseases (Kelly, 1952; Haruki, 1959; Bottiger et al., 1964). The concentration of GAG in serum and their excretion in urine are increased in patients with rheumatoid arthritis (Di Ferrante, 1957). The determination of blood and urine GAG have been used in evaluating the activity of rheumatism (Fedorishko, 1968), although serum GAG elevations are also observed in arthrosis (Tuzi and Bonito, 1968), osteoarthritis (Yusipova and Kriuk, 1979) and other rheumatic diseases (Ghirardo and Rovetta, 1968).

Badin et al., (1955) reported a significant ($P < 0.01$) increase in glycosaminoglycans present in the plasma euglobulin fraction to a mean ± S.D. value of 38.8 ± 10.7 µg per 10 ml plasma in 10 rheumatoid arthritis patients as compared to the values of 20.6 ± 5.1 µg/10 ml in 7 control subjects. In another investigation serum GAG were elevated ($P < 0.05$) in 13 rheumatoid arthritis patients to 17.6 ± 4.2 as compared to the values of 14.4 ± 2.1 µg hexuronate/10 ml serum in 13 normal control subjects (Kerby, 1958). In another comparative study (Harris and Fraser, 1969) serum GAG were noted to be higher ($P < 0.05$) at mean ± S.D. values of 32.8 ± 11.2 μg hexuronate/10 ml in 8 rheumatoid patients than the values of 22.8 ± 11.2 in 18 control subjects.

Variations in individual GAG fractions have been reported in rheumatoid arthritis. The ratio of hyaluronic acid to chondroitin 4-sulfate in blood serum was altered and the presence of protein-associated heparin and chondroitin 6-sulfate were noted (Merkur'eva and Bukhtoyarova, 1976). The serum concentrations of chondroitin 4-sulfate, keratan sulfate and dermatan sulfate increased 7.2-, 2- and 4.4- fold, respectively in patients with acute rheumatism as compared to those of normal subjects (Balesnaya, 1977). These increases correlated positively with the activity of the rheumatic process. Contrary to the aforementioned reports of serum GAG elevation, Friman et al. (1977) found no difference in total serum GAG and electrophoretic patterns in 35 patients with active erosive rheumatoid arthritis and 50 control subjects. Measurement of the free GAG directly adsorbable from undiluted plasma on DE-52 ànion-exchange cellulose and the bound GAG adsorbable on AG 1-X2 (Cl^-) resin after proteolysis revealed that increased free GAG was associated with lower bound GAG levels with no net change in total serum GAG. The values of free, protein-bound and total GAG expressed as μg hexuronate/10 ml of plasma were 20.4 ± 4.6, 37.5 ± 10.2 and 57.9 ± 12.0 in the rheumatoid arthritis patients as compared to the normal control values of 16.0 ± 4.3, 41.0 ± 9.3 and 57.1 ± 10.2 μg hexuronate/10 ml.

Changes in connective tissue degrading enzymes of human leukocytes have been noted and were considered of a significance in injury of the joint in rheumatoid arthritis (Oronsky et al., 1973; Oronsky and Perper, 1975; Keiser et al., 1976). Increased hyaluronidase activity was noted in the sera of patients with rheumatoid arthritis (Merkur'eva and Bukhtoyarova, 1976). The activities of lysosomal enzymes, including hyaluronidase, acid phosphatase and acid proteinase, were higher in sera of patients with rheumatic arthritis, podagra osteoarthrosis and Bekhterev's arthritis than in normal subjects in positive correlation with the intensity of inflammation (Fil'chagin et al., 1977). The serum mucopolysaccharide content, acid cathepsins and β-D-glucosidase increased and β-D-galactosidase decreased in the sera of patients with rheumatoid arthritis (Yusipova and Kriuk, 1977). Anti-inflammatory preparations restored the elevated blood and urine GAG to normal levels (Pidemski and Afonina, 1974).

16.7.7 Skin diseases (dermatoses)

The glycosaminoglycan metabolism in the ground substance of the skin is disturbed in a number of conditions affecting the skin (dermatoses) and these changes are reflected in blood in some disease states. Increased serum hexosamine levels were observed in dermatoses (Weber, 1958; Weber et al., 1954). Similar increases were observed in hypertrophy of the skin and subcutaneous tissue due to obstructed blood or lymphatic circulation such as seen in elephantiasis occurring pre- and post- dermoplastic surgery (Kamev, 1969). The glucuronic acid content of the blood plasma was found to increase in dermatoses reflecting an increased metabolic turnover of the acidic glycosaminoglycans in the corium (Weber and Thaesler, 1959). Elevations of mucopolysaccharides were observed in a number of disorders of collagen metabolism including Skleroderma progressiva, Skleroderma cicumscripta, Erythematodes chronicus and Erythematodes acutus (Balbanov and Samsonova, 1965). The elevation of GAG in serum was reflected in the excretion of these macromolecules in urine and therapy with gestagenic substances modified the levels in the two biological fluids (Holzman et al., 1968).

Elevation of GAG in skin, particularly of hyaluronic acid and chondroitin sulfate, elevation in blood (Zimmerman et al., 1965) and slightly increased excretion of GAG in urine (Clausen and Asboe-Hansen, 1966) were noted in mastocytosis. Irradiation has been reported to increase the vascular permeability with consequent elevation of blood mucopolysaccharide levels (Chernov et al., 1965).

16.7.8 Effects of hormones: hormonal disorders

Blood glycosaminoglycan levels and the spectrum of individual glycosaminoglycans are altered in hypo or hyper secretion from the endocrine glands. The corresponding changes can be produced experimentally by removal of a particular gland or by its chemical destruction and by administration of the corresponding hormone preparation (Asboe-Hansen, 1961, 1966). The effects of endocrines on blood elements have been reviewed (Boseila, 1966).

16.7.8.1 Diabetes mellitus

Disturbances of blood glycosaminoglycans have been observed in diabetes in human subjects due to a relative deficiency of the effective insulin as well as in alloxan and steroid diabetes produced artificially in experimental animals. The protein-bound hexosamines, which represent both glycoproteins and glycosaminoglycans together, were increased up to 185 mg/100 ml as compared to the normal values of 103.9 ± 11.6 mg/100 ml in 15 control subjects (Andreani and Gray, 1956). The rise in hexosamine varied directly with the severity of the complications. The values were higher in patients with hyaline vascular changes than in sera of diabetics with no signs of hyaline vascular changes (Tornblom and Nordstrom, 1954). In diabetics with renal complications, the protein-bound hexosamine of serum increased from the normal value of 66-99 mg/100 ml to 145.2 ± 23 mg/100 ml (Winzler, 1965).

Elevations of serum GAG have been reported in diabetes mellitus (Paukman, 1967; Malathy and Kurup, 1973) and in experimentally-induced alloxan and steroid diabetes (Jakab et al., 1966, 1966a). In contradiction with these reports, no increase in mucopolysaccharide or β-glucuronidase in serum was seen in diabetes without clinical evidence of vascular disease or those with atherosclerotic thrombotic occlusion of coronary or cerebral arteries (El-Ebrashy et al., 1973). However, serum mucopolysaccharides in diabetics with retinopathy and serum β-glucuronidase activities in diabetics with microangiopathy were elevated. Similarly, Kanazawa et al. (1977) found no significant difference in serum levels and urinary excretion of acid mucopolysaccharides between nondiabetics and diabetics, between patients with positive and negative aortic arch calcification and before and after glucose administration. In a study (Malathy and Kurup, 1973) carried out in sera from 5 normal subjects and 10 diabetic patients, increases in total GAG and hyaluronic acid with simultaneous decreases in heparan sulfate, chondroitin 4- and 6-sulfates were observed. The dermatan sulfate content of serum decreased in 7 patients but was unchanged in 3 patients. An increase in heparin was also noted. These changes in different fractions of GAG may be related to the vascular complications of diabetics due to the ability of these macromolecules to interact with low density lipoproteins (Nakashima et al., 1975; Donnelly et al., 1978; Dahlen et al., 1978). In fact, a correlation between serum mucopolysaccharide concentration and serum lipid patterns has been observed (Vegh and Bihari-Varga, 1967; Robinson et al., 1975).

The mechanism of elevations in serum GAG in diabetics is not clear, especially when decreased GAG synthesis in tissues and one-third the normal rate of formation of hyaluronic acid and dermatan sulfate in skin occur (Schiller and Dorfman, 1955). Also, cultured skin fibroblasts from both insulin-dependent and non-insulin-dependent diabetics had higher proportions of heparan sulfate in the media relative to other sulfated GAG than did the fibroblasts from normal subjects (Silbert and Kleinman, 1979). However, increased hyaluronidase and β-glucuronidase and increased GAG in aorta and liver have been noted (Padmaji et al., 1978).

16.7.8.2 Thyroid disorders: exophthalmos, cretinism, thyrotoxicosis

Thyroid hormone reduces the synthesis of sulfated GAG and produces a decrease in blood count of basophil leukocytes. Hypersecretion of thyroid hormone in thyrotoxic subjects decreases the sulfated and nonsulfated GAG contents of the connective tissue and the water-binding capacity (Asboe-Hansen, 1961). In hypothyroidism, excessive accumulations of GAG occur in skin, subcutaneous and other tissues producing edema (Bland, 1963). These tissue changes in hypothyroidism are reflected in increased levels of GAG in blood and urine of the (exophthalmos) hypothyroid patients (Winand, 1967). The blood of hypothyroid patients contained more mucopolysaccharides (100 mg hexosamine/100 ml) than normal subjects (83 mg/100 ml), whereas in hyperthyroid patients the values were less at 65 mg hexosamine/100 ml (Krompecher et al., 1961).

In 25 congenitally hypothyroid children showing symptoms of cretinism, a significant increase of serum ($P < 0.05$) and urinary ($P < 0.001$) GAG was noted (Khafagy et al., 1974). The serum GAG values in the patients were 1.06 ± 0.13 mg uronic acid/100 ml as compared to the values of 0.94 ± 0.11 mg uronic acid/100 ml in normal controls. These increased levels were normalized following an adequate thyroid medication. The authors (Khafagy et al., 1974) suggested that the abnormality involved thyrotropin (TSH) causing overproduction of GAG, mainly hyaluronic acid in the connective tissue. The increased synthesis of GAG in granulomas caused by cotton implants has been reported to follow induced hypothyroidism or thyroidectomy in rats and the reverse was observed in hyperthyroidism produced by administration of triiodothyronine (Gries et al., 1966). The increased TSH production seen in thyroidectomized rats may be responsible for increased GAG synthesis.

Thyroidectomy (Boas, 1955) or hypophysectomy increased the plasma hexo samine in rat from the control value of 98 ± 4 mg/100 ml to 122 ± 4 mg/100 ml ($P < 0.02$). The increase with hypophysectomy was a little less than that in thyroidectomy (Boas and Foley, 1955). Blood mucopolysaccharides decreased in rabbits by administration of thyrotropic hormone or thyroid extract (Jozsa et al., 1965). Changes in serum mucopolysaccharides have been noted with variations in adrenal activity or upon administration of corticoids (Diaz-Rubio et al., 1966; Ignesti et al., 1965). Blood mucopolysaccharide and hexosamine levels increased in rabbits given metothyrine and TSH. Thyroid administration reduced these concentrations. However, these values were normal in rabbits given thyroid hormone extract after treatment with metothyrine (Jozsa et al., 1965).

16.7 8.3 Adrenal gland disorders

The serum mucopolysaccharides determined as uronic acid and glycoproteins determined as mannose were both low in the Addison's disease and high in Cushing syndrome patients (Diaz-Rubio et al., 1966). A two day treatment with prednisone decreased the concentration of mucopolysaccharide (uronic acid).

Changes in serum mucopolysaccharides have been noted with variations in adrenal activity or upon administration of corticoids (Diaz-Rubio et al., 1966; Ignesti et al., 1965). Administration of prednisone for two days in subjects with normal mucopolysaccharide levels decreased the hexuronic acid in serum. Hydrocortisone or adrenocorticotropin (ACTH) injections have a similar effect on serum mucopolysaccharides. When rabbits were injected with 25 mg hydrocortisone or 10 units of ACTH for 10 days, the serum protein-bound glucosamine reached a maxima 7 days after the first injection (Kawashima, 1959).

16.7.8.4 Gonadal hormones and Glucagon

Estrogens induce an increase in the content of acid mucopolysaccharides, particularly hyaluronic acid in all connective tissues and progesterone counteracts these effects (Asboe-Hansen, 1961). Changes of these hormones during estrous cycle and pregnancy produce corresponding changes in serum hexosamine and glycosaminoglycan levels (Winzler, 1965). Serum GAG and vascular connective tissue GAG changes in castrated male and female rabbits receiving estrogens have been reported (Lorenzen, 1966). Intravenous estrogen administration produced changes in serum and dermal hyaluronidase (Wayne et al., 1961). Glucagon administration in rabbits produced changes in serum mucopolysaccharides and during hyperglycemia treatment with glucagon there was no parallelism between the modification of serum mucopolysaccharide content and blood sugar level (Giacovazzo and Bianchi, 1958).

16.7.8.5 Parathyroid glands

Administration of parathyroid extracts in rats produced increased sulfomucopolysaccharides or sulphur-containing glycoproteins in the serum and kidney of rat due to release of these macromolecules from bone matrix by depolymerization, liberation into blood and transport of these to kidney in blood (Bradford et al., 1959).

16.7.9 Bilharziasis

The trematod worm, Schistosoma produces not only hematuria and tumor like swelling of the skin (bilharzioma) but also an elevation of serum glycosaminoglycans. In four groups of patients, the serum protein-bound uronic acid, hexosamine, hexose and L-fucose increases correlated with the progress and severity of bilharziasis (Khafagy et al., 1972).

16.7.10 Peritonitis

The serum mucopolysaccharide levels have been reported to alter in experimental acute and chronic peritonitis (Szulc and Majewski, 1968, 1968 a; Szulc et al., 1970, 1970 a). In chronic peritonitis induced in 70 rats by introducing approximately 2.5 g of sterile talcum into the peritoneal cavity, serum acid mucopolysaccharides were measured at 7, 10, 14, 21, 30, 42 and 365 days later. During the first seven days a drop in serum GAG occurred followed by a rise up to the 14th day. Starting from that day the GAG levels in serum steadily decreased. These changes may be related to an index of early injury to the connective tissue; rise in serum GAG being related to increased connective tissue metabolism and vice-versa.

16.7.11 Renal diseases

Since the majority of urinary GAG are derived from plasma by glomerular filteration, it is logical to assume that increased plasma GAG levels could be expected in kidney diseases associated with decreased glomerular filteration. Variations in quantity and ratios of GAG components determined by electrophoretic method varied in acute and various forms of chronic diffuse glomerulonephritis (Dabolina, 1972). The author suggested the determination of serum GAG as a prognostic parameter for patients with diffuse glomerulonephritis. Severe renal failure is accompanied by a marked accumulation of GAG in uremic patients on dialysis and in those not treated by dialysis (Friman et al., 1977 a). The increased concentration of plasma GAG may be a causative factor in the genesis of some of the clinical symptoms associated with uremia. In 35 patients divided into 3 groups according to renal insufficiency based on plasma creatinine levels of 110-400 μM, 400 μM and over 400 μM/l, plasma GAG concentrations were inversely correlated to renal function (Friman et al., 1977 a). The mean concentrations of both free plasma GAG directly adsorbable on DE-52 anion exchange cellulose and the protein-bound plasma GAG adsorbable on AG 1-X2 resin after papain proteolysis were as controls < group 1 < group 2 < group 3 patients.

16.7.12 Cystic fibrosis

Although cystic fibrosis is a generalized abnormality in secretions of the exocrine glands, abnormalities of GAG metabolism in blood have been found in this disorder. A 30 % reduction of β-glucuronidase activity in white cell cultures from patients with cystic fibrosis was observed (Griffin and Gibbs, 1971). The lymphocytes from these patients cultured in serum responded poorly to the stimulation by phytohemagglutinin; the stimulation of β-glucuronidase activity observed being one-tenth of that observed in lymphocytes from normal subjects (Marchi et al., 1973). Other studies have reported an increased accumulation of ^{35}S-labeled heparin or heparin-like material in leukocytes cultured in absence or presence of phytohemagglutinin (Steel and Rodnan, 1971; Caudill et al., 1974). The increased heparin accumulation was due to increased biosynthesis rather than decreased degradation. Similar labeled heparin accumulations occurred in lymphocytes in affected homozygotes and heterozygotes of cystic fibrosis (Robertson et al., 1974). The above mentioned abnormalities of GAG metabolism and lysosomal enzymes in blood cells suggest that cystic fibrosis is a lysosomal disease. Recently, Casola et al., (1979) have reported a systematic study of 8 glycosidases in serum of patients with cystic fibrosis; only α-glycosidase was elevated.

16.7.13 Mental disorders

The serum GAG have been reported to alter in mental retardation associated with mucopolysaccharidoses and some behavioral disorders.

16.7.13.1 Mucopolysaccharidoses-related mental retardation

Neurological and mental deterioration are among the principle clinical manifestations (Suzuki, 1972) of the different variants of the genetic defects of acid mucopolysaccharide metabolism called the mucopolysaccharidoses. Besides the excessive deposition of dermatan sulfate and heparan sulfate in tissues, elevations of serum glycosaminoglycans and increased urinary excretion of these macromolecules are the most common observations and are used for diagnosis of mucopolysaccharidoses. The spectrum of serum GAG components usually corresponds to that in urine and the GAG present in excess in these biological fluids are the same which accumulate in the tissues in a particular variant of these genetic defects of degradation of glycosaminoglycans (see section 16.3). Various studies on blood cells and serum glycosaminoglycans and enzymes in mucopolysaccharidoses have been extensively reviewed (Dorfman and Matalon, 1972, 1976; Spranger, 1972; McKusick et al., 1978; Horwitz, 1979). A survey of mentally backward children using an Alcian blue spot test for excess mucopolysaccharides in serum or plasma and urine was carried out (Carson and Neill, 1962) and positive spot tests were found in mentally retards.

Elevation of glycosaminoglycans in blood of the mucopolysaccharidoses patients is reflected not only in the free plasma but also in the inclusions of the formed elements of the blood. In Sanfilippo syndrome, a faulty degradation of GAG in tissues due to a deficiency of α-L-iduronidase, and an elevation of dermatan sulfate and heparan sulfate in the serum along with an excessive excretion of these macromolecules in 3:1 ratio were observed (Sanfilippo and Good, 1964). In children with Hurler-Hunter syndromes, plasma glucuronic acid at 1-25 mg/100 ml may be present (Calatroni et al., 1969). Along with this rise in total plasma GAG, differences in the free, bound and total glycosaminoglycans of serum in male and female children with mucopolysaccharidosis have been noted. The values of the free, bound and total GAG in the male patients were increased to 22.32 ± 4.87, 53.95 ± 12.30 and 76.27 ± 14.55 from the values in normal male children of 8.40 ± 1.02, 30.95 ± 2.42 and 39.35 ± 2.42 μg hexuronic acid/10 ml, respectively. The corresponding values in female patients were increased to 18.18 ± 10.59, 36.78 ± 8.06 and 54.96 ± 16.21 from the values in normal female children of 11.60 ± 1.78, 30.62 ± 8.06 and 42.22 ± 8.14 μg hexuronic acid/10 ml, respectively (Calatroni et al., 1969

Dekaban and Constantopoulos (1973) have tabulated the values of glycosaminoglycans in body fluids of a total of 14 patients affected with 4 types of mucopolysaccharidoses: types I, II, III and V. The serum GAG measured as μg hexuronic acid/100 ml (carbazole-borate method) were markedly elevated in all the patients as compared to the values of 310-592 (average 525) in 11 controls. In another investigation (Dekaban and Constantopoulos, 1977) in 7 patients with mucopolysaccharidoses types I, II, IIIA and V, the serum GAG levels were 1.5 to 2.5 times higher (at 1950-3450 μg/100 ml) as compared to the average value of 1475 μg/100 ml in 11 controls.

Increased accumulation of GAG in the inclusions of lymphocytes of patients with Gargoylism has been observed with histochemical staining reactions (Mittwoch, 1961). Similarly, increased GAG accumulations in peripheral leukocytes have been demonstrated using fluorescence microspectrophotometry and microspectropolarimetry (West and Lorincz, 1973). The leukocytes from mucopolysaccharidoses patients took up 1000-fold more acridine orange dye than did the cells from normal individuals. Also, the acridine orange-treated leukocytes from the patients gave a peak at 680 nm with a red emission as compared to the emission peak at 540 nm and green fluorescence in cells from normal subjects. Szabo et al., (1973) have described a procedure for investigating metachromatic granules in peripheral leukocytes for diagnosis of mucopolysaccharidosis. Cultures of white blood cells have been used for genetic studies on the human mucopolysaccharidoses (Foley et al., 1969).

16.7.13.2 Cerebral rheumatism

Serum GAG and glycoproteins were elevated in 20 children, 9-15 years old, with different forms of cerebral rheumatism (Aksent'ev and Yasinovskaya, 1961). Highest values were encountered in the vascular form of cerebral rheumatism. In 14 children with chorea, the protein-bound hexosamine elevations returned to normal following therapy.

16.7.13.3 Psychiatric disorders

The serum GAG levels were significantly ($P < 0.001$) elevated to the range of 8.04-15.91 with a mean ± S.D. of 10.5 ± 1.1 μg hexuronic acid/ml in 46 paranoid schizophrenia patients as compared to the control values ($N = 50$) of 2.36-7.84 with a mean ± S.D. of 5.2 ± 1.6 (Varma et al., 1979). The levels returned to slightly above the normal range on partial remission. Serum GAG were also elevated in organic brain syndrome-associated psychosis, manic type of manic-depressive psychosis, schizoaffective psychosis and in different types of personality disorders. However, in catatonic and hebephrenic types of schizophrenia the values were in the upper part of normal range, and were either in the upper normal range or slightly higher in non-psychotic organic brain syndrome. On the other hand, values lower than normal or close to the lower normal range were encountered in depressed type of manic-depressive psychosis.

The serum protein-bound hexosamine was elevated in epilepsy (Tasher and Andreev, 1968). These authors have also reported values of serum hexosamine in other mental diseases. Elevation of serum GAG was also noted in epileptic patients with psychosis (Varma et al., 1979). Since rheumatoid arthritis is absent in schizophrenics (Mellsop et al., 1974), the elevation of GAG in serum may involve some other mechanism, which is not clear. Abnormal leukocytes with altered histochemical properties are present in blood of the schizophrenic patients (Fessel and Hirata-Hibi, 1963), which may contribute to serum GAG.

16.7.13.4 Aspartylglucosaminuria

It is a metabolic disorder which is associated with mental retardation and excretion of over 300 mg/day of a compound, 2-acetamido-1-(β^1-L-aspartamido)-1,2-dideoxy-β-D-glucose (AADG) due to a deficiency of an enzyme involved in the degradation of this compound (Scriver and Rosenberg, 1973). Coarse facial features and other signs are suggestive of a connective tissue disorder. For these reasons GAG metabolism has been studied in these patients.

Serum hyaluronidase and serum and urine β-glucuronidase activities were increased in this disorder. Serum GAG alterations and the presence of an unknown GAG in urine (containing N-acetylneuraminic acid, hexosamine, 1 or 2 uronic acid residues and aspartylglucosamine) on two-dimensional thin layer chromatography of urine were observed (Palo and Savolainen, 1972).

16.7.14 Carcinomas

Serum and urinary glycosaminoglycans are elevated in leukemia (Slater and Lovell, 1961; Medgyessy et al., 1971). Deutsch and Morton (1956) reported the presence of a large quantity of hyaluronic acid in the sera of two patients, one with reticulum cell sarcoma and the other with neuroblastoma, while no hyaluronic was detected in sera from normal subjects. This component moving faster than albumin on electrophoresis was initially established as a hyaluronic acid (of low M_r 58,000, sedimentation constant 2.45, axial ratio 50-60) on the basis of its acidic groups, its sensitivity to hyaluronidase and identification of glucosamine as all of the 34 % hexosamine present (Deutsch, 1957).

The presence of a large excess of glycosaminoglycans, particularly hyaluronic acid such as seen in carcinomas, produce coagulation and hematological disorders (Bagnoud, 1966). On this basis, abnormal components (GAG) have been suspected in patients. Broughton et al. (1970) while investigating a case of neuroblastoma suspected the presence of an abnormal component in blood from raised erythrocyte sedimentation rates, sludging of red cells in whole blood, marked ruleaux formation, an atypical Leishman stain, increased plasma viscosity and a distorted protein electrophoretic pattern. On the basis of precipitation upon acidification, positive Alcian blue spot test and the disappearance of concave bowings in the electrophoretic pattern upon digestion with hyaluronidase, the presence of hyaluronic acid (or chondroitin sulfate) in blood of the patient was suggested (Broughton et al., 1970). Hexuronic acid equivalent to 75 mg hyaluronic acid per 100 ml serum was present. Similarly, hyaluronic acid was isolated from the blood and urine of a three years old child with metastatic nephroblastoma (Morse and Nussbaum, 1967). The serum of the patient gave a precipitate upon acidification and this property disappeared upon incubation of the serum with testicular hyaluronidase. The serum hexuronic acid concentration was equivalent to hyaluronic acid levels of 163-272 mg/100 ml. The isolated glycosaminoglycan contained 40.3 % hexuronic acid and 32.8 % hexosamine and was identified on the basis of several criteria: specific rotation, absence of sulfate in IR spectrum, C/N elemental ratio and presence of glucosamine as the only hexosamine (Morse and Nussbaum, 1967).

In the case of a malignant renal tumor of young children (Wilm's tumor) approximately 20 mg hyaluronic acid and 8 mg sulfated GAG were present in urine and plasma, respectively (Hopwood and Dorfman, 1978). Elevation of GAG also occurred in experimental tumor growth (Skyvova et al., 1973). Bacchus (1965) has reported elevation of serum GAG and seromucoid in various malignant neoplasms, including solid tumors and neoplasms of the hemopoietic system. The source of elevated serum GAG in various neoplasms is not definitely known. In some cases it has been demonstrated not to arise from synthesis in malignant cells (Broughton et al., 1970). The electron micrographic evidence supports the suggestion by Deutsch (1957) that malignant tumor cells elaborate hyaluronidase which partially depolymerizes the hyaluronic acid in the connective tissue of the host and releases the GAG into the blood circulation. However, the total acidic glycosaminoglycans (consisting of hyaluronic acid, chondroitin 4-sulfate and a small amount of dermatan sulfate) in the serum of tumor-bearing rats was 1/15 of that in tumorigenic ascites. The Tawa sarcoma cells appeared to produce and discharge GAG to the tumorigenic ascites (Kuno, 1973).

In chronic myeloid leukemia, the immature, hyaluronidase-sensitive basophilic granules showed less intense metachromasia than the mature cells (Astaldi et al., 1955). The major glycosaminoglycan component in both normal and leukemic leukocytes was found to be chondroitin 4-sulfate (Olsson and Gardell, 1967; Olsson, 1968). The chondroitin sulfate isolated from leukemic cells had a higher uronic acid/hexosamine ratio but a lower sulfate content than a similar material isolated from normal nonleukemic cells. These differences were traced down by using ^{35}S-sulfate and glucosamine-^{14}C to be due to the facts that the chondroitin 4-sulfate was formed before sulfation occurred and that the leukemic cells more actively incorporated ^{14}C

and ^{35}S labels than did the normal leukocytes (Olsson, 1968; Olsson et al., 1968). Also, Olsson et al. (1968) by gel filtration noted different pools of the chondroitin 4-sulfate proteoglycan with different metabolic activities in leukocytes and in fact there may be separate fast and slow synthesizing pools accounting for the metabolic heterogeneities. The largest molecule also had the highest ^{14}C and ^{35}S radioactivities.

The GAG in peripheral lymphocytes of normals and cancer patients were revealed by histochemical technics and differences in staining were noted (Riesco and Leyton, 1971). The acidic glycosaminoglycans of leukocytes and thrombocytes in patients with leukemia, myeloid leukosis and polycythemia vera have been studied (Kharchenko and Seits, 1970; Kharchenko, 1973; Kharchenko et al., 1969). In short-term suspensions cultures of human bone marrow, virtually all of the $^{35}SO_4^{2-}$ was incorporated into myeloblasts, promyeloblasts and myelocytes suggesting that the majority of GAG synthesis and sulfation occurred in these cells. Radioautography of normal bone marrow cultures labeled with $Na_2{}^{35}SO_4$ indicated that only 2 % of the mature granulocytes contained radioactive granules slightly above the background, while 45 % of the granulocyte precursors were heavily labeled (Lau et al., 1970).

Leukemic cells display continuous synthesis and storage of chondroitin sulfates as opposed to normal leukocytes. The major portion of the GAG recovered from lymphocytes in type B chronic lymphocytic leukemia consisted of chondroitin sulfate and undersulfated chondroitin sulfate and these appeared to be present on the lymphocyte cell surface as the antigenic components (Picard et al., 1978; Capeau et al., 1978). However, these investigators did not detect any hyaluronic acid reported to be present in normal lymphocytes at 2.6 µg/mg of dry cell weight (Anderson et al., 1976). A positive and a significant correlation has been found between cancer curability and the total number of peripheral neutrophils, segmented and nonsegmented (Riesco, 1970). These findings indicate that the immunological activity of the peripheral lymphocytes may be a favorable factor in the treatment of cancer. Similar gel filtration patterns were obtained in short term mucopolysaccharide sulfation into chondroitin sulfate using normal granulocytic precursors and leukemic leukocytes. However, sulfation of mucopolysaccharides was only 50 % of the normal levels in cells from chronic granulocytic and about 15 % in cells of acute granulocytic leukemia or chronic granulocytic leukemia in blastic crisis. Virtually no sulfate incorporation was found in cells from acute lymphocytic leukemia; this fact assists in distinguishing between acute granulocytic leukemia from acute lymphocytic leukemia (Lau et al., 1972).

Alterations in glycosaminoglycans have been observed to occur in blood, urine and tissues in hepatic cancers(Kojima et al., 1975; Anghilery, 1974; Yamamoto and Terayama, 1973). The chemical characteristics of the cellular mucopolysaccharides were determined by cytochemical methods in 30 blood samples from patients with different myelocytic leukemia and in 10 bone marrow samples (Catini et al., 1974). Hyaluronic acid, chondroitin sulfate and heparin were detected. The azurophilic granules contained a chondroitin-like GAG in 50 % of the cases. The presence of a heparin-like GAG in 3.2 % of the cases was related to the basophilic surge. Heparin was detected in 50 % of the cases. These cytochemical characteristics were related to the clinical status of the leukemia patients.

16.7.15 Varicose veins

An increase in glycoproteins and sulfated glycosaminoglycans, from measurements of protein-bound hexosamine and sulfate, was observed in 26 patients with varicose veins as compared to 24 healthy subjects. The hexosamine increased significantly ($P < 0.02$) from 80.9 ± 9.7 mg/100 ml serum to 86.7 ± 6.5 mg/100 ml and sulfate increased ($P < 0.01$) from 5.05 ± 0.49 to 6.03 ± 1.06 mg/100 ml in the patients (Niebes, 1972). In addition to these, three lysosomal GAG hydrolyzing enzymes, β-glucuronidase, β-N-acetylglucosaminidase and arylsulfatase were also increased significantly ($P < 0.001$) and these increases varied with the severity of the varices. Niebes (1972, 1972 a), Niebes and Laszt (1971) and Niebes and Berson (1973) found significant increases in lysosomal enzymes in the varicose vein of man. These enzymatic elevations were suggested to be responsible for the increase of GAG in serum. The flavonoid (+)-catechin inhibited the lysosomal enzymes in vivo in experimental thrombosis of dog's vein (Niebes, 1972 b).

The mean concentration of GAG measured as hexuronic acid was significantly ($P < 0.0005$) increased in 34 patients suffering from serious varicose conditions as compared with serum samples from 30 healthy subjects. The GAG concentration (mean ± S.D.) of 0.602 ± 0.066 mg/100 ml in healthy subjects was elevated to 0.751 ± 0.108 mg/100 ml and these were not different in patients with complications of trophic changes such as phlebitis, thrombosis and ulcers (Niebes and Schifflers, 1975). The increase in serum glycosaminoglycan levels was attributed to the observed increases in neutral sugars (+45 %, $P < 0.25$), hexosamines (+80 %, $P < 0.0025$) and hexuronic acid (+56 %, $P < 0.001$) contents of 12 varicose human saphenous veins as compared to 7 healthy saphenous veins (Niebes and Jegerlehner, 1974).

16.7.16 Cardiovascular disorders

Serum levels and urinary excretion were not different in patients with positive and negative aortic arch calcification in arterioslerosis (Kanazawa et al., 1977). Similar studies (Izawar, 1977) involved relationship of serum and urine in arteriosclerosis. Other studies have reported changes in capillary permeability and GAG changes in patients with arteriosclerosis and treated with ACS and dextran (Bachinskaia, 1971) and in those treated with antireticular cytotoxic serum (Miagkaia et al., 1973). Significantly lower concentrations of heparinoid substances in atherosclerosis (Antonini, 1956) may be responsible for abnormal lipid patterns (Kasatkina et al., 1974). Since the glycosaminoglycans are known to interact with lipoproteins, the lipid patterns change after administration of glucuronyl glucosaminoglycan sulfate (Gallone et al., 1975). Blood coagulation disorders were observed in patients with myocardial infarction and angina pectoris (Olszewski, 1968; Stark and Nikolaeva, 1968) and these were found to be related to blood glycosaminoglycans. In patients with hypertension and chronic insufficiency of cerebral circulation, blood heparin decreased in correlation with the degree of the disorder and blood pressure normalized after heparin treatment (Mitenko and Rogova, 1976).

16.7.17 Other miscellaneous diseases

Serum glycosaminoglycan changes occur in several other disease states, some of which have been reviewed (Brimacombe and Weber, 1964; Hasegana et al., 1978). A nonspecific increase in serum mucopolysaccharides occurred in *magnesium deficiency* in rats (Augusti et al., 1973), in *chronic diseases* (Seibert et al., 1947), in *chronic pneumonia-associated amyloidosis* (Kusaev, 1967) and in *general amyloidosis* (Kusaev, 1968). Heparan sulfate which appears not to be present to a much extent in normal palsma became elevated and detectable in patients with *amyloidosis* (Muir, 1969). Changes in blood mucopolysaccharides and blood coagulation were observed in *hot shop workers* (Mitsenko, 1976), probably due to the hormonal changes induced in this stress. Serum hyaluronic acid levels were elevated 1.35-fold in 27 patients with *chronic suppurative otitis media* without complications and 2.2-4.4 fold in patients with *intracranial complications (brain abscess)*. In the former group the levels normalized in remission while in the latter cases surgical treatment and remission only reduced the elevation to 1.25-2.5 fold above normal (Ostrovskii, 1977). Changes occur in Jaundice (Starosel'skii et al., 1973), in *Curling's ulcer* (Kuo et al., 1973), in *gastric and duodenal ulcers* (Szulc and Szwabe, 1968, 1969), *disorders of mesenchymal tissues* (Friman, 1972) etc. Blood GAG elevation also occurs in *tuberculosis* and the measurements of GAG levels in blood have been suggested as a biochemical criteria for recovery of tuberculosis patients (Blokhin, 1968) and in follow up of treatment of some internal diseases (Sladki and Zak, 1968). The tuberculosis tissue is rich in viscous mucopolysaccharides, which are locally formed and poured into the blood stream to increase the mucopolysaccharide and hexosamine blood levels in active tuberculosis (Berencsi, 1961). Mucopolysaccharide contents of both serum and saliva were altered in *parodontosis* (Pozharitskaia et al., 1975). The leukocytes in *Dyggve-Melchior-Clausen syndrome* patients had increased levels of chondroitin sulfate sulfatase, N-acetyl-6-sulfate sulfatase and sulfamidase and decreased levels of the enzymes arylsulfatase A and B (Rastogi et al., 1977).

16.8 Bibliography

1. Aborg, C. H. and Uvnas, B. (1971): A mucopolysaccharide-protein complex with amine binding properties in rat thrombocytes. Acta Physiol. Scand. 81, 568-570

2. Aksent'ev, S. B. and Yasinovskaya, A. Y. (1961): Changes in mucopolysaccharides in specific manifestations in the course of cerebral rheumatism. Reumatizm, Odessa, Sb. From Zh. Khim. Biokhim 1962, Abstr. No. 16S1306

3. Ali, S. Y. (1969): A quantitative method for estimation of acid polysaccharides in blood. Biochim. Biophys. Acta 177, 641-643

4. Anderson, B. B. and Odell, T. T. (1958): Chemical studies of mucopolysaccharides of rat blood platelets. Proc. Soc. Exptl. Biol. Med. 99, 765-768

5. Anderson, P., Slorach, S. A. and Uvnas, B. (1974): 5-HT storage in rat and rabbit blood platelets: The separation of ATP-containing and sulphomucopolysaccharide-containing granules. Acta Physiol. Scand. 90, 522-532

6. Anderson, L. C., Smith, Q. T. and Shapiro, B. L. (1976): Cellular glycosaminoglycans in lymphocytes from patients with cystic fibrosis. Clin. Chim. Acta 73, 63-65

7. Andreani, D. V. and Gray, C. H. (1956): Serum polysaccharides in diabetes mellitus. Clin. Chim. Acta 1, 7-15

8. Andreev, N. A. (1965): Quantitative determination of hyaluronic acid and of chondroitin sulfates A and C in blood serum. Vop. Med. Khim. 11, 72-75

9. Anghileri, L. J. (1974): Metabolism of acid mucopolysaccharides in hepatoma and in normal liver. Oncology 30, 304-317

10. Antonini, F. M. (1956): Biochemical changes of blood in atherosclerosis and in old age with particular reference to mucopolysaccharides and heparinoid substances. Experientia Suppl. No. 4, 176-180

11. Anzai, T. (1972): Determination of mucopolysaccharides in body fluids. Methods and their significance. Rinsho. Byori, Rinjizokan No. 20, 80-91

12. Asboe-Hansen, G. (1961): Endocrine control of connective tissue. In: L. C. Mills and J. H. Moyer (Eds.). Inflammation and Diseases of Connective Tissue. Saunders, Philadelphia, pp. 38-43

13. Asboe-Hansen, G. (Ed.) (1966): Hormones and Connective Tissue. Williams and Wilkins, Baltimore

14. Astaldi, G., Rondanelli, E. G., Strusselli, E. and Gorini, P. (1955): La myelopenie erythroblastique etude de la moelle osseuse dans la forme acquise. Sang 26, 345-362

15. Augusti, Y., Rayssiguier, Y., Larvor, P. and Durlach, J. (1973): Magnesium deficiency in the rat and variations of protein parameters of blood serum. C. R. Soc. Biol. 167, 1804-1808 (Fr)

16. Bacchus, H. (1960): Serum seromucoid and acid mucopolysaccharide in the Marfan syndrome. J. Lab. Clin. Med. 55, 221-228

17. Bacchus, H. (1965): Serum seromucoid and acid mucopolysaccharide in malignant neoplastic diseases. Cancer 18, 1285-1291

18. Bachinskaia, L. (1971): Vascular permeability and mucopolysaccharide metabolism in arteriosclerosis under the effect of ACS and Dextran. Vrach Delo 11, 31-33 (Rus)

19. Badin, J. and Schubert, M. (1955): Conditions of formation of euglobulin-like-precipitates from serum and chondroitin sulfate. J. Clin. Invest. 34, 1312-1316

20. Badin, J., Schubert, M. and Vouras, M. (1955): Plasma polysaccharide fraction containing uronic acid, in normal subjects and in patients with rheumatoid arthritis. J. Clin. Invest. 34, 1317-1323

21. Bagnoud, F. (1966): Blood coagulation disorders and carcinomas rich in mucopolysaccharides. Thromb. Diath. Hemorrh. 15, 143-160 (Fr)

22. Balabanov, K. and Samsonova, S. (1965): Investigations on mucopolysaccharides in blood serum of some collagenoses. Derm. Wschr. 151, 876-80 (Ger)

23. Balesnaya, A. N. (1977): Glycosaminoglycans and glycoproteins in the serum of rheumatism patients. Zdravookhr. Beloruss. (2), 84-85 (Russ)

24. Bassiouni, M. (1954): The estimation of heparin and similar substances in human blood and tissues using a combined biologic and colorimetric method with paper electrophoretic studies. J. Clin. Pathol. 7, 330-335

25. Bassiouni, M. (1955): Studies of the acid polysaccharides of the white cells in rheumatic and other diseases showing its similarity to the acid polysaccharide of amyloid. Ann. Rheumat. Dis. 14, 288-292

26. Bednarski, W., Figarska, J. and Student, J. (1970): Glycosaminoglycan (acid mucopolysaccharide) level in the serum and their urinary excretion in patients with virus hepatitis. Wiad. Lek. 23, 1795-1799 (Pol)

27. Berencsi, G. (1961): The mucopolysaccharides of tuberculous tissue. Acta Biol. Med. Ger. 6, 441-446 (Ger)

28. Bergendorf, A. and Uvnas, B. (1972): Storage of 5-hydroxytryptamine in rat mast cells. Evidence for an ionic binding to carboxyl groups in a granule heparin-protein complex. Acta Physiol. Scand. 84, 320-331

29. Bjornesjo, K. B., Werner, I. and Odin, L. (1959): The influence of surgery on serum and urine hexosamine, serum mucoprotein, glutamic oxaloacetic transaminase. (GOT) and C reactive protein. Scand. J. Clin. Lab. Invest. 11, 238-244

30. Bland, J. H. (1963): Clinical Metabolism of Body Water and Electrolytes. Saunders, Philadelphia

31. Blokhin, N. N. (1968): Biochemical criteria of recovery of patients from tuberculosis. Probl. Tuberk. 46, 62-69 (Russ)

32. Boas, N. F. (1955): Electrophoretic distribution of hexosamine in plasma proteins of rat following thyroidectomy. Proc. Soc. Exptl. Biol. Med. 90, 4-7

33. Boas, N. F., Bollet, A. J. and Bunim, J. J. (1957): Effect of acute clinical stress on the levels of hexosamine in serum and its excretion in urine. J. Clin. Invest. 34, 782-789

34. Boas, N. F. and Foley, J. B. (1955): Effect of thyroidectomy and hypophysectomy on plasma hexosamine levels in the rat. Endocrinology 56, 305-311

35. Boas, N. F. and Peterman, A. F. (1953): Effects of age, food intake, and stress on plasma hexosamine levels in the rat. Proc. Soc. Exptl. Biol. Med. 82, 19-24

36. Bollet, A. J., Saraydarian, M. W. and Simpson, W. F. (1957): Acid mucopolysaccharides in normal serum. J. Clin. Invest. 36, 1328-1332

37. Boseila, A.-W. A. (1966): Endocrine influence on blood elements. In: G. Asboe-Hansen (Ed.). Hormones and Connective Tissue. Williams and Wilkins, Baltimore, pp. 38-69

38. Bottiger, L. E., Malmquist, E. and Olhegan, B. (1964): Serum protein-bound carbohydrates in rheumatic disease. Ann. Rheumat. Dis. 23, 480-495

39. Bowness, J. M. and Tan, Y. H. (1968): Chromatographic distinction between hyaluronidases from human serum and ovine testes. Biochim. Biophys. Acta 151, 288-290

40. Bradford, R. H., Howard, R. P., Joel, W. and Shetlar, M. R. (1959): Effects of parathyroid extract on serum and kidney. Biochem. Biophys. Res. Commun. 1, 350-355

41. Brimacombe, J. S. and Weber, J. M. (1964): Mucopolysaccharides in disease. Adv. Clin. Chem. 7, 199-234

42. Broughton, P. M. G., Dykes, J.-R., Holt, S. Ridley, J. W. and Steel, A. E. (1970): Mucopolysaccharide in the blood of a patient with neuroblastoma. J. Clin. Path. 23, 246-248

43. Calatroni, A. (1974): Proceedings: Products of the breakdown of glycosaminoglycans in normal and pathological human plasma and urine. Ital. J. Biochem. 23, 267-268

44. Calatroni, A., Donnelly, P. V. and Di Ferrante, N. (1969): The glycosaminoglycans of human plasma. J. Clin. Invest. 48, 332-343

45. Capeau, J., Picard, J. and Paul-Gardais, A. (1978): Isolation of glycosaminoglycans from lymphocytes in chronic lymphoid leukemia. Biomedicine 28, 124-129

46. Carson, N. A. J. and Neill, D. W. (1962): Metabolic abnormalities detected in a survey of mentally backward individuals in Northern Ireland. Arch. Dis. Child. 37, 505-513

47. Casola, L., DiMatteo, G., Romano, M., Rutigliano, B. and Mastella, G. (1979): Glycosidases in serum of cystic fibrosis patients. Clin. Chim. Acta 94, 83-88

48. Catini, C., Gheri, G. and Milani, A. (1974): Cytochemistry of acid mucopolysaccharides of the granulocytic series in the chronic myeloid leukemia. Haematologica 59, 410-427

49. Caudill, M., Schafer, I. and Stjernholm, R. (1974): Sulfate incorporation of leukocytes from patients with cystic fibrosis. Lancet 1, 32

50. Chernov, G. A., Sheremet, Z. I. and Lenskaya, R. V. (1965): Effect of irradiation on vascular permeability and on blood mucopolysaccharides and serotonin levels. Fed. Proc. (Transl. Suppl.) 24, 974-976

51. Chmiel, J., Blazejewska, B., Suszka, B. and Waliszewski, S. (1975): Determination of plasma acid mucopolysaccharides. Diagn. Lab. 11, 309-318 (Pol)

52. Clausen, V. and Andersen, V. (1963): Acid mucopolysaccharides in human leukocytes. Clin. Chim. Acta 8, 505-512

53. Clausen, J. and Asboe-Hansen, G. (1966): Urinary acid mucopolysaccharides in mastocytosis. A new technique for quantitative estimation. Clin. Chim. Acta 13, 475-483

54. Cosson, A., Maroteaux, P., Tapon, J. et al. (1977): Blood and bone marrow cytology in the diagnosis of illnesses of excess. Forty observations of mucopolysaccharidosis, mucolipidosis and mannosidosis. Nouv. Rev. Fre. Hematol. Blood Cells 18, 205-206 (Fre)

55. Dabolina, T. (1972): Condition of serum acid mucopolysaccharides (glycosaminoglycans) in diffuse glomerulonephritis. Latv. PSR Zinat, Akad Vestis (3) 127-130

56. Dahlen, G., Ericsson, C. and Berg, K. (1978): In vitro studies on the interaction of isolated Lp(a) lipoprotein and other serum lipoproteins with glycosaminoglycans. Clin. Genet. 14, 36-42

57. Daimon, T., Uchida, K. and Mizuhira, V. (1977): Ultrastructural localization of acid protein polysaccharides and calcium in vacuoles of the chicken thrombocyte. Histochemistry 52, 25-32

58. Dekaban, A. S. and Constantopoulos, G. (1973): Mucopolysaccharidoses. Relation of elevated cerebral spinal fluid acid mucopolysaccharides to mental retardation. Arch. Neurol. 28, 385-388

59. Dekaban, A. S. and Constantopoulos, G. (1977): Mucopolysaccharidosis types I, II, IIIA and V. Pathological and biochemical abnormalities in the neural and mesenchymal elements of the brain. Acta Neuropath. (Berlin) 39, 1-7

60. Deutsch, H. F. (1957): Some properties of human serum hyaluronic acid. J. Biol. Chem. 224, 767-774

61. Deutsch, H. F. and Morton, J. I. (1956): Some properties of hyaluronic acid from human serum. Fed. Proc. 15, 242

62. Diaz-Rubio, M., Martinez, A. and Gonzalez Rodriguez, N. (1966): Adrenal activity and effect of corticoids on mucopolysaccharide levels in blood. Rev. Clin. Espan. 101, 177-182 (Span)

63. Di Ferrante, N. (1957): Urinary excretion of acid mucopolysaccharides by patients with rheumatoid arthritis. J. Clin. Invest. 36, 1516-1520

64. Di Ferrante, N. and Rich, C. (1956): The mucopolysaccharide of normal human urine. Clin. Chim. Acta 1, 519-524

65. Dillard, G. H. L., Brecher, G. and Cronkite, E. P. (1951): Separation, concentration and transfusion of platelets. Proc. Soc. Exptl. Biol. Med. 78, 796-799

66. Donnelly, P. V., Di Ferrante, N. and Jackson, R. L. (1978): The interaction of human plasma glycosaminoglycans with plasma lipoproteins II. Hemagglutination studies. Circ. Res. 43, 234-237

67. Dorfman, A. (1979): Mucopolysaccharidoses. In: W. Nelson, V. C. Vaughn, R. J. McKay and R. J. Behrman (Eds.) Nelson Textbook of Pediatrics, 11th Ed., Saunders, Philadelphia, pp. 1845-1848

68. Dorfman, A. and Matalon, R. (1972): The mucopolysaccharidoses. In: J. B. Stanbury, J. B. Wyngaarden and D. S. Fredrickson (Eds.). The Metabolic Basis of Inherited Disease, 3rd Ed. McGraw Hill, New York, Chap. 49, pp. 1218-1272

69. Dorfman, A. and Matalon, R. (1976): The mucopolysaccharidoses. A Review. Proc. Natl. Acad. Sci. U.S.A. 73, 630-637

70. Dryzhova, L. P. (1968): Mucopolysaccharide content of the blood serum in pregnant women and their children in isoantigen incompatibility. Akush. Genekol. (Mosk) 44, 18-21 (Rus)

71. El-Ebrashy, N., Shahin, H., El-Roubi, O., Wasfy, A., Gaber, A. and Abul Fadl, M. (1973): Serum β-glucuronidase activity and mucopolysaccharide levels in diabetes. J. Egypt. Med. Assoc. 55, 821-832

72. Emura, Y. and Fukuda, T. (1973): Microanalysis of serum acid mucopolysaccharides by modified carbazole reaction. J. Jap. Biochem. Soc. 45, 30-36

73. Emura, Y. and Koizumi, T. (1975): A method for the quantitative analysis of acidic glycosaminoglycans in a small amount of serum. Ketsugo Soshiki 7, 51-59 (Japan)

74. Endo, M., Namiki, O. and Yosizawa, Z. (1976): Proteochondroitin sulfate A in the euglobulin fraction of human plasma. J. Biochem. (Tokyo) 79, 5-10

75. Endo, M., Yamamoto, R., Namiki, O., Satake, S. and Yosizawa, Z. (1979): Comparison of glycosaminoglycans (GAG) in normal human plasma and urine. Tohoku J. Exp. Med. 128, 89-99

76. Engelberg, H., Dudley, A. and Freeman, L. (1955): An improved method for the determination of plasma heparin. J. Lab. Clin. Med. 46, 653-656

77. Enomura, Y. and Kurata, T. (1971): Quantitative analysis of serum acid mucopolysaccharide. Jap. J. Clin. Pathol. 19, 350 (Japan)

78. Erickson, R. P., Sandman, R. and Epstein. C. J. (1975): Lack of relationship between blood and urine levels of glycosaminoglycans and lysosomal enzymes. Biochem. Med. 12, 331-339

79. Fedorishko, S. S. (1968): Blood and urine acid mucopolysaccharides in evaluating the activity of rheumatism. Vrach. Delo 6, 18-21

80. Fedorko, M. E. and Morse, S. I. (1965): Isolation, characterization and distribution of acid mucopolysaccharides in rabbit leucocytes. J. Exp. Med. 121, 39-48

81. Fessel, W. J. and Hirata-Hibi, M. (1963): Abnormal leukocytes in schizophrenia. Arch. Gen Psychiat. 9, 601-613

82. Fil'chagin, N. M., Agababova, E. R., Nazarenko, V. M. and Chepoi, V. M. (1977): Lysosomal enzyme activities in the blood serum and synovial fluid in articular cartilage. Ter. Arkh. 49, 120-123 (Russ)

83. Foley, K. M., Danes, B. S. and Bearn, A. G. (1969): White blood cell cultures in genetic studies on the human mucopolysaccharidoses. Science 164, 424-426

84. Freeman, L., Engelberg, H. and Dudley, A. (1954): Plasma heparin levels I. A method for determination of plasma heparin based on anticoagulant activity. Am. J. Clin. Pathol. 24, 599-606

85. Friman, C. (1972): Glycosaminoglycans in the urine and serum of healthy persons and of patients with disorders afflicting tissues of mesenchymal origin. Soc. Sci. Fenn. Comment. Biol. 50, 43

86. Friman, C. and Brunish, R., (1966): A zone electrophoretic study of acid mucopolysaccharides (AMPS) in normal human serum. Proc. Exptl. Biol. Med. 122, 599-601

87. Friman, C., Juvani, M. and Skrifvars, B. (1977): Acid glycosamino-glycans in plasma. II. Findings in rheumatoid arthritis. Scand. J. Rheumatol. 6, 177-182

88. Friman, C., Storgards, E., Juvani, M. and Kock, B. (1977 a): The glycosaminoglycans (mucopolysaccharides) in plasma in patients with renal insufficiency. Clin. Nephrol. 8, 435-439

89. Galambos, J. T. (1975): Connective tissue metabolism and cirrhosis. In: H. Popper and K. Becker (Eds.). Collagen Metabolism in Liver. Proc. Int. Conf. 1973. Stratton, New York, pp. 57-61

90. Galambos, J. T. and Dale, E. (1976): Plasma concentrations of connective tissue compounds. I. The effect of uterine involution. Proc. Soc. Exp. Biol. Med. 151, 366-367

91. Gallone, S., Mangiameli, D., Vanaria, F. et al. (1975): Changes in the lipid pattern after administration of glucuronyl-glucosaminoglycan sulfate (3 GS) in patients with acute coronary syndromes. Boll. Soc. Ital. Cardiol. 20, 309-314 (Ita)

92. Gazith, J. and Schimmelpfennig, K. (1964): Plasma-glycosaminoglycans for pharmacological experiments. Nauyn Schmiedberg's Arch. Pharmacol. 287 Suppl., R64

93. Ghirardo, G. and Rovetta, G. (1968): Behavior of the fractionation curve of blood acid mucopolysaccharide in various rheumatic diseases. Reumatismo 20, 118-120 (Ita)

94. Giacovazzo, M. and Bianchi, P. (1958): Serum mucopolysaccharides in rabbits during and after glucagon treatment. Boll. Soc. Ital. Biol. Sper. 34, 599-601

95. Gibson, R. B., Carr, T. L., Green, S. and Fowler, T. M. (1951): Photometric assay of plasma heparin. Proc. Soc. Exptl. Biol. Med. 79, 577-580

96. Goldstein, G. (1961): Serum β-glucuronidase assay by the phenolphthalein mono-β-glucuronide method. Clin. Chem. 7, 136-142

97. Greiling, H. (1974): Hyaluronic acid. In: H. U. Bergmeyer (Ed.). Methods of Enzymatic Analysis, Academic Press, New York, Vol. 3, pp 1217-1220

98. Gries, G., Lindner, J. and Freytag, G. (1966): Effect of thyroid gland function on mucopolysaccharide metabolism. Med. Pharmacol. Exptl. 14, 276-283

99. Griffin, G. D. and Gibbs, G. F. (1971): Lysosomal enzymes of cultured white cells in cystic fibrosis. Proc. Soc. Exp. Biol. Med. 137, 438

100. Gross, J. I., Mathews, M. B. and Dorfman, A. (1960): Sodim chondroitin sulfate-protein complexes of cartilage. II. Metabolism. J. Biol. Chem. 235, 2889-2892

101. Grudzien, M. (1971): Serum acid mucopolysaccharides in women during labor and puerperium and in the umbilical-cord blood and placental tissue in full-term and prolonged pregnancy. Pol. Tyg. Lek. 26, 947-949 (Pol)

102. Grudzien, M. (1973): Serum acid mucopolysaccharides in women during labor and puerperium and in the umbilical-cord blood and placental tissue following spontaneous and instrumental deliveries. Ann. Univ. Mariae Curie-Sklodowska, Sect. D 28, 263-269 (Pol)

103. Harris, G. and Fraser, J. R. E. (1969): Behavior of serum in the borate modification of the carbazole reaction. Anal. Biochem. 27, 433-438

104. Harris, G. S. and Fraser, J. R. E. (1969 a): Extraction and measurement of glycosaminoglycans in serum and cell culture medium: a system with the use of an aliphatic quaternary ammonium salt. J. Lab. Clin. Med. 74, 527-535

105. Haruki, H. (1959): Clinical and experimental studies on hexosamines in serum and urine. Rheumatism (Tokyo) 2, 46-61 (Jap)

106. Hasegana, O. et al., (1978): Serum glycosaminoglycans in health and disease. Nippon Ronen Igakkai Zasshi 15, 245-250 (Japan)

107. Hata, R.-I. and Nagai, Y. (1978): A low sulfated chondroitin sulfate in rat blood. An acidic glycosaminoglycan with a high metabolic rate. Biochim. Biophys. Acta 543, 149-155

108. Hata, R.-I., Ohkawa, S.-I. and Nagai, Y. (1978): Low sulfated chondroitin sulfate in human blood and urine. Biochim. Biophys. Acta 543, 156-166

109. Holzmann, H., Korting, G. W. and Morsches, B. (1968): On the modification of serum and urinary mucopolysaccharides of patients with scleroderma using gestagen therapy. Arch. Klin. Exp. Derm. 231, 156-160 (Ger)

110. Hopwood, J. J. and Dorfman, A. (1978): Glycosaminoglycan synthesis by Wilm's tumor. Pediatr. Res. 12, 52-56

111. Horn, R. G. and Spicer, S. S. (1964): Sulfated mucopolysaccharide and basic protein in certain granules of circulating heterophils of rabbits during endotoxin induced leukocytes. Am. J. Pathol. 44, 905-919

112. Horn, R. G. and Spicer, S. S. (1964 a): Sulfated mucopolysaccharide and basic protein in certain granules of rabbit leukocytes. Lab. Invest 13, 1-15

113. Horwitz, A. L. (1979): The mucopolysaccharidoses: clinical and biochemical correlations. Am. J. Ment. Defic. 84, 113-123

114. Howell, W. H. (1925): The purification of heparin and its presence in blood. Am. J. Physiol. 71, 553-562

115. Ignesti, C., Franchi, F. and Silvestrini, E. (1965): Influence of hydrocortisone and ACTH on the serum level and urinary excretion of hexosamines and acid mucopolysaccharides. I. Behavior of the nyctohemeral rhythm of neutral and acid glycoproteins in the serum and urine before and after intravenous administration of a single dose of hydrocortisone and ACTH. Riv. Crit. Clin. Med. 65, 650-661 (Ita)

116. Izawar, M. (1977): Relationships of serum and urine acid mucopolysaccharides to aging, glucose tolerance and aortic arch calcification-studies on metabolism, secretion and excretion of glycoprotein in arteriosclerosis. Nippon Ronen. Igakkai Zasshi 14, 268-278 (Japan)

117. Jakab, L., Bretan, M., Feher, J. et al., (1966): Serum mucopolysaccharides in alloxan and steroid diabetes. Acta Med. Acad. Sci. Hung. 22, 53-60

118. Jakab, L., Bretan, M., Feher, J. et al., (1966 a): Serum mucopolysaccharide studies in experimental alloxan and steroid diabetes. Orv. Hetil, 107, 1404-1407 (Hun)

119. Jaques, L. B., Monkhouse, F. C. and Stewart, M. (1949): A method for the determination of heparin in blood. J. Physiol. 109, 41-48

120. Jozsa, L., Perneczky, M., Szedekenyi, G. and Lusztig, G. (1965): The effect of chemical thyroidectomy, thyroid therapy, and thyrotropin on the mucopolysaccharide content of rabbit blood. Endokrinologie 48, 44-51

121. Juvani, M., Friman, C., Ranta, H. and Wegelius, O. (1975): Isolation and characterization of undersulfated chondroitin 4-sulfate from normal human plasma. Biochim. Biophys. Acta 411, 1-10

122. Kamev, M. F. (1969): Blood serum protein-polysaccharide components in patients with elephantiasis before and after radical dermoplastic surgery. Sovet. Med. 32, 95-99 (Rus)

123. Kanazawa, T., Komatsu, T., Izawa, K., Terata, T., Mori, H., Shibutani, K., Izumiyama, S., Oike, Y. and Moteki, H. (1977): Relationship of serum acid mucopolysaccharides (AMPS) and of urine AMPS to age, aortic arch calcification and glucose tolerance. Studies on metabolism and excretion of acid mucopolysaccharides in arteriosclerosis. Hirosaki Igaku 29, 657-670

124. Karcher, K. H. (1966): The importance of biochemical studies of the connective tissue fundamental structure during radiotherapy. I. Electrophoretic demonstration of acid mucopolysaccharides in serum, urine and tissue. Strahlentherapie 129, 80-94 (Ger)

125. Kasatkina, L. V., Shtel'makh, M. I. and Kovaleva, O. N. (1974): Age-specific changes of glycosaminoglycan and lipid content of blood in coronary arteriosclerosis. Kardiologiia 14, 103-106 (Rus)

126. Kawashima, S. (1959): Mucopolysaccharide II. Factors influencing serum mucopolysaccharide levels with special reference to pituitary and adrenocortical hormones. Ann. Paediat. Japon. 5, 884-902

127. Keiser, H., Greenwald, R. A., Feinstein, G. et al. (1976): Degradation of cartilage proteoglycan by human leukocyte granule neutral proteases- a model of joint injury. II. Degradation of isolated bovine nasal cartilage proteoglycan. J. Clin. Invest. 57, 625-632

128. Kelly, V. C. (1952): Acute phase reactants II. Serum hexosamine in patients with rheumatic fever and related diseases. J. Pediat. 40, 413-422

129. Kerby, G. P. (1955): The occurrence of acid mucopolysaccharides in human leucocytes and urine. J. Clin. Invest, 34, 1738-1743

130. Kerby, G. P. (1958): Distribution of acid mucopolysaccharides in Cohn fraction of human plasma proteins. J. Clin. Invest. 37, 678-681

131. Kerby, G. P. (1958 a): The effect of inflammation on the hexuronate-containing polysaccharides of human plasma. J. Clin. Invest. 37, 962-964

132. Kerby, G. P. and Langley, N. M. (1959): Human thrombocytes and leucocytes as a source of some of the acid mucopolysaccharides of serum. J. Lab. Clin. Med. 53, 708-713

133. Khafagy, E. Z., Kolb, A., Gabr, M. and Osman, H. G. (1974): Serum and urinary acid mucopolysaccharides in cretinism. Gaz. Egypt Paediatr. Assoc. 22, 163-168

134. Khafagy, E. Z., Osman, H. G., El-Raziky, El. H. and Shalaby, F. H. (1972): Evaluation of the level of mucopolysaccharides in serum in bilharziasis. Clin. Chim. Acta 40, 371-375

135. Kharchenko, M. F. (1973): Glycosaminoglycans of human thrombocytes in normal conditions, in chronic myeloid leukosis, myelofibrosis and polycythemia vera. Vopr. Med. Khim. 19, 613-617 (Rus)

136. Kharchenko, M. F. and Seits, I. F. (1968): Acid mucopolysaccharides in human leukocytes. Biokhimiya 33, 43-47 (Rus)

137. Kharchenko, M. F. and Seits, I. F. (1970): Quantitative levels and turnover rate of acid mucopolysaccharides in the leukocytes of leukemia patients. Vop. Onkol 16, 14-19 (Rus)

138. Kharchenko, M. F., Seits, I. F. and Blinova, A. I. (1969): Acid mucopolysaccharides of leukocytes in patients with leukemia and polycythemia Vera. Probl. Gemat. 14, 35-37 (Rus)

139. King, J. S. Jr. and Boyce, W. H. (1963): High molecular weight substances in human urine. Charles C. Thomas, Springfield, Chap. 7, pp. 106-118

140. Kojima, J., Nakamura, N., Kanatawi, M. et al., (1975): The glycosaminoglycans in human hepatic cancer, Cancer Res. 35, 532-547

141. Krompecher, St., Fornet, B., Balogh, G., Olah, E. H. and Laszlo, M. B. (1961): Mucopolysaccharide and thyroid function. Nature 189, 670

142. Kuno, H. (1973): Comparative study on the components of acid mucopolysaccharides in Tawa sarcoma cells and tumorigenic ascites. Shika Igaku 36, 121-139 (Jap)

143. Kuo, T. P., Linn, T. F. and Chen, S. H. (1973): The serum mucoprotein in Curling's ulcer. J. Formosan Med. Assoc. 72, 199-207

144. Kupchella, C. E., Heckman, J. R., Lega, R. D. and Plamp, C. E. (1977): The influence of cold-torpor and water deprivation on plasma and kidney glycosaminoglycans in emydine turtles. Comp. Biochem. Physiol. A 56, 255-259

145. Kusaev, V. I. (1967): Mucopolysaccharides and mucoproteins of the blood serum in patients with chronic pneumonia complicated by general amyloidosis. Klin. Med. (Moskva) 45, 113-121 (Russ)

146. Kusaev, V. I. (1968): Mucopolysaccharides and mucoproteins in the blood serum of patients with secondary general amyloidosis and controlled primary diseases. Probl. Tuberk. 46, 41-48 (Russ)

147. Lau, P., Cornwell, C. G. and Williams, W. J. (1970): Mucopolysaccharide synthesis by human bone marrow in short-term suspension cultures. J. Lab. Clin. Med. 76, 739-746

148. Lau, P., Gottlieb, A. J. and Williams, W. J. (1972): Mucopolysaccharide sulfation in normal and leukemic leukocytes. Blood 40, 725-732

149. Leaback, D. H. (1970): The metabolic hydrolysis of hexosaminide linkages. In: W. H. Fishman (Ed.). Metabolic Conjugation and Metabolic Hydrolysis. Vol. 2. Academic Press, New York, pp. 443-517

150. Loginov, A. S., Privalenko, M. N. and Isakova, Z. S. (1972): Serum mucoid substances (glycoproteins and glycosaminoglycans) in patients with chronic liver diseases. Ter. Arkh. 44, 51-54 (Rus)

151. Lorenzen, I. (1966): Vascular connective tissue under the influence of oestrogens. I. Biochemical and morphological studies of aortae of untreated female and male rabbits, and aortae of castrated and sham-operated female rabbits. Acta Endocr. (Kobenhavn) 52, 565-572

152. MacPherson, G. G. (1972): Synthesis and localization of sulphated mucopolysaccharide in megakaryocytes and platelets of the rat, an analysis by electron-microscope autoradiography. J. Cell. Sci. 10, 705-717

153. Malathy, K. and Kurup, P. A. (1973): Levels of glycosaminoglycans in the serum of diabetic patients. Indian J. Biochem. Biophys. 10, 132-133

154. Marchi, A. G., Marchi, M., Mastella, G. and Nordio, S. (1973): Abnormal β-glucuronidase activity in cystic fibrosis. Helv. Paediatr. Acta 28, 427-435

155. McCluskey, R. T. and Thomas, L. (1958): The removal of cartilage matrix, in vivo, by papain. Identification of crystalline papain protease as the cause of the phenomenon. J. Exptl. Med. 108, 371-384

156. McDonald, T. P. (1966): Effects of vitmain A on mucopolysaccharides of rat blood platelets. Am. J. Physiol. 210, 807-816

157. McKusick, V. A. (1972): Heritable disorders of connective tissue. C. V. Mosby, St. Louis

158. McKusick, V. A., Neufeld, E. F. and Kelly, T. E. (1978): The mucopolysaccharide storage diseases. In: J. B. Stanbury, J. B. Wyngaarden and D. S. Fredrickson (Eds.). The Metabolic Basis of Inherited Disease. McGraw-Hill Book Company, New York, Chap. 53, pp. 1282-1307

159. Medgyessy, G., Szirmai, E. and Guttsches, O. (1971): Diphenylamine reaction of serum in clinical medicine and chemistry. Acta Med. Okayama 25, 205-215

160. Mellsop, G. W., Koadlow, L. and Wittingham, S. (1974): Absence of rheumatoid arthritis in schizophrenia. Aust. N.Z. J. Med. 4, 247-252

161. Merkur'eva, R. V. and Bukhtoyarova, F. G. (1976): Glycosaminoglycans in the blood serum of patients with rheumatoid arthritis. Vopr. Med. Khim. 22, 291-296 (Russ)

162. Miagkaia, I. P., Bachinskaia, L. I. U. (1973): Mucopolysaccharides of blood serum and capillary permeability in patients with arteriosclerosis treated with antireticular cytotoxic serum. Ter. Arkh. 45, 70-73 (Rus)

163. Mitensko, M. D. (1976): Characteristics of changes in humoral indicators mucopolysaccharide metabolism and in blood coagulation of workers of hot shops and practical importance of detection of these changes. Ter. Arkh. 48, 91-96 (Rus)

164. Mitensko, M. D. and Rogova, T. S. (1976): Characteristics of mucopolysaccharide metabolism indexes in hypertensive disease with a cerebral syndrome. Klin. Med. (Moscow) 54, 62-67 (Rus)

165. Mitra, S. K. and Blau, K. (1978): An improved determination of total glycosaminoglycans in body fluids by formation of complexes with quinacrine: changes in amniotic fluid total glycosaminoglycans during normal pregnancies and in pregnancies at risk for mucopolysaccharidoses. Clin. Chim. Acta 89, 127-134

166. Mitruka, B. M. (1971): Biochemical aspects of Diplococcus pneumoniae infections in laboratory animals. Yale J. Biol. Med. 44, 253-264

167. Mittwoch, U. (1961): Inclusions of mucopolysaccharide in the lymphocytes of patients with gargoylism. Nature 191, 1315-1316

168. Morganstern, E. and Weber, E. (1974): Localization of polysaccharide proteins and lipids in human and pig blood platelets. Histochemistry 40, 69-88

169. Morse, B. S. and Nussbaum, M. (1967): The detection of hyaluronic acid in the serum and urine of a patient with nephroblastoma. Amer. J. Med. 42, 996-1002

170. Mosher, R. E. (1962): Interferences in the colorimeteric determination of glucuronic acid in blood serum. Clin. Chem. 8, 378-388

171. Muir, H. (1969): The structure and metabolism of mucopolysaccharides (glycosaminoglycans) and the problem of mucopolysaccharidoses. Am. J. Med. 47, 673-690

172. Murata, K. (1974): Acidic glycosaminoglycans in normal and connective tissue disease states: with special reference to enzymatic studies of chondroitin sulfate isomers at the disaccharide subunits levels. In: Y. Otaka (Ed.). Biochemistry and Pathology of Connective Tissue. Igaku Shoin, Tokyo, Chap. 5, pp. 48-84

173. Murata, K. (1974 a): Acidic glycosaminoglycans in human platelets and leukocytes: The isolation and enzymatic characterization of chondroitin 4-sulfate. Clin. Chim. Acta 57, 115-124

174. Murata, K. and Horiuchi, Y. (1977): Molecular weight-dependent distribution of acidic glycosaminoglycans in human plasma. Clin. Chim. Acta 75, 59-69

175. Murata, K., Ogura, T. and Okuyama, T. (1972): Studies on the disaccharide units of chondroitin sulfates in human plasma. Biochem. Med. 6, 223-230

176. Murata, K. and Okuyama, T. (1974): Acidic glycosaminoglycans in leukocytes. Application of enzymic methods. Connect. Tissue Res. 2, 101-109

177. Nakashima, Y., Di Ferrante, N., Jackson, R. L. and Pownal, H. J. (1975): The interaction of human plasma glycosaminoglycans with plasma lipoproteins. J. Biol. Chem. 250, 5386-5392

178. Neri, G., Di Ferrante, N. and Hrgovcic, R. (1971): Measurement of plasma glycosaminoglycans. Extension of the method. Clin. Chim. Acta 35, 359-362

179. Niebes, P. (1972): Determination of enzymes and degradation products of glycosaminoglycan metabolism in the serum of healthy and varicose subjects. Clin. Chim. Acta 42, 399-408

180. Niebes, P. (1972 a): Influence des flavonoides sur le metabolisme des mucopolysaccharides dans la paroi veineuse. Angiologica 9, 226-234

181. Niebes, P. (1972 b): Influence des flavonoides sur le metabolisme des mucopolysaccharides dans la paroi veineuse. Fourth Int. Symp. Angiologica Sanatoriana Fribourg-Nyon

182. Niebes, P. and Berson, I. (1973): Determination of enzymes and degradation products of mucopolysaccharide metabolism in the serum of healthy and varicose subjects. Bibl. Anat. 11, 499-506

183. Niebes, P. and Jegerlehner, M. L. (1974): Differences in the composition of glycosaminoglycans (GAG) between healthy and varicose human saphenous veins. IV. Europ. Sympos. on Connective Tissue Res., Padova, Italy, September 8-12, 1974, Tipo Lito Poligrafica Moderna, P. 55

184. Niebes, P. and Laszt, L. (1971): Recherches sur l'activité des enzymes dans le métabolisme des mucopolysaccharides des veins saphenes humaines saines ou variqueses. Angiologica 8, 7-16

185. Niebes, P. and Schifflers, M. H. (1975): Micromethod for the determination of glycosaminoglycans in the serum. Results from the serum of healthy and varicose subjects. Clin. Chim. Acta 62, 195-202

186. O'Brien, J. R. (1972): Platelet membrane mucopolysaccharides: some relevant observations. Ann. N. Y. Acad. Sci. 201, 316-328

187. Odell, T. T. and Anderson, B. (1957): Isolation of a sulfated mucopolydaccharide from the blood patelets of rats. Proc. Soc. Exptl. Biol. Med. 94, 151-156

188. Odell, T. T. Jr. and McDonald, T. P. (1964): Two mechanisms of sulfate-S^{35} uptake by blood patelets of rats. Am. J. Physiol. 206, 580-584

189. Olsson, I. (1968): Biosynthesis of glycosaminoglycans (mucopolysaccharides) in leukemic myeloid cells. Biochim. Biophys. Acta 165, 324-334

190. Olsson, I. (1969): Chondroitin sulfate proteoglycan of human leukocytes. Biochim. Biophys. Acta 177, 241-249

191. Olsson, I. (1969 a): Intracellular distribution and sites of synthesis of glycosaminoglycans (mucopolysaccharides) in human leukocytes. Exp. Cell. Res. 54, 314-317

192. Olsson, I. (1969 b): Intracellular transport of glycosaminoglycans (mucopolysaccharide) in human leukocytes. Exp. Cell. Res. 54, 318-324

193. Olsson, I. (1972): Subcellular sites for synthesis of glycosaminoglycans (mucopolysaccharides) of rabbit bone marrow cells. Exp. Cell. Res. 70, 173-184

194. Olsson, I. and Gardell, S. (1967): Isolation and characterization of glycosamingolycans from human leukocytes and platelets. Biochim. Biophys. Acta 141, 348-357

195. Olsson, I. and Gardell, S. (1971): Biosynthesis of glycosaminoglycans in leukocytes. Biochim. Biophys. Acta 237, 203-213

196. Olsson, I., Gardell, S. and Thunell, S. (1968): Biosynthesis of glycosaminoglycans (mucopolysaccharides) in human leukocytes. Biochim. Biophys Acta 165, 309-323

197. Olszewski, W. (1968): Disorders of blood coagulation and in mucopolysaccharides in patients with myocardial infarction and angina pectoris before and after heparin loading. Pol. Tyg. Lek. 23, 1963-1966 (Pol)

198. Orenstein, N. S., Galli, S. J., Dvorak, A., Silbert, J. and Dvorak, H. F. (1978): Sulfated glycosaminoglycans of guinea pig basophilic leukocytes. J. Immunol. 121, 586-592

199. Oronsky, A. L., Ignarro, L. and Perper, R. J. (1973): Release of cartilage mucopolysaccharide degrading protease from human leukocytes. J. Exp. Med. 138, 461-472

200. Oronsky, A. L. and Perper, R. J. (1975): Connective tissue-degrading enzymes of human leukocytes. Ann. N.Y. Acad. Sci. 256, 233-253

201. Ostrovskii, I. I. (1977): Changes in blood serum hyaluronic acid in chronic suppurative otitis media and otogenic intracranial complications. Vestn. Otorinolaringol. (5), 26-29 (Rus)

202. Padmaji, G., Chempakam, B. and Kurup, P. A. (1978): Insulin and metabolism of glycosaminoglycans in rabbit. Acta Diabetol. Lat. 15, 16-23

203. Palo, J. and Savolainen, H. (1972): Serum and urinary glycopeptides and glycosaminoglycans in aspartylglucosaminuria. Clin. Chim. Acta 36, 431-437

204. Paukman, L. I. (1967): Mucopolysaccharides in diabetes mellitus. Probl. Endokr. (Moskwa) 13, 9-13 (Rus)

205. Picard, J., Capeau, J. and Paul-Gardais, A. (1978): Glycosaminoglycans on the surface of chronic lymphocytic leukemia B-Cells. Protides Biol. Fluids 25, 735-738

206. Pidemski, E. L. and Afonina, T. D. (1974): Effect of some anti-inflammatory preparations on the mucopolysaccharide content in the blood and liver homogenates. Patol. Fiziol Eksp. Ter. 14, 74-76 (Rus)

207. Pozharitskaia, M. M., Groshikov, M. I., Starosel'tseva, L. K. et al. (1975): Mucopolysaccharide content in saliva and blood serum in parodontosis. Stomatologia (Mosk) 54, 25-28 (Rus)

208. Rastogi, S. C., Clausen, J., Melchior, J. C., Dyggve, H. V. and Jensen, G. E. (1977): Lysosomal (leucocyte) proteinase and sulfatase levels in Dyggve-Melchior-Clausen (DMC) syndrome. Acta Neurol. Scand. 56, 389-396

209. Riesco, A. (1970): Five-year cancer cure: Relation to total amount of peripheral lymphocytes and neutrophils. Cancer 25, 135-140

210. Riesco, A. and Leyton, C. (1971): Mucopolysaccharides in peripheral leukocytes of cancer patients. Brit. J. Cancer 25, 284-290

211. Robertson, J. A., Segal, D. J., Chernick, B. and McCoy, E. E. (1974): Increased uptake of ^{35}S-heparin by lymphocytes from patients with cystic fibrosis. Lancet 1, 1256

212. Robinson, R. W., Likar, I. N., Likar, L. J. (1975): Glycosaminoglycans and Arterial Disease. Monographs on Atherosclerosis, Vol. 5, S. Karger, Basel 136 pp.

213. Salkie, M. L. (1971): Substrate specificity of the glycosaminoglycuronoglycan-degrading enzymes of human serum. Enzyme 12, 409-416

214. Sanfilippo. S. J. and Good, R. A. (1964): A laboratory study of the Hurler syndromes. (Abstract). Amer. J. Dis. Child. 102, 140

215. Scharrer, I. and Breddin, K. (1972): Mucopolysaccharides and platelet aggregation. Acta Univ. Carol. (Med. Monogr.) (Praha) 53, 27-31

216. Schiffler, D. (1957): Histochemical study of senile platelets by use of hyaluronidase and lysozyme. Boll. Soc. Ital. Biol. Sper. 33, 144-146

217. Schiller, S. (1958): The isolation of chondroitinsulfuric acid from normal human plasma. Biochim. Biophys. Acta 28, 413-416

218. Schiller, S. and Dewey, K. F. (1956): Isolation of chondroitinsulfuric acid from normal human plasma. Fed. Proc. 15, 348 Abstract

219. Schiller, S. and Dorfman, A. (1955): The biosynthesis of mucopolysaccharides in the skin of alloxan-diabetic rats. Biochim. Biophys. Acta 16, 304-305

220. Scriver, C. R. and Rosenberg, L. E. (1973): Amino Acid Metabolism and its Disorders. W. B. Saunders, Philadelphia, pp. 445-446

221. Seibert, F. B., Seibert, M. V., Atno, A. J. and Campbell, H. W. (1947): Variation in protein and polysaccharides content of sera in chronic disease. J. Clin. Invest. 26, 90-102

222. Selroos, O. (1972): Thrombocytosis in rheumatoid arthritis. Scand. J. Rheumatol. 1, 136-140

223. Shetlar, M. R., Knobloch, W. H., Richmond, V., Shetlar, C. L. and Everett, M. R. (1953): Effect of fever on serum polysaccharide level. Proc. Soc. Exptl. Biol. Med. 83, 75-77

224. Silbert, C. K. and Kleinman, H. K. (1979): Studies of cultured human fibroblasts in diabetes mellitus. Changes in heparan sulfate. Diabetes 28, 61-64

225. Singh, J., Di Ferrante, N., Gyorkey, F. and Wilson, N. (1977): Plasma glycosaminoglycans in normal individuals of various age. Atherosclerosis 28, 319-324

226. Skyvova, M., Kocent, A. and Vermousek, I. (1973): Host-tumour relationship. XXXI. Acid glycosaminoglycans in the plasma and in the ascitic fluid of rats during experimental tumour growth. Neoplasma 20, 181-188

227. Sladki, E. and Zak, E. (1968): Mucopolysaccharides metabolism in connective tissue, on the origin of some internal diseases and contemporary possibilities of treatment. Przegl. Lek. 24, 461-464 (Pol)

228. Slater, T. F. and Lovell, D. (1961): Serum diphenylamine reaction in patients with leukemia. Experientia 17, 272-274

229. Smith, H. and Collins, R. J. (1977): Population of lymphocytes in human blood distinctive in morphology and other characteristics. Clin. Pathol. 30, 243-249

230. Spranger, J. W. (1972): The systemic mucopolysaccharidoses. Ergeb. der Med. Kinderheilk. 32, 165-265

231. Stark, V. M. and Nikolaeva, L. F. (1968): Comparative studies of mucopolysaccharides and indices of blood coagulation and anti-coagulation systems in coronary atherosclerosis. Kardiologiia 8, 14-20 (Rus)

232. Sarosel'skii, D. V., Nechaev, S. B., Pankratov, S. M. et al. (1973): Method of determining the summary quantity of acid mucopolysaccharides (amino polysaccharides) in the blood serum in jaundice. Lab. Delo 3, 149-151 (Rus)

233. Stary, Z. (1959): Mucopolysaccharides and glycoproteins. Ergeb. Physiol. Chem. Exptl. Pharmakol. 50, 175-408

234. Stary, Z. and Yuvanidis, M. (1953): Über den Hexuronsäuregehalt der Serumproteine. Biochem. Zeitschr. 324, 206-213

235. Steele, M. W. and Rodnan, J. B. (1971): Incorporation of heparin S^{35} by cultured leucocytes as a diagnostic tool in cystic fibrosis (CF). Pediatr. Res. 5, 393

236. Suschke, J. (1973): Acid mucopolysaccharides in the urine, amniotic fluid and serum. Fortschr. Med. 91, 485-487 (Ger)

237. Suzuki, K. (1972): Neurological aspects of mucopolysaccharidoses. In: A. Lajtha (Ed.). Handbook of neurochemistry, Vol. VII, Plenum Press, New York, pp. 17-32

238. Szabo, L., Laszlo, A., Puljer, I. et al. (1973): A new method for the study of metachromatic granuli of peripheral leukocytes and its importance in the diagnosis of mucopolysaccharidosis. Orv. Hetil. 114, 988-990 (Hun)

239. Szulc, H. and Majewski, C. (1968): Serum level of acid mucopolysaccharides in experimental chronic peritonitis (1968): Patol. Pol. 19, 55-66

240. Szulc, H. and Majewski, C. (1968 a): Blood Serum Acid Mucopolysaccharides in Experimental Chronic Peritonitis. Pol. Med. J. 7, 1457-67

241. Szulc, H. and Szwabe, E. (1968): Behavior of acid mucopolysaccharides in blood serum in peptic ulcer of the stomach and duodenum after surgery. Pol. Przegl. Chir. 40, 52-58 (Pol)

242. Szulc, H. and Szawabe, E. (1969): Behavior of acid mucopolysaccharides in the blood in gallstones, gastric of duodenal ulcer, in stomach or colon cancers following surgery. Przegl. Lek. 25, 694-698

243. Szulc, H., Szwabe, E. and Majewski, C. (1970): Serum proteins and acid mucopolysaccharides in acute experimental peritonitis. I. Proteinogram. Pat. Pol. 21, 19-25 (Pol)

244. Szulc, H., Szwabe, E. and Majewski, C. (1970 a): Serum protein and acid mucopolysaccharides in acute experimental peritonitis. IV. Relation between acid mucopolysaccharide and the alpha 1-alpha 2 index. Patol. Pol. 21, 299-305 (Pol)

245. Taniguchi, N., Moriya, N. and Nanba, I. (1974): Isolation and preliminary chracterization of glycosaminoglycans in human plasma. Clin. Chim. Acta 50, 319-328

246. Taniguchi, N., Okuda, N. and Nanba, I. (1974 a): Characterization of glycosaminoglycans in rabbit plasma and leukocytes at disaccharide subunit level. Biochim. Biophys. Acta 354, 130-138

247. Taniguchi, N., Nanba, I. and Koizumi, S. (1974 b): Characterization of glycosaminoglycans in human leukocytes: Enzymatic subunit assay of isomeric chondroitin sulfates. Biochem. Med. 11, 217-226

248. Tasher, T. and Andreev, B. (1968): Value of serum glucoproteids in mental diseases. Folia Medica 10, 317-321

249. Tornblom, N. and Nordstrom, K. (1954): Serum polysaccharides and hyaline vascular changes in diabetes mellitus. Acta Endocrinologica 17, 426-481

250. Tuzi, T. and Bonito, G. (1968): Changes in the total mucopolysaccharide level of the blood in healthy subjects as related to age, and in subjects with arthrosis. Reumatismo 20, 512-513 (Ita)

251. Varma, R. S., Varma, R., Allen, W. S. and Wardi, A. H. (1974): Urinary excretion of acid mucopolysaccharides in schizophrenia. Biochem. Med. 11, 358-369

252. Varma, R. S., Varma, R. and Hoshino, A. Y. (1979): Serum acid mucopolysaccharide (Glycosaminoglycans) levels in psychiatric patients. Neurosci. Lett. 14, 349-354

253. Vegh, M. and Bihari-Varga, M. (1967): Correlation of serum lipid pattern with the concentration of mucopolysaccharides in normal and hyperlipemic human sera. Clin. Chim. Acta 16, 371-375

254. Wasteson, A., Hook, M. and Westermark, B. (1976): Demonstration of a platelet enzyme, degrading heparan sulphate. FEBS. Lett. 64, 218-221

255. Wasteson, Å. and Wessler, E. (1971): Molecular size and xylose content of urinary glycosaminoglycans. Biochim. Biophys. Acta 252, 13-17

256. Wayne, L., Coots, M., Glueck, H. I., Baum, G. L., Pecora, L. and Putnam, L. (1961): The effect of intravenous estrogen on tissues and capillaries as measured by coagulation tests, intradermal hyaluronidase, hyaluronidase serum inhibitor and pulmonary diffusion studies. J. Lab. Clin. Med. 58, 970

257. Weber, G. (1958): Vergleichende Untersuchungen über das quantitative Verhalten proteingebundener Kohlenhydrate im Blutserum bei Dermatosen. Acta derm-venereol. (Stockholm) 38, 127

258. Weber, G., Braun-Falco, O. and Thaesler, G. (1954): Zur Frage des Verhaltens proteingebundener Polysaccharide im Serum bei Dermatosen. Arch. Derm. Syph. (Berlin) 198, 634-651

259. Weber, G. and Thaesler, G. (1959): On the "free" glucuronic acid in blood serum in dermatoses. Minerva Derm. (Torino) 34, 554-556 (Ger)

260. West, S. S. and Lorincz, A. E. (1973): Fluorescent molecular probes in fluorescence microspectrophotometry and microspectropolarimetry. In: A. A. Thaer and M. Sernetz, M. (Eds.). Fluorescent Techniques in Cell Biology (Proc. Conf. Quant. Fluoresc. Tech. Appl. Cell Biol., 1972, Publ. 1973). Springer, New York, pp. 395-407

261. Winand, R. J. (1967): Increased urinary excretion of acidic mucopolysaccharides in exophthalmos. J. Clin. Invest. 47, 2563-2568

262. Winzler, R. J. (1965): Glycoproteins and glycosaminoglycans in plasma and in some other body fluids. In: E. A. Balazs and R. W. Jeanloz (Eds.). The Amino Sugars, Vol. IIA, Academic Press, New York, Chap 34, pp. 337-352

263. Wong, K. P. (1970): A new fluorometric assay of β-glucuronidase in tissues and serum. Anal. Biochem. 35, 35-45

264. Yamamoto, K. and Terayama, H. (1973): Comparison of cell coat acid mucopolysaccharides of normal liver and various ascites hepatoma cells. Cancer Res. 33, 2257-2264

265. Yusipova, N. A. and Kriuk, A. S. (1977): Hydrolase and carbohydrate biopolymers in blood serum during rheumatoid arthritis. Zdravookhr. Beloruss. 3, 16-20 (Rus)

266. Yusipova, N. A. and Kriuk, A. S. (1979): Articular cartilage, blood serum glycosaminoglycans and glycoprotein in osteoarthritis. Clin. Chim. Acta 94, 9-22

267. Zeya, H. and Spitznagel, J. K. (1971): Isolation of polymorphonuclear leukocyte granules from rabbit bone marrow. Lab. Invest. 24, 237-245

268. Zeya, H. and Spitznagel, J. K. (1971 a): Characterization of cationic protein-bearing granules of polymorphonuclear leukocytes. Lab. Invest. 24, 229-236

269. Zimmer, J. G., McAllister, B. M. and Demis, D. J. (1965): Mucopolysaccharides in mast cells and mastocytosis. J. Invest. Dermatol. 44, 33-37

Chapter 17

Glycosaminoglycans of Urine

17.1 Introduction

The finding of an increased acidic glycosaminoglycan excretion in Hurler's syndrome (Dorfman and Lorincz, 1957) and further developments helped bring about the realization that other mucopolysaccharidoses are also connective tissue disorders. The glycosaminoglycans, present in tissues as proteoglycans, are extensively involved in the maintenance of all tissue structures. The metabolism of proteoglycans is influenced in physiological and pathological conditions affecting connective tissues all over the body, which is reflected in the changes in the spectrum of glycosaminoglycans in blood circulation and urine. These realizations have prompted investigations of abnormal excretion of glycosaminoglycans in urine in disease states in which the proteoglycans in the connective tissue anywhere in the body may be directly or indirectly involved. The glycosaminoglycans excreted in urine are not exactly the same as they occur in the tissue. In tissues the GAG are present natively as very high molecular weight proteoglycans and their aggregates (section 1.7), while urine contains only low (8,000-12,000) molecular weight polymers. During their release from the connective tissue into the blood stream and their excretion in urine, these polymers might have undergone some modifications of the chemical structures, either as a result of depolymerization or desulfation or both. Thus the glycosaminoglycans present in urine urine are degradation products of tissue proteoglycan and contain a few amino acids with intact carbohydrate-peptide linkages in most of the GAG species (Wasteson and Wessler, 1971). This suggests that proteolysis was involved in their release from the tissue proteoglycans. Some of the proteoglycans are degraded to such a higher degree (oligosaccharides) that they are not recognizable as GAG in urine by precipitation procedures and also, these are lost during dialysis following their isolation from an anion-exchange column.

Of the 250 mg of the GAG disposed of daily in an adult (Leaback, 1970) only 5-10 mg are excreted in urine, the remaining being degraded in the body. The urine is not the major route for disposal of GAG released from tissue proteoglycans catabolism. Evidence for this is provided by studies on excretion of injected GAG and proteoglycans in experimental animals like guinea pigs (Revell and Muir, 1972). When chondroitin 4-sulfate was injected intravenously in rat, most of it was found to be degraded in the liver (Wood et al., 1973). The injected heparin appeared in urine as the modified polymer called *uroheparin*. When heparan N-[^{35}S]sulfate was administered to rats intravenously or through ureter cannulae, substantial desulfation occurred in vivo in liver, kidney, spleen, pancreas or gastrointestinal tract with elimination of labeled sulfate in urine. Unlabeled heparin administered simultaneously interfered with in vivo desulfation of the labeled heparan sulfate, indicating that both heparan sulfate and heparin have the common fate and undergo desulfation (Perry et al., 1977). When chondroitin 4-sulfate proteoglycan was injected into guinea pigs, a portion of it appeared in urine as chondroitin 4-sulfate (Wood et al., 1973) suggesting that the proteoglycan was cleaved in vivo either at the glycopeptide linkage or the proteolysis of its protein moiety had taken place. When chondroitin 4-sulfate was administered to guinea pigs (Revell and Muir, 1972) or to rats intravenously (Wood et al., 1973), this glycosaminoglycan was excreted in the urine partly in the undegraded form. It has been established that the protein and glycosaminoglycan moieties of the tissue proteoglycans are synthesized and degraded in parallel (Gross et al., 1960). Both injection of a proteolytic enzyme like papain (Bryant et al., 1958) or hyaluronidase cause degradation of the tissue proteoglycans with the release of glycosaminoglycans in blood circulation and increased excretion in urine. The glycosaminoglycans accumulated in tissues or those present in excess in urine of mucopolysaccharidoses patients are degraded forms of GAG with smaller (about 1/3 the normal size, 2,700-5,500 M_r for heparin) molecular weights and are deficient in the carbohydrate-protein linkage constituents. Chemically, these represent segments of a single parent molecule.

The urinary GAG excretion varies with the age and these age-related differences are in turn related to the GAG metabolism in tissue proteoglycans. For this reason, uronic acid and all bound carbohydrates are higher in infants and young children, when the formations of bone, cartilage and other connective tissue are at their peaks (Stidworthy et al., 1958; Masuda et al. 1978).

There are age- and sex-related differences in excretion. Therefore, it is important to bear these facts in mind while interpreting any qualitative or quantitative changes observed in a physiological or a pathological state. Also, as will be described in later sections, the GAG in urine are present as complexes with glycoprotein and as free GAG, both of which occur as high molecular weight and low molecular weight uronic acid-containing species. The relative proportions expressed as GAG-glycoprotein complexes/ free GAG and high molecular weight GAG/low molecular weight GAG may change in various disease states.

17.2 Source of urinary glycosaminoglycans

The small amounts of GAG (2.4-6.2 mg uronic acid) excreted daily in urine are not derived to a significant extent from the lower urinary tract (Kerby, 1954). A part of the heparan sulfate present in urine may originate from the renal tissue (Wasteson and Wessler, 1971). The general consensus is that the majority of the GAG in urine are derived from filtration of plasma in the kidney glomerulous. This is supported by several pieces of evidence. The connective tissue disorders producing elevation of GAG in plasma are also accompanied by a corresponding elevation in urine. Papain injection produced elevation of GAG both in plasma and urine (Bryant et al., 1958). When ^{35}S-labeled chondroitin sulfate was injected intravenously in guinea pigs, about 30-40 % of it appeared in urine in 30 minutes (Revell and Muir, 1972). In rabbits, 5-10 % of it appeared in urine in the next 2-4 days after injection (Smith and Kerby, 1960). In renal conditions (renal insufficiency), a decreased glomerular filtration was accompanied by a retention of GAG in plasma and increased levels of GAG in urine (Friman et al., 1977).

From the foregoing information it would appear that as long as the glomerular filtration is normal a direct relationship exists between GAG levels in blood and urine. Whereas the relationship is true in general, such a relationship may not exist in some situations (see section 16.3).

17.3. Analysis and fractionation procedures

All the methods of quantitation and fractionation of glycosaminoglycans in urine are based upon the polyanionic character and the differential distribution of charges on the GAG fractions. These methods have already been described in Chapter 2. In general, the quantitation involves precipitation of the GAG in urine followed by quantitation of hexuronic acid in the precipitate by spectrophotometry using a colorimetric reaction (Chapter 2, section 2). Fractionation can be carried out on this precipitate by elution with varying salt concentrations from a cellulose column or by anion-exchange chromatography of the precipitated material freed from the precipitating agent. The majority of the quantitation and screening procedures use cetylpyridinium salts first used for urine GAG by Di Ferrante and Rich (1956, 1956 a).

The quaternary ammonium salts (CPC and CTAB) have been considered superior to other precipitating agents for urinary GAG. However, there are some problems associated with their use; temperature, salt concentration, and presence of toluene influence the amount of GAG precipitated from urine (Di Ferrante, 1967). Some of these problems can be resolved by using optimum conditions, including dilution of urine if specific gravity is greater than 1.020. Also, glycoproteins are coprecipitated, but these can be separated by dissolving the precipitate in 0.59 M NaCl (Di Ferrante and Popenoe, 1957). If CPC is used in excess, keratan sulfate may be redissolved and appears in the supernatent fraction (Dorfman and Matalon, 1972). Incomplete precipitation of low molecular weight glucuronic acid-containing compounds and losses occur the same way as in dialysis. However, failure to dialyze concentrated urine samples results in incomplete precipitation of urinary GAG with cationic agents (Linker et al., 1969). These losses due to incomplete precipitation or dialysis may be prevented by isolation of urinary GAG by gel filtration on Sephadex columns.

Single dimensional electrophoresis on cellulose acetate using 0.1 M HCl or 0.1 M barium acetate buffer (Wessler, 1970), copper acetate, barbiturate buffer (Varma et al., 1974) separates the component GAG into various fractions which can be quantified by densitometric scanning of the electropherograms after staining with a basic dye. Alternatively, the dye bound to strips corresponding to different fractions, measured spectrophotometrically, gives a quantitation of the various fractions (Kimura et al., 1974). The monosaccharides present in strips cut for different GAG fractions may be quantitated by gas chromatography (Lehtonen et al., 1966; O'Brien et al., 1973). Two-dimensional electrophoresis in 0.1 M pyridine-0.47 M formic acid buffer of pH 3 in the first dimension and 0.1 M barium acetate pH 8 in the second direction separates chondroitin 4-sulfate from chondroitin 6-sulfate and hyaluronic acid from chondroitin (Hata and Nagai, 1972). Thin layer chromatographic methods, even in the presence of irregular spots, have been described for fractional quantitation (Samuels et al., 1974). The urine samples from positive screening tests can be evaluated by two dimensional electrophoresis on cellulose acetate after precipitation with cetylpyridinium chloride (Abeling et al., 1974). Preparative separation of individual GAG components in the precipitated material can be carried out by electrophoresis on agar gel. (Taniguchi, 1972). Thin-layer chromatography on precoated cellulose plates developed with 0.04 M ammonium formate (pH 4.3)-isopropanol, 3:2 v/v, followed by staining with 1 % Alcian blue in 9 % acetic acid separated GAG into three groups: chondroitin 4- and 6-sulfates R_F 0.3-0.38; dermatan sulfate R_F 0.4-0.47; and heparan sulfate plus heparin R_F 0.7-0.85 (Teller and Ziemann, 1969).

Chromatography of the isolated GAG on thin-layer and column chromatography (Humbel et al., 1969; Taniguchi, 1970; Taniguchi et al., 1975) have been used for molecular weight determination and differential diagnosis of mucopolysaccharidoses. Fractionation of urinary GAG can be carried out by anion-exchange chromatography (see Chapter 2), using different anionic materials, including DEAE-Sephadex A-50 (Masuda et al., 1975 a).

Enzymatic analyses of various fractions have used digestion with hyaluronidase followed by separation and identification of the undigested material by electrophoresis (Varma et al., 1974; Taniguchi et al., 1975) or by gel filtration (Allalouf and Ber, 1970). Dermatan sulfate and heparan sulfate are recovered in the hyaluronidase-resistant fraction. The chondroitin sulfate isomers present in urine can be determined by digestion with bacterial enzymes chondroitinase ABC and AC (Murata et al., 1973; Murata 1974). Partition fractionations of complexes between GAG and quaternary ammonium salts have been carried out in two phase systems of butanol and aqueous salt (Hurst et al., 1977).

Keratan sulfate present in urine cannot be quantitated by carbazole reactions since it lacks hexuronic acid. Rapid and specific characterization of keratan sulfate could be carried out by thin-layer chromatography on silica gel or cellulose (Humbel, 1974). Direct quantitation of galactose content of urinary GAG has been used as a measure of keratan sulfate-like substances in urine (Robertson and Harvey, 1972). Galactose content determination in selectively eluted keratan sulfate fraction from Dowex 1-X2 column is a better procedure (Robertson and Harvey, 1972; Humbel, 1974).

Optical rotatory dispersion and circular dichroism of the GAG fractions isolated from urine of mucopolysaccharidoses patients have been described (Stone et al., 1970). Methods for isolation of urinary GAG using cesium chloride gradient centrifugation and gel chromatography have been reported (Kleine et al., 1972). The molecular sizes of GAG components of urine have been differentiated and characterized by thin-layer gel filtration on Sephadex G-200 (Taniguchi et al., 1975 a), gel column chromatography on Sephadex G-200 (Constantopoulos, 1968) or from the xylose contents of the isolated fractions (Wasteson and Wessler, 1971).

The ultraviolet optical rotatory dispersion and circular dichroism spectra of dermatan sulfate isolated from the urine of mucopolysaccharidoses patients was different from those of heparan sulfate. This provided an optical method for distinguishing among the Sanfilippo and Hurler variants (Stone et al., 1970). The heparan sulfate and dermatan sulfate isolated from urine of patients with mucopolysaccharidoses have been characterized by nuclear magnetic resonance spectroscopy (Tira et al., 1979). The GAG isolated from urine may be fractionated by preparative electrophoresis or isoelectric focusing (see Chapter 2, Section 2.5). Methods for microdetermination and fractionation using cetylpyridinium chloride or cetyltrimethylammonium bromide precipitation, proteolytic digestion and anion exchange chromatography on small columns have been reported (Kao and Leslie, 1974; Applegarth et al., 1975).

Several screening procedures, including one using commercially available mucopolysaccharidosis paper test, have been described for an early diagnosis of mucopolysaccharidoses (Pennock, 1969; Rezvani et al., 1973; Valdivieso et al., 1973; Meaney, 1974; Lewis et al., 1975; Rufini et al., 1979). Based on the pattern of mucopolysacchariduria, the various positive cases can be classified (Kaplan, 1969) into different types of mucopolysaccharidoses.

17.4 Glycosaminoglycans of normal urine

Normal men and women on the average excrete glycosaminoglycans (GAG) equivalent to 2.3-5.2 mg glucuronic acid per 24 hour with individual variations. The values of GAG excretion in different age groups of males and females, as reported by different investigators, differ to some extent. These are shown in Table 17.1. In general, the males excrete more GAG than the females of a corresponding age group.

All the species of GAG present in the ground substance of the connective tissue are excreted in normal human urine. However, unlike the GAG present in the connective tissue as proteoglycans, the material excreted in urine consists of depolymerized polymers which lack their peptidic moiety. A large portion of these polymers are nondialyzable and by gel filtration on Sephadex G-200 elute at a volume corresponding to an average molecular weight of 8,000 (Constantopoulos, 1968). Based on the xylose content, the molecular weights of the GAG in normal urine were 10,000-12,000 (Wasteson and Wessler, 1971).

The low molecular weight dialyzable hexuronic acid-containing material far exceeds the nondialyzable material and it escapes detection in methods involving dialysis and is not precipitated by precipitating agents (Di Ferrante et al., 1972). The ratio of the polymeric, nondialyzable GAG to the low molecular weight degradation product is the *degradation index* or *degradation ratio* and it gives information about the degradation capacity of the body. In normal subjects the index is less than unity, while in genetic mucopolysacchariduria (in mucopolysaccharidoses) its value is higher than unity (Di Ferrante et al., 1972). The degradation ratio for urinary GAG from 92 normal individuals from birth to 25 years averaged 0.42 and was not influenced by ages of 2-20 years, sex or puberty. In contrast, the degradation ratios for 8 patients with genetic mucopolysaccharidoses were greater than 1.0 (Douglas et al., 1973). Since the degradation ratio was constant and same when determined on a random urine sample or a 24-hour urine collection, the use of degradation ratio has been recommended for diagnosis, especially when a 24-hour urine collection is not available.

17.5 Nondialyzable, polymeric and dialyzable, low molecular weight fractions

The glycosaminoglycans in urine are present both as dialyzable and nondialyzable fractions. The dialyzable fraction far exceeds the nondialyzable fraction and consists of low molecular weight glucuronic acid-containing degradation products of tissue proteoglycans (McGale, 1969). The nondialyzable fraction consists of glycosaminoglycans complexed with glycoproteins (Kao et al., 1967) and free unbound GAG. The major portion (two thirds) consists of chondroitin 4-/6-sulfate, and the remaining one-third consists of a mixture of heparan sulfate and keratan sulfate (Wessler, 1971). Traces of nonsulfated GAG components, hyaluronic acid and chondroitin, are also present in a ratio of 2:1 (Wasteson and Wessler, 1971). The individual fractions determined by ion-exchange chromatography of normal urine showed chondroitin 4-sulfate 58.5 %, heparan sulfate 27.5 % and dermatan sulfate 14 % (Teller, 1967). In contrast with this, other investigators have detected no keratan sulfate in normal urine, with possible presence of a little of hyaluronic acid (Kao et al., 1967; Goldberg and Cotlier, 1972) and dermatan sulfate which comprised 1 % of the GAG excreted in normal urine (Winand and Lambotte, 1971). Analysis on a single, random urine specimen was not reliable, whether expressed as milligram per unit volume or milligram per

17.1. Daily Urinary Glycosaminoglycan Excretion in Children and Adults (as mg hexuronic acid/24 hr, unless otherwise stated)

Method used	Adult Males	Adult Females	Children	Males and Females	Reference
Benzidine precipitation	4.9±1.3	3.0±0.6	–	–	Cradock and Kerby (1955)
Cetyltrimethyl ammonium bromide (CTAB) pptn.	4.4–7.3 (Av. = 6.0, n = 6)	2.7–5.4 (Av. = 3.8, n = 9)	–	–	Di Ferrante and Rich (1956)
CTAB precipitation, carbazole reaction	–	–	4.1–14.7 (Av. = 7.4, 25 boys, 7–16 yrs) 4.8–8.1 (Av. = 5.8, 4 girls, 7–8 yrs)	–	Rich et al. (1957)
CTAB precipitation	6.88±5.1	6.57±2.9	13.77±7.0 (3–15 yrs)	–	Loewi (1959)
Dialysis, ethanolacetate ppn. proteolysis	–	–	–	5–10 mg (as GAG)	Berenson and Dalferes (1965)
CPC precipitation	–	–	–	7.1±1.8 (as GAG)	Thompson et al. (1966)
ECTEOLA-cellulose column	5.7±1.2 (n = 16)	4.1±1.1 (n = 15)	–	–	Di Ferrante (1967)
CPC precipitation	5.3±1.1 (n = 16)	3.7±0.9 (n = 15)	–	–	Di Ferrante (1967)
CTAB pptn., carbazole reaction	–	–	5.01±0.18 (2–13 yrs)	–	Khafagy et al. (1974)
CTAB pptn. carbazole, dialysis, AG1 X-2	5.82±0.30 (S.E.) n = 60, 30–45 yrs	–	–	–	Varma et al. (1974)

gram creatinine. The ratio of milligram of urinary GAG per gram of creatinine was fairly constant in the same subject only when calculated from a complete 24-hour urine collection (Di Ferrante and Lipscomb, 1970).

The ratio of high molecular weight to low molecular weight GAG has been found to be constant, whether determined on a 24-hour collection or a sporadic urine specimen (Di Ferrante et al., 1972; Lakatos and Di Ferrante, 1974). This provides another approach for obviating the necessity of 24-hour urine collections.

The high molecular weight, polymeric, nondialyzable glycosaminoglycans and the low molecular weight, dialyzable degradation products can be determined separately by precipitation of the former with cetylpyridinium chloride and recovering the low molecular weight fraction from the supernatent by sequential ion-exchange chromatography through Dowex 50 and AG 1-X2 (Cl^-) resins (Di Ferrante et al., 1972; Varma et al., 1979). Another method (Lakatos and Di Ferrante, 1974) involved a tandem arrangement of two syringe barrels, the upper one packed with ECTEOLA-cellulose adsorbed the polymeric GAG, while the lower one packed with AG 1-X2 selectively, removed the hexuronic acid-containing low molecular weight oligosaccharide fraction. A modification used pentanol for extraction of cetylpyridinium chloride in the supernatent obtained after polymeric GAG precipitation and the low molecular weight fraction was recovered from cetylpyridinium chloride-free supernatent by using anion-exchange chromatography (Vitiello et al., 1977).

17.6 Relationship between plasma, urine and amniotic fluid levels

Since the majority of the urinary GAG originate as a glomerular ultrafiltrate, the fluctuations of plasma GAG levels and types are correspondingly reflected in urine. However, under certain circumstances there may not be any relationship between blood and urinary GAG (see Chapter 16, Section 16.3; Chapter 17, Section 17.10.4). The fetal urine contributes to the amniotic fluid GAG and this contribution increases as the pregnancy progresses (Chapter 11, Section 11.3).

17.7 Relationship between glycosaminoglycan and 17-hydroxycorticosteroid excretion

The excretion of 17-hydroxycorticosteroids is increased during stress situations. Since the GAG also vary in response to different stressors, some workers have investigated the correlation between the 17-hydroxycorticosteroids and GAG excretion in urine. Mikolajczyk (1967) and Mikolajczyk et al. (1968) studied the involvement of the endocrine system in rats exposed to simulated high altitude environment by exposing them to a low air pressure. Significantly increased urinary GAG excretion with corresponding changes in urinary steroids was noted. In hypophysectomized and gonadectomized rats 17-ketosteroid excretion was decreased. Kantorovich et al. (1976) noted a triphasic change in diuresis in healthy flat-land inhabitants exposed to high altitude hypoxia for 50 days. The 17-hydroxycorticosteroid excretion increased significantly during the first 20-30 days followed by a sharp decrease which continued through a 20-day deadaptation period.

Kastelan et al. (1970) found that the daily excretion of 17-hydroxycorticosteroids and GAG paralleled each other in human males during two simulated Apollo lunar flights, each of 14 days duration. A direct relationship may not exist in all stress situations. Thus, whereas the 17-hydroxycorticosteroids were elevated in both after anaphylactic shock and after sensitization to allergic reactions, the GAG excretion decreased in the former case and increased in latter situation (Nagai et al., 1960).

17.8 Urinary glycosaminoglycans and renal lithiasis

Several reports have claimed that the urinary GAG act as protective colloids against kidney stone formation. Also, the increased excretion of depolymerized GAG by subcutaneous injection of hyaluronidase stopped formation of urinary concrements in persons predisposed to urinary lithiasis (for review see Butt, 1956). Maclagan and Anderson (1958)' also reported that the protective colloid efficiency increased with the amount of mucosubstances present and was found to be low in 5 out of 25 cases of urinary calculosis and high in cases of proteinuria. Some other reports have contradicted the claim of the protective action of urinary GAG against renal lithiasis, since they found increased mucopolysaccharide excretion in these patients (King and Warnock, 1956; Hesse et al., 1975).

The ability of the urinary GAG to act as protective colloids may just be only one of the several factors influencing the pathogenesis of urolithiasis; other predisposing factors being high pH (Vermeulen et al., 1951), high specific gravity (Grove et al., 1950) and high concentrations of magnesium, calcium or phosphate ions in urine (Vermeulen et al., 1951 a).

Histochemical study (Watanabe, 1972) showed that glycosaminoglycans, in particular the sulfated ones, participated in the formation of urinary stones as bridging agents for the crystalline components. However, recent work (Ito and Shimazaki, 1978) based on in vitro studies on urinary calcium oxalate crystal growth suggested that the inhibitory activity of urinary GAG, especially of chondroitin sulfate, was too low to account for the observed inhibition of urolithiasis. It was suggested that factors other than pepsinogen and chondroitin sulfates may be the inhibitors involved. More recently, Sallis and Lumley (1979) reported the presence of an unknown potent inhibitor of calcium oxalate crystal formation in vitro. It had little hexosamine and hexuronic acid and a molecular weight of 10,000. Thus it appeared that besides the urinary GAG fractions isolated on vinylite support, unknown inhibitors and other factors may be important in the etiology of urolithiasis. Diuretics affect urolithiasis by changing the specific gravity of urine and cast excretion (Imhof, 1972). The significance of kidney GAG (Kiseleva and Blagodarov, 1973) and other views about renal calculi have been described (Germinale and Sommo, 1968; Kalandadse, 1970).

17.9 Effects of physiological factors

17.9.1 Diurnal variation and circadian rhythm

The excretion of glycosaminoglycans shows a wide daily variation (Spranger et al., 1967). Madsen and Linker (1969) have reported a greater variation in day to day excretion of GAG. Besides this, the GAG excretion is also influenced by a circadian rhythm, since the chondroitin sulfate/heparan sulfate in normal subjects showed a marked peak at 1800-2200 hours (Scott and Newton, 1975). Because of the individual excretion rate variations, diurnal variation, variations in urine volumes due to aqueous intakes and the problems of 24-hour specimen collections, several investigators have tried to relate the GAG excretion to creatinine to serve as a better index for comparison between normal and abnormal urine samples. However, there are problems with this too; the creatinine excretion varies with the body weight and muscle mass and also the creatinine clearance is affected slightly by dietary protein intake and by diurnal variations (Addis et al., 1951). To obviate some of these problems, the results of analysis have been expressed in several different ways like milligram uronic acid per gram creatinine per kilogram body weight (Goldberg and Cotlier, 1972). Di Ferrante and Lipscomb (1970) have critically evaluated the relationship between GAG and creatinine excretion and have concluded that because of variations in excretions at different times of the day the GAG analysis on a random urine sample was not reliable whether expressed as milligram per unit volume or as milligrams per gram creatinine. However, the milligrams of GAG excreted per gram creatinine excretion were fairly constant within the same subject when these analyses were carried out on a 24-hour urine collection.

17.9.2 Sex-related variations

The glycoproteins and glycosaminoglycans excreted in urine were found to show a tendency to be larger in males than in females, yet there was no significant difference in the relative amounts of the two macromolecules (Endo et al., 1974; Cortinovis et al., 1962).

Higher total daily excretion of uronic acid containing GAG occurred in males than in females (Di Ferrante, 1967). However, when expressed in terms of milligram uronic acid per kilogram body weight per day, no significant differences between males and females were noted (Masuda et al., 1978). Moreover, Thompson et al. (1966) did not find any significant difference between males and females and reported a value of 7.1 mg (calculated as GAG) in 24 hour urine of both sexes.

17.9.3 Age-related variations from birth to senescence

The GAG excretion in human urine varies qualitatively and quantitatively in various ages of development. In the preterm newborns the distribution of various GAG fractions was: chondroitin 4- and 6-sulfates 60 %, heparan sulfate 20 %, chondroitin 12 % and traces of hyaluronic acid, dermatan sulfate and keratan sulfate (Coppa et al., 1978). The GAG excretion in neonates has been investigated by several other authors (Pennock et al., 1971; Scatena et al., 1977). Premature neonates excreted an excess of a high molecular weight (50,000) chondroitin, which constituted 21.5 % and 14 % of the total GAG excreted at 10 and 11 days after birth, respectively. The excretion of this high molecular weight GAG fraction decreased in the extrauterine life to that in the urine of fullterm neonates due to increased sulfation (Tanaka et al., 1979). In premature infants, chondroitin 6-sulfate, chondroitin 4-sulfate and chrondoitin were the major urinary components. The percentage of chondroitin 6-sulfate remained high till the age of 10 years but decreased with advancing age thereafter, while chondroitin 4-sulfate increased with increasing age to a maximum at 40 years of age (Tanaka et al., 1975). The identification and characterization of GAG in normal infants and those with mucopolysacchariduria have been reported (Taniguchi, 1973).

The healthy infants on the average excreted GAG equivalent to 1.63 mg glucuronic acid per 24 hour. This excreted GAG was distributed among the various fractions as keratan sulfate 0.41 mg, hyaluronic acid plus chondroitin 0.1 mg, heparan sulfate 0.19 mg, chondroitin 4-sulfate 0.20 mg, chondroitin 6-sulfate 0.25 mg, dermatan sulfate 0.14 mg and heparin 0.11 mg in the 24 hour urine (Gadamska, 1977). The pattern of urinary GAG excretion in normal male children showed components with variable sulfate contents (Orii, 1968; Orii et al., 1972). In the normal male children, chondroitin

4- and 6-sulfates comprised 80 % of the total GAG excreted, the remaining 20 % consisted of keratan sulfate 8 %, chondroitin 6 %, heparan sulfate 5 %, dermatan sulfate 2 % and hyaluronic acid 0.5 % (Orii, 1968).

Infants and children excreted more GAG in urine than the adults (Pennock et al., 1973). The amounts of GAG and glycoproteins in the 24-hour urine of children increased with age, but remained constant after puberty. The relative amounts of GAG and glycoproteins, expressed as mg/24 hr urine/ kg body weight, increased with age in younger children and then decreased to an almost constant level in adults. The 10-14 year old group, probably represented the period of transition between child and adult (Endo et al., 1974).

Not only the GAG excretion was much higher in children than the adults (Rich et al., 1957) but also the children excreted higher amounts of glycoproteins. The amounts of both glycoproteins and GAG increased with age after 8 years (Kusuhara, 1978). Fractionation on cellulose acetate of GAG from children showed that chondroitin sulfate > heparan sulfate plus keratan sulfate > hyaluronic acid. Hyaluronic acid did not change with age, while heparan sulfate plus keratan sulfate fractions were the lowest at 11 years and increased afterwards with age. Chondroitin sulfate was the highest at 11 years and then decreased gradually with age (Kusuhara, 1978).

The average cetylpyridinium chloride precipitate/creatinine ratios were higher in young children and gradually decreased with age. A value of 170 in 10-months-old children decreased to 40 in the 14 year old children and during this age-span the uronic acid/creatinine ratios decreased from 13.8 to 5.4 (Gadowska and Borzymowska, 1977). The non-sulfated GAG fraction eluted with up to 0.8 M NaCl from a Dowex 1-X2 column consisted of 8.5 % and 26.3 % of the total urinary GAG in normal male children and normal adult urine, respectively. Multiple sample analysis gave an average nonsulfated GAG excretion of 6.6 % and 20.2 % in children and adults, respectively (Orii, 1968). The relative proportions of the electrophoretic fractions on cellulose acetate varied with age. In childhood the chondroitin sulfate fraction was high and thereafter decreased with age. On the other hand, the fractions with mobilities corresponding to heparan sulfate and keratan sulfate were low in childhood (Shichijo et al., 1978).

Determination of the GAG in urine samples from 7 infants, 7 small children, 7 preadolescents, 7 adolescents and 9 adults revealed that whereas chondroitin 4-sulfate remained constant throughout life the heparan sulfate fraction decreased and dermatan sulfate increased with increasing age (Teller and Krueger, 1967). Peaks of the absolute amounts of heparan sulfate and dermatan sulfate were found during puberty, when the proportion of chondroitin 4-sulfate to the total GAG excreted was also the highest (Teller and Krueger, 1970). Keratan sulfate was present only in traces at all stages. The urinary GAG determined as glucuronic acid were the highest (6 mg/24 hr) and declined continuously to 3 mg and 2 mg in men and women, respectively, at age 70 (Cortinovis et al., 1962). Tanaka et al. (1975) have also reported a similar decrease with increasing age to 80 years in human males.

Masuda et al. (1978) determined glycopeptides and GAG excretion in urine from newborns to 68 years old normal human subjects. When the results were expressed on the basis of milligram uronic acid excretion per kilogram body weight per day, significant higher excretion occurred in infants and children below 5 years than in older children and adults but no difference was noted with age or sex after the age of 7 years.

Differential determination of GAG fractions revealed an increase in excretion of chondroitin 4- and 6-sulfates and a decrease in heparan sulfate during the course of growth (Spranger et al., 1967). The excretions of hydroxyproline, GAG and creatinine were correlated with weight and height in children and increased with age in different age groups as 1-3 < 4-7 < 8-11 < 12-15 years with maximum values at 13 years (Shamkhalova et al., 1979).

In normal adult urine, the GAG are present both as free GAG (20-30 %) and as complexes with glycoproteins (70-80 %). A higher percentage of GAG fraction bound to glycoproteins was present in urine with increasing ages of 6-70 (Kao et al., 1967). Similar complexes have been reported in the urine of normal children (Orii, 1970). The heparinoid substances quantitated colorimetrically by reaction with toluidine blue showed variations in infants, children, men and women. The amount of dye-binding compared to heparin was 11 ± 14 heparinoid units in 1-8 days old infants whereas for 2-12 year old children it was 80 ± 13 heparinoid units (Rosenfeld et al., 1961). The

GAG precipitable from the urine at acid and alkaline pH varied with age (Teller and Muller, 1974). The changes in the connective tissue with aging appear to be reflected in the hydroxyproline and GAG excretion with aging (Katsunuma et al., 1975).

The relative proportion of urinary GAG resistant to testicular hyaluronidase, heparan sulfate and dermatan sulfate (Allalouf and Ber, 1970, 1970 a), or those resistant to chondroitinase ABC increased with advancing age. The increase in the excretion of materials resistant to chondroitinase ABC was characteristic of aged individuals (Taniguchi, 1972). In healthy 3-83 years old males, the hyaluronidase-resistant fraction represented 6-45 % and it increased with advancing age (Allalouf et al., 1970).

17.9.4 Fasting and starvation

The excretion rate of both dialyzable and non-dialyzable amino sugars in urine decreased in fasted guinea pigs (Boas, 1965). Several investigators, including us, have noted a trend that GAG in urine drop in human subjects under fasting or starvation conditions. The effect on GAG excretion may be similar in anorexia nervosa. This fact is very important to keep in mind while interpreting the results of the study on the GAG excretion in a disease in which the patients might tend to starve as a side effect. The GAG changes caused by starvation in rats are not altered by intraperitoneal administration of physiological saline (Ber and Allalouf, 1969).

In rats, starvation for 1-3 days interspersed with 1-15 days of feeding and with or without intraperitoneal administration of physiological saline revealed variations in urinary GAG excretion in different phases (Ber and Allalouf, 1969). The GAG excretion decreased with increasing saline administration during feeding periods but were always greater than those obtained during starvation periods. However, fluid loading had no significant effect on GAG excretion during the starvation periods.

17.9.5 Therapeutic proteolytic enzyme injections

Therapeutic injection of papain for enzymic dissolution of chondromucoprotein in the area of prolapsed intervertebral disc (herniation) resulted in loss of cartilage rigidity. A moderate temporary increase occurred in urinary GAG, which was qualitatively different from the GAG in normal urine in having a high proportion of chondroitin 6-sulfate (Stern et al., 1968). Similarly, the enzyme pronase has been administered in experimental arteriosclerosis and blood pressure (Nakamura et al., 1966).

17.9.6 Effect of hyaluronidase injection

Injection of commercially available testicular hyaluronidase during the five days caused a three-fold increase in the excretion of urinary GAG. Upon stopping the injection a sharp drop to below the initial excretion level was seen. A second injection again produced a rise in excretion of low molecular weight (5,000) components due to the breakdown of the larger molecules (M_r 11,000-13,000), so that the ratio of high molecular weight GAG/low molecular weight degradation product (= fast electrophoretic component/slow electrophoretic component) decreased. Along with this, improvement in the Hurler/Hunter patients was noted (Dietrich et al., 1972). Besides the treatment of mucopolysaccharidoses, the hyaluronidase injections have been used for stopping concrement formation in individuals predisposed to urolithiasis (Butt, 1956; see also Section 17.8).

17.9.7 Effect of surgical stress

A large temporary increase in the nondialyzable polyuronides occurred in surgical stress (Rosenfeld, 1971; Stern et al., 1968). The GAG excreted postsurgically consisted mainly of a component with electrophoretic mobility of hyaluronic acid. However, this component differed from hyaluronic acid or chondroitin in having uronic acid contents twice as much as hexosamine and a slight susceptibility to hyaluronidase digestion (Stern et al., 1968). The postoperative urinary GAG excretion increase was maximum on the second postoperative day and was related to the age for the same surgical procedure (Rosenfeld, 1971). Arthrotomy and mastoidectomy produced the least increase in GAG excretion, while the increase in thyroidectomy and herniorrhaphy was more than this but less than those with major surgical procedures such as gastrectomy, intestinal resection, cholecystectomy, which produced the maximum increase in excretion of GAG.

17.9.8 Effect of physical effort

Physical activity, such as strenuous physical exercise, in normal humans influences the elimination of urinary GAG probably by interfering with the metabolism of a fundamental substance of connective tissue (Rigillo et al., 1969). Significantly increased excretion of GAG occurred in rats exposed to a simulated altitude environment created by exposure to a low air pressure of 500 mm Hg (Mikolajczyk et al., 1968). These changes were mediated through the effect of the endocrine system on connective tissue components.

17.9.9 Allergic sensitization

Urinary GAG excretion rose after sensitization to allergic reactions and fell after anaphylactic shock, whereas the excretion of 17-hydroxycosticosteroids was elevated in both situations (Nagai et al., 1960). Similarly, after histamine-induced shock glucosamine and glycoprotein decreased while urinary GAG increased in rabbits.

17.9.10 Frostbite and extremes of temperature

Urinary GAG excretion is altered on exposure to extremes of temperature. The GAG content of urine was significantly increased in patients with II and III degree frostbite especially at 3 and 7 days after the frostbite. However, the GAG excretion did not differ from the controls in advanced cases (III-IV degree) of frostbite (Zorin and Zorina, 1978). Similarly, connective tissue metabolism was found to be altered in hot-shop workers and led to GAG alterations in body fluids (see Chapter 16, section 16.7.16).

17.9.11 Sonic vibrations

Vibrations induce changes in the metabolism of the connective tissue, although no perceptible clinical changes may be observed. The daily urinary excretion was increased to 8.3 mg glucuronic acid as compared to 3.2 mg in controls (Mikolajczyk and Izycki, 1974). The effect of sound vibrations on GAG of the inner ear fluid has been described in Chapter 10, section 10.9.

17.9.12 Histamine shock and experimental inflammation

Histamine-induced shock increased GAG excretion in rabbits, whereas allergic sensitization increased and shock decreased the GAG excretion in guinea pigs (Nagai et al., 1960). An increase in GAG excretion occurred in early stages of experimentally induced inflammation (Bazin et al., 1968; Koizumi et al., 1969). The increase in excretion of GAG in urine of rats preceded an increase in serum haptoglobulin in aseptic inflammation. Administration of ACTH with the inflammatory agent further aggravated these changes and prolonged the duration of the inflammatory focus (Alimova et al., 1975).

17.9.13 Pregnancy

The quantity of the non-dialyzable hexosamine increased significantly and consistently during the last half of pregnancy (King and Warnock, 1956). The toluidine-binding material measured in 24 hour urine and expressed as *heparinoid units* (amount of dye-binding as compared to heparin) was increased from a basal value of 16 ± 13 units per hour to higher values in the last trimester of pregnancy (Rosenfeld et al., 1961). While the urinary excretion of histamine was significantly increased in women with late pregnancy toxemia, the 24 hour urinary GAG excretion was not increased (Anton, 1971). The total GAG excretion increased during pregnancy from the antepartum levels and this increase persisted in the postpartum periods (Thompson et al., 1974). The postpartal increase was probably related to changes involved in involution of the uterus.

The GAG excretion in 24-hour urine was increased beyond the 20th week of pregnancy and in the postpartum period in 59 out of the 96 pregnancies (Thompson et al., 1974). However, infection in the obstetric and gynecological patient caused increased GAG excretion at any stage (Ledger et al., 1974).

In most young women, dermatan sulfate is not detected in nonpregnant or pregnant state, while young adult males excrete an appreciable amount of dermatan sulfate in urine. This is probably related to the inhibitory action of estrogen on synthesis of dermatan sulfate demonstrated in rat skin (Kuarstein et al., 1963).

17.9.14 Infants with retarded growth, weight deficit or accelerated growth

Measurement of GAG in the 24-hour urine in normally grown near term, preterm and growth retarded infants showed that the GAG excretion on the third day was correlated with the rate of intrauterine growth, gestational age, birth weight and body length of the infants (Klujber and Sulyok, 1972). A predominance of nonsulfated GAG (hyaluronic acid) was found in all the groups studied up to the thirtieth day of extrauterine, postnatal life, in contrast with children and adults where chondroitin sulfate is the predominant GAG. On the third postnatal day the urine contained hyaluronic acid, chondroitin sulfate and chondroitin-heparin fractions. The 10- and 20-day old infants with increasing body weights excreted significantly more hyaluronic acid and chondroitin than those with no change or decreased weights after birth (Klujber and Sulyok, 1973).

Rich et al. (1957) investigated the GAG excretion in normal children and children with accelerated growth. Infants under weight by 20-49 % in the age group 1-10 months, who were hospitalized for illnesses like diarrhoea or pneumonia, excreted increased amounts (2.90 mg/24 hr) of GAG as compared to 1.47 mg in normal infants. The increased GAG excretion was probably related to their decreased corticoid excretion. Large doses of prednisolone significantly lowered their elevated GAG levels (Lipinska-Piotrowska, 1965).

17.9.15 Effect of plasma, blood or leukocyte infusion

Plasma and the white cells contain the enzymes for degradation of GAG (sections 16.5 and 16.6). These have been administered to patients with mucopolysaccharidosis for supplementing the enzymes deficient in these patients (Di Ferrante et al., 1971; Dean and Muir, 1973; Raine, 1975). Such treatments bring about degradation of GAG accumulated as proteoglycans in the tissues with the resultant increase in the low molecular weight fraction and a decrease in the high molecular weight fraction (Lambotte et al., 1971; Lambotte and Winand, 1973). In some cases, improvements in patients are noted along with these urinary GAG changes. In Hurler-Hunter syndromes, the GAG excreted in urine have molecular weights higher than those of the normal urine so that the high M_r/low M_r GAG ratio (degradation ratio) is about 1.0 as compared to 0.4 for normal urine (Douglas et al., 1973). Only a partial success has been achieved with these plasma and leukocyte infusions. Other approaches like hyaluronidase injection (Nader et al., 1974) have also been used. Hyaluronidase injection also produced increased excretion of GAG in patients with Hunter and Hurler syndromes.

17.9.16 Effects of chemicals and drugs

Drugs and chemicals can affect the synthesis or breakdown of the protein core and GAG chains in the tissue proteoglycans. Indirect evidence for this comes from the effects observed in cultured cells. These changes induced in the proteoglycan metabolism are reflected in the blood levels and excretion of GAG in urine.

The intravenous injection of chondroitin sulfate into rabbits followed by the determination of urinary GAG by cetyltrimethylammonium bromide precipitation gave a recovery of 5-10 % of the injected dose. Subcutaneous injection of the chelating agent ethylenediaminetetraacetate (EDTA) into rabbits caused a great but variable increase in GAG excretion (Smith and Kerby, 1960).

The antirheumatic agents inhibit the GAG biosynthesis. Some of these agents are salicylate, phenylbutazone, prednisone, indomethacin, clofezone, flufenamic acid, cortisone, etc. Indirectly, these drugs inhibited collagen formation by Carrageenan in rat (Brettschneider et al., 1976). Acetylsalicylate, N-acetyl-p-aminophenol and phenylbutazone but not phenacetin significantly increased GAG excretion in urine by their effects on GAG in the intercellular matrix of connective tissue (Lehtonen et al. 1967).

Prolonged exposure of guinea pigs to nitrous oxide produced focal emphysema and increased hydroxyproline and GAG excretion in urine (Kosmider et al., 1972). The alterations in GAG metabolism and GAG excretion in urine affected by teratogens have been reviewed (Kochar and Larsson, 1977).

Long-term administration of hydralazine to guinea pigs produced a collagen-like syndrome. In these experimentally-diseased guinea pigs, urinary excretion of GAG was increased along with an elevation of seromucoid and protein-bound hexosamine in serum (Kucharz and Drozdz, 1976).

17.9.17 Effects of vitamins

Urinary GAG excretion in children aged 8-15 years increased from 2.95 mg glucuronic acid per day to 3.84 mg per day following administration of 15,000 units/day of vitamin D_3 for 7 days (Gebala and Gadamska, 1977). The excretion of keratan sulfate and glycoproteins in the urine of infants with rickets before and after administration of vitamin D_3 was studied (Gadamska, 1977 a).

The excretion of GAG was below normal in juvenile cases of vitamin A deficiency (Mohanram and Reddy, 1971). Both total uronic acid and sulfated GAG were low and the chondroitin sulfate excreted was of a lower sulfate content. Vitamin A therapy restored the GAG spectrum of urine to that of normal subjects (Cherian et al., 1970). Other studies (Chandrasekharan, 1977, 1977 a) reported no difference in total urinary GAG excretion and sulfated GAG in 182 children aged 6-12 years with varying degrees of hypovitaminosis A measured by serum retinol levels. No correlation was found between urinary GAG and the degree of hypovitaminosis A. However, since urinary creatinine was low, the sulfated GAG-to-creatinine ratio was higher in the deficient group (Chandrasekharan, 1977). The contradictory reports of urinary GAG in vitamin deficiency are difficult to explain in view of the effect of vitamin A in tissues to increase ^{35}S-incorporation into GAG (Barker and Kennedy, 1969).

When large doses of vitamin A (4,000 units /kg/day, for 4-14 months) were given to 2 Hurler, 1 Hunter, 2 Sanfilippo and 2 Morquio-Brailford patients, instead of a useful response adverse manifestations of hepatosplenomegaly and neurological regression increased. Increased urinary excretion of GAG in response to vitamin A was observed in only 3 patients out of 7 (Madsen and Linker, 1969). Onisawa and Lee (1970) confirmed these negative effects of vitamin A and also of vitamin C in mucopolysaccharidoses patients.

Vitamin C is an important factor in the metabolism of GAG affecting the sulfated and non-sulfated GAG differentially. In scorbutic guinea pigs, ^{35}S incorporation was decreased while hyaluronic acid production was increased (Friberg and Ringertz, 1954; Slack, 1958). Because of this effect in guinea pig and a similar increased synthesis of GAG in cultured Hurler's fibroblasts, an ascorbic acid-free (scorbutic) diet was given for one year to an infant with Hurler's syndrome. However, the urinary GAG excretion did not change as expected. No clinical improvement was seen but the patient did not develop scurvy (DeJong et al., 1968). Probably the Hurler's patients are resistant to scurvy.

17.10 Alterations in pathological states

17.10.1 Pneumonia and other infections

The daily urinary excretion of total GAG in 7-14 year old children with chronic pneumonia of over 4 years duration increased 1.5 fold to 7.38-8.91 (mean 7.8) mg uronic acid/24 hr as compared to 4.19-5.41 (mean 5.1) mg uronic acid/24 hr in normal chldren. Hyaluronic acid, heparan sulfate, dermatan sulfate and total sulfate increased, whereas the chondroitin sulfate levels decreased (Kucinskiene and D'yachkova, 1976).

Variations in urinary excretion occur in children with *measles* (Lipinska-Piotrowska et al., 1968). The GAG excretion in urine was increased in obstetric and gynecological patients (Ledger et al., 1974).

Increased urinary GAG excretion occurred in *obstetric and gynecological patients* affected by *infections* (Ledger et al., 1974). Similarly, the GAG excretion patterns were found to change in *acute respiratory infections* in small children (Tebenchuk, 1971). The urine from *influenza patients* contained cells with inclusion bodies and these inclusions stained for GAG with Alcian blue (Shiina et al., 1979).

17.10.2 Rheumatic diseases

Several reports have indicated increased excretion of GAG in patients with active untreated rheumatoid arthritis. Administration of sodium salicylate or cortisone to these patients decreased the excretion of GAG to normal, which by paper electrophoresis, paper chromatography and enzymatic analysis were found to consist of a mixture of chondroitin sulfate and hyaluronate (Di Ferrante, 1957, 1957 a). The GAG in the Astrup fraction increased from 3.7 mg in 10 normal individuals to 5.5 mg glucuronic acid per 24 hours in 9 patients (Bonomo and Cenetelli, 1957). The elevation occurred in only 20 % of the rheumatoid arthritis patients when the fraction precipitable with amino-acridine was measured (Bitter et al., 1970). Similarly, Loewi (1959) observed inconsistent elevation and in the children patients the excretion was even less than in normal controls. Other investigators have confirmed increased excretion (11.65 mg GAG/day) of the nondialyzable GAG as compared to (7.1 mg/day) in normal urine (Thompson et al., 1966). Urinary GAG elevation in these patients has been found to be associated with correspondingly increased serum GAG levels. The excretion of keratan sulfate was raised in 13 out of 30 cases of rheumatoid arthritis (Robertson and Harvey, 1972). In these patients the increase occurred in both the sulfated and nonsulfated glycosaminoglycans measured by anion-exchange chromatography (Wessler, 1967). The ratios of mean GAG per gram creatinine in the urine of 200 patients with rheumatic diseases were increased as compared to the normal average value of 4.3 in 43 normal subjects (Acquaviva et al., 1970). The ratios in rheumatoid polyarthritis, tuberculous osteoarthritis, gout and Paget's disease were 7.5, 9.49, 8.58 and 11.15, respectively. However, the ratios were close to normal in postinfectious rheumatism and reflex algodystrophies.

Methods for analysis and urinary GAG levels in males, females, and children have been reviewed (Bollet, 1967, 1975; Ksiezny, 1978). The values increased in males, females and children with rheumatoid arthritis to 9.0, 16.0 and 6.0 mg glucuronic acid per day as compared to the values in the corresponding normal subjects at 6.0, 3.7 and 3-23 mg glucuronic acid per day. Delbrueck et al. (1967) have also reported significantly increased excretion rates of GAG in rheumatoid arthritis. It is interesting to note that rheumatoid arthritis is absent in schizophrenics (Mellsop et al., 1974), who have also been reported to excrete decreased amounts of glycosaminoglycans (Varma et al., 1974).

By using electrophoretic analysis, glycosaminoglycans with abnormally high and low sulfate contents could be detected in urine of rheumatoid arthritis patients (Friman, 1972). Hyaluronic acid levels increased in urine with a simultaneous decrease in hyaluronic acid content of synovial tissue of arthritic joints (Merkur'eva and Bukhtoyarova, 1976).

17.10.3 Skin diseases (dermatoses)

17.10.3.1 Psoriasis

The levels of both sulfated and nonsulfated glycosaminoglycans were increased in psoriasis and some other dermatoses (Mier and Urselmann, 1970). A transient 4-fold increase in total GAG excretion, most of which was low sulfated heparan sulfate with absence of chondroitin sulfate, was noted during the acute psoriatic erythrodermic stage. The increased excretion of low sulfated heparan sulfate and GAG correlated with the erythrodermic reaction, decreasing in subacute state and normalizing with disappearance of erythroderma (Friman et al., 1974).

17.10.3.2 Scleroderma and lichen sclerosus

The reports on the GAG in urine of patients with scleroderma have been conflicting. Ohlenschlaeger and Friman (1968) found no consistent difference in the daily urinary excretion expressed as uronic acid/m^2 body surface and uronic acid/creatinine ratios in 30 patients and normal subjects, in addition to similar electrophoretic pattern. Similarly, Thompson et al. (1966) found no significant increase. However, Wessler (1967) reported increased levels of both sulfated and nonsulfated GAG in diffuse scleroderma. The GAG levels in urine of patients with scleroderma showed elevations to 7.95 as compared to the normal value of 4.38 for uronic acid/gram creatinine (Acquaviva et al., 1970). The increase was related to the development of the illness and the relative increase of dermatan sulfate was observed in the sclerotic skin in succession of the perivascular edema (Ishikawa, 1974). Sasaki (1975) reported abnormalities in the total and propanol-insoluble fractions of GAG from patients with systemic scleroderma and these changes were reversed in improvement brought about by treatment with asiaticoside, D-thyroxine or human placenta preparation. Kreysel et al. (1977) from analysis using Bio-Gel P-2 and Dowex 1-X2 chromatography found increased levels of glycopeptides, chondroitin or a low sulfated chondroitin sulfate but no correlation between GAG excretion and progressive scleroderma. On the other hand, Hardy et al. (1971) found increased GAG excretion in 34 out of 40 scleroderma patients and greatly elevated carbazole/creatinine ratios correlated with the severity of the disease; 4 out of 5 patients with ratios over 10:1 died from rapidly progressing scleroderma.

The urine from a patient with wide-spread lichen sclerosus et atrophicus contained an increased amount of hyaluronic acid and low sulfated chondroitin. The dermal tissue of the patient contained only one-half the hyaluronic acid content of the skin of healthy adults (Yoshima et al., 1978). Ishikawa et al (1975) reported induction of experimental scleroderma in mice by intraperitoneal injections of crude GAG obtained from the urine of a patient with systemic scleroderma.

17.10.3.3 Urticaria pigmentosa (mastocytosis)

A pathological increase of GAG excretion in urine was demonstrated in 61 % of patients with this condition using cetyltrimethylammonium bromide precipitation (Clausen and Asboe-Hansen, 1966). An increased hyaluronic acid excretion was previously reported in Bullous diffuse mastocytosis and the excreted GAG were characterized by infra-red spectra (Asboe-Hansen, 1962). Piper et al. (1967) also identified the GAG in slightly increased excretion by benzidine precipitation and infra-red spectra. Elevated levels of hyaluronic acid and dermatan sulfate in urine and serum were present in a 51 year old male patient with mastocytosis (Asboe-Hansen and Clausen, 1964). These GAG originated from tissue mast cells and treatment with the mast cell-active substance Polymyxin B normalized the electrophoretic pattern. Hyaluronic acid increased in urine and other biological fluids of mastocytosis patients (Asboe-Hansen and Clausen, 1964 a).

The GAG excretion in urine was increased in the keratan sulfate fraction in 4 out of 6 patients with dermatomyositis (Robertson and Harvey, 1972).

17.10.4 Renal disorders

Statistically significant increase in glycosaminoglycan excretion in urine was observed in patients with subacute glomerulonephritis (Galletti and Stancari, 1957). Contralateral shifts in the levels of mucoproteins in blood and urine occurred in 32 children with a nephrotic form of diffuse glomerulonephritis and the changes in these biological fluids were related to the state of filtration capacity of kidneys (Klyukina et al., 1968).

An undersulfated chondroitin sulfate has been reported to occur in the sera of the aged and it appeared in urine when tubular absorption was impaired in these aged (Ohkawa et al., 1972). Excessive mucopolysacchariduria (average 14.7 mg GAG/24 hr) was observed in renal diseases of a variety of etiologies as compared to a mean normal GAG excretion of 7.1 mg (Thompson et al., 1966). The daily rhythm of excretion of GAG in the urine in patients with acute nephritis differed from that in healthy persons. The involvement of distal convoluted tubules increased diuresis with a concomitant increase of GAG levels in urine and decrease in hyaluronidase activity (Morgorian, 1968). Alterations in the spectrum of GAG of urine also occur in children with kidney diseases (Mogorian, 1970) and in renal tuberculosis (Mares and Vulcan, 1965).

Kasanen et al. (1964) investigated the GAG excretion in patients with renal diseases with and without abuse of phenacetin-containing analgesics.

17.10.5 Muscle disorders

The urinary concentration of GAG is increased in various conditions affecting the muscle, including dermatomyositis and polymyositis (Murata et al., 1970). Another study (Robertson and Harvey, 1972) reported an increased keratan sulfate excretion in some cases of dermatomyositis. Out of the four unrelated patients with myotonic dystrophy, a slight increase in urinary GAG occurred in two, while the other two had normal values (Hayashi et al., 1978). Hyaluronic acid was observed in urine of all patients and one patient showed an increased excretion of heparan sulfate. Increased excretion of dermatan sulfate and hyaluronic acid occurred in 60 patients with different forms of neuromuscular disorders. Since these changes were apparent immediately and simultaneously with marked clinical symptoms in cases of denervation amyotrophy, urinary GAG analysis was suggested for differential diagnosis and differential drug therapy of progressive muscular dystrophy and denervation amyotrophy (Stinikov and Temin, 1979).

Extensive ectopic ossification of the soft tissue occurs in myositis ossificans progressiva. Parathyroid infusion in a 17 year old male patient increased the urinary values by 40 %, which returned to the base line value in 48 hours. A second infusion elicited no further response suggesting that such infusions had no therapeutic potential (Bland et al., 1973).

17.10.6 Otosclerosis

The formation of spongy bone in the capsule of the labyrinth of the ear (otosclerosis) has been considered a local manifestation of a generalized mesenchymopathy. Results of hexosamine and glycosaminoglycans excretion in the otosclerotic patients, however, suggested that the histochemical changes observed in the ground substance might well be confined to the otosclerotic focal areas and were not part of the generalized mesenchymopathy. Schoenheyder et al. (1966) found no variations from normal in the urinary excretion of amino acids and hexosamines. Similarly, Soifer et al. (1970) failed to find any difference in total GAG excretion and electrophoretic patterns between 36 otosclerotic and 38 non-otosclerotic male patients.

17.10.7 Stein-Leventhal syndrome

This syndrome is characterized by sterility resulting from secondary amenorrhea, anovulation and polycystic ovaries. The hormonal disturbances in women with this syndrome produce changes in urinary mucoprotein and 17-ketosteroid fractions (Jablonski et al., 1969), but not in pregnancy toxemia.

17.10.8 Homocystinuria

The urinary excretion of GAG increased in 3 out of 4 children (2.5-9 years old) with homocystinuria (Barashnev and Semyachkina, 1979). The GAG excreted consisted of hyaluronic acid, heparan sulfate and chondroitin sulfate as in children with Marfan's syndrome. However, the relative amounts of those components were different in homocystinuria and Marfan syndrome.

17.10.9 Marfan syndrome and Marfanoids

A 2 to 4-fold increase in the urinary glycosaminoglycans was noted in Marfan syndrome patients. By electrophoretic analysis, hyaluronic acid and chondroitin-4- and 6-sulfates were found to be the major GAG excreted in these patients (Berenson and Dalferes, 1965). An earlier report (Berenson and Serra, 1959) had described an essentially normal distribution between chondroitin 4-sulfate, chondroitin 6-sulfate and dermatan sulfate in urine.

Goldberg and Cotlier (1972) found a positive urine mucopolysaccharide-cetyltrimethylammonium bromide precipitation screening test in 4 out of 8 subjects who had the clinical characteristics of the Marfan syndrome, including congenital lens abnormalities and cataracts and some showed abnormal arm span to height ratios. Although the screening test was positive, the four patients showed no abnormalities with respect to qualitative and quantitative glycosaminoglycan excretion when expressed in terms of creatinine excretion. These authors considered these patients with normal urinary GAG but with characteristic clinical features as variants of Marfan syndrome and called them *Marfanoids*.

The elevation of glycosaminoglycans in urine may be secondary to the reported increase of these macromolecules in sera of these patients (Chapter 16, section 16.7.4). Hyaluronic acid and chondroitin sulfate are elevated in the biological fluids, probably as a result of a disturbance of these GAG components in the tissues. Support for this has been provided by the fact that the cultured fibroblasts from these patients contained large amounts of hyaluronic acid (Matalon and Dorfman, 1968; Lamberg and Dorfman, 1973). Tortorolo et al. (1965) have critically reviewed the GAG excretion in this syndrome. By using the Dorfman's turbidity screening test, the absorbance was higher (0.035-0.042) in 3 out of 6 Marfan patients as compared to (0.002-0.015) that in normals (Romano et al., 1966).

17.10.10 Rothmund Thomson syndrome

This rare hereditary disorder is characterized by accelerated aging changes in skin, subcutaneous tissue, muscle, bone and joint, which suggest involvement of the connective tissue. In a case of Rothmund Thomson syndrome associated with osteosarcoma in a 19 year old Japanese boy, the urinary GAG increased 2-3 times the normal values (Tokunaga et al., 1978). The major GAG fractions in urine of this patient consisted of partially degraded heparan sulfate and chondroitin sulfate, which eluted as the 0.75 M and 1.0 M NaCl fractions from Dowex 1 column, respectively. A metabolic disorder of GAG was suggested, since urinary excretion in two cases of osteosarcoma was within the normal range.

17.10.11 Weber Christian disease

This connective tissue disorder is of an unknown etiology and is characterized by recurrent fever and severe inflammation of nodular, nonsuppurative panniculitis in subcutaneous tissue. A three-fold increase in chondroitinase - digestible and resistant GAG was observed in a 28 year old female patient whose brother had died with this disease at 28 years of age. Digestion with chondroitinase ABC and cellulose acetate electrophoresis revealed that the increase in urinary GAG was mainly due to heparan sulfate, although chondroitin 4- and 6-sulfates also increased whereas dermatan sulfate remained negligibly small (Murata et al., 1973).

17.10.12 Lowe syndrome

The Lowe syndrome is characterized by mental retardation, congenital cataracts and glaucoma, and failure of renal tubular reabsorption. Because of these triad symptoms it is also called Cerebro-Oculo-Renal Syndrome. Defective ossification resulting in underdeveloped skeleton suggested connective tissue abnormalities and for this reason the GAG and glycoproteins were studied. Suschke and Murken (1969) noted an elevated excretion of GAG in urine. The GAG excreted in excess in urine of a patient were identified as undersulfated 4- or 6-sulfate and heparan sulfate (Nagata et al., 1972). Other workers (Akasaki et al., 1978) reported 4-5 times higher sialic acid in unidentified high molecular weight glycoproteins and as free sialic acid, but the most remarkable finding was excretion of undersulfated chondroitin 4-sulfate. The occurrence of undersulfated chondroitin 4-sulfate in urine was attributed to a disorder in sulfation of GAG in GAG-synthesizing cells in Lowe syndrome and may be, to a lesser extent, due to the failure of renal tubular absorption. Thus, whereas the major GAG components in normal urine consisted of chondroitin 4- and 6-sulfates, in Lowe syndrome only chondroitin 4-sulfate was excreted as the principal component and that too was undersulfated.

17.10.13 Dyggve-Melchior-Clausen syndrome

The patients with this syndrome have a uniform symptomatology consisting of retarded growth (short stature), skeletal (radiological) abnormalities and mental retardation, probably resulting from an abnormality of a specific sulfatase (Rastogi et al., 1977). The patients have been found to have abnormal excretion of urinary glycosaminoglycans, which consisted of hyaluronic acid and oversulfated chondroitin 4- and 6-sulfates and undersulfated forms of keratan sulfate and heparan sulfate. This syndrome resembles Morquio syndrome in some clinical features but is a distinct syndrome.

17.10.14 Werner's syndrome

Increased hyaluronic acid was excreted in the urine of patients with this hereditary condition. Of the total GAG 16 % was represented by hyaluronic acid which was isolated in the 0.8 M and 0.6 M NaCl fractions from Sephadex G-100 column and had a maximum molecular weight of 360,000 (Goto and Murata, 1978). Hyaluronuria is a characteristic feature of the urinary GAG in Werner's syndrome. But see schizophrenia, section 17.10.30.4.

17.10.15 Hepatic Disorders

The excretion of glycosaminoglycans was elevated in pathological conditions affecting liver (Thompson et al., 1966). Using the benzidine precipitation method, the urinary GAG levels were within the normal range in acute hepatitis, but were decreased in primary hepatoma and in patients with terminal cirrhosis in whom liver was shrunken. In experimental liver damage produced in rats with carbon tetrachloride, the urinary GAG increased after 40 days. Similarly, urinary GAG excretion was found to be markedly increased in chronic hepatitis, florid cirrhosis with an enlarged liver and in obstructive jaundice (Kawata et al., 1961).

The glycosaminoglycans appear to play an important role in early stages of liver disease. A comparison of GAG in normal subjects and patients with liver diseases and belonging to the ages 2-75 years revealed higher than normal GAG in urine of patients; the greatest difference being in the older age group (Yoshida, 1962). Whereas the GAG excretion decreased in urine of patients with terminal cirrhosis and primary hepatoma, marked increases ocurred in obstructive jaundice and chronic hepatitis. Oral administration of chondroitin sulfate or glucuronic acid rapidly increased urinary GAG in normal subjects (Morrison, 1974) but no such effect was observed in hepatic cirrhosis patients (Yoshida, 1962). Kupchella and Tamburro (1978) observed a significancantly elevated GAG excretion in hepatitis, cirrhosis and metastatic cancer of liver. The clinical significance of urinary GAG changes in virus hepatitis has been discussed (Stempien et al., 1966).

17.10.16 Mongolism (Down's syndrome)

In this genetic disorder the quantities of GAG and sialic acid containing glycoproteins were elevated as compared to those of healthy children. However, no definite qualitative difference was found between the GAG isolated from the urine of 2-6 years old healthy children and three mongoloids of ages 4, 5 and 6 years. In both cases the major urinary GAG was chondroitin 4-sulfate in addition to heparan sulfate and hyaluronic acid in lesser amounts (Kamikubo, 1965, 1965 a). In contrast to this, a significantly decreased excretion of hydroxyproline and GAG was noted in 28 infants and children with Down's syndrome (Klujber and Mehes, 1973). This suggested that the connective tissue metabolism was abnormal in mongoloids, but fractionation of GAG revealed no characteristic excretion pattern.

17.10.17 Lupus erythematosus

Using cetyltrimethylammonium bromide precipitation, higher than normal levels of glycosaminoglycans were reported in 5 women with systemic lupus erythematosus (Di Ferrante et al., 1957). The levels increased in the patients to 5-16 mg glucuronic acid per day as compared to the values of 2.7-5.4 mg glucuronic acid per day in normal women (Bollet et al., 1975). Values increased to 2-3 times both in the patients with or without the associated renal damage. Prednisone or ACTH decreased the elevated excretion to some extent (Di Ferrante et al., 1957). However, the marked elevation of GAG excretion appeared to be little affected by the clinical course and the salicylate or cortisone therapy did not return the GAG excretion to normal levels. Further investigations (Wessler, 1967; Bitter, 1971) established that both the sulfated and nonsulfated GAG increased in the urine of these patients.

17.10.18 Effects of hormones: Hormonal disorders

The changes in excretion of GAG in urine upon administration of a hormone are, in general, a reflection of the effects of the hormones on the connective tissue proteoglycans.

17.10.18.1 Adrenocortical hormones

In human subjects receiving $Na_2{}^{35}SO_4$ intravenously, prednisone did not alter the total 24 h urinary GAG excretion and the GAG fractions with slower turnover rates. However, the excretion of the nondialyzable ^{35}S was depressed during the first day of prednisone administration (Galambos, 1964). Other reports suggest that the adronocortical hormones inhibit GAG biosynthesis and have been used for therapy of Hurler syndrome patients who showed a decreased excretion of GAG with some clinical benefits (Wolfson et al., 1963). Large doses of prednisone produced an irregular decrease in GAG excretion, a decrease in dermatan sulfate/heparan sulfate ratios and modification of GAG excretion pattern (Onisawa and Lee, 1970). Steroids also had a similar effect in mucopolysaccharidoses patients. Thyroid hormone and glucosteroids also changed the dermatan sulfate/heparan sulfate ratio and glucocorticoids decreased the GAG excretion.

17.10.18.2 Thyroid hormone disorders

The urinary GAG levels in 24 hour urine are related to functional state of the thyroid gland. Hypofunction (hypothyroidism) raises the excretion levels due to an increased biosynthesis of GAG in the tissues. Hypothyroid children excreted an excess of hyaluronic acid. Thyroid hormone administration in hypothyroid children produced a drop in the GAG excretion (Khafagy et al., 1974). Similar changes in GAG excretion were observed in girls with juvenile goiter (Gebala et al., 1973).

Administration of thyroid hormone to myxedematous patients reduced the total content of fluid in the body with a concomitant increase in the hexosamine and electrolytes excretion in urine. These urinary and tissue changes indicated a breakdown of the water and electrolyte-binding (Chapter 3) glycosaminoglycans in the ground substance (Wiener et al., 1955).

The glycosaminoglycan excretion was significantly ($P < 0.001$) increased to 8.33 ± 1.14 mg uronic acid/24 hr in 25 cretins aged 2-13 years as compared to normal control values of 5.01 ± 0.18 mg uronic acid/day (Khafagy et al., 1974). Adequate replacement therapy with thyroid hormone resulted in drop to normal levels. Post-treatment values were similar or lower than values for normal controls. Serum GAG were similarly affected.

17.10.18.3 Parathyroid and osteoporosis

Infusion of parathyroid extract in a 17 year old boy, with myositis ossificans progressiva caused an increase in urinary excretion of hexosamine, glycosaminoglycans and hydroxyproline. The second infusion was ineffective (Bland et al., 1973). A single dose of parathyroid hormone enhanced urinary excretion of hydroxyproline, GAG and pyrophosphate in urine suggesting that these changes were due to increased bone resorption (Muhlbach et al., 1971). The GAG increase occurred simultaneously in blood and urine (Shetlar et al., 1961). Similar changes in hydroxyproline, GAG and pyrophosphate excretion occurred in senile osteoporosis and secondary osteoporosis (Muhlbach et al., 1970).

17.10.18.4 Sex (gonadal) steroids

The variations in amounts and types of urinary GAG in aging and the differences in excretion between males and females could be viewed in terms of the levels of sex steroids in males and females at various age groups. This is because of the fact that estrogens and testosterone affect the metabolism of GAG in various tissues. During the normal menstrual cycle, the disappearance of hyaluronic acid in sexual skin of baboon correlated with increased hyaluronic acid in urine during the deturgence phase (Naidoo and Qureshi, 1974). Perineal deturgence during the menstrual cycle in the baboon was accompanied by a marked transient increase in urine volume and urinary GAG and the hyaluronic acid causing increase in total GAG originated in the perineum (Naidoo, 1971).

17.10.18.5 Effect of hypophysectomy

The GAG levels in 24 hour urine from 10 hypophysectomized rats were decreased during the 8 consecutive days after hypophysectomy as compared to the average of levels during 15 prehypophysectomized days (Mikolajczyk et al., 1968 a). Uronic acid 135 ± 38 µg/24 hr decreased to 80 ± 42/24 hr.

17.10.18.6 Effects of heparin and biogenic amines

Administration of heparin to growing rats produced an elevation of GAG in urine. Similarly, injections of biogenic amines, histamine or serotonin influenced the urinary levels of GAG (Mikolajczyk, 1967).

17.10.18.7 Diabetes mellitus

The excretion of glycosaminoglycans, measured in normal urine as 3-4.9 mg glucuronic acid/24 hr by benzidine precipitation, was found slightly increased in diabetics with normal kidney function but was decreased in those with renal impairment (Craddock and Kerby, 1955). In all diabetic patients, this increase was reflected in higher glucosamine/galactosamine ratios in the Astrup fraction, which by electrophoresis was separated into a glucosamine-containing fraction (hyaluronic acid) and galactosamine-containing chondroitin sulfate (Azerad et al., 1962). In uncomplicated diabetes the glucosamine/galactosamine ratio increased to 20/80 due to an increase in hyaluronic acid/chondroitin sulfate ratio (Azerad et al., 1962 a) and in glomerulosclerosis complication it rose to 34/66 as compared to 15/85 in normal subjects. A gradual increase of total GAG and hyaluronic acid excreted in urine occurred in parallel with the deterioration of vascular changes in diabetes and these changes were particularly significant in the stage of Kimmelstiel-Wilson glomerulosclerosis (Azerad et al., 1962 a). Galletti and Stancari (1957) and Delbrueck et al. (1967) confirmed the elevation of GAG in urine of diabetic patients. A similar study on GAG excretion in diabetic children has been reported (Valuiskova, 1973). Besides the disturbed metabolism of GAG in tissues, biochemical alterations also occur in renal glomerular basement membrane of diabetics (Spiro, 1973).

17.10.19 Cystic fibrosis

Slight elevations of GAG excretion were found in cystic fibrosis patients receiving gentamicin and abnormal distribution of GAG returned to normal in some of the patients after the drug therapy was discontinued (Constantopoulos et al., 1971). In another study (Kollberg et al., 1973) elevations in the urinary excretion of hexosamine, sialic acid and fucose were noted and these appeared to correlate with the age and severity of the disease.

17.10.20 Dyslipemia

The urinary glycosaminoglycans have been found to vary with the plasma lipid patterns. In a comparative study, the urinary GAG excretion was 3.5 mg glucuronic acid/24 hr in 11 normolipemic subjects with total cholesterol 260 mg/dl and Beta lipoprotein 35 %. The GAG excretion decreased to 1.82 with the increases of total cholesterol to 357 mg/dl and Beta lipoprotein to 55 % in 14 dyslipemic subjects (Cortinovis et al., 1961). Although the exact mechanism of the relationship is not clear, the interaction of lipoproteins and glycosaminoglycans in plasma (see Chapters 1 and 16, sections 1.10 and 16.5) may be involved in this.

17.10.21 Amyloidosis

The urinary GAG excretion in 11 patients with hereditary primary amyloidosis was not different from normals (Jackson et al., 1960). Similarly, Loewi (1959) found no abnormalities of urinary GAG excretion in primary and secondary amyloidosis. On the contrary, Suzuki (1962) reported that like the Hurler's and Hunter's syndrome patients, the urine and tissue of amyloidosis patients contained an excess of heparan sulfate and dermatan sulfate. This partly corresponded to the changes in plasma in these patients, since heparan sulfate not present to a much extent in normal plasma became clearly detectable in these patients (Muir, 1969).

17.10.22 Bilharziasis

A comparative investigation in 15 normal subjects and 54 bilharziasis patients revealed an elevation of urinary GAG in patients with active bilharziasis infection and in those bilharziasis patients with clinical complications of hepatosplenomegaly and ascites. However, no elevation occurred in those patients with clinical complications of clinical hepatosplenomegaly plus ascites and edema (Khafagy et al., 1975). In the patients with GAG elevation, heparan sulfate fraction was increased. These alterations indicated that in bilharzial infection connective tissue metabolism was affected (Khafagy et al., 1975 a).

17.10.23 Aspartylglucosaminuria

Besides the excretion of an excess of a aspartylglucosamine in urine due to the absence of a lysosomal hydrolytic enzyme, the urine of these patients by two-dimensional thin-layer chromatography or Sepraphore electrophoresis showed the presence of a diacidic compound containing aspartylglucosamine,

1-2 uronic acid residues, hexosamine and sialic acid (Palo and Savolainen, 1972). Besides this, serum hyaluronidase and β-glucuronidase activities of serum and urine were also increased (Palo and Savolainen, 1972).

17.10.24 Kwashiorkor (protein-calorie malnutrition)

The excretion of total GAG measured as uronic acid was lower than normal in children with protein-caloric malnutrition with and without clinical signs of vitamin A deficiency. This decrease was due to a lower excretion of sulfated GAG and was related to lower serum vitamin A levels in these children. These changes were reversed only when the high protein diets were supplemented with vitamin A for 10 days (Mohanram and Reddy, 1971). Isolation and characterization of the excreted GAG in the Kwashiorkor children revealed the absence of hyaluronic acid, but a low sulfated chondroitin sulfate made 17 % of the total GAG in patients, which was present in normal urine in only a small quantity. The various GAG present were in the increasing order: chondroitin 4-sulfate > chondroitin 6-sulfate > low sulfated chondroitin sulfate > heparan sulfate (Cherian et al., 1970).

17.10.25 Burns of various degrees

The daily urinary excretion of non-dialyzable hexosamine, uronic acid and hexose was increased in burn injuries caused by heat or any cauterizing agent. Dermatan sulfate showed the most consistent elevation and some of this fraction was undersulfated (Shetlar et al., 1969). The urinary GAG elevation correlated with the increasing order of one to five degrees of burns in experimental animals because of a greater damage to the tissues with the increasing degree of burn. In human adults and children blistering observed in second degree and epidermis destruction and exposure of nerve terminals in the third degree burns produced increasing elevations of GAG excretion, respectively (Teller and Mayer, 1973).

The GAG changes observed in urine are related to those occurring in blood and tissues. An increase in GAG in skin of guinea pigs and rats induced by thermal burns or freezing was accompanied by a doubled hexosamine content of blood shortly after injury. These elevations decreased from the third or fourth day following the injury (Berencsi and Krompecher, 1963). Histochemical studies in third and fourth degree burns produced experimentally on the anterior abdominal walls of 18 dogs showed correlation of GAG changes with the degree of burn. Antishock therapy of these dogs with burns antagonized the decrease in permeability of the ground substance in regions of injury (Nikitina et al., 1966). Chemical analysis of the hypertrophic scars also showed elevations of uronic acid, hexosamine, sialic acid and hexose as compared to normal skin (Shetlar et al., 1971).

17.10.26 Ocular disorders

Patients with degenerative myopia (short sightedness) excreted significantly higher GAG in urine equivalent to 18.6 ± 1.7 mg uronic acid per 24 hour as compared to 10.8 ± 1.8 mg uronic acid in the controls. However, these differences were insignificant in patients over 51 years of age (Balacco-Gabrieli et al., 1972). The value of such metabolic studies in myopia has been discussed (Avetisow, 1976).

A three-fold increase in total serum GAG and about two-fold increase of chondroitin, chondroitin 4- and 6-sulfates, keratan sulfate and heparan sulfate fractions of urine occurred in patients with progressive exophthalmos (Winand, 1968). However, these changes were not related to thyroid function.

The importance of urine GAG analysis in eye diseases has been reviewed (Kuhl and Schroder, 1971). The relationship of vitamin A to dark adaptation is well known. Intake of vitamin A increased the sulfated GAG excretion in urine and dark adaptation was found to be delayed in subjects who excreted sulfated conjugates less than 25 % of the total GAG (Koyanagi et al., 1974).

The transparent cornea becomes cloudy in various mucopolysaccharidoses as a result of deposition of abnormal amounts of various GAG. In corneal clouding seen in these syndromes, dermatan sulfate, heparan sulfate or keratan sulfate is present in large amounts in the urine of the patients (McKusick, 1969).

17.10.27 Cardiovascular disorders

A significantly increased urinary glycosaminoglycan excretion was observed in patients with *arteriosclerosis* (Delbrueck et al., 1967, Maas et al., 1968). In *myocardial infarction*, an undersulfated chondroitin sulfate accompanied the major urine components, chondroitin 4- and 6-sulfates (Ohkawa et al., 1977). Other workers (Voigt et al., 1968) have contradicted the previous reports of elevated GAG excretion in arteriosclerosis. In their work no statistically significant differences could be established in the urinary excretion of hexosamine or glucuronic acid between normal and arteriosclerotic subjects.

Statistically significant increases of acidic glycosaminoglycans were observed in urines of patients with *cardiac decompensation* and *arterial hypertension* (Galletti and Stancari, 1957). The urinary concentration of glycosaminoglycans was found to be increased in patients with *aortic syndrome* (Murata et al., 1970).

In *myocardial infarction* the urinary excretion of GAG was elevated but did not correlate with total hydroxyproline. The GAG excretion was the highest (9.87 mg uronic acid/day) between the first and third days of infarction. Thereafter it decreased to 6.67 mg on the sixth day and to 5.53 mg/day between 9th and 22nd day (Rupniewska et al., 1968, 1969). A correlation of glucuronic acid, glycosaminoglycan excretion in 24 hour urine and *coronary disease* was observed (Rupniewska et al., 1968 a). Fedorishko and Padalka (1972) have investigated urinary GAG excretion in *ischemic heart disease*.

17.10.28 Mucolipidoses and related disorders

In various mucolipidoses, simultaneous storage of lipid and polysaccharide material occurs. Thus the mucolipidoses have characteristic deposition of GAG but usually do not show increased elimination of GAG in urine (Orii et al., 1974), as seen in mucopolysaccharidoses. The simultaneous effect on glycolipids and glycosaminoglycans is due to the fact that the enzyme of lysosomal origin deficient in mucolipidoses is effective in degradation of both types of macromolecules.

Philippart (1972) has reviewed the glycolipid and GAG distribution in urine, plasma and tissues in different mucolipidoses. Hooghwinkel (1974) has provided a classification for these disorders.

17.10.28.1 GM_1-gangliosidosis (Mucolipidosis I)

Due to the deficiency of the lysosomal enzyme, β-galactosidase, GM_1 gangliosides, their asialo derivatives and keratan sulfate are not degraded and these accumulate in tissues. The GAG contents of liver are 10-30 times the control values and keratan sulfate is excreted in excess in urine of these patients (Van Hoof, 1973). In this respect, the GM_1-gangliosidosis resembles Morquio's disease but differs in clinical characteristics and by the presence of normal liver β-galactosidase activity in the latter. Robertson and Harvey (1972) reported grossly elevated excretion of keratan sulfate in 2 patients with GM_1-gangliosidosis.

17.10.28.2 Systemic infantile lipidosis

This lysosomal disorder complicated with cerebral accumulation of gangliosides was also associated with a deficiency of the lysosomal enzyme, β-D-galactosidase (EC 3.2.1.23). Keratan sulfate accumulated in tissues as a result of defective breakdown by the deficient enzyme and also keratan sulfate was the predominant GAG excreted in urine (Wolfe et al., 1969).

17.10.28.3 I-cell disease (Mucolipidosis II)

These patients display Hurler-like characteristics with marked skull and skeletal bone deformities but absence of visceromegaly or corneal clouding and only a slight elevation of urinary glycosaminoglycans. The glycosaminoglycans in liver, kidney, brain and spleen are normal or only slightly elevated. Nonmetachromatic inclusions are present in the cells and because of this the name I-cell disease has been given. A number of lysosomal enzymes are deficient. The β-galactosidase deficiency was responsible for mucolipid and glycosaminoglycan accumulation in tissues, slight elevation of total GAG in urine and excessive keratan sulfate excretion in urine (Wolfe et al., 1970). The cultured skin fibroblasts from these patients accumulated keratan sulfate (Callahan et al., 1970).

Coppa et al. (1979, 1979 a) found normal daily urinary GAG excretion in a patient with I-cell disease, but GAG fractions had different elution behavior on Dowex 1-X2 column than the controls and consisted of chondroitin 4- and 6-sulfates 70 %, heparan sulfate 16 %, dermatan sulfate 10 %, keratan sulfate 3 % and traces of hyaluronic acid and chondroitin. The presence of an excess of dermatan sulfate was confirmed by isoelectric focusing of urinary GAG from a patient (Coppa et al., 1979 a).

17.10.28.4 Metachromatic leukodystrophy

In a variant of this genetic disorder called multiple sulfatase deficiency, the patients excreted in the urine an excess of sulfated glycosaminoglycans, notably dermatan sulfate (Austin, 1973). In 3 cases of a variant with dry and scaly skin (Ichthyosis vulgaris), hepatosplenomegaly, gargoylism and corneal opacities were absent. The kidney contained increased amounts of sulfatides and heparan sulfate; some increase in heparan sulfate and chondroitin sulfate in urine was present (Rampini et al., 1970). Increased urinary GAG excretion also occurred in metachromatic leukodystrophy associated with cerebroside sulfatase or arylsulfatase A deficiency (Dulaney and Moser, 1978).

17.10.28.5 Mucosulfatidosis

These patients show the clinical characteristics of grotesque facial appearance, bone malformations, vacuolated cells in peripheral blood and bone marrow. These clinical features and the excretion of an excess of dermatan sulfate and heparan sulfate in urine resemble that in Hurler's syndrome. On the other hand, the excretion of sulfatides in urine and progressive neurological syndrome with impairment of nerve conduction are like those seen in metachromatic leukodystrophy (Van Hoof, 1973 b).

17.10.28.6 Fabry's disease

This is a X-linked recessive sphingoglycolipid storage disorder due to a deficiency of the lysosomal hydrolase α-galactosidase. Multisystem involvement results from storage of glycolipids and glycosaminoglycans in tissues and these include skin lesions, ocular signs, cerebrovascular and cardiovascular conditions and renal impairment. Because of these involvements, abnormal urinary GAG excretion may occur. Matalon et al. (1969) have reported abnormal mucopolysaccharide elevation in cultured skin fibroblasts, but the relative distribution between hyaluronic acid, chondroitin sulfate and dermatan sulfate was normal.

17.10.28.7 Farber's disease

This rare disease is characterized by infiltration process in subcutaneous tissue or tendons near articulation leading to formation of nodules especially near the dorsal surface of joints. After birth a neurologic handicap starts and death results from somatic debility and arthropathy.

Bierman et al. (1966) reported a moderate excessive excretion of dermatan sulfate in the urine of a patient. A moderate excess of dermatan sulfate was also found in the lung, liver and spleen of this patient on autopsy. Wessler (1967) found no abnormality, however, and reported a normal urinary glycosaminoglycan pattern in one child with Farber disease.

17.10.28.8 Fucosidosis

This inborn error of metabolism results from a deficiency of the enzyme α-L-fucosidase (EC 3.2.1.51) in liver and other tissues (Van Hoof, 1973 a). Since keratan sulfate contains α-fucose at the non-reducing terminal end, keratan sulfate failed to be degraded and was excreted in a large excess in urine of two children with fucosidosis (Greiling et al., 1978). Van Hoof (1973 a) reviewed the older literature in four cases of fucosidosis, one of whom excreted mainly chondroitin 4- and 6-sulfates in urine.

17.10.29 Bone diseases (dystoses)

The urinary excretion patterns of individual glycosaminoglycans may be pathognomic of some hereditary bone diseases (Teller, 1967; Merkur'eva, 1972), including dystoses of children (Spranger and Wiedemann, 1966) and acquired diseases of skeletal system (Kyselka et al., 1969) and orthopedic diseases (Kaiser et al., 1969).

17.10.29.1 Diaphyseal aclasis, multiple exostosis and nail-patella syndrome

Using an acidified albumin turbidity test, an elevation of urinary GAG levels was noted in affected children of two families with dominantly transmitted abnormalities of bone and cartilage development, namely, hereditary deforming chondroplasia or diaphyseal aclasis (Lorincz, 1960). Chondroitin 4- or 6-sulfate was the predominant GAG in urine of these patients. Similar elevations were observed in 3 children with osteoarthro-onychodysplasia (nail-patella syndrome) but this was not observed in normal siblings or affected adults (Lorincz, 1961). In growing children of three families affected with hereditary multiple exostoses, chondroitin 4-sulfate was the principal GAG excreted in the urine of 3 children with nail-patella syndrome (Lorincz, 1962). However, Wessler (1967) by anion-exchange chromatography found normal urinary GAG patterns in these patients.

17.10.29.2 Osteogenesis imperfecta

In the genetic disease called Osteogenesis imperfecta, fragility of the bones is an important characteristic, which suggests the possibility of connective tissue abnormality. However, the urinary excretion of GAG was not quantitatively and qualitatively different from normals; only one case showed a slight increase above the normal range (Berggren et al., 1969). However, other workers have noted changes and have discussed the value of the urinary GAG analysis in the differential diagnosis of osteogenesis imperfecta (Torelli, 1967; Beck et al., 1972). Other workers (Valli et al., 1977) have reported a genetic abnormality of proteoglycan biosynthesis rather than its catabolism being responsible for the fragility of bones and other clinical features associated with osteogenesis imperfecta. Although the excretion of total GAG in urine was normal, the spectrum of the GAG fractions isolated from one patient by cellulose column chromatography was shifted from the normal pattern. Treatment with (+)-catechin normalized the GAG spectrum in this patient's urine (Valli et al., 1977).

17.10.29.3 Spondylepiphyseal dysplasia, achondroplasia and athetosis

Spondylepiphyseal dysplasia congenita is a heritable bone dysplasia of the disproportionate short trunk type. The sialic acid and uronic acid contents of the GAG isolated from the urine of a 14 year old female patient were within the normal range. However, the sulfate content (0.5 mg/day) was much lower than (3.5 ± 0.9 mg/day) in 5 normal subjects (Masuda et al., 1975, 1976). Infrared spectra of the GAG showed similar components in patients and normals. Keratan sulfate excretion was raised in 2 out of 6 cases of achondroplasia (Robertson and Harvey, 1972). Keratan sulfate excretion in excess also occurred in patients with athetosis (Maroteaux, 1973). The relationship of keratan sulfate excretion to the constant, slow, writhing movements of the fingers and hands and sometimes of the toes and feet in the athetoid patients is not clear, but it has been considered a new type of mucopolysaccharidosis.

17.10.29.4 Rickets

The glycosaminoglycan excretion in urine of children with rickets was only slightly elevated and that too not in all patients (Christiaens et al., 1964). Similar studies have been reported by others (Fontains et al., 1969).
Also see Section 17.9.17.

17.10.29.5 Pycnodysostosis, polydystrophic dwarfism, Alaskan malamute chondrodysplasia

In patients with pycnodysostosis (Dodge, 1969) the GAG excretion was higher and urine glycoprotein was less than normal. The GAG excreted in urine were highly depolymerized and desulfated (Namiki et al., 1974). Increased urinary excretion of dermatan sulfate and heparan sulfate was reported in a case of polydystrophic dwarfism (Fallis et al., 1968). Similarly, alterations in urinary excretion of hydroxyproline and GAG as uronic acid were observed in Alaskan malamute chondrodysplasia (Hoag et al., 1976). These changes were related to the arrest of the growth in epiphyseal plates of long bones resulting in a stocky dwarfism with short extremities but normal head and trunk.

17.10.29.6 Pachydermoperiostoses and enchondral dystoses

Abnormal excretion of GAG in urine was seen in the autosomal dominantly inherited syndrome called pachydermoperiostosis (Yamato et al., 1974). These changes may be related to the clubbing of the digits and new periosteal bone formation especially over the distal ends of the long bones commonly observed in this disorder. Similar disturbances of GAG metabolism occur in enchondral dysostoses (Kyselka and Schramm, 1968).

17.10.29.7 Brachycephalic "snorter" dwarf cattle

A 30-fold excess of sulfated glycosaminoglycan was observed in brachycephalic "snorter" dwarf cattle. The major GAG component excreted in excess was identified by a combination of various technics to be chondroitin 4-sulfate (Hurst et al., 1975). Earlier the snorter dwarf syndrome was identified in cattle as a naturally occurring hereditary disorder of GAG metabolism resembling Hurler syndrome in human subjects (Lorincz, 1960 a). Tyler et al. (1962) refuted the claim that brachycephalic bovine dwarf was homologous to Hurler syndrome in man, since their analysis showed that the GAG component excreted was chondroitin 4-sulfate and not dermatan or heparan sulfate.

17.10.29.8 Platyspondylisis (Brachyolmia)

This is a congenital disorder due to the absence of the sulfotransferase involved in the transfer of sulfate for biosynthesis of chondroitin 6-sulfate. The clinical features include flattened vertebral bodies in radiograph, growth retardation after the age of 5-6 years, short trunk, and short *bull neck*. These patients excrete undersulfated chondroitin 6-sulfate but the total urinary GAG excretion is not increased. The major GAG present in urine in 4 siblings had a slower electrophoretic mobility and was characterized by digestion with chondroitinases AC and ABC (Dietrich et al., 1972; Mourao et al, 1973).

17.10.30 Mental Disorders

The excretion of the total GAG and the distribution of the GAG components are altered in various mental disorders.

17.10.30.1 Acquired and hereditary nervous diseases

In children with hereditary or acquired nervous diseases, urinary levels of GAG and especially of heparan sulfate were increased as compared to normal children (Bondarenko et al., 1976). Hereditary disorders of the amino acid metabolism such as homocystinuria (section 17.10.8), aspartylglucosaminuria (section 17.10.23), and Down's syndrome (section 17.10.16) are also accompanied by urinary GAG changes. A mass metabolic screening of children with various psychoneurological diseases using the cetyltrimethylammonium bromide screening test has been reported (Hambraeus and Holmgren, 1974).

17.10.30.2 Myoclonus epilepsy of different types

Familial myoclonus epilepsy (Unverricht's syndrome) is characterized by myoclonus, progressive dementia, generalized epileptic seizures and often, intraneuronal inclusion bodies of Lafora (Lafora bodies) composed of glycosaminoglycans and glycoproteins are present. Infante and Rallo (1968) first reported an abnormal, increased excretion of GAG in a Lafora patient. Only the fraction eluted from an anion-exchange column with 1.25 M NaCl corresponding to heparan sulfate was responsible for this increase. Medved et al. (1967) had earlier reported nodular, cutaneous lesions, showing mild hyperkeratosis with moderate acanthosis of the epidermis of a patient with Unverricht's syndrome. Guazzi et al. (1973) investigated the GAG excretion patterns in a 28 year old female born of blood-related wedding, who had myoclonus epilepsy with cherry red spot in the fundus oculi. An increased excretion of 40.2 mg uronic acid per gram of creatinine was found. Anion-exchange chromatography of the isolated material revealed that it contained 36.4 % heparan sulfate, 40.9 % chondroitin 4-sulfate and 22.7 % dermatan sulfate. This patient was then suggested to be a case of mucopolysaccharidosis with a characteristic GAG excretion pattern (Guazzi et al., 1973).

Rallo (1969) reported that the abnormal metabolism of glycosaminoglycans suggested by the altered and increased urinary GAG excretion was related to the etiology of the Unverricht-Lundborg syndrome in the five Valasian families they studied. Rallo et al. (1968) investigated the histological and histochemical analysis of liver biopsy samples in relation to the elevated GAG excretion in the patients.

Poulsen (1975) also confirmed the increased GAG excretion in 3 affected members (of a family of 10) who had progressive familial epilepsy. The increase in GAG was, however, found in 0.65 M and 0.71 M fractions due to some low sulfated GAG. The unaffected members of the family without neurological symptoms had normal GAG excretion. Masuda et al. (1975 b) determined glycopeptides and GAG in the urine of a patient with myoclonus epilepsy of Lafora-body type (Lafora-body disease) and another patient with myoclonus epilepsy of degenerative type. The daily urinary excretion of these macromolecules was not different from those of the normals. However, the sialic acid content of the material isolated from the patients was three times higher than that in normal subjects.

17.10.30.3 Friedreich's ataxia

Friedreich's ataxia is marked by loss of muscular coordination of the extremities followed by paralysis and contractures resulting from the posterior and lateral columns of the spinal cord. In this hereditary spinal ataxia, there was no significant change in excretion of glycosaminoglycans. In 4 patients the μ moles glucuronic acid/mg creatinine ratios ranged from 0.13-0.42 as compared to the normal range of 0.17-0.38 (Bose and Carroll, 1979). Although a β-galactosidase deficiency might be present, yet there was neither an accumulation of GAG in tissues, nor was there any mucopolysacchariduria. This suggested that it was not a lysosomal storage disease.

17.10.30.4 Schizophrenia

A significantly decreased excretion of GAG was noted in 60 schizophrenics as compared to 30 normal subjects matched for age and sex (Varma et al., 1974, 1979, 1979 a). The daily excretion in male schizophrenics expressed as glucuronic acid was 2.13 ± 0.09 (mean ± S.E.) as compared to that of 5.82 ± 0.30 in normal subjects. Also, the spectrum of the GAG excreted in schizophrenics' urine differed in containing chondroitin sulfate 38-43 %, heparan sulfate + keratan sulfate 22-30 % and hyaluronic acid 30-35 % as compared to the normal values in corresponding fractions of 66-75 %, 25-33 % and 0 % (Varma et al., 1974). The glycoprotein excreted in urine of schizophrenics also showed a corresponding decrease and like the glycosaminoglycans had higher glucosamine/galactosamine ratios (Varma et al., 1976, 1976 a). Since the amounts and types of glycoproteins and glycosaminoglycans in urine, in general, reflect the metabolism of these macromolecules in connective tissue, the synchronicity of decreased excretion and of higher glucosamine/galactosamine ratios in both these excreted macromolecules suggested the possibility of an altered metabolism of glucosamine-containing macromolecules in the connective tissue of schizophrenic patients.

17.10.30.5 Mucopolysaccharidoses-related mental retardation

In almost all mental disorders, disturbances of two components, dermatan sulfate and heparan sulfate, are involved. In the various mental retardations associated with various types of mucopolysaccharidoses except the Morquio's syndrome, one or the other or a combination of the dermatan sulfate and heparan sulfate are excreted in different proportions. In the Maroteaux-Lamy syndrome, dermatan sulfate is excreted in excess in urine and the patients do not show mental retardation. Besides the various mucopolysaccharidoses, increased excretion of one or other type of GAG component occurs also in some mucolipidoses (see section 17.10.28), since the enzyme deficient in some mucopolysaccharidoses is also common to glycolipid metabolism. Also see section 17.10.28.

Mucopolysaccharidoses and mucolipidoses

The basic defect in these disorders is the genetic deficiency of one or the other enzyme involved in the degradation of glycosaminoglycans and mucolipids. The nature and extent of the defect of the degradation varies with the deficient enzyme. As a result of the defective degradation, the undegraded or incompletely degraded glycosaminoglycans accumulate in various tissues and an excess of glycosaminoglycans are excreted in urine. The biochemical identification and differentiation of various mucopolysaccharidoses is based on identification and characterization of the relative proportions of the GAG excreted in urine, tissue culture of skin fibroblasts from the patient and the identification of the deficient enzyme. The enzyme deficient and GAG excreted in excess in various classes of mucopolysaccharidoses are given in Table 17.2.

The classification or nosology of the mucopolysaccharidoses (McKusick, 1969) is based on a combination of clinical, genetic, morphologic and biochemical approaches and has permitted firm delineation of several distinct mucopolysaccharidoses. The classification has gone through several revisions with the discoveries of new variants of the previously known mucopolysaccharidoses and new forms of these disorders. The classification presented in Table 17.2 is based mainly on that presented by McKusick et al. (1978). Some new mucopolysaccharidoses forms still remain unclassified.

Table 17.2. Glycosaminoglycan Excretion in Various Mucopolysaccharidoses

Type	Designation	Deficient Enzyme	Excess Glycosamino-glycan Excreted	Mental Retardation (M.R.)
IH	Hurler Syndrome	α-L-Iduronidase	DS : HS = 7 : 3	Present
IS	Scheie Syndrome	α-L-Iduronidase	DS + HS	Normal intelligence
IH/S	Hurler-Scheie Syndrome	α-L-Iduronidase	DS + HS	Intermediate between Hurler and Scheie syndromes (mild M.R.)
IIA	Hunter Syndrome A (severe)	L-Iduronic acid 4-sulfate sulfatase (sulfoiduronate sulfatase)	DS : HS = 1 : 1	Present
IIB	Hunter Syndrome B (mild)	L-Iduronic acid 4-sulfate sulfatase	DS + HS	Fair intelligence
IIC	Hunter Syndrome	L-Iduronic acid 4-sulfate sulfatase	DS + HS	Present
IIIA	Sanfilippo A	Heparan sulfate N-sulfatase (sulfamidase)	HS	Present
IIIB	Sanfilippo B	α-D-Acetamideoxyglucos-idase (N-acetyl-D-glucos-aminidase)	HS	Severe
IIIC	Sanfilippo C	Acetyl-CoA: α-glucos-aminide-N-acetyltransferase	HS	Present
IIID	Sanfilippo D	N-Acetylglucosamine 6-sulfate sulfatase	HS	–
IVA	Morquio syndrome type A	N-Acetylgalactosamine 6-sulfate sulfatase	KS	Normal intelligence
IVB	Morquio-Dale syndrome	β-galactosidase	KS	Normal intelligence
V	Scheie Syndrome	Similar to type I-S	DS	Normal intelligence
VI (severe)	Maroteaux-Lamy syndrome (severe type)	Arylsulfatase B (N-acetylgalactosamine 4-sulfatase)	DS	Normal intelligence
VI (Intermediate)	Maroteaux-Lamy syndrome (intermediate type)	Arylsulfatase B (N-acetylgalactosamine 4-sulfatase)	DS	Normal intelligence
VI (mild)	Maroteaux-Lamy syndrome (mild type)	Arylsulfatase B (N-acetylgalactosamine 4-sulfatase)	DS	Normal intelligence
VII	β-Glucuronidase deficiency (Sly syndrome)	β-Glucuronidase	DS + HS	Variable
VIII	Di Ferrante syndrome	N-Acetylglucosamine 6-sulfate sulfatase	KS + HS	Present

Abbreviations HS = heparan sulfate; DS = dermatan sulfate; KS = keratan sulfate

The literature on biochemical studies of urinary glycosaminoglycan in mucopolysaccharidoses is so vast that it could make a volume of itself. For references to original articles, several excellent reviews are available for consultation (Dorfman and Matalon, 1972, 1976; Groover et al., 1972; Spranger, 1972; Neufeld, 1972; Van Hoof and Hers, 1972; Muir, 1973; Van Hoof, 1973, 1974; Haust, 1974; Hooghwinkel, 1974; Kennedy, 1976; Horwitz, 1979; McKusick et al., 1978).

17.10.31 Carcinomas

Increased urinary excretion of GAG was observed in 7 out of 13 patients afflicted with a variety of malignancies, including lymphoma, leukemia, multiple myeloma and metastatic cancers (Rich and Myers, 1959). The remaining 6 showed normal values. Whereas in the leukemias most urines showed elevated glycosaminoglycans, in solid tumors this was seen only in 50 % of the cases. Also, there was a relatively higher elevation in urinary GAG in the neoplasms of hemopoietic system than in solid tumors. The values were raised to 12.0-23.0 mg glucuronic acid from the normal adult male values of 4.6-7.5 mg glucuronic acid per day. In adult female leukemic patients the values increased to 21.0 as compared to normal female values of 3.7 mg glucuronic acid per day (Bollet, 1975). Electrophoretic analysis revealed presence of glycosaminoglycans with abnormally high and low sulfate contents in urine from chronic lymphatic leukemia patients (Friman, 1972). Masamune et al. (1959) reported isolation of a cancer-specific *K-mucopeptide* by paper chromatography of urinary polypeptides.

The urinary excretion of GAG in two cases of osteosarcoma was not significantly different from those of the normal controls (Tokunaga et al., 1978).

Hepatic angiosarcoma patients exhibited significantly elevated GAG excretion in urine with the greatest increase in the heparin fraction (Kupchella and Tamburro, 1978). The role of GAG in xanthomatogenesis is suggested by xanthoma production following subcutaneous injection of dextran sulfate. The formation of a dextran sulfate-β-lipoprotein complex may play a role in this process (Kodama and Masuda, 1977).

The tumors of connective tissue are generally correlated with excretion patterns mirroring the mucopolysaccharide compostion of the tissue in which the tumor originates. In case of a metastatic cancer, the cells of the original tissue of origin retain the biosynthesis pattern in the host tissue (see section 13.2). Suschke and Kunze (1971) reported excretion of both hyaluronic acid and dermatan sulfate in the urine of patients with multiple soft tissue tumors.

The implanted granuloma in rats induced increased amounts of GAG in both the granuloma tissue and the urine (Koizumi et al., 1969). However, the relative distribution of sulfated and nonsulfated fractions was different in the tissue and urine. An increase in the urine and plasma of Wilm's tumor patient resulted from increased synthesis by the tumor (Hopwood and Dorfman, 1978). The urine and plasma of the patient contained approximately 20 mg of hyaluronic acid and 8 mg sulfated GAG per 100 ml, respectively. Morse and Nussbaum (1967) reported the occurrence of hyaluronic acid in the urine and blood of a case of nephroblastoma.

The urinary GAG excretion in malignant diseases of the haemopoietic and lymphatic tissues has been studied (Friman and Juvani, 1975). A small increase occurred in all cases but the maximum increase in excretion of GAG was found in chronic lymphatic leukemics, who also excreted a larger amount of dermatan sulfate, than the controls. The 7 patients with Hodgkin's disease did not have an increased GAG excretion. The total GAGs were calculated as hexuronic acid in CTAB precipitates and were electrophoresed in 0.5 lithium acetate on cellulose acetate. Besides chronic lymphatic leukemics, 6 patients with mycosis fungoides also had larger than normal amounts of dermatan sulfate in their urines. In myeloid leukemia, the qualitative GAG distribution was similar to those in controls.

17.11 Bibliography

1. Abeling, N. G., Wadman, S. K. and Van Gennip, A. H. (1974): Two-dimensional electrophoresis of urinary mucopolysaccharides on cellulose acetate after N-cetylpyridinium chloride (CPC) precipitation: a method suitable for the routine laboratory. Clin. Chim. Acta 56, 297-303

2. Acquaviva, P., Garrigues, J. C., Frezet, P. and Recordier, A. M. (1970): Acid mucopolysaccharides. Urinary determination in rheumatology. Marseille Med. 107, 467-472 (Fr)

3. Addis, T., Barrett, E., Poo, L. J., Ureen, H. J. and Lippman, R. W (1951): The relation between protein consumption and diurnal variations of the endogenous creatinine clearance in normal individuals. J. Clin. Invest. 30, 206-209

4. Akasaki, M., Fukui, S., Sakano, T., Tanaka, T., Usui, T. and Yamashina, I. (1978): Urinary excretion of a large amount of bound sialic acid and of undersulfated chondroitin sulfate by patients with the Lowe syndrome. Clin. Chim. Acta 89, 119-125

5. Alimova, M. M., Koroleva, T. S. and Dumnova, A. G. (1975): Effect of ACTH on the excretion of acid glycosaminoglycans in urine and on the content of haptoglobin in blood serum of rats in acute aseptic inflammation. Vopr. Med. Khim. 21, 581-586 (Rus)

6. Allalouf, D. A. and Ber, A. (1970): Gel filtration of testicular hyaluronidase digests of urinary glycosaminoglycans. Biochim. Biophys. Acta 208, 141-143

7. Allalouf, D. and Ber, A. (1970a): Pattern of urinary glycosaminoglycan excretion in relation to age of man. Biochim. Biophys. Acta 201, 61-68

8. Allalouf, D. A., Wasserman, L. and Ber, A. (1970): Pattern of urinary glycosaminoglycans excretion in relation to age of man. II. Isolation, proportions and properties of the fraction resistant to testicular hyaluronidase. Biochem. Med. 4, 476-485

9. Anton, W. (1971): Pathogenesis of late pregnancy toxemia. Zentralbl. Gynaekol. 93, 445-452

10. Applegarth, D. A., Bozoian, G. and Lowry, R. B. (1975): Analysis of urinary mucopolysaccharides using small ion exchange columns. Clin. Biochem. 8, 151-160

11. Asboe-Hansen, G. (1962): Bullous diffuse mastocytosis with urinary hyaluronate excretion. Acta Dermatovener. 42, 21-27

12. Asboe-Hansen, G. and Clausen, J. (1964): Mastocytosis (urticaria pigmentosa) with urinary excretion of hyaluronic acid and chondroitin sulfuric acid. Am. J. Med. 36, 144-150

13. Asboe-Hansen, G. and Clausen, J. (1964 a): Urinary excretion of acid mucopolysaccharides in 12 cases of mastocytosis (urticaria pigmentosa). J. Invest. Dermatol. 43, 81-85

14. Austin, J. (1973): Metachromatic leukodystrophy. In: H. G. Hers and F. Van Hoof (Eds.). Lysosomes and Storage Diseases. Academic Press, New York, Chapter 18

15. Avetisow, E. S. (1976): The metabolic values of acid mucopolysaccharides in myopia. Monatsbl. Augenheilkd. 168, 750-754 (Ger)

16. Azerad, E., Morard, J. C., Ghata, J. and Lubetzki, J. (1962): The acid mucopolysaccharides of urine (Astrup fraction) in diabetes and its complication; determination, fractionation, and significance. Pathol. Biol. Semaine Hop. 10, 217-230

17. Azerad, E., Morard, J. C. and Ghata, J. (1962 a): Quantitative and qualitative analysis of acid mucopolysaccharides excreted by the diabetic patients. Vnitrni Lekar. 8, 256-262

18. Balacco-Gabrieli, C., De Vietra, T. and D'Ammacco, F. (1972): Urinary mucopolysaccharides in degenerative myopia. Boll. Soc. Ital. Biol. Sper. 48, 715-717

19. Barashnev, Y. I. and Semyachkina, A. N. (1979): State of connective tissue in homocystinuria. Vopr. Med. Khim. 25, 238-242

20. Barker, S. A. and Kennedy, J. F. (1969): Human skin acidic mucopolysaccharides and clinical conditions. Life Sci., Part II, 8, 989-994

21. Bazin, S., Delaunay, A., Allain, J. C. and Briquelet, N. (1968): Biochemistry of inflammation. VII. Modification of acid mucopolysaccharide levels in rapidly developing inflammation. Ann. Inst. Pasteur 115, 899-907

22. Beck, O., Schramm, G. and Kyselka, R. (1972): Does the determination of acid mucopolysaccharides and hydroxyproline contribute to the differential diagnosis of fibrous dysplasia (Jaffe-Lichtenstein) and osteogenesis imperfecta congenita? Beitr. Orthop. Traumatol. 19, 740-742

23. Ber, A. and Allalouf, D. A. (1969): Effects of starvation and of intraperitoneal administration of physiological saline on the levels of urinary glycosaminoglycans in rats. Med. Exp. 19, 194-200

24. Berencsi, G. and Krompecher, St. (1963): Thermal injuries and formation of mucopolysaccharides in the skin. Nature 197, 191-192

25. Berenson, G. S. and Dalferes, E. R., Jr. (1965): Urinary excretion of mucopolysaccharides in normal individuals and in the Marfan syndrome. Biochim. Biophys. Acta 101, 183-192

26. Berenson, G. S. and Serra, M. T. (1959): Mucopolysaccharides in urine from patients with Marfan's syndrome. Fed. Proc. 18, 190

27. Berggren, L., Wessler, E. and Wennerstrom, J. (1969): Intraocular pressure and excretion of mucopolysaccharides in osteogenesis imperfecta. Acta Ophthal. (Kobenhavn) 47, 122-128

28. Bierman, S. M., Edginton, T. S., Newcomer, V. D. and Pearson, C. M. (1966): Farber's disease: a disorder of mucopolysaccharide metabolism with articular respiratory and neurologic manifestations. Arthritis Rheum. 9, 620-630

29. Bitter, T. (1971): Systemic lupus erythematosus antinuclear serology and cell mediated immunity in the light of clinico-hematologic diagnostic criteria. Preliminary communication. Schweiz. Med. Wochenschr. 101, 181-186

30. Bitter, T., Siegenthaler, P., Depreux, T. and Martin, E. (1970): Excretion in the urine of aminoacridine precipitable polyuronides (acid mucopolysaccharides) in patients with rheumatoid arthritis. Ann. Rheum. Dis. 29, 427-433

31. Bland, J. H., Kirschbaum, B., O'Connor, G. T., Whorton, E. (1973): Myositis ossificans progressiva. Effect of intravenously given parathyroid extract on urinary excretion of connective tissue components. Arch. Intern. Med. 132, 209-212

32. Boas, F. (1965): Aminosugar-containing compounds in urine. In: E. A. Balazs and R. W. Jeanloz (Eds.). The Amino Sugars, Vol. IIA, Academic Press, New York, Chap. 21, pp. 95-115

33. Bollet, A. J. (1967): Urinary excretion of products of connective tissue metabolism. In: A. S. Cohen (Ed.). Laboratory Diagnostic Procedures in the Rheumatic Diseases. Little, Brown and Company, Boston, Chap. 10, pp. 304-331

34. Bollet, A. J. (1975): Urinary excretion of products of connective tissue metabolism. In: A. S. Cohen (Ed.). Laboratory Diagnostic Procedures for Rheumatic Diseases, 2nd Ed., Little, Brown and Company, Boston, pp. 375-392

35. Bondarenko, E. S., Shamkhalova, U. G. and Shvabrina, T. F. (1976): Urinary excretion of glycosaminoglycans in normal children and in children with some nervous system disorders. Neuropatol. Psikhiatr. im S. S. Korsakova 76, 1509-1513 (Rus)

36. Bonomo, E. and Cenetelli, P. (1957): Urinary glycoproteins in rheumatic diseases and in some other morbid conditions. Reumatismo 9, 98-113

37. Bose, S. and Carroll, M. (1979): Friedreich's ataxia. A lysosomal storage disease? Biochem. Soc. Trans. 7, 419-420

38. Brettschneider, I., Pravis, R., Musilova, J. and Musilova, M. (1976): Effects of some antirheumatic agents on connective tissue. Arzneim. Forsch. 26, 846-848

39. Bryant, J. H., Leder, I. G. and Stetten, Jr., D. (1958): Release of chondroitin sulfate from rabbit cartilage following the intravenous injection of crude papain. Arch. Biochem. Biophys. 76, 122-130

40. Butt, A. J. (1956): Pathogenesis of renal lithiasis: A working hypothesis. In: A. J. Butt (Ed.). Etiologic Factors in Renal Lithiasis. Thomas, Springfield, Chap. 17, pp. 320-358

41. Callahan, J. W., Pinsky, L. and Wolfe, L. S. (1970): GM_1-gangliosidosis (type II): Studies on a fibroblast cell strain. Biochem. Med. 4, 295-316

42. Chandrasekharan, N. (1977): Urinary excretion of mucopolysaccharides in vitamin A deficiency states in children. IRCS Med. Sci. Libr. Compend 5, 396

43. Chandrasekharan, N. (1977 a): Changes in serum proteins and the excretion of acid mucopolysaccharides in vitamin A deficiency states in children. Proc. Malays. Biochem. Soc. Conf. 4, 227-235

44. Cherian, R., Chandrasekaran, E. V. and Bachhawat, B. K. (1970): Glycosaminoglycans of human urine 3. Isolation and characterization of urinary glycosaminoglycans in normal and kwashiorkor children. Indian J. Biochem. 7, 174-177

45. Christiaens, L., Fontaine, G., Cuvelier, R., Walbaum, R., Farriaux, J. P. and Duquesnoy, C. (1964): L'excretion urinaire des mucopolysaccharide acides (M.P.S.A.) dans le rachitisme commun de l'enfant. Presse Med. 72, 3073

46. Clausen, J. and Aboe-Hansen, G. (1966): Urinary acid mucopolysaccharides in mastocytosis. A new technique for quantitative estimation. Clin. Chim. Acta 13, 475-483

47. Constantopoulos, G. (1968): Hunter-Hurler syndrome: gel filtration and dialysis of urinary acid mucopolysaccharides. Nature 220, 583-585

48. Constantopoulos, G., Dekaban, A. S., Lapey, A. and di Sant'Agnese, P. A. (1971): Urinary acid mucopolysaccharides in cystic fibrosis. J. Pediat. 78, 806-811

49. Coppa, G. V., Gabrielli, O., Maiorana, A., Sani, S., Benincasa, A., Belsito, M. and Montali, U. (1978): Undersulfated urinary glycosaminoglycans in preterm newborns. Biol. Neonate 33, 18-24

50. Coppa, G. V., Maiorana, A., Gabrielli, O. and Sani, S. (1979): Glycosaminoglycans from urine and tissues in mucolipidoses II (I-cell disease). Clin. Chim. Acta 95, 135-137

51. Coppa, G., Sani, S., Paris, D., Gabrielli, O. and Maiorana, A. (1979 a): Isoelectric focusing of serum lysosomal enzymes and characterization of urinary and tissutal glycosaminoglycans in "I-cell" disease (mucolipidosis II). In: B. Berra and P. Durand (Eds.). Perspect. Inherited Metab. Dis. 2 (Glycoproteins Glycosaminoglycans Health Dis.) Edition Ermes, Milan, pp. 149-164

52. Cortinovis, A., Silenzi, C. and Gambini, A. (1962): Glycoprotein metabolism in relation to age and sex. Acta Gerontol. 12, 235-239

53. Cortinovis, A., Silenzi, C. and Tavazzani, C. (1961): Behaviour of neutral glycoproteins (total protein hexoses) of serum and of urinary acid mucopolysaccharides in dyslipemic and normolipemic subjects. Rass. Fissiopatol. Clin. Terap. 33, 288-294

54. Craddock, J. G., Jr. and Kerby, G. P. (1955): Urinary excretion of acid mucopolysaccharides by diabetic patients. J. Lab. Clin. Med. 46, 193-198

55. Dean, M. F. and Muir, H. (1973): Mobilization of glycosaminoglycans by plasma infusion in mucopolysaccharidosis type III-two types of response. Nature New Biol. 243, 143-146

56. DeJong, B. P., Robertson, W. Van B. and Schafer, I. A. (1968): Failure to induce scurvy by ascorbic acid depletion in a patient with Hurler's syndrome. Pediatrics 48, 889-903

57. Delbrueck, A., Grimme, H. and Wette, K. (1967): Excretion of acid mucopolysaccharides in urine. II. Excretion rates during various pathological conditions in connective tissues. Z. Klin. Chem. 5, 10-16

58. Dietrich, C. P., Mourao, P. A. S. and Toledo, S. P. A. (1972): A new mucopolysaccharidosis characterized by excretion of chondroitin sulfate C of low sulfate content. In: R. Piras and H. G. Pontis (Eds.). Biochemistry of the Glycosidic Linkage. Vol. 2, Academic Press, New York, pp. 751-758

59. Dietrich, C. P., Nader, H. B., Mourao, P. A. and Toledo, S. A. P. (1972): The effect of hyaluronidase in the degradation and excretion of mucopolysaccharides, in Hunter's and Hurler's syndromes (gargoylism). In: R. Piras and H. G. Pontis (Eds.). Biochemistry of the Glycosidic Linkage: An Integrated View. PAABS Sympos. Vol. 2, Academic Press, New York, pp. 741-749

60. Di Ferrante, N. (1957): Urinary excretion of acid mucopolysaccharides by patients with rheumatoid arthritis. J. Clin. Invest. 36, 1516-1520

61. Di Ferrante, N. (1957 a): Urinary excretion of acid mucopolysaccharides by patients with rheumatoid arthritis. Fed. Proc. 16, 172-173

62. Di Ferrante, N. (1967): The measurement of urinary mucopolysaccharides. Anal. Biochem. 21, 98-106

63. Di Ferrante, N. and Lipscomb, H. S. (1970): Urinary glycosaminoglycans versus creatinine excretions: A used and abused parameter. Clin. Chim. Acta 30, 69-72

64. Di Ferrante, N., Neri, G. and Neri, M. E. (1972): Measurement of urinary glycosaminoglycans with quaternary ammonium salts. Extension of the method. Connect. Tissue Res. 1, 93-101

65. Di Ferrante, N., Nichols, B. L., Donnelly, P. V. and Neri, G. (1971): Induced degradation of glycosaminoglycans in Hurler's and Hunter's syndromes by plasma infusion. Proc. Nat. Acad. Sci. U.S. 68, 303-307

66. Di Ferrante, N. and Popenoe, E. A. (1957): Hexosamine-containing proteins co-precipitated with urinary mucopolysaccharides. Scand. J. Clin. Lab. Invest. 10: Supp. 31, 268-269

67. Di Ferrante, N. and Rich, C. (1956): The determination of acid aminopolysaccharide in urine. J. Lab. Clin. Med. 48, 491-494

68. Di Ferrante, N. and Rich, C. (1956 a): The mucopolysaccharide of normal human urine. Clin. Chim. Acta 1, 519-524

69. Di Ferrante, N., Robbins, W. C. and Rich, C. (1957): Urinary excretion of acid mucopolysaccharides by patients with lupus erythematosus. J. Lab. Clin. Med. 50, 897-900

70. Dodge, J. A. (1969): Pycnodysostosis. Develop. Med. Child Neurol. 11, 103-104

71. Dorfman, A. and Lorincz, A. E. (1957): Occurrence of urinary acid mucopolysaccharides in the Hurler syndrome. Proc. Natl. Acad. Sci. U.S. 43, 443-446

72. Dorfman, A. and Matalon, R. (1972): The mucopolsaccharidoses. In. J. B. Stanbury, J. B. Wyngaarden and D. S. Fredrickson (Eds.) The Metabolic Basis of Inherited Disease. 3rd Ed., McGraw Hill, New York, pp. 1218-1272

73. Dorfman, A. and Matalon, R. (1976): The mucopolysaccharidoses (a review). Proc. Natl. Acad. Sci. USA 73, 630-637

74. Douglas, C., Nowak, J. and Danes, B. S. (1973): Mucopolysaccharides in urine during normal human development. Pediatr. Res. 7, 724-727

75. Dulaney, J. T. and Moser, H. W. (1978): Sulfatide lipidosis: metachromatic leukodystrophy. In: J. B. Stanbury, J. B. Wyngaarden and D. S. Fredrickson (Eds.). The Metabolic Basis of Inherited Disease, 4th Ed. McGraw-Hill, New York, Chap. 38, pp. 770-809

76. Endo, M., Namiki, O. and Yosizawa, Z. (1974): Excretion of glycosaminoglycans and glycoproteins in normal urine with age. Tohoku J. Exp. Med. 113, 65-75

77. Fallis, N., Barnes, F. L. and Di Ferrante, N. (1968): A case of polydystrophic dwarfism with urinary excretion of dermatan sulfate and heparan sulfate. J. Clin. Endocr. 28, 26-33

78. Fedorishko, S. S. And Padalka, M. N. (1972): Acid mucopolysaccharide excretion in ischemic heart disease. Vrach. Delo 9, 8-10 (Rus)

79. Fontains, G., Biserte, G., Havez, R. et al. (1969): Urinary amino acids and acid mucopolysaccharides in common rickets in children. Folia Clin. Int. (Barc.) 19, 69-75 (Spa)

80. Friberg, U., Ringertz, N. R. (1954): Uptake of radio sulfate and radiophosphate in various tissues of normal and scorbutic guinea pigs. Exp. Cell. Res. 6, 527-528

81. Friman, C. (1972): Electrophoretic analysis of urinary glycosaminoglycans (GAG) in acquired diseases afflicting tissues of mesenchymal origin. Scand. J. Clin. Lab. Invest. 29, Suppl. 123, 12

82. Friman, C. and Juvani, M. (1975): Urinary excretion of glycosaminoglycans in malignant diseases of the haemopoietic and lymphatic tissues. Acta Med. Scand. 198, 115-122

83. Friman, C., Juvani, M. and Johansson, E. (1974): Transiently increased urinary excretion of low-sulfated heparan sulfate in psoriatic erythroderma associated with benign gammopathy. Clin. Chim. Acta 57, 103-107

84. Friman, C., Storgards, E., Juvani, M. and Kock, B. (1977): The glycosaminoglycans (mucopolysaccharides) in plasma in patients with renal insufficiency. Clin. Nephrol. 8, 435-439

85. Gadamska, T. (1977): Studies on the behavior of fractions of glycosaminoglycans (acid mucopolysaccharides) in the urine of healthy infants. Pediatr. Pol. 52, 1089-1093

86. Gadamska, T. (1977 a): Studies on fraction I of glycosaminoglycans (keratan sulfate and glycoproteins) in the urine of infants with rickets prior to and following administration of vitamin D-3. Pediat. Pol. 10, 1095-1100 (Pol)

87. Gadowska, M. and Borzymowska, J. (1977): Normal acid mucopolysaccharide excretion values in the urine of a random sample of healthy Polish children. Prezgl. Lek. 34, 401-405 (Pol)

88. Galambos, J. T. (1964): The effect of prednisone on glycoprotein sulfate metabolism in normal men. Metab. Clin. Exptl. 13, 437-444

89. Galletti, F. and Stancari, V. (1957): Studi sull'escrezione urinaria dei glicoproteidi. La reazione di donaggio in condizioni normali e patologiche. Clinica (Bologna) 16, 419

90. Gebala, A. and Gadamska, T. (1977): Studies on the effect of vitamin D_3 on the urinary excretion of glycosaminoglycans in healthy children aged 8-15 years. Pediatr. Pol. 52, 1119-11121 (Pol)

91. Gebala, A., Tarkowska, A., Gadamska, T. et al. (1973): Acid mucopolysaccharide excretion and thyroid function tests in girls with juvenile goiter. Pediatr. Pol. 48, 1101-1105 (Pol)

92. Germinale, T. and Sommo, G. (1968): Current etiopathogenetic views concerning urinary calculosis. Pathologica 59, 433-440 (Ita)

93. Goldberg, J. M. and Cotlier, E. (1972): Specific isolation and analysis of mucopolysaccharides (glycosaminoglycans) from human urine Clin. Chim. Acta 41, 19-27

94. Goto, M. and Murata, K. (1978): Urinary excretion of macromolecular acidic glycosaminoglycans in Werner's syndrome. Clin. Chim. Acta 85, 101-106

95. Greiling, H., Stuhlsatz, H. W., Cantz, M. and Gehler, J. (1978): Increased urinary excretion of keratan sulfate in fucosidosis. J. Clin. Chem. Clin. Biochem. 16, 329-334

96. Groover, R. V., Burke, E. C., Gordon, H. and Berden, W. F. (1972): The genetic mucopolysaccharidoses. Seminars Hematol. 9, 371-402

97. Gross, J. I., Mathews, M. D. and Dorfman, A. (1960): Sodium chondroitin sulfate-protein complexes of cartilage, II. Metabolism. J. Biol. Chem. 235, 2889-2892

98. Grove, W. J. Vermeulen, C. W., Goetz, R. and Ragins, H. D. (1950): Experimental urolithiasis. II. The influence of urine volume upon calculi experimentally produced upon foreign bodies. J. Urol. 64, 549-554

99. Guazzi, G. C., Ghetti, B., Barbieri, F. and Cecio, A. (1973): Myoclonus epilepsy with cherry-red spot in adult: A peculiar form of mucopolysaccharidosis. Acta Neurol. (Napoli) 28, 542-549

100. Hambraeus, L. and Holmgren, G. (1974): Urinary metabolic screening in children with psycho-neurological diseases. Evaluation of different screening tests. Uppsala J. Med. Sci. 15, 1-23

101. Hardy, K. H., Rosevear, J. W., Sams, W. M. and Winkelmann, R. K. (1971): Scleroderma and urinary excretion of acidic glycosaminoglycans. Mayo Clin. Proc. 6, 119-127

102. Hata, R. and Nagai, Y (1972): Rapid and micro method for separation of acidic glycosaminoglycans by two-dimensional electrophoresis. Anal. Biochem. 45, 462-468

103. Haust, M. D. (1974): The genetic mucopolysaccharidoses (GMS). Part II. Int. Rev. Exp. Pathol. 12, 251-314

104. Hayashi, S., Kimura, A., Nagai, H. and Tsurumi, K. (1978): Studies on urinary glycosaminoglycans and hydroxyproline from patients with myotonic dystrophy. Fukushima J. Med. Sci. 25, 37-44

105. Hesse, A., Hartmann, U., Schneider, H. J. and Horn, G. (1975): Mucopolysaccharide excretion in urinary calculus disease. Z. Urol. Nephrol. 68, 401-405 (Ger)

106. Hoag, G. N., Brown, R. G., Smart, M. E. et al. (1976): Alaskan malamute chondrodysplasia. II. Urinary excretion of hydroxyproline, uronic acid and acid mucopolysaccharides. Growth 40, 13-18

107. Hooghwinkel, G. J. M. (1974): Classification of sphingolipidoses and mucopolysacchridoses. In: J. M. Tager, G. J. M. Hooghwinkel and W. Th. Daems (Eds). Enzyme Therapy in Lysosomal Storage Diseases. North-Holland Publishing Company, Amsterdam, pp. 13-24

108. Hopwood, J. J. and Dorfman, A. (1978): Glycosaminoglycan synthesis by Wilm's tumor. Peditr. Res. 12, 52-56

109. Horwitz, A. L. (1979): The mucopolysaccharidoses: Clinical and biochemical correlations. Am. J. Ment. Defic. 84, 113-123

110. Humbel, R. (1974): Identification and quantitation of keratan sulfate in urine. Clin. Chim. Acta 52, 173-177

111. Humbel, R., Marchal, C. and Fall, M. (1969): Differential diagnosis of mucopolysaccharidosis by means of thin-layer chromatography of urinary acidic glycosaminoglycans. Helv. Paediat. Acta 24, 648-650

112. Hurst, R. E., Cezayirli, R. C. and Lorincz, A. E. (1975): Nature of the glycosaminoglycanuria (mucopolysacchariduria) in brachycephalic "snorter" dwarf cattle. J. Comp. Pathol. 85, 481-486

113. Hurst, R. E., Jennings, G. C. and Lorincz, A. E. (1977): Partition techniques for isolation and fractionation of urinary glycosaminoglycans. Anal. Biochem. 79, 502-512

114. Imhof, P. R. (1972): Importance of cast excretion following administration of diuretics. Praxis 61, 1059-1060 (Ger)

115. Infante, F. and Rallo, E. (1968): Les mucopolysaccharides urinaires dans l'epilepsie myoclonique progressive familiale (Types Lafora et Lundborg). In: H. Peeters, (Eds.) Protides of the Biological Fluids. Vol. 16. Elsevier, Amsterdam, pp. 611-618

116. Ishikawa, H. (1974): Glycosaminoglycans of the skin and urine in patients with progressive scleroderma. In: Y. Otaka (Ed.). Biochemistry and Pathology of Connective Tissue, Igaku Shoin Ltd., Tokyo, Chap. 7, pp. 97-109

117. Ishikawa, H., Suzuki, S., Horiuchi, R. and Sato, H. (1975): An approach to experimental scleroderma, using urinary glycosaminoglycans from patients with systemic scleroderma. Acta Dermatovener. (Stockholm) 55, 97-107

118. Ito, H. and Shimazaki, J. (1978): High molecular weight substances which inhibit calcium oxalate crystal growth. 4. Studies on the role of urinary pepsinogen and mucopolysaccharides as inhibitors of calcium oxalate crystal growth. Nippon Hinyokika Gakkai Zasshi 69, 1355-1359

119. Jablonski, K., Bauer, A., Karolak, A. and Dobryszycka, W. (1969): Urinary mucoprotein fractions and 17-ketosteroid in healthy women and in patients with the Stein-Leventhal syndrome. Archivum Immunologiae et therapiae experimentalis 17, 253-260

120. Jackson, C. E., Block, W. D. and Ratliff, W. C. (1960: Serum hexosamine content and urinary acid mucopolysaccharide excretion in hereditary primary amyloidosis. J. Lab. Clin. Med. 56, 544-546

121. Kaiser, G., Kystelka, R., Muhlbach, R. et al. (1969): The significance of the excretion of amino acids and acid mucopolysaccharides in the urine in orthopedic diseases. Beitr. Orthop. Trauma 16, 545-552

122. Kalandadse, N. I. (1970): Mechanism of the formation of renal calculi in the light of histochemical studies. Z. Urol. Nephrol. 63, 9-13

123. Kamikubo, K. (1965): Studies on acid mucopolysaccharides in urine. I. Urinary acid mucopolysaccharides in mongolism (Down's syndrome). Fukushima J. Med. Sci. 12, 89-101

124. Kamikubo, K. (1965 a): Studies on acid mucopolysaccharides in the urine. I. On urinary acid mucopolysaccharides in mongolism. Acta Paediat. Jap. 69, 904-909

125. Kantorovich, I. N., Kadyrkulova, M. T. and Fedorova, L. P. (1976): Urinary excretion of adrenaline, noradrenaline, and 17-hydroxycorticosteroids during individual adaptation to high-altitude. Tr. Kirg. Gos. Med. Inst. 110, 110-121

126. Kao, K. Y. T., Hizer, C. A. and McGavack, T. H. (1967): Urinary protein and carbohydrate. III. Age differences in acid mucopolysaccharides in human beings. Proc. Soc. Exp. Biol. Med. 125, 1040-1045

127. Kao, K.-Y. T. and Leslie, J. G. (1974): Micro fractionation and determination of urinary glycosaminoglycans. Biochem. Med. 9, 317-326

128. Kaplan, D. (1969): Classification of the mucopolysaccharidoses based on the pattern of mucopolysacchariduria. Am. J. Med. 47, 721-729

129. Kasanen, A., Nanto, V. and Lehtonen, A. (1964): Urinary acid mucopolysaccharides in renal patients with and without abuse of Phenacetin-containing analgesics. Ann. Med. Int. Fenn. 53, 33-38

130. Kastelan, M., Di Ferrante, N. and Lipscomb, H. S. (1970): Relation between urinary 17-hydroxycorticoids and glycosaminoglycans. Experientia 1970, 937-939

131. Katsunuma, H., Kobayashi, H., Sano, Y. et al. (1975): Connective tissue and aging. (4) Changes in urinary acid mucopolysaccharides and hydroxyproline metabolism with aging. Jpn. J. Geriatr. 12, 135-139 (Jap)

132. Kawata, H., Koizumi, T., Wada, R. and Yoshida, T. (1961): Urinary acid mucopolysaccharide excretion in liver damage. Gastroenterology 40, 507-512

133. Kennedy, J. F. (1976): Chemical and biochemical aspects of the glycosaminoglycans and proteoglycans in health and disease. Advan. Clin. Chem. 18, 1-101

134. Kerby, G. P. (1954): The excretion of glucuronic acid and acid mucopolysaccharides in normal human urine. J. Clin. Invest. 33, 1168-1174

135. Khafagy, E. X., Kolb, A., Garb, M. and Osman, H. G. (1974): Serum and urinary acid mucopolysaccharides in cretinism. Gaz. Egypt Paediatr. Assoc. 22, 163-168

136. Khafagy, E. Z., El-Gohary, A., Shalaby, F. Y. and Osman, G. (1975): Urinary excretion of acidic glycosaminoglycans in bilharziasis. Acta Med. Sci. Hung. 31, 25-29

137. Khafagy, E. Z., Osman, H. G., El-Raziky, E. H. et al. (1975 a): Significance of abnormalities in urinary neutral mucopolysaccharides in bilharziasis. Egypt J. Biharz. 2, 111-116

138. Kimura, A., Tsurumi, K. and Ogane, S. (1974): Urinary glycosaminoglycans. Simple method for the determination of urinary acidic glycosaminoglycans by electrophoresis on cellulose acetate paper. Application to normal and pathology urine samples. In: Y. Otaka (Ed.). Biochemistry and Pathology of Connective Tissue. Igaku Shoin, Tokyo, pp. 36-47

139. King, J. S. and Warnock, N. H. (1956): Some clinical conditions affecting urinary excretion of non-dialyzable hexosamine. Soc. Exptl. Biol. Med. 92, 369-371

140. Kiseleva, A. F. and Blagodarov, V. N. (1973): Significance of kidney mucopolysaccharides in the pathogenesis of urolithiasis. Vrach. Delo 7, 49-52 (Rus)

141. Kleine, T. O., Muske, J. and Timmers, H. W. (1972): Isolation of glycosaminoglycans from human urine using gel chromatography and CsCl-gradient centrifugation. Hoppe Seylers Z. Physiol. Chem. 353, 1536 (Ger)

142. Klujber, L. and Mehes, K. (1973): Urinary hydroxyproline and glycosaminoglycan excretion in Down's syndrome. Z. Kinderheilk. 114, 205-209

143. Klujber, L. and Sulyok, E. (1972): Urinary glycosaminoglycan excretion in normally grown and growth retarded neonates. Acta Paediat. (Budapest) 13, 81-87

144. Klujber, L. and Sulyok, E. (1973): Urinary glycosaminoglycan exretion in normally growing and growth retarded neonates. II. Quantitative distribution of glycosaminoglycans. Acta Paediat. (Budapest) 14, 209-220

145. Klyukina, S. S., Fokeeva, V. V. and Matveev, M. P. (1968): Mucoproteins in children with a nephrotic form of diffuse glomerulonephritis. Pediatriia 47, 1-18 (Rus)

146. Kochar, D. M. and Larsson, K. S. (1977): Alterations in the metabolism of glycosaminoglycans and collagen. In: J. Wilson and F. C. Fraser (Eds). Handbook of Teratology Vol. 2, Plenum Press, New York, pp. 231-269

147. Kodama, H. and Masuda, T. (1977): Xanthoma formation by intracutaneous dextran sulfate injection and the role of acid mucopolysaccharides in xanthomatogenesis, Nippon Hifuka Gakki Zasshi 87, 299-303

148. Koizumi, T., Mukuda, T., Nakamura, N. and Abe, H. (1969): Urinary and tissue acid mucopolysaccharides in rat with connective tissue proliferation. Clin. Chim. Acta 26, 477-480

149. Kollberg, H., Lundblad, A. and Ekbohm, G. (1973): Studies in cystic fibrosis. Urinary excretion of hexosamine, sialic acid and fucose. Acta Paediat. Scand. 62, 279-288

150. Kosmider, S., Ludyga, K., Misiewicz, A., Drozdz, M. and Sagan, J. (1972): Experimental and clinical investigations on the emphysemaforming action of nitrogen oxides. Zentralbl. Arbeitsmed. Arbeitsschutz 22, 362-368

151. Koyanagi, T., Nakahara, K., Takanohashi, T., Oikawa, K. and Akazawa, N. (1974): Effects of administration of vitamins and methionine on the dark adaptation, urinary sulfate, and creatinine in adults and children. Eiyo To Shokuryo 27, 211-219

152. Kreysel, H. W., Kraus, H., Stuhlsatz, H. W., Greiling, E., Nissen, H. P. and Kimmig, J. (1977): Quantitative isolation (column chromatographic separation) of acid mucopolysaccharides from urine in progressive scleroderma. Arch. Dermatol. Res. 259, 11-19 (Ger)

153. Ksiezny, S. (1978): Determination of the content and composition of glycosaminoglycans (acid mucopolysaccharides) excreted with the urine. Reumatologia 16 (2), 273-280

154. Kuarstein, B., Nordiso, H. and Shult-Haudt (1963): An effect of estrogen on chondroitin sulfate of rat skin connective tissue. Arch. Endocr. 44, 209-215

155. Kucharz, E. and Drozdz, M. (1976): Serum glycoprotein level and urinary glycosaminoglycan excretion in drug induced collagen-like syndrome in guinea pigs. Acta Biol. Med. Ger. 35, 1541-1545

156. Kucinskiene, Z. and D'yachkova, A. Ya. (1976): Study of glycosaminoglycan metabolism in the urine of older children with chronic pneumonia and their nearest relatives. Pediatriya (Moscow) 5, 31-33 (Rus)

157. Kuhl, W. and Schroder, K. D. (1971): Importance of mucopolysaccharide determination in urine in eye diseases. Z. Aerztl. Fortbild. (Jena) 65, 155-157 (Ger)

158. Kupchella, C. E. and Tamburro, C. H. (1978): Urinary and tissue glycosaminoglycan patterns in hepatic angiosarcoma. In: H. E. Nieburgs (Ed.), Detection and Prevention of Cancer, Marcel Dekker, Inc., New York, Part 1, Vol. 1, pp. 915-926

159. Kusuhara, T. (1978): Chemical patterns of urinary glycopeptide and glycosaminoglycans elicited during growth of infants and children. Kurume Igakkai Zasshi 41, 233-247 (Japan)

160. Kyselka, R. and Schramm, G. (1968): Mucopolysaccharide studies in enchondral dysostoses. Beitr. Orthop. Trauma 15, 665-667 (Ger)

161. Kyselka, R., Schramm, G. and Beck, O. (1969): Quantitative and qualitative changes of acid mucopolysaccharides in urine in diseases of the skeletal system. Beitr. Orthop. Trauma 16, 552-558 (Ger)

162. Lakatos, M. and Di Ferrante, N. (1974): Simple measurement of urinary glycosamoniglycans and degradation products. Biochem. Med. 9, 256-260

163. Lamberg, S. I. and Dorfman, A. (1973): Synthesis and degradation of hyaluronic acid in the cultured fibroblasts of Marfan's syndrome. J. Clin. Invest. 52, 2428-2433

164. Lambotte, C. and Winand, R. (1973): Treatment of mucopolysaccharidoses with perfusions of plasma or leukocytes. Acta Univ. Carol. (Med. Monogr.) (Praha) 56, 191-195 (Fre)

165. Lambotte, C., Winand, R., Chantraine, J. M., Bierthog, G. and Leruth-Fafchamps, Y. (1971): Treatment of mucopolysaccharidoses with human plasma perfusions. Acta Paediatr. Belgica 25, 297-309

166. Leaback, D. H. (1970): The metabolic hydrolysis of hexosaminide linkages. In: W. H. Fishman (Ed.). Metabolic Conjugation and Metabolic Hydrolysis. Vol. 2, Academic Press, New York, pp. 443-517

167. Ledger, W. J., Thompson, G. R. and Vries, L. De (1974): The influence of infection in obstetric and gynecologic patients upon increased urinary mucopolysaccharide excretion. Am. J. Obstet. Gynecol. 120, 407-410

168. Lehtonen, A., Karkkainen, J. and Haahti, E. (1966): Gas-chromatographic characterization of the elctrophoretically separated fractions of acid mucopolysaccharides. J. Chromatogr. 24, 179-182

169. Lehtonen, A., Nanto, V. and Kasanen, A. (1967): The effects of some anti-inflammatory agents on the urinary excretion of acid mucopolysaccharides in the rat. Ann. Med. Exp. Biol. Fenn 45, 32-34

170. Lewis, P. W., Kennedy, J. F. and Raine, D. N. (1975): The laboratory diagnosis of the mucopolysaccharidoses. I: J. B. Holton and J. T. Ireland (Eds.). Inborn Errors of Skin, Hair and Connective Tissue. University Park Press,Baltimore, pp. 229-244

171. Linker, A., Evans, L. R. and Madsen, J. A. (1969): Problems in the analysis of urinary mucopolysaccharide excretion. Biochem. Med. 2, 448-456

172. Lipinska-Piotrowska, I. (1965): Mucopolysaccharides and glucocorticoids in infants with weight deficit. Pediat. Pol. 40, 1231-1237

173. Lipinska-Piotrowska, I., Deloff, M. and Mikulowska, H. (1968): Urinary excretion of acid mucopolysaccharides and creatinine clearance in children with measles. Wiad. Lek. 21, 1717-1720 (Pol)

174. Loewi, G. (1959): Urinary excretion of acid polysaccharide in rheumatoid arthritis and other diseases. Ann. Rheumatic Diseases 18, 239-243

175. Lorincz, A. E. (1960): Urinary acid mucopolysaccharides in hereditary deforming chondrodysplasia (diaphysial aclasis). Fed. Proc. 19, 148

176. Lorincz, A. E. (1960 a): "Snorter" dwarf cattle: a naturally occurring heritable disorder of acid mucopolysaccharide metabolism which resembles the Hurler syndrome. A. M. A. J. Diseases Children 100, 488-489

177. Lorincz, A. E. (1961): Heritable disorders of acid mucopolysaccharide metabolism in humans and in snorter dwarf cattle. Ann. N.Y. Acad. Sci. 91, 644-658

178. Lorincz, A. E. (1962): Mucopolysacchariduria in children with hereditary arthro-osteo-onychondysplasia. Fed. Proc. 21, 173

179. Maas, W., Klempien, E. J. and Utermann, D. (1968): Studies on values and composition of the mucopolysaccharide fraction in urine of healthy persons and of those with arteriosclerosis. Klin. Wschr. 44, 1058-1062 (Ger)

180. Maclagan, N. F. and Anderson, A. J. (1958): Colloidal properties of urinary mucopolysaccharides. In: G. W. E. Wolstenholme and M. O'Connor (Eds.). Ciba Found. Symp. on the Chemistry and Biology of Mucopolysaccharides. Little Brown and Company, Boston, pp. 268-281

181. Madsen, J. A. and Linker, A. (1969): Vitamin A and mucopolysaccharidosis: A clinical and biochemical evaluation. Pediatrics 75, 843-852

182. Mares, V. and Vulcan, R. (1965): Contribution to the study of mucopolysacchrides in renal tuberculosis. Ann. Histochim. 10, 53-58 (Fr)

183. Maroteaux, P. (1973): Un nouveau type de mucopolysaccharidose avec athetose et elimination urinaire de keratan-sulfate. Nouv. Presse Med. 2, 975-979

184. Masamune, H., Hakamori, S. and Sugo, T. (1959): Urinary mucopolypeptides and mucoproteins in cancer. II. An acid mucopolypeptide characteristic of cancer. Tohoku J. Exptl. Med. 69, 371-381

185. Masuda, H., Shichijo, S., Imai, H., Takeuchi, M., Shetlar, M. R. (1978): Urinary glycopeptides and glycosaminoglycans in children and adults as affected by age. Texas Rep. Biol. Med. 36, 151-156

186. Masuda, H., Shichijo, S. and Takeuchi, M. (1975): Glycosaminoglycan from the urine of spondyloepiphyseal dysplasia congenita. Clin. Chim. Acta 65, 149-151

187. Masuda, H., Shichijo, S. and Takeuchi, M. (1975 a): Fractionation of urinary glycopeptide by DEAE Sephadex A-50 column chromatography. Kurume Med. J. 21, 111-115

188. Masuda, H., Shichijo, S. and Takeuchi, M. (1976): Study on the urinary glycosaminoglycan of spondyloepiphyseal dysplasia congenita. Jpn. J. Clin. Pathol. 24, 65-67 (Jap)

189. Masuda, H., Takeuchi, M., Kuinitake, A. and Anraku, S. (1975 b): Urinary glycopeptides in myoclonus epilepsy (Lafora-type epilepsy). Rinsho Byori 23, 555-558 (Jap)

190. Matalon, R. and Dorfman, A. (1968): The accumulation of hyaluronic acid in cultured fibroblasts of the Marfan syndrome. Biochem. Biophys. Res. Commun. 32, 150-154

191. Matalon, R., Dorfman, A., Dawson, G. and Sweeley, C. C. (1969): Glycolipid and mucopolysaccharide abnormality in fibroblasts of Fabry's disease. Science 164, 1522-1533

192. McGale, E. H. (1969): Protein-carbohydrate compounds in human urine Advances Carbohyd. Chem. 24, 435-452

193. McKusick, V. (1969): The nosology of the mucopolysaccharidoses. Am. J. Med. 47, 730-747

194. McKusick, V. A., Neufeld, E. F. and Kelly, T. E. (1978): The mucopolysaccharide storage diseases. In: J. B. Stanbury, J. B. Wyngaarden and D. S. Fredrickson (Eds.). The Metabolic Basis of Inherited Disease. 4th Ed., McGraw-Hill, New York, Chap. 53, pp. 1282-1307

195. Meaney, M. F. (1974): Increased accuracy and precision in screening for mucopolysaccharides. N. Z. J. Med. Lab. Technol. 28, 29-34

196. Medved, A., Petersen, W. C. and Johnson, R. V. (1967): Cutaneous findings in Unverricht's syndrome (progressive familial myoclonic epilepsy). Arch. Derm. 95, 206-209

197. Mellsop, G. W., Koadlow, L., Syme, J. and Whittingham, S. (1974): Absence of rheumatoid arthritis in schizophrenia. Aust. N. Z. J. Med. 4, 247-252

198. Merkur'eva, R. V. (1972): Comparative study of urinary excretion of glycosaminoglycans and glycopeptides in bone diseases. Ortop. Travmatol. Protez. 33, 86-88 (Rus)

199. Merkur'eva, R. V. and Bukhtoyarova, F. G. (1976): Glycosaminoglycans in the blood serum of patients with rheumatoid arthritis. Vopr. Med. Khim. 22, 291-296 (Rus)

200. Mier, P. D. and Urselmann, E. (1970): Urinary acid mucopolysaccharide excretion in skin diseases. Dermatologica 141, 34-37

201. Mikolajczyk, H. (1967): Mucopolysaccharide levels in the urine of growing rats treated with histamine, 5-hydroxytryptamine or heparin. Endokr. Pol. 18, 197-204 (Pol)

202. Mikolajczyk, H. (1967 a): Changes in atmospheric pressure and mucopolysaccharide levels in the urine of hypophysectomized-gonadectomized rats treated with ACTH, FSH, ACTH + FSH or acetylcholine. Endokr. Pol. 18, 205-13 (Pol)

203. Mikolajczyk, H., Allalouf, D. and Ber, A. (1968): The effect of simulated altitude (500 mm Hg) on urine acid mucopolysaccharides excretion in normal rats. Int. J. Biometeor. 12, 283-287

204. Mikolajczyk, H., Allalouf, D. and Ber, A. (1968 a): Effect of hypophysectomy on the excretion of acid mucopolysaccharides in rat urine. Endocrinology 83, 861-863

205. Mikolajczyk, H. and Izycki, J. (1974): Urinary excretion of glycosaminoglycans in workers exposed to vibrations. Med. Pr. 25, 555-561

206. Mogorian, P. D. (1968): The role of the tubular apparatus of the kidneys in the excretion of acid mucopolysaccharides in children with acute nephritis. Pediatriia 47, 18-22 (Rus)

207. Mogorian, P. D. (1970): Urinary acid mucopolysaccharides in kidney diseases in children. Pediatriia 49, 23-26 (Rus)

208. Mohanram, M. and Reddy, V. (1971): Urinary excretion of acid mucopolysaccharides in kwashiorkor and vitamin A deficient children. Clin. Chim. Acta 34, 93-96

209. Morrison, R. I. G. (1974): The enzymatic degradation of proteoglycans. In: R. Fricke and F. Hartman (Eds.). Connective Tissues: Biochemistry and Pathophysiology, Springer Verlag, New York, pp. 150-157

210. Morse, B. S. and Nussbaum, M. (1967): The detection of hyaluronic acid in the serum and urine of a patient with nephroblastoma. Am. J. Med. 42, 996-1002

211. Mourao, P. A. S., Toledo, S. P. A., Nadler, H. B. and Dietrich, C. P. (1973): Excretion of chondroitin sulfate C with low sulfate content by patients with generalized platyspondyly (Brachyolmia) Biochem. Med. 7, 415-423

212. Muhlbach, R., Schmidt, U. J., Bruschke, G. et al. (1970): Renal hydroxyproline, acid mucopolysaccharide and pyrophosphate elimination in senile osteoporosis and secondary forms of osteoporosis. Dtsch. Gesundheitsw. 25, 2154 (Ger)

213. Muhlbach, R., Trzenschik, K., Brueschke, G., Schmidt, U. J. and Stecher, K. (1971): Effect of parathormone on bone metabolism and the significance of laboratory diagnostic parameter for the detection of parathyroid gland hyperfunction. Deut. Gesundheitsw. 26, 1587-1593 (Ger)

214. Muir, H. (1969): The structure and metabolism of mucopolysaccharides (glycosaminoglycans) and the problem of mucopolysaccharidoses. Am. J. Med. 47, 673-690

215. Muir, H. (1973): Structure and enzymic degradation of mucopolysaccharides. In: H. G. Hers and F. Van Hoof (Eds.). Lysosomes and Storage Diseases, Academic Press, New York, pp. 79-104

216. Murata, K. (1974): Acidic glycosaminoglycans in normal and connective tissue diseased states; with special reference to enzymatic studies of chondroitin sulfate isomers at disaccharide subunits. In: Y. Otaka (Ed.). Biochemistry and Pathology of Connective Tissue. Igaku Shoin, Tokyo, Chap. 5, pp. 48-84

217. Murata, K., Ishikawa, T. and Oshima, Y. (1970): Enzymatic studies of urinary chondroitin sulfates in normal and systemic connective tissue disease states. Clin. Chim. Acta 28, 213-222

218. Murata, K., Ukiyama, Y. and Horiuchi, Y. (1973): Metabolic changes of urinary glycosaminoglycans in Weber Christian disease. Clin. Chim. Acta 49, 129-139

219. Nader, H. B., Mourao, P. A. S., Toledo, S. P. A. and Dietrich, C. P. (1974): Mucopolysaccharide degradation and excretion after hyaluronidase injection in patients with Hunter's and Hurler's syndromes. Clin. Chim. Acta 50, 245-255

220. Nagai, H., Usui, T., Horie, K. and Kawashima, S. (1960): Congenital nonspecific defense factors in vivo. II. Behavior of congenital nonspecific defense factors in allergic reactions, particularly in anaphylactic shock. Ann. Paediat. Japan. 6, 622-637

221. Nagata, T., Sakuma, T., Yamauchi, A., Nagai, H., Nagasawa, S., Haginoya, K., Sue, A., Ogawa, S., Yamagata, Y. and Fukuchi, S. (1972): Nippon Jinzo Gakkaishi 3, 147-159

222. Naidoo, S. S. (1971): Excretion of urinary glycosaminoglycans during the menstrual cycle in the baboon. Proc. Soc. Exp. Biol. Med. 136, 1014-1019

223. Naidoo, S. S. and Qureshi, A. M. (1974): Origin of glycosaminoglycans in baboon urine during the menstrual cycle. Can. J. Physiol. Pharmacol. 52, 206-214

224. Nakamura, M., Arakawa, K., Ishihara, Y. et al. (1966): Effects of exogenous proteinase (pronase) on experimental arteriosclerosis and blood pressure. Jap. Circ. J. 30, 1645-1650

225. Namiki, O., Tanabe, S. and Morisaki, N. (1974): Urinary acid mucopolysaccharides and glycoproteins in pycnodystosis. Kesugo Soshiki, 6, 91-101 (Jap)

226. Neufeld, E. F. (1972): Mucopolysaccharidoses. In: A. Dorfman (Ed.). Antenatal Diagnosis, University of Chicago Press, Chicago, pp. 217-225

227. Nikitina, T. D., Balanchuk, V. K., Golovina, A. A., Khal'zov, V. L. and Gudimova, N. S. (1966): Changes in mucopolysaccharide complex in third and fourth degree burns under the influence of some antishock agents. In: A. A. Tustanovskii (Ed.). Soedin. Tkan. Norme Patol., Mater. Soveshich. Novosibirsk, USSR, pp. 209-211

228. O'Brien, J. F., Gerritsen, T. and Helmuth, A. C. (1973): Identification of acid mucopolysaccharides by gas chromatographic analysis of component sugars after chemical depolymerization. Anal. Biochem. 65, 465-479

229. Ohkawa, S.-I., Hata, R., Nagai, Y. and Sugiura, M. (1972): Urinary excretion of acidic glycosaminoglycans in the aged. J. Biochem. (Tokyo) 72, 1495-1501

230. Ohkawa, S. I., Sugiura, M., Hata, R. and Nagai, Y (1977): Acidic glycosaminoglycans in urine, serum and myocardium of aged patients with myocardial infarction. J. Mol. Cell. Cardiol. 9, 541-550

231. Ohlenschlaeger, K. and Friman, C. (1968): A normal urinary excretion of acid mucopolysaccharides in generalized scleroderma. Scand. J. Clin. Lab. Invest. 21, 364-368

232. Onisawa, J. and Lee, T.-Y. (1970): Increased urinary excretion of chondroitin sulfate A and C in Hunter's syndrome. Biochim. Biophys. Acta 208, 144-147

233. Orii, T. (1968): The urinary acid mucopolysaccharides in normal male children. Hoppe-Seyler's Z. Physiol. Chem. 349, 816-822

234. Orii, T. (1968 a): Nonsulfated acid mucopolysaccharide excretion in normal male children and adults. Biochim. Biophys. Acta 170, 204-206

235. Orii, T. (1970): Urinary acid mucopolysaccharide-peptide complexes in children. 1. Urinary acid mucopolysaccharide-peptide complexes in normal children. Acta Paediat. Jap. 74, 127-136 (Jap)

236. Orii, T., Chiba, H., Minami, R. et al. (1974): Hyperglycopeptiduria in genetic mucolipidoses. Tohoku J. Exp. Med. 112, 373-380

237. Orii, T., Minami, R. and Nakao, T. (1972): Pattern of urinary acid mucopolysaccharide excretion in normal male children. Tohoku J. Exp. Med. 108, 369-375

238. Palo, J. And Savolainen, H. (1972): Serum and urinary glycopeptides and glycosaminoglycans in aspartylglucosaminuria. Clin. Chim. Acta 36, 431-437

239. Pennock, C. A. (1969): A modified screening test for glycosaminoglycan excretion. J. Clin. Path. 22, 379-380

240. Pennock, C. A., Wharton, B. A. and White, F. (1971): Urinary glycosaminoglycan excretion in the neonatal period. Acta Paediatr. Scand. 60, 299-300

241. Pennock, C. A., White, F., Murphy, D. et al. (1973): Excess of glycosaminoglycan excretion in infancy and childhood. Acta Paediatr. Scand. 62, 481-491

242. Perry, M. A., Powell, G. M., Wusteman, K. S. and Curtis, C. G. (1977): The catabolism of intravenously injected heparan N-[^{35}S] sulfate in the rat. Biochem. J. 166, 373-379

243. Philippart, M. (1972): Glycolipid, mucopolysaccharide and carbohydrate distribution in tissues, plasma and urine from glycolipidoses and other disorders. Complex nature of the accumulated substances. Advan. Exp. Med. Biol. 25, 231-254

244. Piper, H. G., Hadlich, J. and Wurbach, G. (1967): On the identification of, in urticaria pigmentosa, excreted acid mucopolysaccharides using infrared spectroscopy. Arch. Klin. Exp. Derm. 228, 332-336 (Ger)

245. Poulsen, J. H. (1975): The excretion patterns of urinary glycosaminoglycans in a family with progressive familial myoclonus epilepsy. Acta Neurol. Scand. 52, 216-230

246. Raine, D. N. (1975): Treatment of other inherited metabolic disorders. In: D. N. Raine (Ed.) The Treatment of Inherited Metabolic Disease. Medical and Technical Publ. Co., Lancaster, pp. 263-280

247. Rallo, E. (1969): Familial progressive myoclonic epilepsy (Unverricht-Lundborg syndrome in five Valasian families). A contribution to the research of its metabolic origin through the study of urinary excretion of mucopolysaccharides. Acta Genet. Med. Gemellol (Roma) 18, 231-270 (Fre)

248 Rallo, E., Martin, F., Beaumonoir, A. and Klein, D. (1968): Progressive malignant myoclonic epilepsy (Lafora's disease). Clinical study of a case. Acta Neurol. Belg. 68, 356-359 (Fre)

249. Rampini, S. and Clausen, J. (1967): Farber's disease (disseminated lipogranulomatosis). Clinical picture and summary of the chemical findings. Helv. Paediat. Acta 22, 500-515

250. Rampini, S., Isler, W., Baerlocher, K., Bishoff, A., Ulrich, J. and Pluss, H. (1970): Die Kombination von metachromatischer Leukocystrophie und Mukopolysaccharidose als selbstständiges Krankheitsbild (Mukosulfatidose). Helv. Pediat. Acta 25, 436-461 (Ger)

251. Rastogi, S. C., Clausen, J., Melchior, J. C., Dyggve, H. V. and Jensen, G. E. (1977): Lysosomal (Leucocyte) proteinase and sulfatase levels in Dyggve-Melchior-Clausen (DMC) syndrome. Acta Neurol. Scandinav. 56, 389-396

252. Revell, P. A. and Muir, H. (1972): The excretion and degradation of chondroitin 4-sulfate administered to guinea pigs as free chondroitin sulfate and as proteoglycans. Biochem. J. 130, 597-606

253. Rezvani, I., Collip, P. J., DiGeorge, A. M. (1973): Evaluation of screening tests for urinary mucopolysaccharides. Pediatrics 52, 80-84

254. Rich, C., Di Ferrante, N. and Archibald, R. M. (1957): Acid mucopolysaccharides excretion in the urine of children. J. Lab. Clin. Med. 50, 686-691

255. Rich, C. and Myers, W. P. L. (1959): Excretion of acid mucopolysaccharides in the urine of patients with malignant neoplastic diseases. J. Lab. Clin. Med. 54, 223-228

256. Rigillo, N., Torelli, M. and Caringella, A. (1969): Variations of the urinary excretion of mucopolysaccharides in normal humans, following physical exercise. Boll. Soc. Ital. Biol. Sper. 45, 1321-1322

257. Robertson, W. V. B. and Harvey, J. (1972): Determination of galactose in urinary acidic glycosaminoglycans as a measure of keratan sulfate-like substance in urine. Biochem. Med. 6. 246-256

258. Romano, C., Tortorolo, G. and Segni, G. (1966): Mucopolysaccharides in urine in the pediatric age. I. The turbidimetric test according to the Dorfman method as a screening test. Minerva Pediat. 18, 153-159

259. Rosenfeld, L. (1971): Urinary excretion of constituents of acid mucopolysaccharides and other carbohydrate polymers in man following surgery. Surgery 70, 373-384

260. Rosenfeld, L., Neuhof, H. and Mestel, A. L. (1961): Determination of heparinoid substances in urine with a dye-binding technique. Proc. Soc. Exptl. Biol. Med. 107, 317-321

261. Rufini, S., Ghebregzabher, M., Castellucci, G. and Lato, M. (1979): A new screening test for the diagnosis of mucopolysaccharidoses. Clin. Chim. Acta 95, 443-446

262. Rupniewska, Z. M., Widomska-Czekajska, T., Miturzynska-Stryjecka, H. and Bielak, Z. (1968): Urinary excretion of acid mucopolysaccharides and total hydroxyproline in myocardial infarctions. Pol. Tyg. Lek. 23, 2001-2002 (Pol)

263. Rupniewska, Z. M., Bielak, Z., Widomska-Czekajska, T. and Miturzynska-Stryjecka, H. (1968 a): Urinary 24-hour excretion of glucuronic acid in coronary disease and its correlation with the excretion of acid mucopolysaccharides. Pol. Tyg. Lek. 23, 1919-1922 (Pol)

264. Rupniewska, Z. M., Widomska-Czekajska, T., Miturzynska-Stryjecka, H. and Bielak, Z. (1969): Urinary elimination of acid mucopolysaccharides and total hydroxyproline in myocardial infarction. Pol. Med. J. 8, 1064-1068

265. Sallis, J. D. and Lumley, M. F. (1979): On the possible role of glycosaminoglycans as natural inhibitors of calcium oxalate. Invest. Urol. 16, 296-299

266. Samuels, S., Fisher, C., Micca, P. S. and Veliz, G. (1974): Urinary glycosaminoglycans and the direct quantitation of irregular spots on thin-layer chomatograms. J. Chromatogr. 92, 461-464

267. Sasaki, S. (1975): Urinary acidic glycosaminoglycans in patients with systemic scleroderma and its changes by clinical improvement treated with a few drugs. Rinsho Kagaku Shimpojumu 15, 25-29 (Jap)

268. Scatena, P., Ranno, O. and Giordano, P. (1977): Mucopolysaccharides in urine in the newborn. Minerva Pediatr. 29, 779-781 (Ita)

269. Schoenheyder, F., Zimmermann-Nielsen, C. and Anderson, H. C. (1966): Urinary excretion of amino acids and hexosamine in otosclerotic patients. Arch. Otolaryng. 84, 495-496

270. Scott, J. E. and Newton, D. J. (1975): The recovery and characterization of acid glycosaminoglycans in normal human urine. Influence of a circadian rhythm. Connect. Tissue Res. 3, 157-164

271. Shamkhalova, V. G., Ananenko, A. A. and Vel'tischev, Yu. E. (1979): Importance of certain biochemical indexes in a complex evaluation of the physical development of children of different age groups. Fiziol. Che. 5, 249-253 (Rus)

272. Shetlar, M. R., Bradford, R. H., Joel, W. and Howard, R. P. (1961): Effects of parathyroid extract on glycoprotein and mucopolysaccharide components of serum and tissue. In: R. O. Greep and R. V. Talmage (Eds.). The Parathyroids. Charles C. Thomas, Springfield, pp. 144-155

273. Shetlar, M. R., Dobrkovsky, M., Linares, H., Villarante, R., Shetlar, C. L. and Larson, D. L. (1971): Hypertrophic scar. Glycoprotein and collagen components of burn scars. Proc. Soc. Exp. Biol. Med. 138, 298-300

274. Shetlar, M. R., Van Lier, P., Hern, D. L. and Lewis, S. R. (1969): Uronic acid-containing polypeptides from the urine of burned patients. Proc. Soc. Exp. Biol. Med. 132, 999-1002

275. Shichijo, S., Masuda, H., Takeuchi, M., Imai, H. and Yamashita, F. (1978): Excretion patterns of glycosaminoglycans in the urine of normal children and adults. The Kurume Med. J. 25, 11-16

276. Shiina, Y., Kera, K. and Hatakeyama, R. (1979): Morphological and cytochemical characteristics of the inclusion bodies excreted in urine of patients with influenza. Eisei Kensa 28, 126-134

277. Slack, H. G. B. (1958): Connective tissue growth stimulated by carrageenin. 3. The nature and amount of polysaccharide produced in normal and scorbutic guinea pigs and metabolism of a chondroitin sulfate fraction. Biochem. J. 69, 125-134

278. Smith, W. S. and Kerby, G. P. (1960): Urinary excretion of acid mucopolysaccharides by rabbits injected with chelating agents and with chondroitinsulfate. Proc. Soc. Exptl. Biol. Med. 103, 562-566

279. Soifer, N., Waltner, J. G., Endahl, G. L. and Holdsworth, C. E. (1970): Biochemical studies of otosclerosis: urinary acid mucopolysaccharides. Arch. Klin. Exp. Ohren Nasen Kehlkopf Heilkd. 197, 1-6

280. Spiro, R. G. (1973): Biochemistry of the renal glomerular basement membrane and its alterations in diabetes mellitus. New Eng. J. Med. 288, 1337-1342

281. Spranger, J. W. (1972): The systemic mucopolysaccharidoses. Ergeb. Med. Kinderheilk. 32, 165-265

282. Spranger, J., Todt, H. and Wiedemann, H. R. (1967): Comparison of urine mucopolysaccharides in children and adults. Clin. Chim. Acta 17, 142-146

283. Spranger, J. and Wiedemann, H. R. (1966): Studies on mucopolysaccharide excretion in healthy children and in those with dysostosis. Ann. Paediat. (Basel) 206, 342-355 (Ger)

284. Stempien, R., Schmidt, E. and Tomaszewska, L. (1966): Studies on urinary excretion of acid mucopolysaccharides in the course of virus hepatitis and their clinical significance. Pol. Tyg. Lek. 21, 1686-1689 (Pol)

285. Stern, I. J., Cosmas, F. and Smith, L. (1968): Urinary polyuronide excretion in man after enzymic dissolution of the chondromucoprotein of the intervertebral disk or surgical stress. Clin. Chim. Acta 21, 181-190

286. Stidworthy, G., Masters, Y. E. and Shetlar, M. R. (1958): The effect of aging on the mucopolysaccharide composition. J. Gerontol. 13, 10-13

287. Stinikov, V. F. and Temin, P. A. (1979): State of contractile tissue in progressive muscular distrophy. Zh. Nevropathol. Psikhiatr. im. S. S. Korsakova 79, 272-276

288. Stone, A. L., Constantopoulos, G., Sotsky, S. M. and Dekaban, A. (1970): Optical rotatory dispersion of mucopolysaccharides. IV. Optical rotatory dispersion and circular dichroism of glycosaminoglycans and heparan sulfate fractions from the urine of patients with mucopolysaccharidosis (Hurler syndrome). Biochim. Biophys. Acta 222, 79-89

289. Suschke, H. J. and Kunze, D. (1971): Zur Behandlung von Durchfallerkrankungen bei Säuglingen and Kleinkindern mit einem neuen Darmtherapeutikum. Med. Monatsschr. 25, 240-241

290. Suschke, J. and Murken, J.-D. (1969): Über die Ausscheidung saurer Mucopolysaccharide bei einem Kind mit Lowe-Syndrom. Monatsschr. Kinderheilk. 117, 412-414

291. Suzuki, S. (1962): Introduction to the chemistry and biochemistry of mucopolysaccharides. In: F. Egami and Y. Oshima (Eds.). Biochemistry and Medicine of Mucopolysaccharides. Maruzen, Tokyo, pp. 1-11

292. Tanaka, Y., Takazono, I., Ohashi, K. and Tanaka, M. (1979): Urinary glycosaminoglycans in premature and fullterm neonates. Kurume Med. J. 26, 35-44

293. Tanaka, Y., Takazono, I., Yasuoka, C., Iwatani, E. and Gore, I. (1975): The pattern of urinary chondroitin sulfate and chondroitin excretion with age. Kurume Med. J. 22, 153-157

294. Taniguchi, N. (1970): Thin-layer gel filtration of urinary acid mucopolysaccharides in Hunter's syndrome and in normal individuals. Clin. Chim. Acta 30, 801-808

295. Taniguchi, N. (1972): Age differences in the pattern of urinary glycosaminoglycan excretion in normal individuals. Clin. Chim. Acta 37, 225-233

296. Taniguchi, N. (1973): Glycosaminoglycan and infants. Ryoiki Zokan 99, 181-207 (Jap)

297. Taniguchi, N., Koizumi, S., Masaki, K. and Kobayashi, Y. (1975): Diagnosis of genetic mucopolysaccharidoses: Electrophoretic and enzymatic characterization of urinary glycosaminoglycans. Biochem. Med. 14, 241-249

298. Taniguchi, N., Miyawaki, T., Moriya, N. and Okuda, N. (1975 a): Molecular size difference of urinary heparan sulfates from normal individuals and genetic mucopolysaccharidoses. Clin. Chim. Acta 60, 363-369

299. Tebenchuk, G. M. (1971): Mucopolysaccharides in urine in acute respiratory infections in small children. Pediatr. Akush. Ginekol. 2, 22-24 (Ukr)

300. Teller, W. M. (1967): Urinary excretion patterns of individual acid mucopolysaccharides. Nature 213, 1132-1133

301. Teller, W. and Krueger, Ch. (1967): Age-dependent changes of acid mucopolysaccharide excretion patterns in human urine. Experientia 23, 908-909

302. Teller, W. M. and Krueger, C. (1970): Fractionation of acid glycosaminoglycans in the urine of normal people of different ages. Klin. Wochenschr. 48, 688-693

303. Teller, W. M. and Mayer, P. (1973): Excretion of acid glycosaminoglycans in the urine of adults and children with second and third degree burns. Helv. Paediatr. Acta 28, 373-382 (Ger)

304. Teller, W. M. and Muller, S. (1974): Age-dependent changes of glycosaminoglycans precipitable from human urine at acid and alkaline pH. Klin. Wochenschr. 52, 1165-1172

305. Teller, W. M. and Ziemann, A. (1969): Thin-layer chromatography of urinary acid glycosaminoglycans as a screening procedure for mucopolysaccharides. Horm. Metab. Res. 1, 32-35

306. Thompson, G. R., Castor, C. W., Ballantyne, L. M. and Prince, R. K. (1966): The excretion of nondialyzable urinary mucopolysaccharide in rheumatic and other systemic disease states. J. Lab. Clin. Med. 68, 617-627

307. Thompson, G. R., De Vries, L. and Ledger, W. J. (1974): Effect of pregnancy upon urinary mucopolysaccharide excretion. Am. J. Obstet. Gynecol. 119, 1069-1072

308. Tira, M. E., Calatroni, A., Balduini, C., Torri, G., Moretti, R. and Casu, B. (1979): Structure of urinary glycosaminoglycans. Chemical and protein NMR characterization of heparan sulfates and dermatan sulfates from mucopolysaccharidoses type II, III, VI. In: B. Berra et al. (Eds.). Perspectives in Inherited Metabolic Diseases 2 (Glycoproteins Glycosaminoglycans Health Dis.) Editions Ermes, Milan, pp. 165-185

309. Tokunaga, M., Wakamatsu, E. and Yosizawa, Z. (1978): Increased excretion of urinary glycosaminoglycans in a case of Rothmund Thomson Syndrome. Tohoku J. Exp.Med. 126, 1-6

310. Torelli, M. (1967): Research on urinary acid mucopolysaccharides in osteogenesis imperfecta. Clin. Pediat. (Bologna) 49, 310-317 (Ita)

311. Tortorolo, G., Romano, C. and Segni, G. (1965): Mucopolysaccharide urinary excretion and Marfan's syndrome. Minerva Pediat. 17, 1897-1903

312. Tyler, W. S., Gregory, P. W. and Meyer, K. (1962): Sulfated mucopolysaccharides of urine from brachycephalic bovine dwarfs. Am. J. Vet. Res. 23, 1109-1110

313. Valdivieso, F., Martinez-Valverde, A., Maties, M. and Ugarte, M. (1973): Early diagnosis of hypermucopolysacchariduria. Clin. Chim. Acta 44, 357-360

314. Valli, M., Tenni, R., Curzi, L. and Lenzi, L. (1977): Metabolic aspects of connective tissue in osteogenesis imperfecta. Boll. Soc. Ital. Biol. Sper. 53, 805-810 (Ita)

315. Valuiskova, R. P. (1973): Excretion of acid mucopolysaccharides with urine in children with diabetes mellitus. Pediatriia 52, 59-60 (Rus)

316. Van Hoof, F. (1973): GM_1-gangliosidosis. In: H. G. Hers and F. Van Hoof (Eds.). Lysosomes and Storage Diseases. Academic Press, New York, Chap. 12

317. Van Hoof, F. (1973 a): Fucosidosis. In: H. G. Hers and F. Van Hoof (Eds.). Lysosomes and Storage Diseases. Academic Press, New York, Chap. 10

318. Van Hoof, F. (1973 b): Mucopolysaccharidoses. In: H. G. Hers and F. Van Hoof (Eds.). Lysosomes and Storage Diseases. Academic Press, New York, Chap. 8, pp. 217-259

319. Van Hoof, F. (1974): Mucopolysaccharidoses and mucolipidoses. In: D. N. Raine (Ed.). Molecular Variants in Disease. Brit. Med. Assn. London, pp. 64-93

320. Van Hoof, F. and Hers, H. G. (1972): Mucopolysaccharidoses as lysosomal diseases. Advan. Exp. Med. Biol. 19, 211-223

321. Varma, R., Hoshino, A. Y. and Mesmer, R. (1979): Excretion of acidic glycosaminoglycans and their degradation products in urine of schizophrenic patients. Res. Commun. Psychol. Psychiat. Behav. 4, 299-310

322. Varma, R. S., Varma, R., Allen, W. S. and Wardi, A. H. (1974): Urinary excretion of acid mucopolysaccharides in schizophrenia. Biochem. Med. 11, 358-369

323. Varma, R., Varma, R. S., Allen, W. S. and Wardi, A. H. (1976): Protein bound rhamnose in urine of schizophrenic patients. Frontiers Psychiat. 6, 3

324. Varma, R. S., Varma, R. and Hoshino, A. Y. (1979 a): Urinary macromolecules and diagnosis of schizophrenia. Lab Management (New York) 17 (7), 35-41

325. Varma, R. S., Varma, R. and Mesmer, R. (1976 a): Urinary glycoproteins in schizophrenia. Biochem. Med. 15, 296-305

326. Vermeulen, C. W., Goetz, R., Ragins, H. D. and Grove, W. J. (1951): Experimental urolithiasis: Prevention of magnesium ammonium phosphate calculi by reducing the magnesium intake or by feeding on aluminium gel. J. Urol. 66, 5-11

327. Vermeulen, C. W., Ragins, H. D., Grove, W. J. and Goetz, R. (1951 a): Experimental urolithiasis III. Prevention and dissolution of calculi by alteration of urinary pH. J. Urol. 66, 1-5

328. Vitiello, F., Carlomagno, S. and Di Benedetta, C. (1977): Urinary acid mucopolysaccharides determination: a modification of the method. Biochem. Med. 17, 187-192

329. Voigt, K. D., Utermann, D. and Apostolakis, M. (1968): Excretion of acid mucopolysaccharides in the urine of normal and arteriosclerotic subjects. Colloq. Int. Cent. Nat. Rech. Sci. No. 169, 821-834

330. Wardi, A. H., Allen, W. S., Varma, R. and Varma, R. S. (1974): A simple and direct method for the microestimation of glycosaminoglycans in urine. Biochem. Med. 11, 93-97

331. Wasteson, A. and Wessler, E. (1971): Molecular size and xylose content of urinary glycosaminoglycans. Biochim. Biophys. Acta 252, 13-17

332. Watanabe, T. (1972): Histochemical studies on mucosubstances in urinary stones. Tohoku J. Exptl. Med. 107, 345-357

333. Wessler, E. (1967): Determination of acidic glycosaminoglycnas (mucopolysaccharides) in urine by an ion exchange method. Application to "collagenoses", gargoylism, the nail-patella syndrome and Farber's disease. Clin. Chim. Acta 16, 235-243

334. Wessler, E. (1970): Electrophoretic separation of glycosaminoglycans excreted in excess in different kinds of mucopolysaccharidoses. In: E. A. Balazs (Ed.). Chemistry and Molecular Biology of the Intercellular Matrix Vol. 2, Academic Press, London, pp. 895-901

335. Wessler, E. (1971): The nature of the non-ultrafilterable glycosaminoglycans of normal human urine. Biochem. J. 122, 373-384

336. Wiener, R., Iannaccone, A., Eisenberg, J., Griboff, S. I., Ludwig, A. W. and Soffer, I. J. (1955): Influence of hormone therapy on body fluids, electrolyte balance, and mucopolysaccharides in myxedema. J. Clin. Endocrinol. Metab. 15, 1131-1140

337. Winand, R. J. (1968): Increased urinary excretion of acidic mucopolysaccharides in exophthalmos. J. Clin. Invest. 47, 2563-2568

338. Winand, R. and Lambotte, C. (1971): Biochemical study of mucopolysaccharidosis. Rev. Med. Liege 26, 278-284

339. Wolfe, J. S., Callahan, J., Andermann, F., Scriver, C. R., Fawcett, J. S. and Pinsky, L. (1969): Systemic late infantile lipidosis: Beta-galactosidase deficiency with cerebral GM_1 gangliosidosis and accumulation and excretion of a keratan sulfate. Neurology 19, 305

340. Wolfe, L. S., Callahan, J., Fawcett, J. S., Andermann, F. and Scriver, C. R. (1970): GM_1-gangliosidosis without chondrodystrophy or visceromegaly. ß-galactosidase deficiency with gangliosidosis and the excessive excretion of a keratan sulfate. Neurology 20, 23-44

341. Wolfson, S. L., Davidson, E., Harris, J. S., Kahanana, L. and Lorincz, A. E. (1963): Long term corticosteroid therapy in Hurler syndrome. Amer. J. Dis. Child. 106, 3-10

342. Wood, K. M., Wusteman, F. S. and Curtis, C. G. (1973): The degradation of intravenously injected chondroitin 4-sulfate in the rat. Biochem. J. 134, 1009-1013

343. Yamato, K., Handa, S., Yamakawa, T. et al. (1974): The urinary glycosaminoglycans in pachydermoperiostosis. Jap. J. Exp. Med. 44, 19-24

344. Yoshida, Y. (1962): Changes in metabolism of connective tissue components in liver damage. In: F. Egami and Y. Oshima (Eds.) Biochemistry and Medicine of Mucopolysaccharides, Maruzen, Tokyo, pp. 273-285

345. Yoshima, H., Shinkai, H., Sano, S. and Homma, M. (1978): Acidic glycosaminoglycans in skin and urine of a patient with widespreaded lichen sclerosus et atrophicus. Arch. Dermatol. Res. 263, 59-66

346. Zorin, N. A. and Zorina, R. M. (1978): Determination of glycosaminoglycans urinary excretion under normal conditions and after frostbite. Lab. Delo. (6) 340-343 (Rus)

Chapter 18

Secretions of the Skin

18.1 General Views

The sweat glands and the sebacous glands in human skin are clearly distinguishable from each other; the sweat gland is tubular and produces the watery secretion, sweat, while the sebaceous gland is an acinous gland and discharges a fatty substance, sebum. Two types of sweat glands are present in human skin: Apocrine and eccrine glands. The apocrine glands arise from hair follicles, are functional during infancy and secrete the sweat by a combination of necrobiotic discharge and simple secretion. The eccrine glands develop directly from the epidermis, become functional much later and discharge their secretion without loss of their cellular material (Kuno, 1956): The fine structures of the human sebaceous gland (Ellis and Henrickson, 1963; Bell, 1974) and human eccrine sweat gland (Ellis, 1962) have been studied by electron microscopy.

The sebaceous glands are located in skin throughout the body, but are probably absent in the palm and sole. Ectopic sebaceous glands are present in the mucosa of the cheeks, gingiva, tongue and the vermillion surface of the upper lip (Montagna, 1963). The sebaceous glands may actually be considered appendages of the hair follicles and these follow the course of differentiation of hair. Their secretion, sebum, is a holocrine secretion formed by complete disintegration of the cells and release of the stored lipids. The cells lost in this way are replaced by the mitotic activity of the undifferentiated cells. In the fetus the largest sabaceous glands are present on the chest and face and their secretion along with desquamated epithelial cells forms the *vernix caseosa*, the *cheesy varnish* which covers the skin of the fetus. In adults, the secretion of the sebaceous glands present in the oral cavity may modify the chemical composition of saliva. In the external auditory meatus, the cerumen is the combined secretion of sebaceous glands and the ceruminous glands. The physiological role of sebum in man (Kligman, 1963) is not known, although in aquatic birds it acts as a non-wetting agent and as a pheromonal agent in some mammals. Sebum may also act as an emolient, emulsifier, lubricant, fungicidal and bacteriostatic agent (Wheatley, 1953; Kligman, 1963).

The most important function of sweat is thermoregulation although it does help to eliminate urea, electrolytes, etc. to some extent. The majority of the sweat is secreted by the well-developed eccrine sweat glands spread over the general body surface. Alterations in sweat gland function are produced by pharmacological agents. The diuretics, Diamox, furoseamide and ethacrynic acid and ouabain reduce sweat formation (Slegers, 1972). The eccrine sweat glands are simple tubes that extend downward from the epidermis into about the midway in the dermis, or deeper. Following toluidine blue staining of these glands in thick section of collagenase-digested skin, two types of cells can be identified. The larger, basal, acidophilic cells are the *clear cells* which secrete the water and salt fractions of the sweat (Montagna, 1962; Montagna et al., 1963). The *dark cells* are small, superficial, displaced toward the lumen and contain basophilic granules of RNA and mucopolysaccharide nature (Formisano and Lobitz, 1957). These dark cells secrete mucinous substances in sweat. During profuse sweating these mucopolysaccharide granules are decreased and depleted in the dark cells and are repleted during recovery (Dobson, 1960).

The secretion of the eccrine gland is controlled by cholinergic and adrenergic nerves (Robertshaw, 1977) and adrenergic and cholinergic blocking agents can inhibit apocrine and eccrine sweating (Goodall, 1970). Deoxycorticosterone or corticotrophin (ACTH) markedly depresses the sodium and chloride contents of sweat.

18.2 Glycosaminoglycans in normal sweat

A considerable amount of information is available on the chemical composition of sweat (Robinson and Robinson, 1954; Carruthers, 1962). However, very little information is available about the macromolecules present in sweat. Human sweat contains proteins at a concentration of 40 mg per 100 ml of sweat. The first indication of the possible presence of sulfated glycosaminoglycans in sweat was suggested by the observation of Wohnlich (1951) that the wrinkling of skin on soaking hand in water may be due to the extraction of macromolecules from the skin. Concentration of the water extract of the hand and arm (pre-washed with soap and water) gave three fractions: a glycogen fraction, a sialic acid-containing glycoprotein representing 20% of glycogen-free material, and the remaining 80% of the glycogen-free material containing bound hexosamine, sulfate and sulfuric acid. From these results, Wohnlich (1951) suggested the presence of sulfated glycosaminoglycans originating from the stratum corneum, sweat or sebaceous gland secretion.

Jirka and Kotas (1957) reported the presence of mucoproteins as a physiological component of both the apocrine sweat from axilla and eccrine sweat from the backs of human subjects. The mucoprotein contents of apocrine and eccrine sweat preparations from the same subject were not different. This material was soluble in 20% sulfosalicylic acid, was polarographically active and upon hydrolysis gave reaction for hexosamine. The value of hexosamine contributed by this mucoprotein was calculated from the difference in hexosamine contents before and after hydrolysis.

In thick sections of collagenase-digested skin stained with toluidine blue, *dark cells* have been observed to contain basophilic granules which are resistant to both ribonuclease and diastase but are PAS-positive (Montagna et al., 1962). The granules of this material in dark cells are rapidly depleted during profuse sweating (Formisano and Lobitz, 1957) and mucoprotein can be detected in small quantities in the sweat.

Pallavicini et al. (1963) failed to detect any glucuronic acid in carbohydrate-protein complexes in sweat collected upon thermal stimulation in normal subjects and cystic fibrosis patients. The material obtained by precipitation of sweat with 10 volumes of ethanol followed by reprecipitation with 5% trichloroacetic acid gave in supernatent negative results for sulfate by the chloranilate method and by the carbazole reaction for uronic acid. These results indicated the absence of glycosaminoglycans, while the presence of glycoproteins in this material was suggested by the detection of sialic acid, fucose, mannose, glucose, and galactose upon paper chromatography of its acid hydrolysate. The inability to detect hexuronic acid and sulfate may be due to the loss of GAG in the trichloroacetate precipitate, as has been shown for GAG of serum (see section 16.5). Pallavicini et al. (1963) did not check the trichloroacetate precipitate for uronic acid. A better scheme for analysis is proposed in Fig. 18.1.

Seutter and Mali (1965) presented evidence for the presence of GAG in human sweat. By two-dimensional thin-layer chromatography on silica gel plates developed in one direction with chloroform-methanol-17 % ammonia (40.:40:20) and with butanol-acetic acid-water (60:20:20) in the other direction a compound (R_F 0.32, 0.42) was isolated. By infra-red analysis this was shown to be like the desulfated heparin.

Sweat gland has been considered a mucus-secreting gland like any other mucus gland of the body (Seutter et al., 1970). From human crude sweat obtained from the axilla and chest after thermal stimulation, a polysaccharide was obtained. This polysaccharide contained hexosamine and hexuronic acid and gave an infra-red spectrum for sulfate groups at 1240 cm^{-1} and sulfate was also present upon sulfur element analysis. In some patients with light sensitization a deviation in the infra-red spectra for sulfate groups in the polysaccharide was noted.

In a previous publication these authors (Seutter et al., 1966) reported the presence of mucopolysaccharides in two light-sensitive patients which gave an infrared absorption at 867 cm^{-1} which was absent in mucopolysaccharide isolated from sweat of normal subjects. Upon grinding the KBr pellet, this absorption band was replaced by a new infra-red band at 848 cm^{-1}. Also, this material isolated by two-dimensional chromatography did not give the reaction for free peroxide (-O-O-H) group with 10 % w/v titanyl sulfate in 30 % (v/v) sulfuric acid suggesting that the mucopolysaccharide contained a stable dialkyl peroxide group. This was confirmed by preparing a glycosaminoglycan with the aforementioned properties by reacting normal sweat GAG with a weak alkaline solution of 30 % (w/v) hydrogen peroxide in the dark.

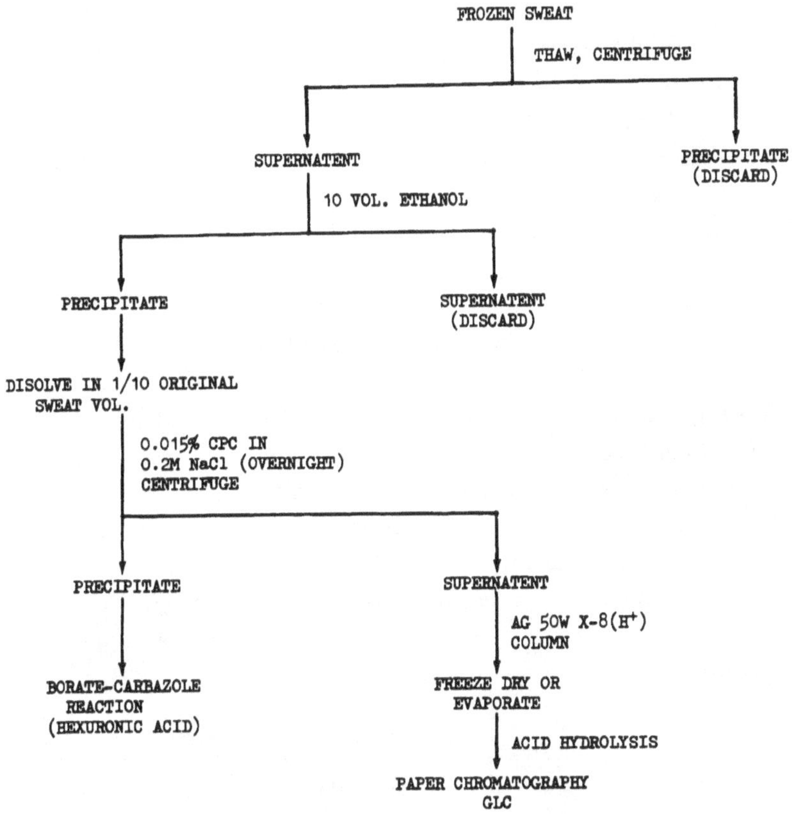

Fig. 18-1 SCHEME FOR ANALYSIS OF CARBOHYDRATE COMPONENTS IN SWEAT

The polysaccharides in the crude sweat are present partly bound to lipids, which are derived in vivo from vitamin K (Seutter and Sutorius, 1971). The origin of these lipids from vitamin K was confirmed by pyrolysis and menadione -^{14}C studies. The release of these vitamin K derivative-bound polysaccharides was increased in the sweat of the Rhesus monkey in the middle of menstruation.

18.3 Glycosaminoglycans of sweat glands in pathological states

Defects in sweat glands have been demonstrated primarily by histochemical tests in several disease conditions affecting skin and in lysosomal storage diseases. Quantitative chemical analysis for hyaluronic acid and dermatan sulfate performed on biopsies from 9 controls and 30 patients revealed a striking decrease in hyaluronic acid with smaller reduction in dermatan sulfate in lesions of discord lupus erythematosus and scleroderma. In dystrophic epidermolysis bullosa, a very low dermatan sulfate level was seen. Changes also occur in several other dermatoses (Hollman et al., 1971).

The normal and abnormal variations in eccrine sweat gland distribution have been reviewed (Willis et al., 1973). Storage of glycosaminoglycans occurs in lysosomes within the eccrine sweat glands, as in other tissues, of patients with fucosidosis (Durand et al., 1969), mucopolysaccharidoses (Belcher, 1972, 1973) and cystic fibrosis of the pancreas (Munger et al., 1961). In 4 patients suffering from Hurler's syndrome, peculiar granular vacuolization of the cytoplasm of epidermal cells and fibrocytes of the dermis was noted and the material within the vacuolated cells stained specifically for mucopolysaccharides (Hambrick and Scheie, 1962). These diffuse vacuoles in the eccrine glands in the skin of patients with Hurler's syndrome were smaller than those in normal sweat glands. Belcher (1972) reported membrane-limited vacuoles within the cells of the eccrine glands which closely resembled mucopolysaccharides-loaded lysosomes identified in other organs of patients with genetic mucopolysaccharidoses. The cells of the proximal duct contain less vacuoles than the distal ducts of the eccrine glands (Belcher, 1973). Defects in eccrine function occur in fucosidosis, glycogen storage diseases, gangliosidoses and cystic fibrosis. The abnormalities studied have been limited to only the electrolytes in sweat and glycosaminoglycans abnormalities in sweat have not been investigated despite a strong histochemical evidence of glycosaminoglycan storage within the lysosomes of the eccrine sweat glands. Even the electrolyte (sodium) levels were normal in patients with genetic mucopolysaccharidoses in contrast to other storage diseases, although mucopolysaccharide storage in lysosomes of tissues and eccrine glands occurs in all these disorders. Active sodium reabsorption normally occurs in the proximal portion of the coiled segment of the eccrine duct, the absence of defect in sodium reabsorption and normal sodium levels in sweat of Hunter-Hurler patients may be related to the absence of excess glycosaminoglycan storage in this coiled segment of the eccrine duct.

Skin biopsies of mucopolysaccharidoses patients have disclosed vacuoles in dermal fibroblasts, epithelial cells of the sweat glands and epidermis. The stored mucopolysaccharides in vacuoles more likely accumulate from intrinsic cell processes rather than from endocytosis of the extracellular material by the cells in the sweat gland coil (Spicer et al., 1974). Sulfated glycosaminoglycans detected by dialysed iron reaction have been investigated at the light and electron microscope levels in patients with Hurler's or Hunter's disease. These sulfated complex carbohydrates were present in both normal and patients' sweat glands on the surface of the intercellular canaliculi and central lumen. However, the secretory granules of the dark cells in normal sweat glands were not stained by dialysed iron, but mucopolysaccharide reacted to this stain in lined and partially filled vacuoles of dark cells showing a laminar distribution in vacuoles of clear cells in sweat glands of the patients (Spicer et al., 1978).

Cystic fibrosis, which is a generalized exocrinopathy, also affects the sweat gland secretion. In this condition the eccrine sweat production rate is normal but the sweat contains abnormally high electrolyte concentrations. The abnormally increased sodium chloride concentration in sweat is responsible for heat prostration and shock secondary to higher salt loss in hot weather in such patients (Dobson, 1962; Dobson and Cage, 1964).

The exact cause of the increased salt concentration in sweat is not known but the fluid formed in the body of the sweat gland is normal and during the passage through the sweat duct to the skin a mucus secretory product inhibits electrolyte reabsorption in the cystic fibrosis patients (Kaiser et al., 1971). A rise of uronic acid content in skin (Wilson, 1972) and glycosaminoglycan accumulation studied by chemical analysis in fibroblasts (Matalon and Dorfman, 1968; Danes and Bearn, 1969), by metachromasia in fibroblasts (Danes and Bearn, 1968, 1969 a) and by abnormal turnover of ^{35}S-labeled compounds (Wiesmann and Neufeld, 1970) may be due to low β-glucuronidase activity in skin (Wilson, 1972 a), in epidermis and sweat gland tissue (Gibbs and Griffin, 1970). Electron microscopic studies on cultured skin fibroblasts have shown no qualitative differences in the number of metachromatic granules between normal subjects and cystic fibrosis patients, but increased production of mucopolysaccharides has been considered to be the basic defect in cystic fibrosis (Munger et al., 1961; Danes and Bearn, 1969). In this context, the role of glycosaminoglycans present on the lining membrane of the secretory cells (Ellis, 1962, 1964) is obvious from the effect of the enzyme hyaluronidase on increased excretion of ^{24}Na in sweat following its intradermal injection. Similar changes appearing in this part of the gland may be responsible for the high sodium concentration in sweat of cystic fibrosis patients (Giec et al., 1968).

The GAG composition of human sweat, although not studied, may vary in those conditions which alter the GAG in skin, including myxedema (Watson and Pearce, 1947), Unverricht's syndrome (Nedved et al., 1967) and other connective tissue related dermatoses (Asboe-Hansen, 1954; Barker and Kennedy, 1969), diabetes (Malathy and Kurup, 1972), and genetic diseases (Landing and Wells, 1969).

18.4 Variations with age and sex

Since the information on the glycosaminoglycan composition is incomplete and scanty, no report has appeared on the variation of GAG in this biological fluid. Even the nature of GAG in sweat is not clear; the GAG present being regarded as desulfation product of heparin (Seutter and Mali, 1965).

The changes observed in sweat, if any, may be related to those in skin. The amount of sweating in the aged is much less than in younger human subjects, because both the number of sweat glands and the sweat production by the single sweat glands are reduced. This is related to the histological findings in the aged, which include involution of the secretory coils, irregularities in size and distribution of secretory epithelial cells in the eccrine sweat glands (Montagna, 1965). No definite and specific aging changes have been reported in apocrine sweat glands. There are a few reports of partial or complete atrophy of these glands but these findings are inconclusive (Asboe-Hansen, 1969). Also see addendum on page 606.

The texture, appearance and elasticity of the skin is different in males and females and in human subjects of different ages. These sex-related changes are the attributes of the gonadal hormones. The hexosamine content of skin is higher in male rats than in females. The GAG in dermis falls from 722 µg uronic acid/g dry weight at birth to 226 µg/g at ages between 67-96 years; along with this hexosamine/hexuronic acid ratios change from 5.8 in newborns to 10.2 in old age (Fleischmajer et al., 1973). Changes in hyaluronic acid and dermatan sulfate levels occur during aging.

18.5 Glycosaminoglycans in sebaceous secretions

The chemistry of sebum has been reviewed (Wheatley, 1953; Nikkari, 1974). It is primarily a waxy material containing different types of lipids. Glycosaminoglycans or hexuronic acid-containing material have not been reported in sebaceous secretion of skin per se. However, the cerumen in the external auditory meatus, which as mentioned before (section 18.1) is partly a sebaceous secretion, has been reported to contain 0.1% glucuronic acid and 3.5% sulfate (Masuda et al., 1978, 1979). The presence of the enzyme β-glucuronidase in sebum has also been reported (Serri and Huber, 1963). The sebum content and the enzymic activity diminishes in senescence. Clinically, atrophy of the sebaceous glands occurs as senile xeroderma (Korting, 1973). In the hereditary disorder called Icthyosis vulgaris, scales formed on the skin contain significantly lower glycosaminoglycans than those present in epidermis of normal men and the chondroitin sulfate-like fraction was the most affected (Masuda et al., 1976).

18.6 Hormonal control of sebaceous secretion

The measurement of the activity and hormonal control of the sebaceous glands have been reviewed (Hodgson-Jones et al., 1952; Strauss and Pochi, 1963; Ebling, 1963, 1974; Schuster and Thody, 1974). Hormone-induced changes occur in these glands starting from their inception in the fetus through the declining sexual years and senescence (Korting, 1973), including pathogenesis of acne at puberty (Montagna et al., 1974), seborrhoea and psoriasis (Fleischmajer and Blumenkrantz, 1970; Fleischmajer et al., 1970). The sebaceous glands are androgen-sensitive where the dihydrosterone combines with a receptor-protein to increase the sebum synthesizing activity of the glands (Adachi, 1974), which is demonstrable in the subcellular organelles (Bell, 1974). The antiandrogens suppress the sebaceous activity either by inhibiting the 5α-reductase dependent conversion of testosterone into dihydrotestosterone (Hsia and Voigt, 1974) or by inhibiting the binding of dihydrosterone to the nuclear protein (Adachi, 1974). Estrogens reduce the size of the sebaceous glands (Ebling, 1963) and have been used as a therapy for acne vulgaris. Factors of pituitary origin have been shown to control the sebaceous gland activity in rats (Ebling, 1974; Schuster and Thody, 1974). Prolactin, growth hormone, ACTH and TSH stimulate these glands directly or indirectly. The sebaceous activity and melanocyte stimulating hormone are increased in Parkinson's disease (Schuster and Thody, 1974) and these changes are reversed, though not uniformly, by L-dopa (Kohn et al., 1973). In castrated male rats, progesterone increases the size of the sebaceous glands (Ebling, 1963).

Addendum

The acidic GAG concentration of sweat from 5 patients with epidermolysis bullosa hereditaria was not different from those in normals, but the excretion of GAG in urine was increased 5-fold (Ref. 73). In comparison to normals, hyaluronic acid and dermatan sulfate were slightly increased in fibroblasts cultured from patients with epidermolysis bullosa dystrophica dominans (Ref. 74).

18.7 Bibliography

1. Adachi, K. (1974): Receptor proteins for androgen in hamster sebaceous glands. J. Invest. Dermatol. 62, 217-223

2. Asboe-Hansen, G. (1954): Some systemic connective tissue disorders pertaining to dermatology. In: G. Asboe-Hansen (Ed.). Connective Tissue in Health and Disease. Ejner Munksgaard, Copenhagen, pp. 275-295

3. Asboe-Hansen, G. (1969): Aging of skin. In: A. Engel and T. Larsson (Eds.). Thule Intern. Symp. on Aging of Connective and Skeletal Tissue. Nordiska Bokhandelns Forlag, Stockholm, pp. 235-262

4. Barker, S. A. and Kennedy, J. F. (1969): Human skin acidic mucopolysaccharides and clinical conditions. Life Sci. Part II 8, 989-994

5. Belcher, R. W. (1972): Ultrastructure and cytochemistry of lymphocytes in the genetic mucopolysaccharidoses. Arch. Pathol. 94, 511-518

6. Belcher, R. W. (1973): Ultrastructure and function of eccrine glands in the mucopolysaccharidoses. Arch. Pathol. 96, 339-341

7. Bell, M. (1974): A comparative study of the ultrastructure of the sebaceous glands of man and other primates. J. Invest. Dermatol. 62, 132-143

8. Carruthers, C. (1962): Biochemistry of Skin in Health and Disease. Charles C. Thomas, Springfield, pp. 152-154

9. Danes, B. S. and Bearn, A. G. (1968): A genetic cell marker in cystic fibrosis of the pancreas. Lancet 1, 1061-1063

10. Danes, B. S. and Bearn, A. G. (1969): Cystic fibrosis: Distribution of mucopolysaccharides in fibroblast culture. Biochem. Biophys. Res. Commun. 36, 919-924

11. Danes, B. S. and Bearn, A. G. (1969 a): Cystic fibrosis of the pancreas, a study in cell culture. J. Exp. Med. 129, 775-793

12. Dobson, R. L. (1960): The effect of repeated episodes of profuse sweating on the human eccrine gland. J. Invest. Dermatol. 35, 195-198

13. Dobson, R. L. (1962): The correlation of structure and function in the human eccrine sweat gland. In: W. Montagna, R. A. Ellis and A. F. Silver (Eds.). Advances in Biology of Skin, Vol. 3, Eccrine Sweat Glands and Eccrine Sweating, Pergamon Press, Chap. IV, pp. 54-75

14. Dobson, R. L., Cage, G. W. (1964): The eccrine sweat gland in systemic diseases. In: P. A. Di Sant'Agnese (Ed.). Research on Pathogenesis of Cystic Fibrosis of the Pancreas. NIH, Bethesda, pp. 57-66

15. Durand, P., Borrone, C. and Della-Cella, G. (1969): Fucosidosis. J. Pediatr. 75, 665-674

16. Ebling, J. F. (1963): Hormonal control of sebaceous glands in experimental animals. In: W. Montagna et al., Eds. (1963), Chap. 13, pp. 200-219

17. Ebling, J. F. (1974): Hormonal control and methods of measuring sebaceous gland acitivity. J. Invest. Dermatol. 62, 161-171

18. Ellis, R. A. (1962): The fine structure of the eccrine sweat glands. In: W. Montagna, R. A. Ellis and A. F. Silver (Eds.). Advances in Biology of Skin, Vol. 3 Eccrine Sweat Glands and Eccrine Sweating, Pergamon Press, New York, Chap. 3, pp. 30-53

19. Ellis, R. A. (1964): The fine structure of salt secreting epithelia. In: P. A. Di Sant'Agnese (Ed.). Research on Pathogenesis of Cystic Fibrosis of the Pancreas. NIH, Bethesda, 1

20. Ellis, R. A. and Henrickson, R. C. (1963): The ultrastructure of the sebaceous glands of man. In: W. Montagna et al. (1963), Chap. 6, 94-109

21. Fleischmajer, R. and Blumenkrantz, N. (1970): Fractionation of glycosaminoglycans from psoriatic skin. J. Invest. Dermatol. 55, 274-276

22. Fleischmajer, R., Burgoyne, A. and Piamphongsant, T. (1970): Dermal glycosaminoglycans in psoriasis. J. Invest. Dermatol. 54, 316-318

23. Fleischmajer, R., Perlish, J. F. and Bashey, R. I. (1973): Aging of human dermis. In: L. Robert and B. Robert (Eds.). Frontiers of Matrix Biology. 1. Ageing of Connective Tissue. Skin, S. Karger, Basel, pp. 90-106

24. Formisano, V. and Lobitz, W. C., Jr. (1957): "The Schiff-positive, non-glycogen material" in the human eccrine glands. I. Histochemistry. A. M. A. Arch. Dermatol. 75, 202-209

25. Gibbs, G. E. and Griffin, G. D. (1970): Beta-glucuronidase in skin components of children with cystic fibrosis. Science 167, 993-994

26. Giec, L., Kumaszka, F. and Zmudzinski, J. (1968): Effect of hyaluronidase on the excretion of sodium in thermal sweat. Pol. Med. J. 7, 315-321

27. Goodall, M. (1970): Innervation and inhibition of eccrine and apocrine sweating in man. J. Clin. Pharmacol. J. New Drugs 10, 235-246

28. Hambrick, G. W. and Scheie, H. G. (1962): Studies of the skin in Hurler's syndrome. Arch. Dermatol. 85, 455-471

29. Hodgson-Jones, I. S., Mackenna, R. M. B. and Wheatley, V. R. (1952): The study of human sebaceous activity. Acta Derm.-Venerol. 32 (Suppl. 29), 155-160

30. Hollman, E. P., Mier, P. D., Van de Staak, W. J., Urselmann, E. and Warndorff, J. A. (1971): Cutaneous acid mucopolysaccharides in some dermatoses. Br. J. Dermatol. 85, 421-423

31. Hsia, S. L. and Voigt, W. (1974): Inhibition of dihydrosterone formation: an effective means of blocking androgen action in hamster sebaceous glands. J. Invest. Dermatol. 60, 224-227

32. Jirka, M. and Kotas, J. (1957): The occurrence of mucoproteins in human sweat. Clin. Chim. Acta 2, 292-296

33. Kaiser, D., Drack, E. and Rossi, E. (1971): Inhibition of net sodium transport in single sweat glands by sweat of patients with cystic fibrosis of the pancreas. Pediatr. Res. 5, 167-172

34. Kligman, A. M. (1963): The uses of sebum? In: W. Montagna et al., Eds. (1963), Chap. 7, pp. 110-123

35. Kohn, S. R., Pochi, P. E., Strauss, J. S. (1973): Sebaceous gland secretion in Parkinson's disease during L-dopa treatment. J. Invest. Dermatol. 60, 134-136

36. Korting, G. W. (1973): The skin in old age and its clinical aspects. In: H. G. Vogel (Ed.). Connective Tissue and Ageing, Excerpta Medica, Amsterdam, pp. 236-240

37. Kuno, Y. (1956): Human Perspiration, Charles C. Thomas, Springfield, Chap. 2, pp. 42-97

38. Landing, B. H. and Wells, T. R. (1969): Sweat gland anatomy in genetic diseases. J. Chron. Dis. 21, 703-713

39. Malathy, K. and Kurup, P. A. (1972): Metabolism of glycosaminoglycans in alloxan diabetic rats. I. changes in tissue glycosaminoglycans. Diabetes 21, 1162-1167

40. Masuda, H., Shichijo, S., Goya, T. and Takeuchi, M. (1978): Isolation and characterization of a glycopeptide from cerumen. Kurume Med. J. 25, 203-205

41. Masuda, H., Shichijo, S. and Takeuchi, M. (1976): Isolation and characterization of the glycosaminoglycans from the scale of ichthyosis vulgaris. Biochem. Med. 15, 22-27

42. Masuda, H., Shichijo, S. and Takeuchi, M. (1979): Carbohydrate composition of glycopeptides from the human cerumen. Biochem. Med. (in press)

43. Matalon, R. and Dorfman, A. (1968): Acid mucopolysaccharides in cultured fibroblasts in cystic fibrosis of the pancreas. Biochem. Biophys. Res. Commun. 33, 954-958

44. Montagna, W. (1962): The Structure and Function of Skin, 2nd Ed. Academic Press, New York, pp. 341-345

45. Montagna, W. (1963): The sebaceous glands in man. In. W. Montagna et al., Eds. (1963), Chap. 2, pp. 19-31

46. Montagna, W. (1965): Morphology of the ageing skin. The cutaneous appendages. In: W. Montagna (Ed.), Advances in Biology of Skin. Vol. 6, Ageing, Pergamon Press, New York

47. Montagna, W., Bell, M. and Strauss, J. S., Eds. (1974): Sebaceous Glands and Acne Vulgaris (Advances in Biology of Skin, Vol. 14). Special issue of J. Invest. Dermatol. 62 (3), 119-339

48. Montagna, W., Ellis, R. A. and Silver, A. F. (Eds.) (1962): Eccrine Sweat Glands and Eccrine Sweating. Vol. 3 of Advances in Biology of Skin, Pergamon Press, New York, 1962

49. Montagna, W., Ellis, R. A. and Silver, A. F., Eds. (1963): The Sebaceous Glands (Advances in Biology of Skin, Vol. 4), Pergamon Press, Oxford

50. Munger, B. L., Brusilow, S. W. and Cooke, R. E. (1961): An electron microscopic study of eccrine sweat glands in patients with cystic fibrosis of the pancreas. J. Pediatr. 59, 497-511

51. Nedved, A., Peterson, W. C., Jr. and Johnson, R. V. (1967): Cutaneous findings in Unverricht's syndrome. Progressive familial myoclonic epilepsy. Arch. Derm. (Chicago) 95, 206-209

52. Nikkari, T. (1974): Comparative Chemistry of Sebum. J. Invest. Dermatol. 62, 257-267

53. Pallavicini, J. C., Gabriel, O., di Sant'Agnese, P. A. and Bukirk, E. R. (1963): Isolation and characterization of carbohydrate-protein complexes from human sweat. Annals N. Y. Acad. Sci. 106, 330-338

54. Robertshaw, D. (1977): Neuroendocrine control of sweat glands. J. Invest. Dermatol. 69, 121-129

55. Robinson, S. and Robinson, A. H. (1954): Chemical composition of sweat. Physiol. Revs. 34, 202-220

56. Schuster, S. and Thody, A. J. (1974): The control and measurement of sebum secretion. J. Invest. Dermatol. 62, 172-190

57. Serri, F. and Huber, W. M. (1963): The development of sebaceous glands in man. In: W. Montagna et al. (1963), Chap. 1, pp. 1-18

58. Seutter, E. and Mali, J. W. H. (1965): Mucopolysaccharides in sweat. Clin. Chim. Acta 12, 17-21

59. Seutter, E. and Sutorius, A. H. M. (1971): Vitamin K derivatives of some skin-mucins. Int. J. Vitam. Nutr. Res. 41, 57-67

60. Seutter, E., Sutorius, A. H. M. and Govaert, F. J. (1966): Light sensitization and peroxides in sweat. Clin. Chem., 12, 728-731

61. Seutter, E., Trijbels, J. M. F., Sutorius, A. H. M. and Urselmann, E. J. M. (1970): Sweat gland as a mucous gland. Dermatologica 141, 397-408

62. Slegers, J. F. G. (1972): Alterations in sweat gland function produced by pharmacological agents. Adv. Biol. Skin 12, 477-493

63. Spicer, S. S., Garvin, A. J., Wohltmann, H. J. and Simson, J. A. V. (1974): The ultrastructure of the skin in patients with mucopolysaccharidoses. Lab. Invest. 31, 488-502

64. Spicer, S. S., Garvin, A. J., Simson, J. A. V. and Wertelecki, V. (1978): Cytochemistry of the skin of patients with mucopolysaccharidoses. Histochem. J. 10, 137-150

65. Strauss, J. S. and Pochi, P.E. (1963): The hormonal control of human sebaceous glands. In: W. Montagna et al., Eds. (1963): Chap. 14, pp. 220-253

66. Watson, E. M. and Pearce, R. H. (1947): The mucopolysaccharide content of skin in localized (pretibial) myxedema. Am.J. Clin. Pathol. 17, 507-512

67. Wheatley, V. R. (1953): Some aspects of the nature and function of sebum. St. Bartholomews Hosp. J. Jan., 1953

68. Wiesmann, V. and Neufeld, E. F. (1970): Metabolism of sulfated mucopolysaccharide in cultured fibroblasts from cystic fibrosis patients. J. Pediatr. 77, 685-690

69. Willis, I., Harris, D. R. and Moretz, W. (1973): Normal and abnormal variations in eccrine sweat gland distribution. J. Invest. Dermatol. 60, 98-103

70. Wilson, R. G. (1972): Glycosaminoglycans in the skin of patients with cystic fibrosis of the pancreas. Clin. Chim. Acta 36, 189-193

71. Wilson, R. G. (1972 a): The activities of some acid hydrolases in the skin of patients with cystic fibrosis of the pancreas. Clin. Chim. Acta 36, 113-117

72. Wohnlich, H. (1951): Mucine der menschlichen Epidermis, ihre Abtrennung von Glykogen. Biochem. Z. 322, 76-84

Note

Masuda et al., 1979 (Ref. 42) later appeared as:

Shichijo, S., Masuda, H. and Takeuchi, M. (1979): Carbohydrate composition of glycopeptides from the human cerumen. Biochem. Med. 22, 256-263

73. Ratzenhofer, E., Legenstein, E., Meznik, E. and Lubec, G. (1979): Excretion of acidic mucopolysaccharides in the urine and sweat in epidermolysis bullosa hereditaria. Z. Hautkr. 54, 987-992 (Ger)

74. Igarashi, M. and Seiji, M. (1980): Acidic glycosaminoglycans in cultured skin fibroblasts of a patient with epidermolysis bullosa dystrophica dominans. Tohoku J. Exp. Med. 130, 207-208

Chapter 19

Other Miscellaneous Body Fluids

19.1 Milk

This is the secretion of pH 6.6-7.0 from the mammary glands in mammals. Milk contains oligosaccharides, glycoproteins, immunoglobulins and glycosaminoglycans in addition to proteins (lactalbumin, lactglobulin and casein). Human milk contains only 1.5 % protein in comparison to 3.5 % in cow's milk. The colostrum (milk produced for a few days after the parturition) differs from the ordinary milk in having higher protein and immunoglobulin contents.

The milk fat globule membrane (MFGM) contains glycoproteins (1), acidic glycoproteins (2) and acidic glycosaminoglycans. The acidic glycoproteins have been extracted and isolated from MFGM by phenol extraction and sephadex G-200 column chromatography. The glycoprotein "B" isolated from bovine MFGM contains sialic acid, mannose, galactose, glucose, glucosamine and galactosamine (1).

The membranes of the fat globules of cow's milk were found to contain a total of GAGs equivalent to a uronic acid content of 163 μg/100 mg dry weight. The major GAG was hyaluronic acid which accounted for 62.5 % of the total uronic acid in cow's MFGM. The remainder 37.5 % uronic acid consisted of sulfated GAGs (chondroitin 4-sulfate, chondroitin 6-sulfate, dermatan sulfate and heparan sulfate) with different degrees of sulfation (3).

Another investigation (4) on the MFGM, isolated from bovine and human milk, reported contradictory finding that heparan sulfate was the major GAG in both human and bovine MFGM and not hyaluronic acid as reported earlier (3). The GAGs in this study were eluted from a Dowex-1 column with 1.0 *M* and 3.0 *M* NaCl and were characterized by cellulose acetate electrophoresis, digestion with chondroitinases and nitrous acid degradation. Whereas the protein content of cow's milk is known to be 3-fold higher than that of human milk, this study (4) showed that the total GAGs (mostly heparan sulfate were 5-10 fold higher in human MFGM. The cow's milk, however, contained larger quantities of chondroitin 4- and 6-sulfates than were present in human milk.

The MFGM preparation from bovine colostrum had a similar GAG composition as the normal bovine MFGM. Although heparan sulfate was the major GAG in normal human MFGM, the human colostrum lacked heparan sulfate. This suggested that GAG changes take place during the lactation periods following parturition in human subjects (4).

Milk contains the GAG-degrading enzyme ß-glucuronidase. Determination of this enzyme, by using phenolphthalein-ß-D-glucuronide as the substrate, showed that its activity was dependent upon the lactation stage. It is higher in colostral milk than in normal milk (5), which may be related to GAG metabolism in the lactating mammary gland at different stages. In milk from mastitis patients; this enzyme activity was high and there was a narrow relation of this enzyme and the leukocyte count (5).

19.2 Interstitial Fluid

This fluid lies between the cells and is about 12 liters in volume in a human adult. The interstitial fluid volume is the difference in the volumes of the total extracellular fluid (measured as the inulin space) and plasma (measured as the I^{131}-space). The ratios of the volumes of interstitial fluid to plasma vary from 3.0 to 4.0. It originates as a dialysate of plasma through the capillary endothelium and its volumes vary with the osmotic pressure of plasma proteins and other factors. In various tissues of the body, this fluid intercommunicates with plasma and lymph. As the blood flows through the capillaries, a substantial portion of plasma diffuses back and forth into the interstitial fluid. At the same time, a part of the interstitial fluid flows out of the interstitial spaces as lymph appearing in the lymphatic capillaries. For this reason the composition of interstitial fluid, lymph and plasma are similar except that protein in the first two is lower (2 %) and replaces chloride ions and in plasma protein averages 7 %. The average interstitial pressure is negative and is -6.0 to - 7.0 mm/Hg. The negative interstitial pressure implies a state of dehydration. Conditions which cause an increase in this pressure to positive produce an edema (6).

Although the quantity of the interstitial fluid is very high, only nanolitres of the fluid can be aspirated with difficulty. Several procedures have been used for sampling of the interstitial fluid and include the direct aspiration with a microneedle, sampling of fluid in chronically-implanted perforated capsules, with a liquid paraffin technique and the wick-cathetar technic. Although this fluid's composition varies with the sampling technic used (7) and with the tissue sampled, the mucopolysaccharides (glycosaminoglycans, GAGs) are consistently present in the form of a "gel". The GAG contents of the interstitial space vary from 0.1-1.0 % and are 0.6 % in skin and muscle, where 2 % plasma proteins are also present (8). The osmotic interaction of GAGs and proteins results in osmotic pressures, which are more than the algebraic sum of the individual osmotic pressures of these two macromolecules. This is because a mixture of plasma proteins and GAGs acts as a two-phase system in which the gel-like GAG phase is associated with a fluid volume from which the protein has been excluded.

Between the GAG-gels is present the free fluid containing the plasma proteins. The two phases are in equilibrium and exert osmotic pressures of their own (8). The physical chemical bases of the interstitial phenomenon, in terms of the GAGs, have been reviewed (9, 10). A solution composed of 1.6 % plasma proteins, 20 % collagen, and 0.4 % hyaluronic acid in Ringer's Solution approximates the composition of the interstitial fluid and yields higher oncotic pressures (11).

The relationship between composition and pressure of the interstitial fluid (or interstitial space) is determined by the changes in the properties of the GAGs making the gel, which result in increases or decreases of its water content. The major GAG in this gel is the high molecular weight hyaluronic acid. This gel serves as the "intercellular matrix or cement", also called the "ground substance", which holds together and binds cells into tissues and organs. Because of the location of the interstitial fluid between the cells, all the nutrients going from the blood capillaries into the cells and the metabolites and secretions leaving the cell on way to blood have to pass through this fluid. For these reasons, this fluid serves as the internal environment, the "milieu interieur" of the cells, whose homeostasis is very essential for the growth, multiplication and maintenance of the life of the cell. The glycosaminoglycans in the jelly-like matrix in this fluid serve to control the intercellular permeability by acting as a selective barrier to large molecules (e. g. proteins), bacteria and viruses and give immunity against infections (6). The high osmotic pressure contributed by glycosaminoglycans in this fluid produces water retention and turgence of the tissues. Depolymerization of the hyaluronic acid gel in some disease states such as rheumatoid arthritis or bacterial and viral infections (which liberate hyaluronidase) results in loss of this selective permeability of the interstitial fluid, deturgence of the tissue and increases the susceptibility to infections.

19.3 Lymph

This fluid is derived from the interstitial fluid. It represents that portion of the interstitial fluid which flows out of the interstitial space into the minute lymphatic capillaries accompanying the blood vessels in all tissues of the body. Since it is derived from the interstitial fluid, its composition differs from that of plasma but is similar to that of the interstitial fluid. The water compartment in the interstitial fluid and lymph together are 120 ml/kg body or 20 % of the body water. The chemical composition varies with the tissue from which sample is taken. Pulmonary lymph and the pulmonary interstitial fluid virtually have the same composition (6).

This fluid contains leukocytes which can be the source of GAGs and enzymes in lymph. Several enzymes are present (12). The significance of lymph and lymph flow for clinical enzymology has been discussed (13). Factors which affect the ground matrix and permeability of blood/lymph capillaries produce an alteration in GAG composition of the lymph. When hydrocortisone was administered to dogs or guinea pigs. the GAG, enzyme and protein contents rose 4 hours later in the lymph and serum (14). This may be attributable to an enhanced resorption capacity of the lymph capillaries and an alteration in the vascular permeability.

According to some investigators, the interstitial fluid and lymph are considered virtually identical (15), since the lymphatics drain the interstitial space and the two fluids resemble in protein composition. The GAG composition of lymph is close to that in plasma. Also, it is necessary to keep in mind that the former fluid is present in the interstitial space and it has two phases, while the latter fluid is a free solution in ducts. It can be surmised that the GAG present as a sieve-like gel in the interstitial fluid will behave physicochemically different than the GAGs present in the lymph as a free solution (6). In lymph from lung, the proteins are only 70 % of that in plasma. In some animals like sheep, the lymph is easier to collect by micropuncture. In the sheep with pulmonary edema, the protein composition is identical in the simultaneously sampled interstitial and lymph fluids from the lungs. However, it should be remembered that modification of the lymph composition can take place in the lymph nodes, where plasma and lymph are in close relationship (6).

19.4 Cartilage Fluid

A fluid aspirated in vivo by micropuncture of the articular and growth plate cartilage is actually the interstitial fluid of the cartilage (16). This cartilage fluid has been studied extensively, since it helps a more specific sampling of the biochemical processes of the cells at the sites of active mineralization of the epiphyseal and metaphyseal cartilage than can be studied in sliced or ground whole cartilage. This fluid is of extracellular origin.

The fluid (5-40 nanoliters) is usually aspirated by micropuncture of the distal fluid hypertrophic cell zone of the upper tibial growth plates in the anesthesized rat or the growth part of the weight-bearing articular cartilage in rat/rabbit. Electron microscopy of the aspirated fluid shows the presence of proteins, glycoproteins, hyaluronidase-digestible matrix granules, collagen fibrils and matrix vesicles. The articular cartilage fluid differs from the fluid of growth plate cartilage in having a less alkaline pH and no nucleation factor for calcification. The growth cartilage fluid contains an organic calcification factor (a protein or glycoprotein) and also a mineral growth inhibitor, which is a sulfated proteoglycan aggregate (17, 18) and is present in this fluid at a concentration of 9.5 mg/ml. The organic calcification (nucleation) factor (a protein or glycoprotein) promotes mineral growth at calcium-phosphate product concentration over 2.6 mM and acts as a mineral ion reservoir. The sulfated proteoglycan, only in aggregated form (80S units) in the cartilage fluid inhibits the mineral growth by preventing the spontaneous conversion of the amorphous calcium phosphate into hydroxylapatite. Degradation of this high molecular weight with hyaluronidase or its degradation with lysozyme releases the inhibition and helps the spontaneous transformation of the amorphous calcium phosphate into hydroxylapatite for endochondral calcification (19).

The cartilage micropuncture fluid collected from a rachitic growth cartilage of rat on the average contained 2.0-2.5 mg/ml hexuronic acid, 2.1 mg/ml hexosamine and 0.8 mg/ml sulfate. From these values. the total proteoglycan (PG) present in the cartilage fluid, calculated as hexuronate X 4.55, was 9.6 mg/ml (20).

Investigations carried out in the cartilage fluid, aspirated in vivo from the upper tibial growth plate in rats with experimentally-induced rickets, support the aforementioned role of proteoglycan aggregates in cartilage calcification (21). Rickets induced in rats with phosphonate (EHDP = di - Na (1-hydroxyethylidene) -1, 1-bisphosphonate) interfered with the mineralization process in cartilage by decreasing calcium-phosphate concentration in cartilage fluid to half that in serum, by preventing disaggregation of the high molecular weight proteoglycans and by eliminating lysozyme activity in the cartilage. In comparison to EHDP-induced rickets, the changes observed were different in rickets induced in rats with vitamin D deficient-low phosphate (VDP) diet; the PG concentration was 20 % of the initial concentration as in normal control rats and calcium-phosphate concentrations were too low to sustain mineral growth. During healing from both EHDP - and VDP - induced rickets, the PG aggregation increased in the cartilage fluid, as evidenced by the decrease in the sedimentation coefficient of the cartilage fluid from 124S to 50S. This is consistent with the role of PG aggregates in cartilage calcification (21). Two recent reviews provide more details of the aforementioned glycoprotein and proteoglycans in the calcification of cartilage and bone (22, 23).

In hyperparathyroidism occurring secondarily to renal insufficiency, a fluid rich in GAGs accumulates as a cyst in bone associated with a marked generalized tissue calcification (p. 391).

Bibliography

1. Horowitz, M. I. and Pigman, W. (1978): The Glycoconjugates, Vol. II, Academic Press, New York, p. 361

2. Kobata, A. (1977): Milk glycoproteins and oligosaccharides. In: M. I. Horowitz and W. Pigman (Eds.) The Glycoconjugates. Vol. I, Academic Press, New York, pp. 423-440

3. Lis, D. and Monis, B. (1978): Glycosaminoglycans of the fat globule membrane of cow milk. J. Supramol. Structure 8, 173-176

4. Shimzu, M., Uryu, N., Yamauchi, K. (1981): Presence of heparan sulfate in the fat globule membrane of bovine and human milk. Agricul. Biol. Chem. 45, 741-745

5. Kiermeir, F. and Guell, J. (1968): Biological effects on the B-glucuronidase in milk. Z. Lebensm.-Unters. Forsch. 138, 205-208

6. Prichard, C. P. (1982): Edema of the Lung, Charles C. Thomas, Springfield, p. 12, 13, 108, 123

7. Haljamae, H., Linde, A. and Amundson, B. (1974): Comparative analysis of the capsular fluid and interstitial fluid. Am. J. Physiol. 227, 1199-1205

8. Wiederhielm, C. A., Fox, J. R. and Lee, D. R. (1976): Ground substance mucopolysaccharides and plasma proteins: their role in capillary water balance. Am. J. Physiol. 230, 1121-1125

9. Granger, H. J. (1981): Physicochemical properties of the extracellular matrix. In: A. R. Hargens (Ed.): Tissue Fluid Pressure and Composition, William and Wilkins, Baltimore, pp. 43-61

10. Comper, W. D. and Laurent, T. C. (1978): Osmotic interaction of plasma proteins with interstitial macromolecules. Physiol. Reviews 58, 255-315

11. Wiederhielm, C. A. and Black, L. L. (1976): Osmotic interaction of plasma proteins with interstitial macromolecules. Am. J. Physiol. 231, 638-641

12. Friedel, R., Bode, R., Trautschold, I. and Mattenheimer, H. (1976): Distribution and transport of cell enzymes within the extracellular space. II. The lymph as a distribution space for cell enzymes. J. Clin. Chem. Clin. Biochem. 14, 119-128

13. Szabo, G. (1978): Enzymes in tissue fluid and lymph. Lymphology 11, 147-155

14. Minnebaev, M. M. (1970): Changes in aldolase and transaminase activities and of the level of mucopolysaccharides in the lymph and serum following administration of hydrocortisone. Probl. Endokrinol. 16, 62-65

15. Renkin, E. M. (1979): Lymph as a measure of the composition of interstitial fluid. In: A. P. Fishman and E. M. Renkin (Eds.) Pulmonary Edema, Williams and Wilkins, Baltimore, p. 145

16. Linn, F. C. and Sokoloff, L. (1965): Movement and composition of interstitial fluid of cartilage. Arthritis Rheum. 8, 481-494

17. Cuervo, L. A., Pita, J. C. and Howell, D. S. (1973): Inhibition of calcium phosphate mineral growth by proteoglycan aggregate in a synthetic lymph. Calcif. Tissue Res. 13, 1-10

18. Urist, M. D. (1976): Biochemistry of Calcification. In: Biochemistry and Physiology of Bone, 2nd Ed., Vol. 4, pp. 1-57

19. Pita, J. C., Muller, F. J. and Howell, D. S. (1975): Disaggregation of proteoglycan aggregate during endochondral calcification. In: P. M. C. Burleigh and A. R. Poole (Eds.) Dynamics of Connective Tissue Macromolecules, North-Holland Publ., Amsterdam, pp. 247-258

20. Pita, J. C. and Howell, D. S. (1978): Microbiochemical studies of cartilage. In: Sokoloff (Ed.) The Joints and Synovial Fluid, Vol. I, Academic Press, New York, pp. 273-330

21. Howell, D. S., Muniz, O. E., Blanco, L. N. and Pita, J. C. (1980): A micropuncture study of growth cartilage in phosphonate (EHDP) induced rickets. Calcif. Tissue Int. 30, 35-42

22. De Bernard, B., Camerotto, R., Stagni, N., Vittur, F. and Zanetti, M. (1981): Role of a glycoprotein and proteoglycans in the calcification of cartilage and bone. In: H.-G. Schweiger (Ed.) Proc. Second International Congress of Cell Biology, Springer, Berlin, pp. 983-992

23. Bonucci, E. (1981): The origin of matrix vesicles and their role in the calcification of cartilage and bone. In: H.-G. Schweiger (Ed.) Proceedings Second International Congress of Cell Biology, Springer, Berlin, pp. 993-1003

Subject Index

Kolibri 360 GmbH
Ferdinand-Jühlke-Straße 7
99095 Erfurt, Deutschland
produktsicherheit@kolibri360.de